Lecture Notes in Computer Science 9849

Commenced Publication in 1973
Founding and Former Series Editors:
Gerhard Goos, Juris Hartmanis, and Jan van Leeuwen

Editorial Board

David Hutchison
  Lancaster University, Lancaster, UK
Takeo Kanade
  Carnegie Mellon University, Pittsburgh, PA, USA
Josef Kittler
  University of Surrey, Guildford, UK
Jon M. Kleinberg
  Cornell University, Ithaca, NY, USA
Friedemann Mattern
  ETH Zurich, Zurich, Switzerland
John C. Mitchell
  Stanford University, Stanford, CA, USA
Moni Naor
  Weizmann Institute of Science, Rehovot, Israel
C. Pandu Rangan
  Indian Institute of Technology, Madras, India
Bernhard Steffen
  TU Dortmund University, Dortmund, Germany
Demetri Terzopoulos
  University of California, Los Angeles, CA, USA
Doug Tygar
  University of California, Berkeley, CA, USA
Gerhard Weikum
  Max Planck Institute for Informatics, Saarbrücken, Germany

More information about this series at http://www.springer.com/series/7407

Raffaele Cerulli · Satoru Fujishige
A. Ridha Mahjoub (Eds.)

Combinatorial Optimization

4th International Symposium, ISCO 2016
Vietri sul Mare, Italy, May 16–18, 2016
Revised Selected Papers

 Springer

Editors
Raffaele Cerulli
University of Salerno
Fisciano
Italy

Satoru Fujishige
Kyoto University
Kyoto
Japan

A. Ridha Mahjoub
LAMSADE, Université Paris-Dauphine
Paris
France

ISSN 0302-9743   ISSN 1611-3349   (electronic)
Lecture Notes in Computer Science
ISBN 978-3-319-45586-0   ISBN 978-3-319-45587-7   (eBook)
DOI 10.1007/978-3-319-45587-7

Library of Congress Control Number: 2016949112

LNCS Sublibrary: SL1 – Theoretical Computer Science and General Issues

© Springer International Publishing Switzerland 2016
This work is subject to copyright. All rights are reserved by the Publisher, whether the whole or part of the material is concerned, specifically the rights of translation, reprinting, reuse of illustrations, recitation, broadcasting, reproduction on microfilms or in any other physical way, and transmission or information storage and retrieval, electronic adaptation, computer software, or by similar or dissimilar methodology now known or hereafter developed.
The use of general descriptive names, registered names, trademarks, service marks, etc. in this publication does not imply, even in the absence of a specific statement, that such names are exempt from the relevant protective laws and regulations and therefore free for general use.
The publisher, the authors and the editors are safe to assume that the advice and information in this book are believed to be true and accurate at the date of publication. Neither the publisher nor the authors or the editors give a warranty, express or implied, with respect to the material contained herein or for any errors or omissions that may have been made.

Printed on acid-free paper

This Springer imprint is published by Springer Nature
The registered company is Springer International Publishing AG Switzerland

Preface

This volume contains the full-papers presented at ISCO 2016, the 4th International Symposium on Combinatorial Optimization, held in Vietri Sul Mare (Italy) during May 16–18, 2016. ISCO 2016 was followed by the Spring School on "Extended Formulations for Combinatorial Optimization" given by Volker Kaibel and Samuel Fiorni. ISCO is a biennial symposium. The first event was held in Hammamet, Tunisia, in March 2010, the second in Athens, Greece, in April 2012, and the third in Lisbon, Portugal, in March 2014. The symposium aims to bring together researchers from all the communities related to combinatorial optimization, including algorithms and complexity, mathematical programming, operations research, stochastic optimization, graphs, and combinatorics. It is intended to be a forum for presenting original research on all aspects of combinatorial optimization, ranging from mathematical foundations and theory of algorithms to computational studies and practical applications, and especially their intersections. In response to the call for papers, ISCO 2016 received 98 fullpaper submissions. Each submission was reviewed by at least three reviewers, with at least two of them belonging to the Program Committee (PC). The submissions were judged on their originality and technical quality and the PC had to discuss in length the reviews and make tough decisions. As a result, the PC selected 38 fullpapers to be presented at the symposium, giving an acceptance rate of 39 % (57 short papers were also selected from both regular and short submissions). Four eminent invited speakers, R. Ravi (Carnegie Mellon University), András Frank (Egerváry Research Group, Eövös University Budapest), Adam N. Letchford (Lancaster University), and Volker Kaibel (Otto-von-Guericke University, Magdeburg), gave talks at the symposium. The revised versions of the accepted full-papers, as well as the abstracts of the invited talks, are included in this volume. We would like to thank all the authors who submitted their work to ISCO 2016, and the PC members and external reviewers for their excellent work. We would also like to thank our invited speakers as well as the speakers of the Spring School for their exciting lectures. They all greatly contributed to the quality of the symposium. Finally, we would like to thank the Organizing Committee members for their dedicated work in preparing this conference, and we gratefully acknowledge our sponsoring institutions for their assistance and support.

July 2016   Raffaele Cerulli
                                                                             Satoru Fujishige
                                                                             A. Ridha Mahjoub

Organization

Program Committee

Edoardo Amaldi	DEI, Politecnico di Milano, Italy
Francisco Barahona	IBM Research, USA
Mourad Baïou	LIMOS - Université Blaise Pascal, France
Daniel Bienstock	Columbia University, USA
Francesco Carrabs	University of Salerno, Italy
Raffaele Cerulli	University of Salerno, Italy
Laureano Escudero	Universidad Rey Juan Carlos, Spain
Matteo Fischetti	University of Padua, Italy
Pierre Fouilhoux	Laboratoire LIP6, Université Pierre et Marie Curie, France
Satoru Fujishige	RIMS, Kyoto University, Japan
Ricardo Fukasawa	University of Waterloo, Canada
Takuro Fukunaga	National Institute of Informatics, Japan
Naveen Garg	IIT Delhi, India
Monica Gentili	University of Salerno, Italy
Bruce Golden	University of Maryland, USA
Laurent Gourvès	lamsade, France
Luis Gouveia	University of Lisbon, Portugal
Mohamed Haouari	Qatar University, Qatar
Hiroshi Hirai	University of Tokyo, Japan
Giuseppe Italiano	University of Rome "Tor Vergata", Italy
Imed Kacem	LCOMS - University of Lorraine, France
Volker Kaibel	Otto-von-Guericke Universitaet Magdeburg, Germany
Naoyuki Kamiyama	Kyushu University, Japan
Shuji Kijima	Kyushu University, Japan
Yusuke Kobayashi	University of Tsukuba, Japan
Martine Labbé	Université Libre de Bruxelles, Belgium
Gilbert Laporte	HEC Montréal, Canada
Leo Liberti	LIX, Ecole Polytechnique, France
Andrea Lodi	DEI, University of Bologna, Italy
Marco Lübbecke	RWTH Aachen University, Germany
Nelson Maculan	Federal University of Rio de Janeiro (UFRJ), Brazil
Ali Ridha Mahjoub	LAMSADE, University Paris-Dauphine, France
Carlo Mannino	Sintef ict, Norway
Francois Margot	Carnegie Mellon University, USA
Silvano Martello	University of Bologna, Italy
Thomas McCormick	Sauder School of Business, UBC, Canada
Ioannis Milis	Athens University of Economics and Business, Greece

Kiyohito Nagano	Future University, Japan
Yoshio Okamoto	University of Electro-Communications, Japan
Gianpaolo Oriolo	Università di Roma "Tor Vergata", Italy
Vangelis Paschos	LAMSADE, University Paris-Dauphine, France
Nancy Perrot	Orange Labs, France
Franz Rendl	University of Klagenfurt, Austria
Giovanni Rinaldi	CNR, Italy
Juan José Salazar González	Universidad de La Laguna, Spain
Marc Sevaux	Lab-STICC, Université de Bretagne-Sud, France
Douglas Shier	Clemson University, USA
Akiyoshi Shioura	Tokyo Institute of Technology, Japan
Maria Grazia Speranza	University of Brescia, Italy
Kenjiro Takazawa	Hosei University, Japan
Shinichi Tanigawa	Kyoto University, Japan
Paolo Toth	DEIS, University of Bologna, France
Chefi Triki	University of Salento, Italy and Sultan Qaboos University, Oman
Eduardo Uchoa	Universidade Federal Fluminense, Brazil
Francois Vanderbeck	University of Bordeaux, France
Hande Yaman	Bilkent University, Turkey
Peng-Yeng Yin	National Chi Nan University, Taiwan
Yuichi Yoshida	National Institute of Informatics, Japan

Additional Reviewers

Alvarez-Miranda, Eduardo
Amanatidis, Georgios
Andrade, Rafael
Aristotelis, Giannakos
Assunção, Lucas
Belmonte, Rémy
Ben-Ameur, Walid
Bendali, Fatiha
Benhamiche, Amal
Boeckenhauer, Hans-Joachim
Bonomo, Flavia
Bulteau, Laurent
Cai, Shaowei
Carlinet, Yannick
Carrabs, Francesco
Casel, Katrin
Catanzaro, Daniele

Cerqueus, Audrey
Cornaz, Denis
Cunha, Alexandre
D'Ambrosio, Ciriaco
Derrien, Alban
Dias, Gustavo
Faenza, Yuri
Fampa, Marcia
Felici, Giovanni
Firsching, Moritz
Fotakis, Dimitris
Gaudioso, Manlio
Grappe, Roland
Imahori, Shinji
John, Maximilian
Karrenbauer, Andreas
Katsikarelis, Ioannis
Kawase, Yasushi
Koca, Esra

Koichi, Shungo
Lampis, Michael
Letsios, Dimitrios
Lhouari, Nourine
M.S., Ramanujan
Mahey, Philippe
Martin, Sébastien
Martinez, Leonardo
Marín, Alfredo
Mendez-Diaz, Isabel
Miyashiro, Ryuhei
Miyazawa, Flavio K.
Mkrtchyan, Vahan
Moura, Pedro
Murota, Kazuo
Mömke, Tobias
Naghmouchi, Mohamed Yassine
Nannicini, Giacomo

Nasini, Graciela
Neto, Jose
Nobili, Paolo
Oliveira, Daniel
Ozbaygin, Gizem
Pahl, Julia
Papadigenopoulos,
  Vasileios-Orestis
Pessoa, Artur
Picouleau, Christophe

Pêcher, Arnaud
Raiconi, Andrea
Raidl, Günther
Rossi, Fabrizio
Rudolph, Larry
Sadykov, Ruslan
Sikora, Florian
Smriglio, Stefano
Soma, Tasuku
Srivastav, Abhinav

Stamoulis, Georgios
Sukegawa, Noriyoshi
Tural, Mustafa Kemal
Umetani, Shunji
Urrutia, Sebastián
van Renssen, André
Ventura, Paolo
Zhou, Yuan
Zois, Georgios

Abstracts

New Graph Optimization Problems
in NP∩co-NP

András Frank

Egerváry Research Group, Eötvös University Budapest
frank@cs.elte.hu

We show that the following three problems in graph theory belong to **NP∩co-NP**.

1. Wang and Kleitman (1972) characterized degree-sequences of simple k-connected undirected graphs. We solve the corresponding problem for digraphs.
2. Edmonds (1973) characterized digraphs admitting k disjoint spanning arborescences of given root, and his result could be extended to the case when there is no prescription for the localization of the roots. Here we exhibit a much more general result that characterizes digraphs admitting k disjoint branchings with specified sizes $\mu_1, \mu_2, \dots, \mu_k$.
3. Ryser (1958) solved the maximum term rank problem which consisted of characterizing the row-sums and column-sums of $(0, 1)$-matrices with term-rank at least μ, or equivalently, characterize the degree-sequences of simple bipartite graphs with matching number at least μ. Recently, it turned out that the maximum term rank problem, though not particularly difficult, is not tractable with network flow or matroid techniques since the weighted version is **NP**-complete. Yet, we found a necessary and sufficient condition for the existence of a simple bipartite graph with matching number at least μ such that the degree of each node lies between specified lower and upper bounds.

As a major novelty, we show that these three apparently quite distant problems stem out from one common root: a general theorem on covering a supermodular function by a minimal *simple* digraph. Since the corresponding weighted optimization version includes **NP**-complete problems, the new results are certainly out of the range of classic general frameworks such as the one of submodular flows.

In the talk, I outline first the origin and the history of optimization problems concerning optimal coverings of supermodular functions and exhibit then the new developments giving rise to the characterizations indicated above. Finally, some open problems are sketched that are hopeful to be attacked successfully with the new approach.

Describing Integer Points in Polyhedra

Volker Kaibel

Otto-von-Guericke University, Magdeburg
kaibel@ovgu.de

Linear mixed integer models are fundamental in treating combinatorial problems via Mathematical Programming. In this lecture we are going to discuss the question how small such formulations one can obtain for different problems. It turns out that for several problems including, e.g., the traveling salesman problem and the spanning tree problem, the use of additional variables is essential for the design of polynomial sized integer programming formulations. In fact, we prove that their standard exponential size formulations are asymptotically minimal among the formulations based on incidence vectors only. We also treat bounds for general sets of 0/1-points and briey discuss the question for the role of rationality of coefficients in formulations.

Some Hard Combinatorial Optimization Problems from Mobile Wireless Communications

Adam N. Letchford

Lancaster University
a.n.letchford@lancaster.ac.uk

In the past decade, a revolution in telecommunications has been taking place. There has been an inexorable trend towards mobile wireless communications, in which there are a large number of *portable devices* (such as smartphones) scattered across a geographical region. Each such region is divided into a number of so-called *cells*. Each cell contains a powerful transmitter called a *base station*. When they wish to send or receive data, the portable devices have to send requests to one or more nearby base stations.

It turns out that mobile wireless communications are a rich source of new and difficult combinatorial optimisation problems. These include strategic problems, such as where and when to locate new base stations, tactical problems, such as how much power to give to each base station, and operational problems, such as how to assign incoming user requests to the available frequency bands.

In this talk, we focus on operational problems associated with so-called *orthogonal frequency-division multiple access* (OFDMA) systems. In these systems, there are a large number of channels available, each of which can be allocated to at most one user. On the other hand, a user can be assigned to more than one channel. The rate at which data is transmitted over a given channel is a nonlinear function of the power allocated to that channel, the bandwidth of the channel, and the noise associated with the channel. So one faces the problem of simultaneously assigning channels to users and allocating the available power to the channels. This leads to several different combinatorial optimization problems, depending on the particular objective in question, the side-constraints imposed, and the time-horizon of interest.

We show that some of these joint channel assignment and power allocation problems can be tackled successfully via mixed-integer linear programming, especially if one uses clever pre-processing tricks, strong cutting planes, and symmetry-breaking techniques. On the other hand, some of the problems still present a formidable challenge.

Improved Approximations for Graph-TSP in Regular Graphs

R. Ravi

Carnegie Mellon University
ravi@andrew.cmu.edu

A tour in a graph is a connected walk that visits every vertex at least once, and returns to the starting vertex. We describe improved approximation results for a tour with the minimum number of edges in regular graphs. En route we illustrate the main ideas used recently in designing improved approximation algorithms for graph TSP.

Contents

On the Finite Optimal Convergence of Logic-Based Benders' Decomposition in Solving 0–1 Min-Max Regret Optimization Problems with Interval Costs

Lucas Assunção[1(✉)], Andréa Cynthia Santos[2], Thiago F. Noronha[3], and Rafael Andrade[4]

[1] Departamento de Engenharia de Produção, Universidade Federal de Minas Gerais, Avenida Antônio Carlos, 6627, Belo Horizonte, MG 31270-901, Brazil
`lucas-assuncao@ufmg.br`
[2] ICD-LOSI, UMR CNRS 6281, Université de Technologie de Troyes, 12, rue Marie Curie, CS 42060, 10004 Troyes Cedex, France
`andrea.duhamel@utt.fr`
[3] Departamento de Ciência da Computação, Universidade Federal de Minas Gerais, Avenida Antônio Carlos, 6627, Belo Horizonte, MG 31270-901, Brazil
`tfn@dcc.ufmg.br`
[4] Departamento de Estatística e Matemática Aplicada, Universidade Federal do Ceará, Campus do Pici, BL 910, Fortaleza, CE 60455-900, Brazil
`rca@lia.ufc.br`

Abstract. This paper addresses a class of problems under interval data uncertainty composed of *min-max regret* versions of classical 0–1 optimization problems with interval costs. We refer to them as *interval 0–1 min-max regret* problems. The state-of-the-art exact algorithms for this class of problems work by solving a corresponding mixed integer linear programming formulation in a Benders' decomposition fashion. Each of the possibly exponentially many Benders' cuts is separated on the fly through the resolution of an instance of the classical 0–1 optimization problem counterpart. Since these separation subproblems may be NP-hard, not all of them can be modeled by means of linear programming, unless P = NP. In these cases, the convergence of the aforementioned algorithms are not guaranteed in a straightforward manner. In fact, to the best of our knowledge, their finite convergence has not been explicitly proved for any interval 0–1 min-max regret problem. In this work, we formally describe these algorithms through the definition of a logic-based Benders' decomposition framework and prove their convergence to an optimal solution in a finite number of iterations. As this framework is applicable to any interval 0–1 min-max regret problem, its finite optimal convergence also holds in the cases where the separation subproblems are NP-hard.

L. Assunção—Partially supported by the Coordination for the Improvement of Higher Education Personnel, Brazil (CAPES).
L. Assunção—The author thanks Vitor A. A. Souza and Phillippe Samer for the valuable discussions throughout the conception of this work.

© Springer International Publishing Switzerland 2016
R. Cerulli et al. (Eds.): ISCO 2016, LNCS 9849, pp. 1–12, 2016.
DOI: 10.1007/978-3-319-45587-7_1

1   Introduction

Robust Optimization (RO) [14] has drawn particular attention as an alternative to stochastic programming [22] in modeling uncertainty. In RO, instead of considering a probabilistic description known *a priori*, the variability of the data is represented by deterministic values in the context of *scenarios*. A *scenario* corresponds to a parameters assignment, *i.e.*, a value is fixed for each parameter subject to uncertainty. Two main approaches are adopted to model RO problems: the *discrete scenarios model* and the *interval data model*. In the former, a discrete set of possible scenarios is considered. In the latter, the uncertainty referred to a parameter is represented by a continuous interval of possible values. Differently from the discrete scenarios model, the infinite many possible scenarios that arise in the interval data model are not explicitly given. Nevertheless, in both models, a classical (*i.e.*, parameters known in advance) optimization problem takes place whenever a scenario is established.

The most commonly adopted RO criteria are the *absolute robustness* criterion, the *min-max regret* (also known as *robust deviation* criterion) and the *min-max relative regret* (also known as *relative robustness* criterion). The absolute robustness criterion is based on the anticipation of the worst possible conditions. Solutions for RO problems under such criterion tend to be conservative, as they optimize only a worst-case scenario. On the other hand, the min-max regret and the min-max relative regret are less conservative criteria and, for this reason, they have been addressed in several works (*e.g.*, [3,5,17,18,20]). Intuitively speaking, the *regret (robust deviation)* of a solution in a given scenario is the cost difference between such solution and an optimal one for this scenario. In turn, the *relative regret* of a solution in a given scenario consists of the corresponding regret normalized by the cost of an optimal solution for the scenario considered. The *(relative) robustness* cost of a solution is defined as its maximum *(relative) regret* over all scenarios. In this sense, the min-max (relative) regret criterion aims at finding a solution that has the minimum (relative) robustness cost. Such solution is referred to as a *robust solution*.

RO versions of several combinatorial optimization problems have been studied in the literature, addressing, for example, uncertainties over costs. Handling uncertain costs brings an extra level of difficulty, such that even polynomially solvable problems become NP-hard in their corresponding robust versions [13,19,20,23]. In this study, we are interested in a particular class of RO problems, namely *interval 0–1 min-max regret* problems, which consist of min-max regret versions of Binary Integer Linear Programming (BILP) problems with interval costs. Notice that a large variety of classical optimization problems can be modeled as BILP problems, including (i) polynomially solvable problems, such as the shortest path problem, the minimum spanning tree problem and the assignment problem, and (ii) NP-hard combinatorial problems, such as the 0–1 knapsack problem, the set covering problem, the traveling salesman problem and the restricted shortest path problem [9]. An especially challenging subclass of interval 0–1 min-max regret problems, referred to as *interval 0–1 robust-hard problems*, arises when we address interval 0–1 min-max regret versions of classical NP-hard combinatorial problems as those aforementioned in (ii).

Aissi et al. [1] showed that, for any interval 0–1 min-max regret problem (including interval 0–1 robust-hard problems), the robustness cost of a solution can be computed by solving a single instance of the classical optimization problem counterpart (*i.e.*, costs known in advance) in a particular scenario. Therefore, one does not have to consider all the infinite many possible scenarios during the search for a robust solution, but only a subset of them, one for each feasible solution. Nevertheless, since the number of these promising scenarios can still be huge, the state-of-the-art exact algorithms for interval 0–1 min-max regret problems work by implicitly separating them on the fly, in a Benders' decomposition [4] fashion (see, *e.g.*, [17,18,21]). Precisely, each Benders' cut is generated through the resolution of an instance of the classical optimization problem counterpart. Notice that, for interval 0–1 robust-hard problems, these separation subproblems are NP-hard and, thus, they cannot be modeled by means of Linear Programming (LP), unless P = NP. In these cases, the convergence of the aforementioned algorithms are not guaranteed in a straightforward manner.

These exact algorithms, which have their roots in *logic-based Benders' decomposition* [11] (also see [6]), have been successfully applied to solve several interval 0–1 min-max regret problems (*e.g.*, [17,19,20]), including interval 0–1 robust-hard problems, such as the robust traveling salesman problem [18] and the robust set covering problem [21]. However, to the best of our knowledge, their convergence has not been explicitly proved for any interval 0–1 min-max regret problem. In this work, we formally describe these algorithms through the definition of a logic-based Benders' decomposition framework and prove their finite convergence to an optimal solution. Precisely, we show, by contradiction, that a new cut is always generated per iteration, in a finite space of possible solutions. As the framework is applicable to any interval 0–1 min-max regret problem, its finite optimal convergence also holds in solving interval 0–1 robust-hard problems, *i.e.*, in the cases where the separation subproblems are NP-hard.

The remainder of this work is organized as follows. The Benders' decomposition method is briefly introduced in Sect. 2, followed by the description of a standard modeling technique for interval 0–1 min-max regret problems (Sect. 2.1). In addition, a generalization of state-of-the-art exact algorithms for interval 0–1 min-max regret problems is devised through the description of a logic-based Benders' decomposition framework (Sect. 2.2). The finite convergence of these algorithms to optimal solutions is proved in the same section, and concluding remarks are given in the last section.

2   A Logic-Based Benders' Decomposition Framework for Interval 0–1 Min-Max Regret Problems

The *classical Benders' decomposition method* [4] was originally proposed to tackle Mixed Integer Linear Programming (MILP) problems of the form P : $\min\{cx + dy : Ax + By \geq b, x \in \mathbb{Z}_+^{n_1}, y \geq \mathbf{0}\}$. In this case, there are n_1

integer variables and n_2 continuous ones, which are represented by the column vectors x and y, respectively, and their corresponding cost values are given by the row vectors c and d. Moreover, b is an m-dimensional column vector, and A and B are $m \times n_1$ and $m \times n_2$ matrices, respectively. Given a vector $\bar{x} \in \mathbb{Z}_+^{n_1}$, the *classical Benders' reformulation* starts by defining an LP *primal subproblem* $PS(\bar{x})$: $\min\{dy : By \geq b - A\bar{x}, y \geq \mathbf{0}\}$ through the projection of the continuous variables y in the space defined by $\bar{x}$. Notice that $PS(\bar{x})$ can be represented by means of the corresponding *dual subproblem* $DS(\bar{x})$: $\max\{\mu(b - A\bar{x}) : \mu B \leq d, \mu \geq \mathbf{0}\}$, where μ is an m-dimensional row vector referred to the dual variables.

Let $EP(\bar{x})$ and $ER(\bar{x})$ be, respectively, the sets of extreme points and extreme rays of a given $DS(\bar{x})$ subproblem. One may observe that the feasible region of $DS(\bar{x})$ does not depend on the value assumed by $\bar{x}$. Thus, hereafter, these sets are referred to as EP and ER for all $\bar{x} \in \mathbb{Z}_+^{n_1}$.

Considering a nonnegative continuous variable ρ, the resolution of each dual subproblem $DS(\bar{x})$ leads to a new linear constraint (i) $\rho \geq \bar{\mu}(b - Ax)$, if $DS(\bar{x})$ has a bounded optimal solution $\bar{\mu} \in EP$, or (ii) $\bar{\nu}(b - Ax) \leq 0$, if DP has an unbounded solution, represented by $\bar{\nu} \in ER$. The Benders' cuts described in (i) are called *optimality cuts*, whereas the ones in (ii) are called *feasibility cuts*. Both types of cuts are used to populate on the fly a *reformulated problem*, defined as:

$$(RP) \quad \min \ \{cx + \rho\} \tag{1}$$
$$s.t. \ \ \rho \geq \bar{\mu}(b - Ax) \quad \forall \, \bar{\mu} \in EP, \tag{2}$$
$$\bar{\nu}(b - Ax) \leq 0 \quad \forall \, \bar{\nu} \in ER, \tag{3}$$
$$\rho \geq 0, \tag{4}$$
$$x \in \mathbb{Z}_+^{n_1}. \tag{5}$$

Let MRP be a relaxed RP problem, called *master problem*, which considers only a subset of the extreme points and extreme rays associated with constraints (2) and (3), respectively. At each iteration of the *classical Benders' decomposition algorithm*, the master problem is solved, obtaining a solution $(\bar{x}, \bar{\rho}) \in \mathbb{Z}_+^{n_1} \times \mathbb{R}_+$. If $(\bar{x}, \bar{\rho})$ is not feasible for the original reformulated problem RP, a corresponding $DS(\bar{x})$ subproblem is solved in order to generate a new Benders' cut, either a feasibility cut or an optimality one. The new cut is added to the master problem MRP and the algorithm iterates until a feasible (and, therefore, optimal) solution for RP is found.

The finite convergence of the classical Benders' decomposition method is guaranteed by the fact that the polyhedron referred to any LP problem can be described by finite sets of extreme points and extreme rays, and that a new Benders' cut is generated per iteration of the algorithm. We refer to [4] for the detailed proof. Methodologies able to improve the convergence of the method were studied in several works (see, *e.g.*, [8, 15, 16]). In addition, nonlinear convex duality theory was later applied to devise a generalized approach, namely *generalized Benders' decomposition*, applicable to mixed integer nonlinear problems [10].

More recently, Hooker and Ottosson [11] introduced the idea of the so-called *logic-based Benders' decomposition*, a Benders-like decomposition approach that is suitable for a broader class of problems. In fact, the latter approach is intended to tackle any optimization problem by exploring the concept of an *inference dual subproblem*. In particular, that is the problem of inferring a strongest possible bound for a set of constraints from the original problem that are relaxed in the master problem. Notice that the aforementioned inference subproblems are not restricted to linear and nonlinear continuous problems. In fact, they can even be NP-hard combinatorial problems (see, *e.g.*, [7]). Therefore, the convergence of the logic-based Benders' decomposition method cannot be showed in a straightforward manner for all classes of optimization problems. As pointed out in [11], the convergence of the method relies on some peculiarities of the (logic-based) Benders' reformulation, such as the way the inference dual subproblems are devised and the finiteness of the search space referred to them.

In the remainder of this section, we describe a framework that generalizes state-of-the-art logic-based Benders' decomposition algorithms widely used to solve interval 0–1 min-max regret problems [17–21]. In addition, we show its convergence to an optimal solution in a finite number of iterations. The framework addresses MILP formulations with typically an exponential number of constraints, as detailed in the sequel.

2.1 Mathematical Formulation

Consider $\mathcal{G}$, a generic BILP minimization problem defined as follows.

$$(\mathcal{G}) \quad \min \ cx \tag{6}$$
$$s.t. \ Ax \geq b, \tag{7}$$
$$x \in \{0,1\}^n. \tag{8}$$

The binary variables are represented by an n-dimensional column vector x, whereas their corresponding cost values are given by an n-dimensional row vector c. Moreover, b is an m-dimensional column vector, and A is an $m \times n$ matrix. The feasible region of $\mathcal{G}$ is given by $\Omega = \{x : Ax \geq b, x \in \{0,1\}^n\}$. We highlight that, although the results of this work are presented by the assumption of $\mathcal{G}$ being a minimization problem, they also hold for interval 0–1 min-max regret versions of maximization problems, with minor modifications.

Now, let $\mathcal{R}$ be an interval min-max regret RO version of $\mathcal{G}$, where a continuous cost interval $[l_i, u_i]$, with $l_i, u_i \in \mathbb{Z}_+$ and $l_i \leq u_i$, is associated with each binary variable x_i, $i = 1, \ldots, n$. The following definitions describe $\mathcal{R}$ formally.

Definition 1. *A scenario s is an assignment of costs to the binary variables, i.e., a cost $c_i^s \in [l_i, u_i]$ is fixed for all x_i, $i = 1, \ldots, n$.*

Let $\mathcal{S}$ be the set of all possible cost scenarios, which consists of the cartesian product of the continuous intervals $[l_i, u_i]$, $i = 1, \ldots, n$. The cost of a solution $x \in \Omega$ in a scenario $s \in \mathcal{S}$ is given by $c^s x = \sum_{i=1}^{n} c_i^s x_i$.

Definition 2. *A solution $opt(s) \in \Omega$ is said to be* optimal *for a scenario $s \in \mathcal{S}$ if it has the smallest cost in s among all the solutions in Ω, i.e., $opt(s) = \arg\min_{x \in \Omega} c^s x$.*

Definition 3. *The* regret (robust deviation) *of a solution $x \in \Omega$ in a scenario $s \in \mathcal{S}$, denoted by r_x^s, is the difference between the cost of x in s and the cost of $opt(s)$ in s, i.e., $r_x^s = c^s x - c^s opt(s)$.*

Definition 4. *The* robustness cost *of a solution $x \in \Omega$, denoted by R_x, is the maximum regret of x among all possible scenarios, i.e., $R_x = \max_{s \in \mathcal{S}} r_x^s$.*

Definition 5. *A solution $x^* \in \Omega$ is said to be* robust *if it has the smallest robustness cost among all the solutions in Ω, i.e., $x^* = \arg\min_{x \in \Omega} R_x$.*

Definition 6. *The* interval min-max regret problem $\mathcal{R}$ *consists in finding a robust solution $x^* \in \Omega$.*

For each scenario $s \in \mathcal{S}$, let $\mathcal{G}(s)$ denote the corresponding problem $\mathcal{G}$ under cost vector $c^s \in \mathbb{R}_+^n$, i.e., the problem of finding an optimal solution $opt(s)$ for s. Also consider y, an n-dimensional vector of binary variables. Then, $\mathcal{R}$ can be generically modeled as follows.

$$(\mathcal{R}) \quad \min \ \max_{s \in \mathcal{S}} \left(c^s x - \overbrace{\min_{y \in \Omega} c^s y}^{(\mathcal{G}(s))} \right) \tag{9}$$

$$\text{s.t. } x \in \Omega. \tag{10}$$

The basic result presented below has been explicitly proved for several interval min-max regret problems (see, *e.g.*, [12,18,23]) and generalized for the case of interval 0–1 min-max regret problems [1] (also see [2]).

Proposition 1 (Aissi et al. [1]). *The regret of any feasible solution $x \in \Omega$ is maximum in the scenario $s(x)$ induced by x, defined as follows:*

$$\text{for all } i \in \{1, \ldots, n\}, \quad c_i^{s(x)} = \begin{cases} u_i, & \text{if } x_i = 1, \\ l_i, & \text{if } x_i = 0. \end{cases}$$

From Proposition 1, $\mathcal{R}$ can be rewritten as

$$(\tilde{\mathcal{R}}) \quad \min \ \left(c^{s(x)} x - \min_{y \in \Omega} c^{s(x)} y \right) \tag{11}$$

$$\text{s.t. } x \in \Omega. \tag{12}$$

One may note that the inner minimization in (11) does not define an LP problem, but a BILP one. Since, in general, there is no guarantee of integrality in solving the linear relaxation of this problem, we cannot represent it by means of extreme points and extreme rays, as in a classical Benders' reformulation [4].

Alternatively, we reformulate $\tilde{\mathcal{R}}$ by adding a free variable ρ and linear constraints that explicitly bound ρ with respect to all the feasible solutions that y can represent. The resulting MILP formulation (see, e.g., [1]) is provided from (13) to (16).

$$(\mathcal{F}) \quad \min \; (\sum_{i=1}^{n} u_i x_i - \rho) \tag{13}$$

$$s.t. \; \rho \leq \sum_{i=1}^{n} (l_i + (u_i - l_i)x_i)\bar{y}_i \quad \forall \bar{y} \in \Omega, \tag{14}$$

$$x \in \Omega, \tag{15}$$

$$\rho \; \text{free}. \tag{16}$$

Constraints (14) ensure that ρ does not exceed the value related to the inner minimization in (11). Note that, in (14), $\bar{y}$ is a constant vector, one for each solution in Ω. These constraints are tight whenever $\bar{y}$ is optimal for the classical counterpart problem $\mathcal{G}$ in the scenario $s(x)$. Constraints (15) define the feasible region referred to the x variables, and constraint (16) gives the domain of the variable ρ. Notice that the feasibility of $\mathcal{F}$ solely relies on the feasibility of the corresponding classical optimization problem $\mathcal{G}$. Thus, for simplicity, we assume that $\mathcal{F}$ is feasible in the remainder of this work.

The number of constraints (14) corresponds to the number of feasible solutions in Ω. As the size of this region may grow exponentially with the number of binary variables, this fomulation is particularly suitable to be handled by decomposition methods, such as the logic-based Benders' decomposition framework detailed below.

2.2 Logic-Based Benders' Algorithm

The logic-based Benders' algorithm here described relies on the fact that, since several of constraints (14) might be inactive at optimality, they can be generated on demand whenever they are violated. In this sense, given a set $\Gamma \subseteq \Omega$, $\Gamma \neq \emptyset$, consider the *relaxed robustness cost* metric defined as follows.

Definition 7. *A solution $opt(s, \Gamma) \in \Gamma$ is said to be Γ-relaxed optimal for a scenario $s \in \mathcal{S}$ if it has the smallest cost in s among all the solutions in Γ, i.e., $opt(s, \Gamma) = \arg\min_{x \in \Gamma} c^s x$.*

Definition 8. *The Γ-relaxed robustness cost of a solution $x \in \Omega$, denoted by R_x^Γ, is the difference between the cost of x in the scenario $s(x)$ induced by x and the cost of a Γ-relaxed optimal solution $opt(s(x), \Gamma)$ in $s(x)$, i.e., $R_x^\Gamma = c^{s(x)}x - c^{s(x)}opt(s(x), \Gamma)$.*

Proposition 2. *For any $\Gamma \subseteq \Omega$, $\Gamma \neq \emptyset$, and any solution $x \in \Omega$, the Γ-relaxed robustness cost R_x^Γ of x gives a lower bound on the robustness cost R_x of x.*

Proof. Consider a set $\Gamma \subseteq \Omega$, $\Gamma \neq \emptyset$, and a solution $x \in \Omega$. According to Proposition 1, the robustness cost of x is given by $R_x = r_x^{s(x)} = c^{s(x)}x - c^{s(x)}opt(s(x))$, where $opt(s(x))$ is an optimal solution for the scenario $s(x)$ induced by x. By definition, the Γ-relaxed robustness cost of x is given by $R_x^\Gamma = c^{s(x)}x - c^{s(x)}opt(s(x), \Gamma)$, where $opt(s(x), \Gamma)$ is a Γ-relaxed optimal solution for $s(x)$. Notice that $c^{s(x)}opt(s(x)) \leq c^{s(x)}x'$ for all $x' \in \Omega$, including $opt(s(x), \Gamma)$. Therefore,

$$R_x^\Gamma = c^{s(x)}x - c^{s(x)}opt(s(x), \Gamma) \leq c^{s(x)}x - c^{s(x)}opt(s(x)) = R_x. \qquad (17)$$

$\square$

Proposition 3. *If $\Gamma = \Omega$, then, for any solution $x \in \Omega$, it holds that $R_x^\Gamma = R_x$.*

Proof. Consider the set $\Gamma = \Omega$ and a solution $x \in \Omega$. In this case, a Γ-relaxed optimal solution $opt(s(x), \Gamma)$ for $s(x)$ is also an optimal solution $opt(s(x))$ for this scenario. Therefore, considering Proposition 1,

$$R_x^\Gamma = c^{s(x)}x - c^{s(x)}opt(s(x), \Gamma) = c^{s(x)}x - c^{s(x)}opt(s(x)) = R_x. \qquad (18)$$

$\square$

Definition 9. *A solution $\tilde{x}^* \in \Omega$ is said to be Γ-relaxed robust if it has the smallest Γ-relaxed robustness cost among all the solutions in Ω, i.e., $\tilde{x}^* = \arg\min_{x \in \Omega} R_x^\Gamma$.*

Considering the relaxed metric discussed above, we detail a logic-based Benders' algorithm to solve formulation $\mathcal{F}$, given by (13)–(16). The procedure is described in Algorithm 1. Let $\Omega^\psi \subseteq \Omega$ be the set of solutions $\bar{y} \in \Omega$ (Benders' cuts) available at an iteration ψ. Also let $\mathcal{F}^\psi$ be a relaxed version of $\mathcal{F}$ in which constraints (14) are replaced by

$$\rho \leq \sum_{i=1}^n (l_i + (u_i - l_i)x_i)\bar{y}_i \quad \forall \bar{y} \in \Omega^\psi. \qquad (19)$$

Thus, the relaxed problem $\mathcal{F}^\psi$, called *master problem*, is defined by (13), (15), (16) and (19). One may observe that $\mathcal{F}^\psi$ is precisely the problem of finding a Γ-relaxed robust solution, with $\Gamma = \Omega^\psi$.

Let ub^ψ keep the best upper bound found (until an iteration ψ) on the solution of $\mathcal{F}$. Notice that, at the beginning of Algorithm 1, Ω^1 contains the initial Benders' cuts available, whereas ub^1 keeps the initial upper bound on the solution of $\mathcal{F}$. In this case, $\Omega^1 = \emptyset$ and $ub^1 := +\infty$. At each iteration ψ, the algorithm obtains a solution by solving a corresponding master problem $\mathcal{F}^\psi$ and seeks a constraint (14) that is most violated by this solution. Initially, no constraint (19) is considered, since $\Omega^1 = \emptyset$. An initialization step is then necessary to add at least one solution to Ω^1, thus avoiding unbounded solutions during the first resolution of the master problem. To this end, it is computed an optimal solution for the worst-case scenario s_u, in which $c^{s_u} = u$ (Step I, Algorithm 1).

Algorithm 1. Logic-based Benders' algorithm.

Input: Cost intervals $[l_i, u_i]$ referred to x_i, $i = 1, \ldots, n$.
Output: $(\bar{x}^*, R^*)$, where $\bar{x}^*$ is a robust solution for $\mathcal{F}$, and R^* is its corresponding
      robustness cost.
$\psi := 1$; $ub^1 := +\infty$; $\Omega^1 := \emptyset$;
Step I. (Initialization)
Find an optimal solution $\bar{y}^1 = opt(s_u)$ for the worst-case scenario s_u;
$\Omega^1 := \Omega^1 \cup \{\bar{y}^1\}$;
Step II. (Master problem)
Solve the relaxed problem $\mathcal{F}^\psi$, obtaining a solution $(\bar{x}^\psi, \bar{\rho}^\psi)$;
Step III. (Separation subproblem)
Find an optimal solution $\bar{y}^\psi = opt(s(\bar{x}^\psi))$ for the scenario $s(\bar{x}^\psi)$ induced by $\bar{x}^\psi$ and
use it to compute $R_{\bar{x}^\psi}$, the robustness cost of $\bar{x}^\psi$;
Step IV. (Stopping condition)
$lb^\psi := u\bar{x}^\psi - \bar{\rho}^\psi$;
if $lb^\psi \geq R_{\bar{x}^\psi}$ then
    $\bar{x}^* := \bar{x}^\psi$;
    $R^* := R_{\bar{x}^\psi}$;
    Return $(\bar{x}^*, R^*)$;
end
else
    $ub^\psi := \min\{ub^\psi, R_{\bar{x}^\psi}\}$;
    $ub^{\psi+1} := ub^\psi$;
    $\Omega^{\psi+1} := \Omega^\psi \cup \{\bar{y}^\psi\}$;
    $\psi := \psi + 1$;
    Go to Step II;
end

After the initialization step, the iterative procedure takes place. At each iteration ψ, the corresponding relaxed problem $\mathcal{F}^\psi$ is solved (Step II, Algorithm 1), obtaining a solution $(\bar{x}^\psi, \bar{\rho}^\psi)$. Then, the algorithm checks if $(\bar{x}^\psi, \bar{\rho}^\psi)$ violates any constraint (14) of the original problem $\mathcal{F}$, *i.e.*, if there is a constraint (19) that should have been considered in $\mathcal{F}^\psi$ and was not. For this purpose, it is solved a *separation subproblem* that computes $R_{\bar{x}^\psi}$ (the actual robustness cost of $\bar{x}^\psi$) by finding an optimal solution $\bar{y}^\psi = opt(s(\bar{x}^\psi))$ for the scenario $s(\bar{x}^\psi)$ induced by $\bar{x}^\psi$ (see Step III, Algorithm 1). Notice that the separation subproblems involve solving a classical optimization problem $\mathcal{G}(\bar{x}^\psi)$, *i.e.*, problem $\mathcal{G}$, given by (6)–(8), in the scenario $s(\bar{x}^\psi)$.

Let $lb^\psi = u\bar{x}^\psi - \bar{\rho}^\psi$ be the value of the objective function in (13) related to the solution $(\bar{x}^\psi, \bar{\rho}^\psi)$ of the current master problem $\mathcal{F}^\psi$. Notice that, considering $\Gamma = \Omega^\psi$, lb^ψ corresponds to the Γ-relaxed robustness cost of $\bar{x}^\psi$. Thus, according to Proposition 2, lb^ψ gives a lower (dual) bound on the solution of $\mathcal{F}$. Moreover, since $\bar{x}^\psi$ is a feasible solution in Ω, its robustness cost $R_{\bar{x}^\psi}$ gives an upper (primal) bound on the solution of $\mathcal{F}$. Accordingly, if lb^ψ reaches $R_{\bar{x}^\psi}$, the algorithm stops. Otherwise, ub^ψ and $ub^{\psi+1}$ are both set to the best upper bound found by the algorithm until the iteration ψ. In addition, a new constraint

(19) is generated from $\bar{y}^\psi$ and added to $\mathcal{F}^{\psi+1}$ by setting $\Omega^{\psi+1} := \Omega^\psi \cup \{\bar{y}^\psi\}$ (see Step IV of Algorithm 1). Notice that the algorithm stops when the value $\bar{\rho}^\psi$ corresponds to the cost of $\bar{y}^\psi = opt(s(\bar{x}^\psi))$ in the scenario $s(\bar{x}^\psi)$, *i.e.*, the optimal solution for $\mathcal{F}^\psi$ is also feasible (and, therefore, optimal) for the original problem $\mathcal{F}$. The convergence of the algorithm is ensured by Proposition 3 and the following results.

Lemma 1. *Every separation subproblem that arises during the execution of Algorithm 1 is feasible.*

Proof. Assuming $\mathcal{F}$ feasible, we must have $\Omega \neq \emptyset$. This implies the existence of at least one feasible solution for every scenario $s \in S$, and, thus, any classical problem $\mathcal{G}$ that arises while executing Algorithm 1 is feasible. □

Proposition 4. *At each iteration $\psi \geq 1$ of Algorithm 1, if the stopping condition is not satisfied, then the resolution of the corresponding separation subproblem leads to a new solution $\bar{y}^\psi \in \Omega \setminus \Omega^\psi$.*

Proof. Consider an iteration $\psi \geq 1$ of Algorithm 1 and assume, by contradiction, that (I) the stopping condition is not satisfied, and (II) the resolution of the separation subproblem of the current iteration ψ does not lead to a solution in $\Omega \setminus \Omega^\psi$. Let $(\bar{x}^\psi, \bar{\rho}^\psi)$ be the solution obtained from the resolution of the corresponding master problem $\mathcal{F}^\psi$. From Lemma 1, the subproblem referred to iteration ψ is feasible, and, thus, its resolution leads to an optimal solution $\bar{y}^\psi = opt(s(\bar{x}^\psi)) \in \Omega$ for the scenario $s(\bar{x}^\psi)$ induced by $\bar{x}^\psi$. From assumption (I), we must have $lb^\psi < R_{\bar{x}^\psi}$. Considering Proposition 1 and letting $\Gamma = \Omega^\psi$, we have that $lb^\psi = R^\Gamma_{\bar{x}^\psi}$, and, moreover,

$$c^{s(\bar{x}^\psi)}\bar{x}^\psi - c^{s(\bar{x}^\psi)}opt(s(\bar{x}^\psi), \Gamma) = R^\Gamma_{\bar{x}^\psi} \tag{20}$$

$$= lb^\psi \tag{21}$$

$$< R_{\bar{x}^\psi} \tag{22}$$

$$= c^{s(\bar{x}^\psi)}\bar{x}^\psi - c^{s(\bar{x}^\psi)}opt(s(\bar{x}^\psi)), \tag{23}$$

where $opt(s(\bar{x}^\psi), \Gamma)$ is a Γ-relaxed optimal solution for $s(\bar{x}^\psi)$, and $opt(s(\bar{x}^\psi))$ is an optimal solution for $s(\bar{x}^\psi)$. From (20)–(23), we obtain

$$c^{s(\bar{x}^\psi)}opt(s(\bar{x}^\psi), \Gamma) > c^{s(\bar{x}^\psi)}opt(s(\bar{x}^\psi)). \tag{24}$$

Notice that, since $\bar{y}^\psi$ is also an optimal solution for $s(\bar{x}^\psi)$, it follows, from (24), that

$$c^{s(\bar{x}^\psi)}opt(s(\bar{x}^\psi), \Gamma) > c^{s(\bar{x}^\psi)}opt(s(\bar{x}^\psi)) = c^{s(\bar{x}^\psi)}\bar{y}^\psi. \tag{25}$$

Nevertheless, as $opt(s(\bar{x}^\psi), \Gamma)$ is a Γ-relaxed optimal solution for $s(\bar{x}^\psi)$, and, from Lemma 1 and assumption (II), $\bar{y}^\psi$ belongs to $\Omega^\psi = \Gamma$, we also have that

$$c^{s(\bar{x}^\psi)}opt(s(\bar{x}^\psi), \Gamma) \leq c^{s(\bar{x}^\psi)}\bar{y}^\psi, \tag{26}$$

which, considering (25), defines a contradiction. □

Theorem 1. *Algorithm 1 solves problem $\mathcal{F}$ at optimality within a finite number of iterations.*

Proof. As Ω is defined in terms of binary variables, it consists of a finite discrete set of solutions. Thus, the convergence of Algorithm 1 is guaranteed by Propositions 3 and 4. $\qquad\square$

3 Concluding Remarks

In this work, we presented the first formal proof of the finite convergence of state-of-the-art logic-based Benders' decomposition algorithms for a class of robust optimization problems, namely interval 0–1 min-max regret problems. These algorithms were generically described by means of a logic-based Benders' decomposition framework, which was proved to converge to an optimal solution in a finite number of iterations.

References

1. Aissi, H., Bazgan, C., Vanderpooten, D.: Min-max and min-max regret versions of combinatorial optimization problems: a survey. Eur. J. Oper. Res. **197**(2), 427–438 (2009)
2. Averbakh, I.: On the complexity of a class of combinatorial optimization problems with uncertainty. Math. Program. **90**(2), 263–272 (2001)
3. Averbakh, I.: Computing and minimizing the relative regret in combinatorial optimization with interval data. Discrete Optim. **2**(4), 273–287 (2005)
4. Benders, J.F.: Partitioning procedures for solving mixed-variables programming problems. Numer. Math. **4**(1), 238–252 (1962)
5. Coco, A.A., Júnior, J.C.A., Noronha, T.F., Santos, A.C.: An integer linear programming formulation and heuristics for the minmax relative regret robust shortest path problem. J. Glob. Optim. **60**(2), 265–287 (2014)
6. Codato, G., Fischetti, M.: Combinatorial Benders' cuts for mixed-integer linear programming. Oper. Res. **54**(4), 756–766 (2006)
7. Côté, J.F., Dell'Amico, M., Iori, M.: Combinatorial Benders' cuts for the strip packing problem. Oper. Res. **62**(3), 643–661 (2014)
8. Fischetti, M., Salvagnin, D., Zanette, A.: A note on the selection of Benders' cuts. Math. Program. **124**(1–2), 175–182 (2010)
9. Garey, M.R., Johnson, D.S.: Computers and Intractability: A Guide to the Theory of NP-Completeness. W. H. Freeman & Co., New York (1979)
10. Geoffrion, A.M.: Generalized Benders decomposition. J. Optim. Theo. Appl. **10**(4), 237–260 (1972)
11. Hooker, J., Ottosson, G.: Logic-based Benders decomposition. Math. Program. **96**(1), 33–60 (2003)
12. Karaşan, O.E., Pinar, M.Ç., Yaman, H.: The robust shortest path problem with interval data. Technical report, Bilkent University, Ankara, Turkey (2001)
13. Kasperski, A.: Discrete Optimization with Interval Data: Minmax Regret and Fuzzy Approach (Studies in Fuzziness and Soft Computing). Springer, Berlin (2008)

14. Kouvelis, P., Yu, G.: Robust Discrete Optimization and Its Applications. Kluwer Academic Publishers, Boston (1997)
15. Magnanti, T.L., Wong, R.T.: Accelerating Benders decomposition: algorithmic enhancement and model selection criteria. Oper. Res. **29**(3), 464–484 (1981)
16. McDaniel, D., Devine, M.: A modified Benders' partitioning algorithm for mixed integer programming. Manag. Sci. **24**(3), 312–319 (1977)
17. Montemanni, R.: A Benders decomposition approach for the robust spanning tree problem with interval data. Eur. J. Oper. Res. **174**(3), 1479–1490 (2006)
18. Montemanni, R., Barta, J., Gambardella, L.M.: The robust traveling salesman problem with interval data. Transp. Sci. **41**(3), 366–381 (2007)
19. Montemanni, R., Gambardella, L.M.: The robust shortest path problem with interval data via Benders decomposition. 4OR **3**(4), 315–328 (2005)
20. Pereira, J., Averbakh, I.: Exact and heuristic algorithms for the interval data robust assignment problem. Comput. Oper. Res. **38**(8), 1153–1163 (2011)
21. Pereira, J., Averbakh, I.: The robust set covering problem with interval data. Ann. Oper. Res. **207**(1), 217–235 (2013)
22. Spall, J.C.: Introduction to Stochastic Search and Optimization: Estimation Simulation and Control. Wiley, New York (2003)
23. Yaman, H., Karaşan, O.E., Pinar, M.Ç.: The robust spanning tree problem with interval data. Oper. Res. Lett. **29**(1), 31–40 (2001)

A Full Description of Polytopes Related to the Index of the Lowest Nonzero Row of an Assignment Matrix

Walid Ben-Ameur, Antoine Glorieux[✉], and José Neto

Samovar UMR5157, Télécom SudParis, CNRS, Universit Paris-Saclay,
9 Rue Charles Fourier, 91011 Evry Cedex, France
{walid.benameur,antoine.glorieux,jose.neto}@telecom-sudparis.eu

Abstract. Consider a $\{0,1\}$ assignment matrix where each column contains exactly one coefficient equal to 1 and let h be the index of the lowest row that is not identically equal to the zero row. We give a full description of the convex hull of all feasible assignments appended with the extra parameter h. This polytope and some of its variants naturally appear in the context of several combinatorial optimization problems including frequency assignment, job scheduling, graph orientation, maximum clique, etc. We also show that the underlying separation problems are solvable in polynomial time and thus optimization over those polytopes can be done in polynomial time.

1 Introduction and Motivations

Let us consider combinatorial optimization problems involving $n \in \mathbb{N}\backslash\{0\}$ variables z^i each of which can be assigned an integer number in $[\![1,k]\!]$ and let $h \equiv \min_{i=1}^{n} z^i$. A natural polytope related to these problems is given by

$$
P = Conv \left\{ \left(\begin{bmatrix} y_k^1 & y_k^2 & \cdots & y_k^n \\ \vdots & \vdots & \ddots & \vdots \\ y_2^1 & y_2^2 & \cdots & y_2^n \\ y_1^1 & y_1^2 & \cdots & y_1^n \end{bmatrix}, h \right) \in M_{k,n} \times \mathbb{N}\backslash\{0\} \,\middle|\, \begin{array}{l} \sum_{l=1}^{k} y_l^i = 1, \ \forall i \in [\![1,n]\!], \\ h = \min_{i \in [\![1,n]\!]} \sum_{l=1}^{k} l y_l^i, \end{array} \right\},
$$

where $M_{k,n}$ is the set of all k-by-n matrices with coefficients in $\{0,1\}$ and the variables y_l^i are interpreted as follows: $y_l^i = 1$ if and only if $z^i = l$.

The matrix (y_j^i) can be seen as a $\{0,1\}$ assignment matrix where each column contains exactly one coefficient equal to 1 while h denotes the index of the lowest row that is not identically equal to the zero row (cf. Fig. 1).

Another variant of P is obtained by considering the index of the highest row that is not identically equal to the zero row. In this case we get the polytope

$$
P' = Conv \left\{ \left(\begin{bmatrix} x_k^1 & x_k^2 & \cdots & x_k^n \\ \vdots & \vdots & \ddots & \vdots \\ x_2^1 & x_2^2 & \cdots & x_2^n \\ x_1^1 & x_1^2 & \cdots & x_1^n \end{bmatrix}, g \right) \in M_{k,n} \times \mathbb{N}\backslash\{0\} \,\middle|\, \begin{array}{l} \sum_{l=1}^{k} x_l^i = 1, \ \forall i \in [\![1,n]\!], \\ g = \max_{i \in [\![1,n]\!]} \sum_{l=1}^{k} l x_l^i, \end{array} \right\}.
$$

© Springer International Publishing Switzerland 2016
R. Cerulli et al. (Eds.): ISCO 2016, LNCS 9849, pp. 13–25, 2016.
DOI: 10.1007/978-3-319-45587-7_2

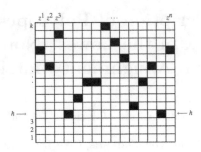

Fig. 1. On this grid representing a k-by-n matrix, each column is a z^i variable and the height of its black cell is the value assigned to it. h thus corresponds to the first non-empty row of the grid starting from the bottom (in this case, $h = 4$).

Polytopes P and P' naturally appear in the context of several combinatorial optimization. Let us for example consider the *minimum-span frequency-assignment problem* which is a variant of the NP-hard *frequency-assignment problem* [8]. Given a simple graph $G = (V, E)$ that is generally called the *interference graph*, the frequency assignment problem consists in assigning a frequency f from a set of available frequencies F to each vertex $v \in V$ in such a way that each pair of antennas $uv \in E$ that may interfere with one another are assigned different frequencies. Frequencies can be seen as ordered integer numbers. To reduce interferences, one might impose stronger constraints: a minimum separation between the frequencies assigned to u and v is required. If frequency i is assigned to u and j is assigned to v, then $|i - j| \geq s_{uv}$ where s_{uv} is a given number. The minimum-span frequency-assignment problem (or MS-FAP) consists in assigning frequencies to nodes taking into account the separation requirements and minimizing the difference between the largest assigned number (frequency) and the smallest assigned number (see, e.g., [7]).

If we consider that $V = \{v_1, \cdots, v_n\}$, $F = [\![1, k]\!]$ where k is an upper bound of the minimum-span, then we obtain the following formulation for MS-FAP

$$\begin{cases} \min g \\ \text{s.t. } x_l^i + x_{l'}^j \leq 1, \ \forall (i, j, l, l') \in [\![1, n]\!]^2 \times [\![1, k]\!]^2 \text{ such that } v_i v_j \in E, |l - l'| < s_{v_i v_j} \\ (x, g) \in P', \ x \in M_{k,n}. \end{cases}$$

where the interpretation of the x variable is the following: $x_l^i = 1$ if and only if the frequency l is assigned to the antenna v_i.

Another example is the *minimum makespan scheduling*, which is a central problem in the scheduling area (see [10]). Given a set J of jobs, a set M of machines that can all process at most one job at a time, and the time $t^{i,j} \in \mathbb{N}$ taken to process job $j \in J$ on machine $i \in M$, the goal of the minimum makespan scheduling problem is to assign a machine $p \in M$ for each job $j \in J$ so as to minimize the *makespan*, i.e. the maximum processing time of any machine. Several approximation schemes have been developed to deal with this NP-hard problem [3], e.g. [4,5]. Since the processing times are integers, the timeline is discretized in identical units of time, e.g. days. We consider here the variant where all the machines in M are identical (or IM-MMS) and preemptions are not allowed. In other words, for any job $j \in J$, $t^{i,j} = t^j$, $\forall i \in M$. In this case, assigning a machine to each job is equivalent to assigning a day d' to be the last day of processing this job, which also determines the first day d of processing and will therefore be processed by a machine free during the period $[d, d']$. Now to make a formulation for IM-MMS with the set of jobs $J = [\![1, n]\!]$ and $m \in \mathbb{N}\backslash\{0\}$ identical machines, we take $k = \sum_{i=1}^{n} t^i$ and the variable $x \in M_{k,n}$ whose interpretation is the following: $x_l^i = 1$ if and only if the processing of the job i ends on the day l. Then we have the following formulation for IM-MMS

$$\begin{cases} \min g \\ \text{s.t. } \sum_{l=1}^{k} l x_l^i \geq t^i, \ \forall i \in [\![1, n]\!], \\ \sum_{i=1}^{n} \sum_{l'=l}^{\min(l+t^i-1,k)} x_{l'}^i \leq m, \ \forall l \in [\![1, k]\!], \\ (x, g) \in P', \ x \in M_{k,n}. \end{cases}$$

For a job $i \in [\![1, n]\!]$, $\sum_{l=1}^{k} l x_l^i \geq t^i$ ensures that its processing ends after enough time has passed for i to be processed, and for a day $l \in [\![1, k]\!]$, $\sum_{i=1}^{n} \sum_{l'=l}^{\min(l+t^i-1,k)} x_{l'}^i \leq m$ ensures that no more than m jobs are being processed. Some additional constraints can be added to this formulation such as a necessary precedence or release time. If we want a job $i \in [\![1, n]\!]$ to be processed before another job $j \in [\![1, n]\!]\backslash\{i\}$ starts processing, we add the constraint $\sum_{l=1}^{k} l x_l^i \leq \sum_{l=1}^{k} l x_l^j - t^j$. If we want a job $i \in [\![1, n]\!]$ to be processed before (resp. on, after) a day $d \in [\![1, k]\!]$, we add the constraint $\sum_{l=1}^{d-1} x_l^i = 1$ (resp. $x_d^i = 1$, $\sum_{l=d+1}^{k} x_l^i = 1$). The objective function can also be any linear function depending on the x and g variables.

Another example is given by the problem of the *most imbalanced orientation of a graph* (or MAXIM) that consists in orienting the edges of a graph such that the minimum over all the vertices of the absolute difference between the outdegree and the indegree of a vertex is maximized (NP-complete) [1]. In other words, for a simple graph $G = (V, E)$, $\text{MAXIM}(G) = \max_{\Lambda \in \overrightarrow{O}(G)} \min_{v \in V} |d_\Lambda^+(v) - d_\Lambda^-(v)|$, where $\overrightarrow{O}(G)$ denotes the set of all the orientations of G and $d_\Lambda^+(v)$ (resp. $d_\Lambda^-(v)$) denotes the outdegree (resp. indegree) of v in G with respect to Λ. Now if we consider the graph G to be arbitrarily oriented and take its incidence matrix $B \in \{-1, 0, 1\}^{|V| \times |E|}$, we can describe an orientation of G with the variable $x \in \{-1, 1\}^{|E|}$ interpreted as follows. For each edge $uv \in E$, its orientation is kept from the original one if $x_{uv} = 1$ and reversed otherwise. Then if we look at the product of B with an orientation vector $x \in \{-1, 1\}^{|E|}$ we obtain $B_v x = d_x^+(v) - d_x^-(v)$, $\forall v \in V$, where B_v denotes the row of B corresponding to node v. In order to make a formulation of MAXIM, we consider indicator variables $t_l^v \in \{0, 1\}$ with $v \in V$ and $l \in [\![-k, k]\!]$, k being the maximum degree of the vertices of G, that have the following interpretation: $t_l^v = 1$ if and only if

$B_v x = d_x^+(v) - d_x^-(v) = l$, and thus we obtain the formulation

$$
\begin{cases}
\max \ h \\
\text{s.t. } h \leq \sum_{l=-k}^{k} |l| t_l^v, \ \forall v \in V, \\
\sum_{l=-k}^{k} t_l^v = 1, \ \forall v \in V, \\
\sum_{l=-k}^{k} l t_l^v = B_v x, \ \forall v \in V, \\
x \in [-1;1]^{|E|}, \ t \in M_{n,2k+1}, \ h \in \mathbb{R}.
\end{cases}
$$

Introducing variables $y_l^v = t_{-l}^v + t_l^v$, $\forall (v,l) \in V \times [\![1,k]\!]$, it becomes

$$
\begin{cases}
\max \ h \\
\text{s.t. } \sum_{l=-k}^{k} l t_l^v = B_v x, \ \forall v \in V, \\
y_l^v = t_{-l}^v + t_l^v, \ \forall (v,l) \in V \times [\![1,k]\!], \\
x \in [-1;1]^{|E|}, \ (y,h) \in P, \ t \in M_{n,2k+1}.
\end{cases}
$$

Considering the last formulation, a polyhedral analysis of the polytope P may be helpful in strengthening the linear relaxation of the original formulation within the framework of a cutting-plane algorithm (see, e.g. [2,11]).

Polytope P also appears in the context of the maximum clique problem. A discretized formulation is proposed in [9] where a variable w_q^i indicates whether the vertex i belongs to a clique of size q. These variables are of course linked to standard vertex variables ($x_i = 1$ if i belongs to the maximum clique). The problem is then equivalent to maximizing q such that $w_q^i = 1$ for some i. This is again related to polytope P.

More generally, several combinatorial optimization problems where discretization techniques are used can benefit from a description of either P or some of its variants.

The rest of the paper is organized as follows. First we give a complete linear description of P in Sect. 2. Then we show that the separation problem with respect to P can be solved in polynomial time in Sect. 3. Finally we give similar results for the polyhedron P' and others in Sect. 4 and conclude in Sect. 5.

2 A Full Description of P

Let us define a set of inequalities that will prove to be an hyperplane representation of P.

$$
\tilde{P} = \begin{cases}
\sum_{l=1}^{k} y_l^i = 1, \ \forall i \in [\![1,n]\!], \\
\sum_{l=2}^{k} \sum_{i=1}^{n} \lambda_l^i y_l^i \geq h - 1, \ \forall \lambda \in \Lambda, \\
\sum_{l=1}^{h_{\max}-1} \sum_{i=1}^{n} (l - h_{\max}) y_l^i + h_{\max} \leq h, \ \forall h_{\max} \in [\![1,k]\!], \\
y_l^i \geq 0, \ \forall (i,l) \in [\![1,n]\!] \times [\![1,k]\!], \ h \in \mathbb{R},
\end{cases}
$$

where

$$
\Lambda = \left\{ \lambda = (\lambda_l^i)_{(i,l) \in [\![1,n]\!] \times [\![1,k]\!]} \in \mathbb{N}^{nk} \ \middle| \ \begin{matrix} \lambda_{l+1}^i \geq \lambda_l^i, \ \forall (i,l) \in [\![1,n]\!] \times [\![1,k-1]\!] \\ \sum_{i=1}^{n} \lambda_l^i = l - 1, \ \forall l \in [\![1,k]\!] \end{matrix} \right\}.
$$

Any element λ of Λ can be constructed as follows: we start with $\lambda_1^i = 0$, $\forall i \in [\![1,n]\!]$, choose an index $i_2 \in [\![1,n]\!]$ and set $\lambda_2^{i_2} = 1$ and $\lambda_2^i = 0$, $\forall i \in [\![1,n]\!]\backslash\{i_2\}$. And we proceed like this for $l = 2, \cdots, k$, we choose an index $i_l \in [\![1,n]\!]$ and set $\lambda_l^{i_l} = \lambda_{l-1}^{i_l} + 1$ and $\lambda_l^i = \lambda_{l-1}^i$, $\forall i \in [\![1,n]\!]\backslash\{i_l\}$.

Lemma 1

$$P \subseteq \tilde{P}$$

Proof. Since P is the convex hull of integer points, it suffices to show that each of those points satisfies all the inequalities in $\tilde{P}$. Let $(y,h) \in P$ be one of those points, that is to say $(y,h) \in M_{k,n} \times \mathbb{N}\backslash\{0\}$, $\sum_{l=1}^k y_l^i = 1$, $\forall i \in [\![1,n]\!]$ and $h = \min_{i \in [\![1,n]\!]} \sum_{l=1}^k l y_l^i$. We firstly show that $\sum_{l=2}^k \sum_{i=1}^n \lambda_l^i y_l^i \geq h - 1$, $\forall \lambda \in \Lambda$. For all $i \in [\![1,n]\!]$, there exists $l_i \in [\![h,k]\!]$ such that $y_{l_i}^i = 1$. Now since for a given $i \in [\![1,n]\!]$, λ_l^i increases with l, we have

$$\sum_{l=2}^k \sum_{i=1}^n \lambda_l^i y_l^i = \sum_{i=1}^n \lambda_{l_i}^i \geq \sum_{i=1}^n \lambda_h^i = h - 1.$$

Now we take $h_{\max} \in [\![1,k]\!]$, and we show that $h_{\max} - \sum_{l=1}^{h_{\max}-1} (h_{\max} - l) \sum_{i=1}^n y_l^i \leq h$. We have

$$h_{\max} - \sum_{l=1}^{h_{\max}-1} (h_{\max} - l) \sum_{i=1}^n y_l^i$$

$$= h_{\max} + \sum_{l=1}^k \min(l - h_{\max}, 0) \sum_{i=1}^n y_l^i$$

$$= h_{\max} + \sum_{l=1}^k \min(l - 1, h_{\max} - 1) \sum_{i=1}^n y_l^i - n(h_{\max} - 1)$$

$$= \sum_{i=1}^n \sum_{l=2}^k \min(l - 1, h_{\max} - 1) y_l^i + 1 - (n-1)(h_{\max} - 1)$$

There exists $i^* \in [\![1,n]\!]$ such that $y_h^{i^*} = 1$, then

$$\sum_{i=1}^n \sum_{l=2}^k \min(l - 1, h_{\max} - 1) y_l^i$$

$$= \sum_{\substack{i=1 \\ i \neq i^*}}^n \sum_{l=2}^k \min(l-1, h_{\max} - 1) y_l^i + \sum_{l=2}^k \min(l-1, h_{\max} - 1) y_l^{i^*}$$

$$\leq \sum_{\substack{i=1 \\ i \neq i^*}}^n \sum_{l=2}^k (h_{\max} - 1) y_l^i + \sum_{l=2}^k (l-1) y_l^{i^*} \leq \sum_{\substack{i=1 \\ i \neq i^*}}^n (h_{\max} - 1) + h - 1$$

$$= (n-1)(h_{\max} - 1) + h - 1.$$

$\square$

Now to prove that P coincides with $\tilde{P}$, we show that all facet-defining inequalities for P are among those defining $\tilde{P}$. Two inequalities are said to be *equivalent* if one can be obtained from the other by multiplying it by a non-zero scalar and adding a combination of equations of the type $\sum_{l=1}^k y_l^i = 1$.

Lemma 2. *Let*

$$\sum_{l=1}^{k}\sum_{i=1}^{n}\beta_l^i y_l^i + \gamma \geq 0, \tag{1}$$

be a facet-defining inequality of P, with $\beta_l^i \in \mathbb{R}$, $\forall (i,l) \in [\![1,n]\!] \times [\![1,k]\!]$, $\gamma \in \mathbb{R}$. Then there exists $(i,l) \in [\![1,n]\!] \times [\![1,k]\!]$ such that (1) is equivalent to $y_l^i \geq 0$.

For an extreme point (y,h) of P and $(\tilde{i},\tilde{l},\tilde{l}') \in [\![1,n]\!] \times [\![1,k]\!]^2$, such that $y_{\tilde{l}}^{\tilde{i}} = 1$ and $\tilde{l} \neq \tilde{l}'$, we denote by $(y_{\tilde{l} \to \tilde{l}'}^{\tilde{i}}, h_{\tilde{l} \to \tilde{l}'}^{\tilde{i}})$ the extreme point (y',h') of P such that $y_l'^i = y_l^i$, $\forall (i,l) \in [\![1,n]\!] \times [\![1,k]\!] \setminus \{(\tilde{i},\tilde{l}),(\tilde{i},\tilde{l}')\}$, $y_{\tilde{l}}'^{\tilde{i}} = 0$, $y_{\tilde{l}'}'^{\tilde{i}} = 1$ and $h' = \min_{i \in [\![1,n]\!]} \sum_{l=1}^{k} l y_l'^i$. For $\tilde{l} \in [\![1,k]\!]$, we denote by $(y_{\to \tilde{l}}, \tilde{l})$ the point of P such that $y_{\tilde{l}}^i = 1$, $\forall i \in [\![1,n]\!]$ and $y_l^i = 0$, $\forall (i,l) \in [\![1,n]\!] \times ([\![1,k]\!] \setminus \{\tilde{l}\})$.

Proof. First, since for each $i \in [\![1,n]\!]$, we have $\sum_{l=1}^{k} y_l^i = 1$, we can replace y_1^i by $1 - \sum_{l=2}^{k} y_l^i$ and get new coefficients $\tilde{\beta}_1^i = 0$ and $\tilde{\beta}_l^i = \beta_l^i - \beta_1^i$, $\forall l \in [\![2,k]\!]$ and a new $\tilde{\gamma} = \gamma + \sum_{i=1}^{n} \beta_1^i$. Hence, without loss of generality, we can assume that $\beta_1^i = 0$ for all $i \in [\![1,n]\!]$. Suppose that the facet defined by (1) is not equivalent to a facet defined by an inequality of the type $y_l^i \geq 0$. If we take $(\tilde{i},\tilde{l}) \in [\![1,n]\!] \times [\![1,k-1]\!]$, we know that there exists an extreme point (y,h) of P saturating (1) and such that $y_{\tilde{l}}^{\tilde{i}} = 1$, otherwise all the extreme points saturating (1) would saturate $y_{\tilde{l}}^{\tilde{i}} \geq 0$ thus contradicting the fact that (1) is facet-defining and not equivalent to $y_l^i \geq 0$ for some $(i,l) \in [\![1,n]\!] \times [\![1,k]\!]$. Since $(y',h') = (y_{\tilde{l} \to \tilde{l}+1}^{\tilde{i}}, h_{\tilde{l} \to \tilde{l}+1}^{\tilde{i}}) \in P$, we have $\sum_{l=1}^{k}\sum_{i=1}^{n}\beta_l^i y_l^i + \gamma = 0$ and $\sum_{l=1}^{k}\sum_{i=1}^{n}\beta_l^i y_l'^i + \gamma \geq 0$ which yields $\beta_{\tilde{l}+1}^{\tilde{i}} \geq \beta_{\tilde{l}}^{\tilde{i}}$. Similarly, taking an extreme point (y,h) of P saturating (1) and such that $y_{\tilde{l}+1}^{\tilde{i}} = 1$ and $(y',h') = (y_{\tilde{l}+1 \to \tilde{l}}^{\tilde{i}}, h_{\tilde{l}+1 \to \tilde{l}}^{\tilde{i}}) \in P$, we have $\sum_{l=1}^{k}\sum_{i=1}^{n}\beta_l^i y_l^i + \gamma = 0$ and $\sum_{l=1}^{k}\sum_{i=1}^{n}\beta_l^i y_l'^i + \gamma \geq 0$, yielding $\beta_{\tilde{l}}^{\tilde{i}} \geq \beta_{\tilde{l}+1}^{\tilde{i}}$. Hence, for all $(i,l) \in [\![1,n]\!] \times [\![1,k-1]\!]$, we have $\beta_l^i = \beta_{l+1}^i$, in other words, $\beta_l^i = 0$, $\forall (i,l) \in [\![1,n]\!] \times [\![1,k]\!]$, and (1) is not facet-defining. $\square$

Lemma 3. *Let*

$$\sum_{l=1}^{k}\sum_{i=1}^{n}\beta_l^i y_l^i + \gamma \geq h, \tag{2}$$

be a facet-defining inequality of P, with $\beta_l^i \in \mathbb{R}$, $\forall (i,l) \in [\![1,n]\!] \times [\![1,k]\!]$, $\gamma \in \mathbb{R}$. Then there exists $\lambda \in \Lambda$ such that (2) is equivalent to $\sum_{l=2}^{k}\sum_{i=1}^{n}\lambda_l^i y_l^i \geq h-1$.

Proof. Again, without loss of generality, we assume that $\beta_1^i = 0$ for all $i \in [\![1,n]\!]$. For an $(\tilde{i},\tilde{l}) \in [\![1,n]\!] \times [\![1,k-1]\!]$, there exists an extreme point (y,h) of P saturating (2) and such that $y_{\tilde{l}}^{\tilde{i}} = 1$. Since $(y',h') = (y_{\tilde{l} \to \tilde{l}+1}^{\tilde{i}}, h_{\tilde{l} \to \tilde{l}+1}^{\tilde{i}}) \in P$, we have $\sum_{l=1}^{k}\sum_{i=1}^{n}\beta_l^i y_l^i + \gamma = h$ and $\sum_{l=1}^{k}\sum_{i=1}^{n}\beta_l^i y_l'^i + \gamma \geq h' \geq h$ which yields $\beta_{\tilde{l}+1}^{\tilde{i}} \geq \beta_{\tilde{l}}^{\tilde{i}}$. Hence for all $i \in [\![1,n]\!]$, $(\beta_l^i)_l$ is increasing with l and therefore non-negative since $\beta_1^i = 0$, $\forall i \in [\![1,n]\!]$.

If we consider the point $(y, h) = (y_{\rightarrow 1}, 1)$, we obtain that $\gamma \geq 1$. If we now consider an extreme point (y, h) of P saturating (2) and such that $y_1^{\tilde{i}} = 1$ for an $\tilde{i} \in [\![1, n]\!]$, we get $\sum_{l=1}^{k} \sum_{i=1}^{n} \beta_l^i y_l^{\prime i} + \gamma = h = 1$. But since both the β_l^i and the y_l^i are non-negative, then so is $\sum_{l=1}^{k} \sum_{i=1}^{n} \beta_l^i y_l^{\prime i}$. Hence $\gamma \leq 1$, yielding $\gamma = 1$.

Considering $(y_{\rightarrow l}, l) \in P$ for $l \in [\![1, k]\!]$, we obtain $\sum_{i=1}^{n} \beta_l^i \geq l - 1$. Let us show by induction on l that $\sum_{i=1}^{n} \beta_l^i = l - 1$, $\forall l \in [\![1, k]\!]$. Our induction is already initialized by $\sum_{i=1}^{n} \beta_1^i = 0$. We suppose that for a $\tilde{l} \in [\![1, k-1]\!]$, we have $\sum_{i=1}^{n} \beta_{\tilde{l}}^i = \tilde{l} - 1$ and show that $\sum_{i=1}^{n} \beta_{\tilde{l}+1}^i = \tilde{l}$. Suppose that all the extreme points (y, h) of P saturating (2) and such that $y_{\tilde{l}+1}^i = 1$ for some $i \in [\![1, n]\!]$ verify $h \leq \tilde{l}$. Then for each $\tilde{i} \in [\![1, n]\!]$, take one of those extreme saturating points such that $y_{\tilde{l}+1}^{\tilde{i}} = 1$ and let $(y', h') = (y_{\tilde{l}+1 \xrightarrow{\tilde{i}} \tilde{l}}, h_{\tilde{l}+1 \xrightarrow{\tilde{i}} \tilde{l}}) \in P$. Since $\beta_{\tilde{l}+1}^{\tilde{i}} \geq \beta_{\tilde{l}}^{\tilde{i}}$, we have $h' - 1 \leq \sum_{l=1}^{k} \sum_{i=1}^{n} \beta_l^i y_l^{\prime i} \leq \sum_{l=1}^{k} \sum_{i=1}^{n} \beta_l^i y_l^i = h - 1$, and since $h \leq \tilde{l}$, we have $h = h'$ and therefore, $\sum_{l=1}^{k} \sum_{i=1}^{n} \beta_l^i y_l^{\prime i} = \sum_{l=1}^{k} \sum_{i=1}^{n} \beta_l^i y_l^i = h - 1$, yielding $\beta_{\tilde{l}+1}^{\tilde{i}} = \beta_{\tilde{l}}^{\tilde{i}}$. Thus $\sum_{i=1}^{n} \beta_{\tilde{l}+1}^i = \sum_{i=1}^{n} \beta_{\tilde{l}}^i = \tilde{l} - 1$ which contradicts $\sum_{i=1}^{n} \beta_{\tilde{l}+1}^i \geq \tilde{l}$. So there exists $\tilde{i} \in [\![1, n]\!]$ and (y, h) an extreme point of P saturating (2) such that $y_{\tilde{l}+1}^{\tilde{i}} = 1$ and $h = \tilde{l} + 1$. We have $\tilde{l} \leq \sum_{i=1}^{n} \beta_{\tilde{l}+1}^i \leq \sum_{l=1}^{k} \sum_{i=1}^{n} \beta_l^i y_l^i = \tilde{l}$, hence $\sum_{i=1}^{n} \beta_{\tilde{l}+1}^i = \tilde{l}$, which concludes our induction.

Now let us show by induction on l that for all $(i, l) \in [\![1, n]\!] \times [\![1, k]\!]$, β_l^i is an integer. This induction is trivially initialized for $\beta_1^i = 0$, $\forall i \in [\![1, n]\!]$. We suppose that for a $\tilde{l} \in [\![1, k-1]\!]$ we have that the $\beta_{\tilde{l}}^i$ for $i \in [\![1, n]\!]$ are integers and we show that the same holds for the $\beta_{\tilde{l}+1}^i$ for $i \in [\![1, n]\!]$. We note $\alpha^i = \beta_{\tilde{l}+1}^i - \beta_{\tilde{l}}^i$, $\forall i \in [\![1, n]\!]$ and for each $i \in [\![1, n]\!]$ we build a new set of inequality coefficients: $\beta_l^{(i),j} = \beta_l^j - \alpha^j + \delta_{i,j}$, $\forall (j, l) \in [\![1, n]\!] \times [\![1, k]\!]$, where $\delta_{i,j}$ equals 1 if $i = j$ and 0 otherwise. Let (y, h) be an extreme point of P and for all $j \in [\![1, n]\!]$, let $l_j \in [\![h, k]\!]$ be such that $y_{l_j}^j = 1$. Then, since $\sum_{i=1}^{n} \alpha^i = 1$, we have for $i \in [\![1, n]\!]$

$$\sum_{l=1}^{k} \sum_{j=1}^{n} \beta_l^{(i),j} y_l^j = \sum_{l=1}^{k} \sum_{j=1}^{n} (\beta_l^j - \alpha^j + \delta_{i,j}) y_l^j$$

$$= \sum_{j=1}^{n} (\beta_{l_j}^j - \alpha^j + \delta_{i,j}) = \sum_{j=1}^{n} \beta_{l_j}^j \geq \sum_{i=1}^{n} \beta_h^i = h - 1,$$

which means that for all $i \in [\![1, n]\!]$, $\sum_{l=1}^{k} \sum_{j=1}^{n} \beta_l^{(i),j} y_l^j + 1 \geq h$ is valid for P. Now since for $(j, l) \in [\![1, n]\!] \times [\![1, k]\!]$,

$$\left(\sum_{i=1}^{n} \alpha^i \beta^{(i)} \right)_l^j = \sum_{i=1}^{n} \alpha^i \left(\beta_l^j - \alpha^j + \delta_{i,j} \right) = \left(\sum_{i=1}^{n} \alpha^i \right) \beta_l^j - \left(\sum_{i=1}^{n} \alpha^i \right) \alpha^j + \sum_{i=1}^{n} \alpha^i \delta_{i,j} = \beta_l^j$$

and $\alpha^i \geq 0$, $\forall i \in [\![1, n]\!]$, (2) is a convex combination of these inequalities. Moreover, if any of the $\beta_{\tilde{l}+1}^i$, $i \in [\![1, n]\!]$ was not an integer, then the convex combination would be non-trivial, which would contradict the fact that (2) is facet-defining. Concluding our induction, we obtain that for all $(i,l) \in [\![1,n]\!] \times [\![1,k]\!]$, β_l^i is an integer. And thus that $\beta \in \Lambda$, i.e. (2) belongs to the set of inequalities defining $\tilde{P}$. □

Lemma 4. *Let*

$$\sum_{l=1}^{k} \sum_{i=1}^{n} \beta_l^i y_l^i + \gamma \leq h, \tag{3}$$

be a facet-defining inequality of P, with $\beta_l^i \in \mathbb{R}$, $\forall(i,l) \in [\![1,n]\!] \times [\![1,k]\!]$, $\gamma \in \mathbb{R}$. Then it is equivalent to $h_{\max} - \sum_{l=1}^{h_{\max}-1}(h_{\max}-l)\sum_{i=1}^{n} y_l^i \leq h$ for some $h_{\max} \in [\![1,k]\!]$.

Proof. Since for each $i \in [\![1,n]\!]$, we have $\sum_{l=1}^{k} y_l^i = 1$, we can replace β_l^i by $\tilde{\beta}_l^i = \beta_l^i - v_i$, for some $v_i \geq 0$ such that $\tilde{\beta}_l^i \leq 0$, $\forall(i,l) \in [\![1,n]\!] \times [\![1,k]\!]$, and γ by $\tilde{\gamma} = \gamma + \sum_{i=1}^{n} v_i$ and thus get new coefficients $\tilde{\beta}$ which are non-positive without changing (3). So without loss of generality, we can assume that β is non-positive and furthermore that γ is minimal for a non-positive β.

For $(\tilde{i}, \tilde{l}) \in [\![1,n]\!] \times [\![2,k]\!]$, we take an extreme point (y,h) of P saturating (3) such that $y_{\tilde{l}}^{\tilde{i}} = 1$ and $(y',h') = (y_{\tilde{l} \xrightarrow{\tilde{i}} \tilde{l}-1}, h_{\tilde{l} \xrightarrow{\tilde{i}} \tilde{l}-1})$. We have $\sum_{l=1}^{k} \sum_{i=1}^{n} \beta_l^i y_l^i + \gamma = h$ and $\sum_{l=1}^{k} \sum_{i=1}^{n} \beta_l^i y_l'^i + \gamma \leq h' \leq h$, which yields $\beta_{\tilde{l}-1}^{\tilde{i}} \leq \beta_{\tilde{l}}^{\tilde{i}}$. In other words, β_l^i is increasing with l, for all $i \in [\![1,n]\!]$. This implies that for all $i \in [\![1,n]\!]$, there exists $l_i \in [\![1,k]\!]$ such that $\beta_l^i = 0$, $\forall l \in [\![l_i, k]\!]$ and $\beta_l^i < 0$ for $l < l_i$ because suppose there exists $i \in [\![1,n]\!]$ for which $\beta_k^i > 0$, then we can replace β_l^i by $\beta_l^i - \beta_k^i$ for all $l \in [\![1,k]\!]$ and subtract β_k^i from γ and thus get new non-positive coefficients $\tilde{\beta}$ with a $\tilde{\gamma} < \gamma$, which contradicts the minimality of γ.

Let $(\tilde{i}, \tilde{l}) \in [\![1,n]\!] \times [\![1,k]\!]$ such that $\beta_{\tilde{l}}^{\tilde{i}} < 0$ (i.e., $\tilde{l} < l_{\tilde{i}}$), we know there exists an extreme point (y,h) of P saturating (3) such that $y_{\tilde{l}}^{\tilde{i}} = 1$. Suppose that $h < \tilde{l}$, we take (y',h') an extreme point of P such that $y_l'^i = y_l^i$, $\forall(i,l) \in [\![1,n]\!] \setminus \{\tilde{i}\} \times [\![1,k]\!]$, $y_{l_{\tilde{i}}}'^{\tilde{i}} = 1$ and obtain $h = \sum_{l=1}^{k} \sum_{i=1}^{n} \beta_l^i y_l^i + \gamma \leq \sum_{l=1}^{k} \sum_{i=1}^{n} \beta_l^i y_l'^i + \gamma \leq h' = h$, that yields $\beta_{\tilde{l}}^{\tilde{i}} = 0$, which is a contradiction. Hence $h = \tilde{l}$, so if we take the extreme point (y',h') of P such that $y_{\tilde{l}}^{\tilde{i}} = 1$, $y_k'^i = 1$, $\forall i \in [\![1,n]\!] \setminus \{\tilde{i}\}$ and $y_l^i = 0$, $\forall(i,l) \in [\![1,n]\!] \times [\![1,k]\!] \setminus (\{(\tilde{i}, \tilde{l})\} \cup \{(i,k), i \in [\![1,n]\!] \setminus \{\tilde{i}\}\})$, we have

$$\tilde{l} = \sum_{l=1}^{k} \sum_{i=1}^{n} \beta_l^i y_l^i + \gamma \leq \sum_{l=1}^{k} \sum_{i=1}^{n} \beta_l^i y_l'^i + \gamma = \beta_{\tilde{l}}^{\tilde{i}} + \gamma \leq h' = \tilde{l}.$$

This gives us that for all $i \in [\![1,n]\!]$, $\beta_l^i = l - \gamma$, $\forall l \in [\![1, l_i - 1]\!]$. Consider the extreme point (y,h) of P such that for all $i \in [\![1,n]\!]$, $y_{l_i}^i = 1$, we have $\gamma \leq \min_{i \in [\![1,n]\!]} l_i$. We call $h_{\max}$ the maximum value of h among the extreme points (y,h) of P saturating (3).

If we take an extreme point (y,h) of P realizing $h_{\max}$, i.e. saturating (3) and such that $h = h_{\max}$, we have $\sum_{l=1}^{k} \sum_{i=1}^{n} \beta_l^i y_l^i + \gamma = h_{\max}$ which, since

$\sum_{l=1}^{k} \sum_{i=1}^{n} \beta_l^i y_l^i$ is non-positive, yields $h_{\max} \leq \gamma$. Now for $\tilde{i} \in [\![1, n]\!]$, there exists an extreme point (y, h) of P saturating (3) such that $y_{l_{\tilde{i}}^{i}-1}^{\tilde{i}} = 1$. Suppose that $h < l_{\tilde{i}} - 1$, we take (y', h') an extreme point of P such that $y_l'^{i} = y_l^i$, $\forall (i, l) \in [\![1, n]\!] \backslash \{\tilde{i}\} \times [\![1, k]\!]$, $y_{l_{\tilde{i}}}^{\tilde{i}} = 1$ and obtain

$$h = \sum_{l=1}^{k} \sum_{i=1}^{n} \beta_l^i y_l^i + \gamma \leq \sum_{l=1}^{k} \sum_{i=1}^{n} \beta_l^i y_l'^{i} + \gamma \leq h' = h,$$

that yields $\beta_{l_{\tilde{i}}^{-1}}^{\tilde{i}} = 0$, which is a contradiction. So $h = l_{\tilde{i}} - 1$, hence $h_{\max} \geq l_{\tilde{i}} - 1$. We obtain $\max_{i \in [\![1,n]\!]} l_i - 1 \leq h_{\max} \leq \gamma \leq \min_{i \in [\![1,n]\!]} l_i \leq \max_{i \in [\![1,n]\!]} l_i$. There are two possibilities, either $\min_{i \in [\![1,n]\!]} l_i = \max_{i \in [\![1,n]\!]} l_i$, or $\max_{i \in [\![1,n]\!]} l_i - 1 = \min_{i \in [\![1,n]\!]} l_i$. Suppose $\max_{i \in [\![1,n]\!]} l_i - 1 = h_{\max} = \gamma = \min_{i \in [\![1,n]\!]} l_i$, then there exists $\tilde{i} \in [\![1, n]\!]$ such that $l_{\tilde{i}} = 1 + h_{\max}$, which implies that $\beta_{h_{\max}}^{\tilde{i}} \neq 0$ and $\beta_{h_{\max}}^{\tilde{i}} = h_{\max} - \gamma = 0$ which is a contradiction. So $\min_{i \in [\![1,n]\!]} l_i = \max_{i \in [\![1,n]\!]} l_i =: L$ and

$$L - 1 \leq h_{\max} \leq \gamma \leq L.$$

Let us assume that $h_{\max} = L - 1$. Then we know that $\gamma < L$, otherwise if $\gamma = L$, the extreme point of P $(y_{\to L}, L)$ would saturate (3) and thus contradict the maximality of $h_{\max}$. We consider the following inequality

$$\sum_{l=1}^{L-1} \sum_{i=1}^{n} (l - L) y_l^i + L \leq h. \tag{4}$$

Let us show that it is a valid inequality for P, that is to say, that every extreme point (y, h) of P verifies it. If $h \geq L$, then $\sum_{l=1}^{L-1} \sum_{i=1}^{n} (l - L) y_l^i = 0$ and we are done. If $h \leq L - 1$, then there exists $\tilde{i} \in [\![1, n]\!]$ such that $y_h^{\tilde{i}} = 1$. Combining this with the validity of (3) implies

$$\sum_{l=1}^{L-1} \sum_{i=1}^{n} (l-L) y_l^i + L = \sum_{l=1}^{L-1} \sum_{i=1}^{n} (l-\gamma) y_l^i + \gamma + \sum_{l=1}^{L-1} \sum_{i=1}^{n} (\gamma-L) y_l^i + L - \gamma \leq h + \gamma - L + L - \gamma = h.$$

Moreover, if (y, h) is an extreme point of P saturating (3), then there exists $\tilde{i} \in [\![1, n]\!]$ such that $y_h^{\tilde{i}} = 1$ and $h \leq h_{\max} = L - 1$ which yields $\sum_{l=1}^{L-1} \sum_{i=1}^{n} (l-\gamma) y_l^i + \gamma = h = \sum_{l=1}^{L-1} \sum_{\substack{i=1 \\ i \neq \tilde{i}}}^{n} (l-\gamma) y_l^i + h - \gamma + \gamma$. So we have $\sum_{l=1}^{L-1} \sum_{\substack{i=1 \\ i \neq \tilde{i}}}^{n} (l-\gamma) y_l^i = 0$ which implies that $y_l^i = 0$, $\forall (i, l) \in ([\![1, n]\!] \backslash \{\tilde{i}\}) \times [\![1, L-1]\!]$. Thus

$$\sum_{l=1}^{L-1} \sum_{i=1}^{n} (l-L) y_l^i + L = \sum_{l=1}^{L-1} \sum_{i=1}^{n} (l-\gamma) y_l^i + \gamma + \sum_{l=1}^{L-1} \sum_{i=1}^{n} (\gamma-L) y_l^i + L - \gamma = h + \gamma - L + L - \gamma = h.$$

Consequently, all points saturating (3) saturate (4), furthermore, $(y_{\to L}, L)$ saturates (4) and not (3). This means that the face of the polyhedron defined by

(3) is strictly contained in the face defined by (4) contradicting the fact that (3) is facet-defining, hence $h_{max} = \gamma = L$ and (3) becomes $\sum_{l=1}^{h_{max}-1} \sum_{i=1}^{n} (l - h_{max}) y_l^i + h_{max} \leq h$ with $h_{max} \in [\![1, n]\!]$. □

Theorem 5

$$P = \tilde{P}$$

Proof. With Lemma 1, we know that $P \subseteq \tilde{P}$. Take any facet-defining inequality of P $\sum_{l=1}^{k} \sum_{i=1}^{n} \beta_l^i y_l^i + \gamma \geq \alpha h$, where $\beta_l^i \in \mathbb{R}$, $\forall (i, l) \in [\![1, n]\!] \times [\![1, k]\!]$, $(\gamma, \alpha) \in \mathbb{R} \times \{-1, 0, 1\}$. If $\alpha = 0$ (resp. $\alpha = 1$, $\alpha = -1$), then Lemma 2 (resp. Lemmas 3, 4) gives us that this inequality is equivalent to one of those defining $\tilde{P}$.

3 Separation Problem

P is defined by n equalities, kn non-negativity constraints, k constraints of type $\sum_{l=1}^{h_{max}-1} \sum_{i=1}^{n} (l - h_{max}) y_l^i + h_{max} \leq h$ and n^{k-1} of inequalities of type $\sum_{l=2}^{k} \sum_{i=1}^{n} \lambda_l^i y_l^i \geq h - 1$. The total number of inequalities is then exponential. However, the following holds.

Theorem 6. *The separation problem which consists in deciding if a vector $(y, h) \in \mathbb{R}^{nk+1}$ is in P, and if not in returning a constraint of P violated by (y, h) can be solved in polynomial time.*

Proof. Let $(y, h) \in \mathbb{R}^{nk+1}$, first, one can verify in linear time if $(y, h) \in [0, 1]^{nk} \times [1, k]$ is such that $\sum_{l=1}^{k} y_l^i = 1$, $\forall i \in [\![1, n]\!]$ and verifies the k inequalities of type $\sum_{l=1}^{h_{max}-1} \sum_{i=1}^{n} (l - h_{max}) y_l^i + h_{max} \leq h$. If not, we return a violated constraint. Otherwise, we build $\tilde{\lambda} \in \Lambda$ as follows: $\tilde{\lambda}_1^i = 0$, $\forall i \in [\![1, n]\!]$, and for $l = 2, \cdots, k$, let $\tilde{i}_l = arg \, min_{i \in [\![1,n]\!]} y_l^i + y_{l+1}^i + \cdots + y_k^i$ and set $\tilde{\lambda}_l^{\tilde{i}_l} = \tilde{\lambda}_{l-1}^{\tilde{i}_l} + 1$ and $\tilde{\lambda}_l^i = \tilde{\lambda}_{l-1}^i$, $\forall i \in [\![1, n]\!] \backslash \{\tilde{i}_l\}$. We will show that if (y, h) satisfies the inequality of P corresponding to $\tilde{\lambda}$, then it satisfies all the inequalities corresponding to an element of Λ. Suppose $\sum_{l=2}^{k} \sum_{i=1}^{n} \tilde{\lambda}_l^i y_l^i \geq h - 1$ and let $\lambda \in \Lambda$ and $(i_2, \cdots, i_k) \in [\![1, n]\!]^{k-1}$ the indices corresponding to the building of λ, i.e. for $l = 2, \cdots, k$, $\lambda_l^{i_l} = \lambda_{l-1}^{i_l} + 1$ and $\lambda_l^i = \lambda_{l-1}^i$, $\forall i \in [\![1, n]\!] \backslash \{i_l\}$. By construction of $i_2, \cdots, i_k$ and $\tilde{i}_2, \cdots, \tilde{i}_k$, we have

$$\sum_{l=2}^{k} \sum_{i=1}^{n} \lambda_l^i y_l^i = \sum_{l=2}^{k} (y_l^{i_l} + y_{l+1}^{i_l} + \cdots + y_k^{i_l}) \geq \sum_{l=2}^{k} (y_l^{\tilde{i}_l} + y_{l+1}^{\tilde{i}_l} + \cdots + y_k^{\tilde{i}_l}) = \sum_{l=2}^{k} \sum_{i=1}^{n} \tilde{\lambda}_l^i y_l^i \geq h - 1,$$

hence the inequality of P corresponding to λ is satisfied. So we can conclude that if (y, h) satisfies the inequality of P corresponding to $\tilde{\lambda}$, then it satisfies all the inequalities of P corresponding to an element of Λ. And since the construction of $\tilde{\lambda}$ is done in polynomial time, the separation problem is indeed polynomial. □

The previous result is very useful in the context of cutting plane algorithms where only violated inequalities are added and not all valid inequalities.

4   Variants

Linear programming formulations aiming to maximize (resp. minimize) the index of the lowest (resp. highest) nonzero row of an assignment matrix are related to polytope Q (resp. Q') described below. Observe that h (resp. g) is only required to be less (resp. more) than or equal to $\min_{i \in [\![1,n]\!]} \sum_{l=1}^{k} l y_l^i$ (resp. $\max_{i \in [\![1,n]\!]} \sum_{l=1}^{k} l x_l^i$).

$$Q = Conv \left\{ \left(\begin{bmatrix} y_k^1 & y_k^2 & \cdots & y_k^n \\ \vdots & \vdots & \ddots & \vdots \\ y_2^1 & y_2^2 & \cdots & y_2^n \\ y_1^1 & y_1^2 & \cdots & y_1^n \end{bmatrix}, h \right) \in M_{k,n} \times \mathbb{N}\backslash\{0\} \left| \begin{matrix} \sum_{l=1}^{k} y_l^i = 1, \ \forall i \in [\![1,n]\!], \\ h \leq \min_{i \in [\![1,n]\!]} \sum_{l=1}^{k} l y_l^i, \end{matrix} \right. \right\}$$

$$Q' = Conv \left\{ \left(\begin{bmatrix} x_k^1 & x_k^2 & \cdots & x_k^n \\ \vdots & \vdots & \ddots & \vdots \\ x_2^1 & x_2^2 & \cdots & x_2^n \\ x_1^1 & x_1^2 & \cdots & x_1^n \end{bmatrix}, g \right) \in M_{k,n} \times \mathbb{N}\backslash\{0\} \left| \begin{matrix} \sum_{l=1}^{k} x_l^i = 1, \ \forall i \in [\![1,n]\!], \\ g \geq \max_{i \in [\![1,n]\!]} \sum_{l=1}^{k} l x_l^i, \end{matrix} \right. \right\}$$

A full description of Q is given below.

Theorem 7

$$Q = \begin{cases} \sum_{l=1}^{k} y_l^i = 1, \ \forall i \in [\![1,n]\!], \\ \sum_{l=2}^{k} \sum_{i=1}^{n} \lambda_l^i y_l^i \geq h - 1, \ \forall \lambda \in \Lambda, \\ y_l^i \geq 0, \ \forall (i,l) \in [\![1,n]\!] \times [\![1,k]\!], \ h \geq 1. \end{cases}$$

□

Proof. It is a simple fact that $h \geq 1$ is the only possible facet of type (3) while the positivity constraints are the only possible facets of type (1). Let us consider an inequality of type (2) defining a facet of Q. Any extreme point of Q saturating such a facet necessarily satisfies $h = \min_{i \in [\![1,n]\!]} \sum_{l=1}^{k} l y_l^i$ implying that it is also a point of P. Using this observation and the fact that Q and P have the same dimension, we deduce that any facet of Q of type (2) is also a facet of P. Using the description of P, we get the result. □

Similarly to Theorem 6 we can deduce that the separation problem with respect to Q is solvable in polynomial time as well.

We can also derive a full description of P' and Q' from the previous results.

Theorem 8

$$P' = \begin{cases} \sum_{l=1}^{k} x_l^i = 1, \ \forall i \in [\![1,n]\!], \\ \sum_{l=1}^{k-1} \sum_{i=1}^{n} \lambda_l^i x_l^i \leq g - k, \ \forall \lambda \in \tilde{\Lambda}, \\ \sum_{l=g_{\min}+1}^{k} \sum_{i=1}^{n} (l - g_{\min}) x_l^i + g_{\min} \geq g, \ \forall g_{\min} \in [\![1,k]\!], \\ x_l^i \geq 0, \ \forall (i,l) \in [\![1,n]\!] \times [\![1,k]\!], \ g \in \mathbb{R}, \end{cases}$$

$$Q' = \begin{cases} \sum_{l=1}^{k} x_l^i = 1, \ \forall i \in [\![1,n]\!], \\ \sum_{l=1}^{k-1} \sum_{i=1}^{n} \lambda_l^i x_l^i \leq g - k, \ \forall \lambda \in \tilde{\Lambda}, \\ x_l^i \geq 0, \ \forall (i,l) \in [\![1,n]\!] \times [\![1,k]\!], \ g \leq k \end{cases}$$

where

$$\tilde{\Lambda} = \left\{ \lambda = (\lambda_l^i)_{(i,l) \in [\![1,n]\!] \times [\![1,k]\!]} \in \mathbb{N}^{nk} \left| \begin{matrix} \lambda_{l+1}^i \leq \lambda_l^i, \ \forall (i,l) \in [\![1,n]\!] \times [\![1,k-1]\!] \\ \sum_{i=1}^{h} \lambda_l^i = k - l, \ \forall l \in [\![1,k]\!] \end{matrix} \right. \right\}.$$

Proof. Take an extreme point (y,h) of P, and let $(x,g) \in M_{k,n} \times \mathbb{N} \backslash \{0\}$ such that $x_l^i = y_{k-l+1}^i$, $\forall (i,l) \in [\![1,n]\!] \times [\![1,k]\!]$ and $g = k - h + 1$, then $g = \max_{i \in [\![1,n]\!]} \sum_{l=1}^{k} l x_l^i$, hence $(x,g) \in P'$. Conversely, any extreme point (x,g) of P' can be obtained from an extreme point of P in this manner. So P' is obtained from P doing the change of variables $x_l^i = y_{k-l+1}^i$, $\forall (i,l) \in [\![1,n]\!] \times [\![1,k]\!]$ and $g = k - h + 1$. Therefore its hyperplane representation is obtained from that of P in Theorem 5. Similarly, Q' is obtained from Q doing the same change of variables and its hyperplane representation is thus obtained from Theorem 7. $\square$

The previous results imply that the separation problems related to P' and Q' can be solved in polynomial time.

5   Conclusion

In this paper we exhibited a family of polyhedra emerging in very diverse combinatorial optimization problems including the most imbalanced orientation of a graph, the minimum span frequency assignment and some scheduling problems. Then a full description of these polyhedra has been derived. We also proved that the separation problems related to these polyhedra can be solved in polynomial time and thus optimization over them can be done in polynomial time.

We think that many combinatorial optimization problems where discretization techniques are used can benefit from the description of the polyhedra introduced in this paper. We are currently carrying out experimentations to study the efficiency of cutting plane algorithms based on these polyhedra. Future work may be directed towards investigations on extensions of the polyhedra we considered here in order to get better approximations while still keeping the feature of computational tractability. One can, for example, study $\{0,1\}$ assignment matrices appended with both h and g (the index of the lowest (resp. highest) nonzero row of the matrix). The related polytope is included in the intersection of P or Q. Some preliminary investigations show that more inequalities are necessary to describe the polytope.

References

1. Ben-Ameur, W., Glorieux, A., Neto, J.: On the most imbalanced orientation of a graph. In: Xu, D., Du, D., Du, D. (eds.) COCOON 2015. LNCS, vol. 9198, pp. 16–29. Springer, Heidelberg (2015)

2. Cook, W., Cunningham, W., Pulleyblank, W., Schrijver, A.: Combinatorial optimization (1997). ISBN: 978-0-471-55894-1
3. Garey, M., Johnson, D.: Computers and Intractability. A Guide to the Theory of NP-Completeness. W.H. Freeman & Co., New York (1990)
4. Hochbaum, D., Shmoys, D.: A polynomial approximation scheme for scheduling on uniform processors: using the dual approximation approach. SIAM J. Comput. **17**(3), 539–551 (1988)
5. Horowitz, E., Sahni, S.: Exact and approximate algorithms for scheduling nonidentical processors. J. ACM **23**(2), 317–327 (1976)
6. Karp, R.: Reducibility among combinatorial problems. In: Miller, R.E., Thatcher, J.W., Bohlinger, J.D. (eds.) Complexity of Computer Computations. Plenum Press, New York (1972)
7. Koster, A.: Frequency assignment - models and algorihtms. Ph.D. thesis, University of Maastricht, The Netherlands (1999)
8. Mann, Z., Szajkó, A.: Complexity of different ILP models of the frequency assignment problem. In: Mann, Z.Á. (ed.) Linear Programming - New frontiers in Theory and Applications, pp. 305–326. Nova Science Publishers, New York (2012)
9. Martins, P.: Extended and discretized formulations for the maximum clique problem. Comput. Oper. Res. **37**(7), 1348–1358 (2011)
10. Vazirani, V.: Minimum makespan scheduling. In: Vazirani, V.V. (ed.) Approximations Algorithms, vol. 10, pp. 79–83. Springer, Heidelberg (2003)
11. Wolsey, L.: Integer Programming (1998). ISBN: 978-0-471-28366-9

On Robust Lot Sizing Problems with Storage Deterioration, with Applications to Heat and Power Cogeneration

Stefano Coniglio[1]([✉]), Arie Koster[2], and Nils Spiekermann[2]

[1] Department of Mathematical Sciences, University of Southampton,
University Road, Southampton SO17 1BJ, UK
s.coniglio@soton.ac.uk
[2] Lehrstuhl II für Mathematik, RWTH Aachen University,
Pontdriesch 14-16, 52062 Aachen, Germany
{koster,spiekermann}@math2.rwth-aachen.de

Abstract. We consider a variant of the single item lot sizing problem where the product, when stored, suffers from a proportional loss, and in which the product demand is affected by uncertainty. This setting is particularly relevant in the energy sector, where the demands must be satisfied in a timely manner and storage losses are, often, unavoidable. We propose a two-stage robust optimization approach to tackle the problem with second stage storage variables. We first show that, in the case of uncertain demands, the robust problem can be solved as an instance of the deterministic one. We then address an application of robust lot sizing arising in the context of heat and power cogeneration and show that, even in this case, we can solve the problem as an instance of the deterministic lot sizing problem. Computational experiments are reported and illustrated.

1 Introduction

Lot Sizing (LS) is a fundamental problem in a large part of modern production planning systems. In its basic version, given a demand for a single good over a finite time horizon, the problem calls for a feasible production plan which minimizes storage, production, and setup costs, also guaranteeing that certain lower and upper bounds on both the production and the amount of good that is stored at each point in time are met.

In the paper, we focus on a generalized variant of lot sizing where the product suffers from proportional losses when stored and the objective function is not necessarily a linear function of the production variables. We also assume uncertain product demands and, consequently, we tackle the problem from a robust optimization perspective. This suits the case of many applications in the energy sector where the product demands of, typically, heat or power, are often not known in advance, especially when the decision maker has to commit to a production plan some time before it becomes operational.

In the paper, we first show that the robust counterpart of this generalized variant of the lot sizing problem can be solved as a special instance of its deterministic counterpart with suitably defined demands and storage upper bounds.

© Springer International Publishing Switzerland 2016
R. Cerulli et al. (Eds.): ISCO 2016, LNCS 9849, pp. 26–37, 2016.
DOI: 10.1007/978-3-319-45587-7_3

We next investigate an application of this result to a production planning problem arising in the context of Combined Heat and Power Production (CHPP), which we also show to be solvable as a special instance of the same deterministic generalized variant of the lot sizing problem that we consider.

The paper is organized as follows. In Sect. 2, we report on some previous work on lot sizing and introduce some relevant robust optimization concepts. Our main contribution, the reduction of the generalized variant of the robust lot sizing problem that we consider to its deterministic counterpart, is outlined in Sect. 3. Section 4 illustrates the application to heat and power cogeneration, while Sect. 5 reports on some computational results and observations. Section 6 concludes the paper with some final comments.

2   Previous Work

In this section, we give a brief account of some relevant works on different versions of the lot sizing problem, also encompassing the uncertain case.

2.1   Deterministic Case

Many variants of the deterministic lot sizing problem have been studied. For an extensive account of the most relevant works, we refer the reader to the monograph [PW06]. The problem is known to be in $\mathcal{P}$ for the case with linear costs, complete conservation (i.e., no losses in the stored product), zero storage lower bounds, nonzero time dependent storage upper bounds, and no production bounds, as shown by Atamtürk and Küçükyavuz in [AK08]. A similar result holds for nonnegative and nondecreasing concave cost functions, complete conservation, production bounds which are constant over time, and unrestricted storage, as shown by Hellion, Mangione, and Penz in [HMP12]. For a polynomial time algorithm for the case with storage losses and nondecreasing concave costs, but no storage or production bounds, see the work of Hsu [Hsu00]. As to $\mathcal{NP}$-hard cases, Florian, Lenstra, and Kan in [FLK80] provide a number of examples. These include the case of linear as well as fixed production costs, no inventory costs, no storage bounds, and no lower production bounds, but nonconstant production upper bounds.

2.2   Uncertain Case

Classical approaches to handle uncertainties in lot sizing are, historically, stochastic in nature, dating back as early as 1960 [Sca60]. The idea is to first assign a probability distribution to the uncertain demand and, then, to solve the problem by looking for a solution of minimum *expected* cost. Unfortunately, as pointed out in [LS05], even when the distribution is estimated within sufficient precision from historical data, such methods can yield solutions which, when implemented with the demand that realizes in practice, can be substantially more costly than those that were predicted with the stochastic approach. Moreover, and regardless of the accuracy of the estimation, these techniques are, in many cases, intrinsically doomed to suffer from the curse of dimensionality [BT06]. Indeed, they usually require a computing time which is, at least, linear in the size of the (discrete) probability space to which the realized demand belongs, which is typically exponential in the size of the instance.

A different option, arguably more affordable from a computational standpoint, is of resorting to a *robust optimization* approach. It corresponds to looking for solutions which are feasible for any realization of the uncertain demand belonging to a given *uncertainty set*, and which also minimize the *worst case* cost. Two seminal papers in this direction are those of Bertsimas and Thiele [BT04,BT06], tackling the uncapacitated lot sizing problem with backlogging and fixed costs, with demands subject to a so-called Γ-robustness model of uncertainty.[1,2] Among other results, they show that the Γ-robust counterpart of the variant of lot sizing they consider can be solved as a version of the deterministic problem with modified demands, also for the case where production bounds are in place. We remark that, for the result of Bertsimas and Thiele to hold, bounds on the storage cannot be enforced. This is an issue in the energy sector, where backlogging is not tolerable as the demand of energy, be it heat or electrical power, must be satisfied when issued.

3   Robust Lot Sizing Under Demand Uncertainty

In this section, we first introduce the generalized variant of lot sizing that we will address. After presenting its robust counterpart, we illustrate how to reduce the robust problem to a special version of the deterministic one and draw some computational complexity considerations.

3.1   Lot Sizing with Deterioration and Storage Bounds

Consider a single product and a time horizon $T = \{1, \ldots, n\}$. For each time step $t \in T$, let $d_t \geq 0$ be the product demand, $q_t \geq 0$ the production variable, c_t^i the inventory cost, $z_t \in \{0, 1\}$ an indicator variable equal to 1 if $q_t > 0$, and $u_t \geq 0$ a variable corresponding to the value of the storage at the end of time period t. Let also $u_0 \geq 0$ be the initial storage value. We assume time dependent bounds $\underline{Q}_t \leq q_t \leq \overline{Q}_t$ on the production and $\underline{U}_t \leq u_t \leq \overline{U}_t$ on the storage. The generalized variant of lot sizing that we consider can be cast as the following

---

[1] In case of backlogging, shortages in the inventory are allowed—or, said differently, unmet demand can be postponed, at a cost, to the future.

[2] When applied to the lot sizing problem, the idea of Γ-robustness is of assuming that the uncertain parameters, i.e., the demand at different time steps, belong to symmetric intervals and that, given an integer Γ, the total number of time steps in which the uncertain demand deviates from its nominal value to either of the extremes of its intervals is bounded by Γ in any constraint of the problem. See [BS03,BS04].

Mixed-Integer Linear Programming (MILP) problem:

$$(\text{LS} - \text{DET}) \quad \min \quad f(q, z) + \sum_{t \in T} c_t^i u_t \tag{1a}$$

$$\text{s.t.} \quad \alpha_t u_{t-1} + q_t = u_t + d_t \qquad \forall t \in T \tag{1b}$$

$$\underline{U_t} \le u_t \le \overline{U_t} \qquad \forall t \in T \tag{1c}$$

$$z_t \underline{Q_t} \le q_t \le z_t \overline{Q_t} \qquad \forall t \in T \tag{1d}$$

$$z_t \in \{0, 1\} \qquad \forall t \in T \tag{1e}$$

$$q_t \ge 0 \qquad \forall t \in T \tag{1f}$$

$$u_t \ge 0 \qquad \forall t \in T. \tag{1g}$$

The generalization goes along two directions. First, we assume that Objective (1a) be the sum of a (general) function $f(q, z)$ of variables q, z and a linear one of u, namely: $\sum_{t \in T} c_t^i u_t$. The classical case is obtained for $f(q, z) = \sum_{t \in T} (a_t q_t + b_t z_t)$, where a_t and b_t are, respectively, the unit production cost and the setup cost at time $t \in T$. Secondly, we assume that, at the end of each time period $t \in T$, a fraction $(1 - \alpha_t)$ of the stored product is lost, as determined by the (possibly time dependent) *conservation factor* $\alpha_t \in (0, 1]$.

We remark that, similarly to the classical case where $\alpha_t = 1$ for all $t \in T$, u_t is uniquely determined as a function of q_t, u_{t-1}, and d_t in every feasible solution (q, z, u) of LS-DET. Indeed, from Constraints (1b), we deduce, by substitution from $t = 1$ to $t = |T|$:

$$u_t := \left(\prod_{k=1}^{t} \alpha_k \right) u_0 + \sum_{i=1}^{t} \left(\prod_{k=i+1}^{t} \alpha_k \right) (q_i - d_i) \qquad \forall t \in T. \tag{2}$$

Note that Eq. (2) is *causal*, as the value of the storage at time t only depends on the demand (and production) at times $1, \ldots, t - 1$. From the equation, it follows that a pair (q, z) suffices to fully characterize a solution to LS-DET, as the (unique) value of the missing vector u can be calculated *a posteriori* via Eq. (2). We call a pair (q, z) satisfying Constraints (1d)–(1f) a *production plan* and denote by $\mathcal{I}_{\text{LS-DET}}(q, z) = (q, z, u)$, with u as in Eq. (2), its *induced solution* to LS-DET.

3.2 Uncertain Demands

Assuming that the uncertain demand vector d takes values in the uncertainty set D, the robust counterpart of LS-DET, i.e., a version of the problem where we look for a solution (q, z, u) which is feasible for *all* realizations $d \in D$, has to be a solution to:

$$(\text{LS} - \text{ROB}) \quad \min \quad (1a) \tag{3a}$$

$$\text{s.t.} \quad \alpha_t u_{t-1} + q_t = u_t + d_t \qquad \forall t \in T, d \in D \tag{3b}$$

$$(1c) - (1g), \tag{3c}$$

where the original Constraints (1b) (which are called Constraints (3b) here) are enforced for all realizations $d_t \in D$. As we will better explain in the following, we remark that, in this case, all the variables are *first stage* variables, as they are required to take a single value independently of the realization of d.

For any nontrivial D, the following holds:

Proposition 1. *If* $\exists \{d^1, d^2\} \subseteq D$ *with* $d^1 \neq d^2$, *then LS-ROB is infeasible.*

Proof. For any value of q, d, Constraints (3b) induce the linear system $Au = q - d + e_1\alpha_1 u_0$, where A is a full rank matrix with the all-one vector as main diagonal and the vector $-\alpha$ as the diagonal below it. Assuming that LS-ROB is feasible, we have $\exists u : Au = q - d^1 + e_1\alpha_1 u_0$ and $Au = q - d^2 + e_1\alpha_1 u_0$. This implies $A^{-1}(q - d^1 + e_1\alpha_1 u_0) = A^{-1}(q - d^2 + e_1\alpha_1 u_0)$, that is, $d^1 = d^2$, a contradiction.   □

It is thus natural, as well as very reasonable in practice, to assume that the value of the storage u_t at time $t \in T$ could be adapted as a function $u_t(d)$ of the demand $d \in D$ which has realized up to time $t - 1$, i.e., that u be a *second stage* variable. A direct formulation for the robust counterpart of LS-DET with a second stage u is thus:

$$(\text{LS} - \text{ROB2}) \quad \min \quad f(q, z) + \eta \tag{4a}$$

$$\text{s.t.} \quad \alpha_t u_{t-1}(d) + q_t = u_t(d) + d_t \qquad \forall t \in T, d \in D \tag{4b}$$

$$\underline{U_t} \leq u_t(d) \leq \overline{U_t} \qquad \forall t \in T, d \in D \tag{4c}$$

$$\eta \geq \sum_{t \in T} c_t^i u_t(d) \qquad \forall d \in D \tag{4d}$$

$$(1\text{d}) - (1\text{f}) \tag{4e}$$

$$u_t(d) \geq 0 \qquad \forall t \in T, d \in D \tag{4f}$$

$$\eta \geq 0. \tag{4g}$$

The newly introduced variable η accounts for the worst case storage cost over all $d \in D$ (a so-called, partial, *epigraph reformulation*). Assuming, as it is the case for a discrete scenario approach, a finite D, LS-ROB2 calls for a vector $u(d) \geq 0$ for each $d \in D$ satisfying Constraints (4b)–(4d). Clearly, LS-ROB2 can be solved directly by employing a mixed-integer linear programming solver, although at the cost of expressing Constraints (4b)–(4d) and (4f) $|D|$ times.

3.3   Solving LS-ROB2 as an Instance of LS-DET

We now present a more efficient way to solve LS-ROB2 as an instance of LS-DET with suitably chosen demand and storage upper bound vectors d and $\overline{U}$.

As for LS-DET, a production plan (q, z) is required to satisfy Constraints (1d)–(1f)—which are condensed, in LS-ROB2, in Constraint (4e). Its induced solution $\mathcal{I}_{\text{LS-ROB2}}(q, z) = (q, z, u(d), \eta)$ to LS-ROB2 can be defined, without loss of generality, as:

$$u_t(d) := \left(\prod_{k=1}^{t} \alpha_k\right) u_0 + \sum_{i=1}^{t} \left(\prod_{k=i+1}^{t} \alpha_k\right)(q_i - d_i) \qquad \forall t \in T, d \in D \tag{5a}$$

$$\eta := \max_{d \in D}\left\{\sum_{t \in T} c_t^i u_t(d)\right\}. \tag{5b}$$

Note that, differently from the case of LS-DET, in this case another induced solution can be constructed by selecting any η satisfying $\eta > \max_{d \in D}\left\{\sum_{t \in T} c_t^i u_t(d)\right\}$. Clearly though, such solution cannot be optimal.

Our main result follows:

Theorem 1. *For an uncertainty set D over which a linear function can be optimized in polynomial time, LS-ROB2 can be polynomially reduced (with respect to production plans) to an instance of LS-DET with $d = d'$ and $\overline{U} = \overline{U}'$ thus defined:*

$$d_t' := \max_{d \in D}\left\{d_t - \sum_{i=1}^{t-1}\left(\prod_{k=i+1}^{t} \alpha_k\right)(d_i' - d_i)\right\} \qquad \forall t \in T \tag{6a}$$

$$\overline{U}_t' := \overline{U}_t - \Delta_t \qquad \forall t \in T \tag{6b}$$

$$\Delta_t := \max_{d \in D}\left\{\sum_{i=1}^{t}\left(\prod_{k=i+1}^{t} \alpha_k\right)(d_i' - d_i)\right\} \qquad \forall t \in T. \tag{6c}$$

Proof. First, note that the values for d_t' and $\overline{U}_t'$ can be computed iteratively, from $t = 1$ to $t = |T|$, in polynomial time due to the assumptions on D. For a given production plan (q, z), adopting $d = d'$ and $\overline{U} = \overline{U}'$, we show that $\mathcal{I}_{\text{LS-ROB2}}(q, z) = (q, z, u(d), \eta)$ is feasible for LS-ROB2 if and only if $\mathcal{I}_{\text{LS-DET}}(q, z) = (q, z, u)$ is feasible for LS-DET. We deduce the following:

$$d_t' \geq 0 \qquad \forall t \in T \tag{7a}$$

$$u_t = \min_{d \in D}\{u_t(d)\} \qquad \forall t \in T \tag{7b}$$

$$u_t + \Delta_t = \max_{d \in D}\{u_t(d)\} \qquad \forall t \in T \tag{7c}$$

$$\eta = \sum_{t \in T} c_t^i u_t + \underbrace{\max_{d \in D}\left\{\sum_{t \in T} c_t^i \sum_{i=1}^{t}\left(\prod_{k=i}^{t} \alpha_k\right)(d_i' - d_i)\right\}}_{\text{const}} \tag{7d}$$

For the derivations, which we omit due to space reasons, we refer the reader to the Online Appendix [CKS16], available from the authors upon request. We are to show that Constraints (4c) are satisfied by $u(d)$ if and only if Constraints (1c) are satisfied by u (all the other constraints are satisfied by definition of production plan). This is shown by observing that, for all $t \in T$, the following holds true:

$$\underline{U}_t \leq u_t(d)\, \forall d \in D \Leftrightarrow \underline{U}_t \leq \min_{d \in D}\{u_t(d)\} \Leftrightarrow \underline{U}_t \leq u_t$$

$$u_t(d) \leq \overline{U}_t\, \forall d \in D \Leftrightarrow \max_{d \in D}\{u_t(d)\} \leq \overline{U}_t \Leftrightarrow u_t + \Delta_t \leq \overline{U}_t \Leftrightarrow u_t \leq \overline{U}_t'.$$

Since Objectives (4a) and (1a) are equal up to a constant additive term, i.e.:

$$f(q,z) + \eta = f(q,z) + \sum_{t \in T} c_t^i u_t + \text{const}, \tag{8}$$

we deduce that a production plan is optimal for LS-ROB2 if and only if it is optimal for LS-DET.   □

Intuitively, the newly defined demand d_t' induces a lower bound on the product which has to be available at time $t \in T$, thus ensuring that every demand $d \in D$ can be met. The value Δ reduces the storage upper bound of the transformed problem to prevent that, when a large production is realized but, suddenly, a large deficit in demand occurs (an event which would result in an overflow of storage), the actual storage upper bound $\overline{U}_t$ is not exceeded.

We remark that the assumptions in Theorem 1 subsume the cases of many widely employed robustness models, including polyhedral uncertainty sets (such as the Γ-robustness one), discrete scenario uncertainty sets, and ellipsoidal uncertainty sets, such as those used in [BTEGN09].

From a computational complexity perspective, the following holds:

Corollary 1. *Given an uncertainty set D over which a linear function can be optimized in polynomial time, LS-ROB2 is in $\mathcal{P}$ (respectively, $\mathcal{NP}$-hard) if and only if the corresponding version of LS-DET is in $\mathcal{P}$ (respectively, $\mathcal{NP}$-hard).*

Proof. We use Lemma 1 to polynomially reduce LS-ROB2 to LS-DET. For the polynomial reduction from LS-DET to LS-ROB2, it suffices to consider that every deterministic problem can be regarded as a robust optimization problem.   □

Note that, as a consequence of Corollary 1, LS-ROB2 is in $\mathcal{P}$ for all the polynomially solvable cases of LS-DET that we reported in Sect. 2, provided that their algorithm allows for the introduction of time dependent upper bounds $\overline{U}$ on u. This is, for instance, the case of the problem studied in [HMP12].

We conclude by noting that our result can be slightly extended as follows:

Remark 1. Theorem 1 and Corollary 1 still hold if we introduce additional constraints on z and q or assumptions on the givens (except for d and $\overline{U}$). They are also valid for $\overline{U}_t = \infty$ and for not necessarily nonnegative demands d.

4   Application to Heat and Power Cogeneration

In this section, we consider an application of the previous results to the case of Combined Heat and Power Production (CHPP). CHPP plants are production units in which the heat that is generated when cooling down the plant is extracted and, at least partially, utilized for heating purposes. The units are equipped with a storage tank where the heat in excess can be temporarily stored, subject to constant (over time) proportional losses due to dissipation effects.

From a production planning perspective, two products and two demands are present: one of heat and one of power.[3] Power in excess or defect with respect

---

[3] Although "electrical energy" would be more precise, we will refer to "power" in the following. Due to the hourly time scale, this quantity is, indeed, a measure of power.

to the given demand can be sold or bought from the power market. Storage is costless, while a cost is incurred for fuel consumption. In principle, we can outline three sources of uncertainty: heat demands, power demands, and power market prices. Among the three, heat demands are, arguably, the most critical ones. This is because poorly estimated heat demands can lead to infeasible production plans by which the storage bounds are violated. Differently, poorly estimated power demands or market prices can only introduce an extra cost into the objective function. For this reason, in the following we will focus solely on uncertain heat demands.

4.1   Problem Formulation

We adopt the same variables as in LS-DET, with q, u, and d^h representing the amount of heat which is, respectively, produced, stored, and required as a demand. For power, we introduce a second production variable p_t, indicating the amount of power that is generated at time $t \in T$, two market variables, p^b and p^s, representing the amount of power that is, at each point in time, bought and sold, and a demand vector d^p. Let c^p be the vector of market prices for both buying and selling a MWh of power and let c^f be the vector of fuel prices. Fuel consumption at time t, as denoted by the variable f_t, is modeled as the linear function $sq_t + hz_t$, where hz_t is a constant term corresponding to the activation of the CHPP unit. We assume that heat and power are produced with a fixed proportion $\rho \in (0,1)$ and that α is constant over the time horizon T.

Let D^h be the uncertainty set for the heat demands. As for LS-ROB2, we assume second stage storage variables $u(d^h)$ as a function of the uncertain heat demand $d^h \in D^h$. We introduce the following robust two-stage MILP formulation:

$$(\text{CHPP} - \text{ROB2}) \quad \min \quad \sum_{t \in T} \left(c_t^p (p_t^b - p_t^s) + c_t^f f_t \right) \tag{9a}$$

$$\text{s.t.} \quad \alpha u_{t-1}(d^h) + q_t = u_t(d^h) + d_t^h \quad \forall t \in T, d^h \in D^h \tag{9b}$$

$$p_t^p + p_t^b = d_t^p + p_t^s \qquad \forall t \in T \tag{9c}$$

$$\underline{U} \leq u_t(d^h) \leq \overline{U} \qquad \forall t \in T, d^h \in D^h \tag{9d}$$

$$z_t \underline{Q} \leq q_t \leq z_t \overline{Q} \qquad \forall t \in T \tag{9e}$$

$$f_t = sq_t + hz_t \qquad \forall t \in T \tag{9f}$$

$$p_t^p = \rho q_t \qquad \forall t \in T \tag{9g}$$

$$q_t, f_t, p_t^p, p_t^b, p_t^s \geq 0 \qquad \forall t \in T \tag{9h}$$

$$u_t(d^h) \geq 0 \qquad \forall t \in T, d^h \in D^h \tag{9i}$$

$$z_t \in \{0,1\} \qquad \forall t \in T. \tag{9j}$$

Given a production plan (q, z), its induced solution to CHPP-ROB2, i.e., $\mathcal{I}_{\text{CHPP-ROB2}}(q, z) = (q, z, f, p^p, p^b, p^s, u(d^h))$, is defined as:

$$u_t(d^h) = \alpha^t u_0 + \sum_{i=1}^{t} \alpha^{t-i} \left(q_i - d_i^h \right) \qquad \forall t \in T \qquad (10a)$$

$$f_t = sq_t + hz_t \qquad \forall t \in T \qquad (10b)$$

$$p_t^p = \rho q_t \qquad \forall t \in T \qquad (10c)$$

$$p_t^b = \max\{d_t^p - p_t^p, 0\} \qquad (10d)$$

$$p_t^s = \max\{p_t^p - d_t^p, 0\}. \qquad (10e)$$

The following holds:

Proposition 2. *CHPP-ROB2 can be solved as an instance of LS-DET.*

Proof. By substitution from Constraints (9g), Constraints (9c) become, for all $t \in T$:

$$p_t^b - p_t^s = d_t^p - \rho q_t$$

which, after substitution in Objective (9a) together with Constraints (9f), yield:

$$\sum_{t \in T} \left(c_t^p (d_t^p - \rho q_t) + c_t^f (sq_t + hz_t) \right) = \sum_{t \in T} \left((c_t^f s - c_t^p \rho) q_t + c_t^f hz_t \right) + \underbrace{\sum_{t \in T} c_t^p d_t^p}_{\text{const}}.$$

By setting:

$$f(q, z) := \sum_{t \in T} \left((c_t^f s - c_t^p \rho) q_t + c_t^f hz_t \right) \qquad (11a)$$

$$\alpha_t := \alpha \quad \forall t \in T \qquad (11b)$$

$$c_t^i := 0 \quad \forall t \in T \qquad (11c)$$

and dropping the constant term, we obtain an instance of LS-ROB2. The corresponding instance of LS-DET is obtained by applying Corollary 1. $\qquad \square$

Unfortunately, we are not aware of any specialized algorithm capable of solving LS-DET with $\alpha_t < 1$ in combination with (constant) lower and upper production bounds. In spite of this, in the next section we will rely on the transformation into LS-DET to solve CHPP-ROB2 via mixed-integer linear programming techniques in a much shorter amount of computing time than when tackling the problem directly in its original form.

5   Computational Results

We report and illustrate a set of computational experiments carried out on a CHPP-ROB2 problem originating within the project *Robuste Optimierung* der *Stromversorgungsysteme* (Robust Optimization of Power Supply Systems), funded by the German *Bundesministerium für Wirtschaft und Energie* (Federal Ministry for Economic Affairs and Energy, BMWi).

We consider a dataset of 232 days, spanning a period of two years (with some missing months). Market prices for the power market are taken from EPEX STOP (the European Power Exchange). The power demand is taken from historical data for the whole country of Germany, downscaled to 50000 households, while the heat demand is taken from historical data of a portion of Frankfurt (of around 50000 households). We assume $s = 1.51$ EUR/MWh, $h = 5.43$ EUR/MWh, and $\rho = 0.4$. The bounds are set to $\underline{U} = 0$ MWh, $\overline{U} = 120$ MWh, $\underline{Q} = 37.5$ MWh, $\overline{Q} = 125$ MWh. We also set $u_0 = 36$ MWh.

We adopt a *discrete scenario* uncertainty set D^h, built from a heat demand forecast provided by our industrial partner ProCom GmbH. The first scenario of D^h is the original forecast for the current day, as produced by ProCom. It is generated in a two-stage fashion, with an autoregressive component and a neural network one, with temperature and calendar events as main influence factors. We then single out the 50 days from the set of historical time series where the corresponding pair of demand and forecast is closest in L1 norm. After computing the forecast error between the two, we create a scenario where such error is added to the forecast demand of the current day (for which the problem is being solved). This way we, intuitively, "learn" the forecast error from historical data and apply it to the current forecast, creating 50 additional scenarios. The general idea is that the forecast error follows certain patterns, so that it is more likely that combinations of the historical errors will also apply to the error of the current day for which CHPP-ROB2 is being solved.

The experiments are run on an Intel i7-3770 3.40 GHz machine with 32 GB RAM using CPLEX 12.6 and AMPL as modeling language. We consider four settings, with a time horizon of, respectively, 24, 48, 72, and 96 h. The total time in seconds to solve all the instances, as well as the corresponding standard deviation, are reported in the following table. In it, as well as in the charts that will follow, CHPP-ROB2 accounts for the problem when solved via the original Formulation (9a)–(9j), whereas LS-DET corresponds the problem solved via Formulation (1a)–(1g) after having been transformed, by applying Theorem 1, into an instance of the deterministic lot sizing problem. Proportional speedup factors are also reported.

	CHPP-ROB2		LS-DET		Speedup	
Horizon	totTime	stdev	totTime	stdev	totTime	stdev
24 h	37.05	0.05	7.96	0.01	4.65	5.00
48 h	140.01	0.17	16.32	0.03	8.58	5.67
72 h	323.95	0.47	34.77	0.15	9.32	3.13
96 h	805.86	2.65	64.49	0.42	12.50	6.31

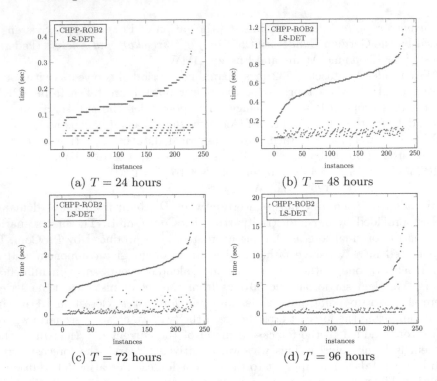

Fig. 1. Computing times when solving CHPP-ROB2 in its original form vs. those when solving it as an instance of LS-DET via the transformation outlined in Theorem 1. Instances are sorted by nondecreasing computing time for CHPP-ROB2.

As the table illustrates, the improvement in computing time achieved when solving the problem as an instance of LS-DET, rather than in its original form, steeply increases with the size of the instances. From an average speedup of 4.65 times for the 24 h instances, we register one of 8.58 times for the 48 h instances, one of 9.32 times for the 72 h instances, and one of 12.5 times for the 96 h instances. On average, the speedup is of 7.69 times. This illustrates that, even in a mixed-integer linear programming setting, the transformation proposed with Theorem 1 allows for a substantial reduction in the computing time. The table also shows that the computing time, if seen as a stochastic process, becomes much more stable when Theorem 1 is employed. Indeed, with its application, we observe a reduction in the standard deviation ranging between 3.13 times for the 72 h instances to 6.31 times for the 96 h ones, with an average reduction of 5.03 times. A visual depiction is reported in Fig. 1.

6   Concluding Remarks

We have considered a generalized variant of lot sizing with proportional storage losses and a nonlinear objective function, showing how, for the case of uncertain demands, the problem can be solved as a special instance of the deterministic one via a polynomial time transformation. We have then considered an application to heat and power cogeneration systems, showing that, when uncertain demands are considered, even that problem can be tackled as a special instance of lot sizing. Computational experiments have shown that our transformation allows for a much shorter computing time even when using a general purpose mixed-integer linear programming solver.

Acknowledgement. This work is supported by the German Federal Ministry for Economic Affairs and Energy, BMWi, grant 03ET7528B.

References

[AK08]   Atamtürk, A., Küçükyavuz, S.: An algorithm for lot sizing with inventory bounds and fixed costs. Oper. Res. Lett. **36**(3), 297–299 (2008)

[BS03]   Bertsimas, D., Sim, M.: Robust discrete optimization and network flows. Mathe. Program. **98**(1–3), 49–71 (2003)

[BS04]   Bertsimas, D., Sim, M.: The price of robustness. Oper. Res. **52**(1), 35–53 (2004)

[BT04]   Bertsimas, D., Thiele, A.: A robust optimization approach to supply chain management. In: Bienstock, D., Nemhauser, G.L. (eds.) IPCO 2004. LNCS, vol. 3064, pp. 86–100. Springer, Heidelberg (2004)

[BT06]   Bertsimas, D., Thiele, A.: A robust optimization approach to inventory theory. Oper. Res. **54**(1), 150–168 (2006)

[BTEGN09] Ben-Tal, A., El Ghaoui, L., Nemirovski, A.: Robust Optimization. Princeton University Press, Princeton (2009)

[CKS16] Coniglio, S., Koster, A.M.C.A., Spiekermann, N.: Online appendix (2016)

[FLK80] Florian, M., Lenstra, J.K., Kan, A.H.G.R.: Deterministic production planning: algorithms and complexity. Manage. Sci. **26**(7), 669–679 (1980)

[HMP12] Hellion, B., Mangione, F., Penz, B.: A polynomial time algorithm to solve the single-item capacitated lot sizing problem with minimum order quantities and concave costs. Euro. J. Oper. Res. **222**(1), 10–16 (2012)

[Hsu00] Hsu, V.N.: Dynamic economic lot size model with perishable inventory. Manage. Sci. **46**(8), 1159–1169 (2000)

[LS05]   Liyanage, L.H., Shanthikumar, J.G.: A practical inventory control policy using operational statistics. Oper. Res. Lett. **33**(4), 341–348 (2005)

[PW06]   Pochet, Y., Wolsey, L.A.: Production Planning by Mixed Integer Programming. Springer Series in Operations Research and Financial Engineering. Springer, New York (2006)

[Sca60] Scarf, H.: The Optimality of (S, s) Policies in the Dynamic Inventory Problem. Mathemtical Methods in the Social Sciences, vol. 1, p. 196. Stanford University Press, New York (1960)

Reducing the Clique and Chromatic Number via Edge Contractions and Vertex Deletions

Daniël Paulusma[1], Christophe Picouleau[2(✉)], and Bernard Ries[3]

[1] Durham University, Durham, UK
daniel.paulusma@durham.ac.uk
[2] CNAM, Laboratoire CEDRIC, Paris, France
christophe.picouleau@cnam.fr
[3] University of Fribourg, Fribourg, Switzerland
bernard.ries@unifr.ch

Abstract. We consider the following problem: can a certain graph parameter of some given graph G be reduced by at least d, for some integer d, via at most k graph operations from some specified set S, for some given integer k? As graph parameters we take the chromatic number and the clique number. We let the set S consist of either an edge contraction or a vertex deletion. As all these problems are NP-complete for general graphs even if d is fixed, we restrict the input graph G to some special graph class. We continue a line of research that considers these problems for subclasses of perfect graphs, but our main results are full classifications, from a computational complexity point of view, for graph classes characterized by forbidding a single induced connected subgraph H.

1 Introduction

When considering a graph modification problem, we usually fix a graph class $\mathcal{G}$ and then, given a graph G, a set S of one or more graph operations and an integer k, we ask whether G can be transformed into a graph $G' \in \mathcal{G}$ using at most k operations from S. Now, instead of fixing a particular graph class, one may be interested in *fixing a certain graph parameter* π. In this setting we ask, given a graph G, a set S of one or more graph operations and an integer k, whether G can be transformed into a graph G' by using at most k operations from S such that $\pi(G') \leq \pi(G) - d$, for some *threshold* $d \geq 0$. Such problems are called *blocker problems*, as the set of vertices or edges involved can be seen as "blocking" some desirable graph property (such as being colorable with only a few colors). Identifying the part of the graph responsible for a significant decrease of the graph parameter under consideration gives crucial information on the graph.

Blocker problems have been given much attention over the last years [1–4,6,7,13,15,16]. Graph parameters considered were the chromatic number, the independence number, the clique number, the matching number and the

D. Paulusma—Author supported by EPSRC (EP/K025090/1).

© Springer International Publishing Switzerland 2016
R. Cerulli et al. (Eds.): ISCO 2016, LNCS 9849, pp. 38–49, 2016.
DOI: 10.1007/978-3-319-45587-7_4

vertex cover number. So far, the set S always consisted of a single graph operation, which was a vertex deletion, edge deletion, edge contraction, or an edge
addition. Here, we consider the chromatic number and the clique number. We
keep the restriction on the size of S and let S consist of an edge contraction or a
vertex deletion. Thus, we continue the research initiated by Bentz et al. [4] and
Diner et al. [7]. In the latter paper, classes of perfect graphs are considered. Here,
we also consider classes of perfect graphs, but in our main results we restrict the
input to graphs that are defined by a single forbidden induced subgraph H, that
is, to so-called H-*free graphs*.

Definitions. The *contraction* of an edge uv of a graph G removes the vertices
u and v from G, and replaces them by a new vertex made adjacent to precisely
those vertices that were adjacent to u or v in G (neither self-loops nor multiple
edges are introduced). Then G can be k-*contracted* into a graph H if G can
be modified into H by a sequence of at most k edge contractions. For a subset
$V' \subseteq V$, let $G - V'$ be the graph obtained from G after deleting the vertices of
V'. Let $\chi(G)$ and $\omega(G)$ denote the chromatic number and the clique number of
G. We now define our two blocker problems formally, where $\pi \in \{\chi, \omega\}$ is the
(fixed) graph parameter:

CONTRACTION BLOCKER(π)
Input: A graph G and two integers $d, k \geq 0$.
Question: Can G be k-contracted into a graph G' such that $\pi(G') \leq \pi(G) - d$?

DELETION BLOCKER(π)
Input: A graph $G = (V, E)$ and two integers $d, k \geq 0$.
Question: Is there a set $V' \subseteq V$, with $|V'| \leq k$, such that $\pi(G - V') \leq \pi(G) - d$?

If we remove d from the input and fix it instead, we call the resulting problems
d-CONTRACTION BLOCKER(π) and d-DELETION BLOCKER(π), respectively.

Relations to known problems. In Sect. 3, we will pinpoint a close relationship
between the blocker problem and the problem of deciding whether the graph
parameter under consideration (chromatic number or clique number) is bounded
by some constant (in order to prove a number of hardness results). We also
observe that blocker problems generalize graph transversal problems. To explain
the latter type of problems, for a family of graphs $\mathcal{H}$, the $\mathcal{H}$-TRANSVERSAL
problem is that of finding a set $V' \subseteq V$ in a graph $G = (V, E)$ of size $|V'| \leq k$ for
some integer k, such that $G - V'$ contains no induced subgraph isomorphic to a
graph in $\mathcal{H}$. By letting, for instance, $\mathcal{H}$ be the family of all complete graphs on
at least two vertices, we find that $\mathcal{H}$-TRANSVERSAL is equivalent to DELETION
BLOCKER(ω) restricted to instances $(G, d = \omega(G) - 1, k)$.

Our Results. In Sect. 2, we introduce some more terminology and give a number
of known results used to prove our results. In Sect. 3, we show how the computational hardness of the decision problems for χ, ω relates to the computational
hardness of the blocker variants. There, we also give a number of additional
results on subclasses of perfect graphs. We need these results for our proofs.
However, these results may be of independent interest, as they continue similar

work on perfect graphs in [7]. In Sect. 4 we present our results for CONTRAC-
TION BLOCKER(π) and d-CONTRACTION BLOCKER(π) for H-free graphs, where
$\pi \in \{\chi, \omega\}$. Amongst others we prove complete dichotomies for all connected
graphs H. In Sect. 5 we perform the same study for DELETION BLOCKER(π) and
d-DELETION BLOCKER(π), where $\pi \in \{\chi, \omega\}$ to obtain complete dichotomies for
all connected graphs H. We conclude our paper in Sect. 6.

2   Preliminaries

All graphs considered are finite, undirected and without self-loops or multiple
edges. The *complement* of G is the graph $\overline{G} = (V, \overline{E})$ with vertex set V and an
edge between two vertices u and v if and only if $uv \notin E$. For a subset $S \subseteq V$,
we let $G[S]$ denote the subgraph of G *induced* by S, which has vertex set S and
edge set $\{uv \in E \mid u, v \in S\}$. We write $H \subseteq_i G$ if a graph H is an induced
subgraph of G. For a vertex $v \in V$, we write $G - v = G[V \setminus \{v\}]$. Recall that for
a subset $V' \subseteq V$ we write $G - V' = G[V \setminus V']$. When we contract an edge uv,
we may also say that a vertex u is *contracted onto* v, and we use v to denote the
new vertex resulting from the edge contraction.

Let $G = (V_G, E_G)$ and $H = (V_H, E_H)$ be two vertex-disjoint graphs. The
disjoint union $G + H$ has vertex set $V_G \cup V_H$ and edge set $E_G \cup E_H$. The disjoint
union of k copies of G is denoted by kG. Let $\{H_1, \ldots, H_p\}$ be a set of graphs.
We say that G is $(H_1, \ldots, H_p)$-*free* if G has no induced subgraph isomorphic to
a graph in $\{H_1, \ldots, H_p\}$. If $p = 1$ we may write H_1-free instead of (H_1)-free.
A subset $C \subseteq V$ is called a *clique* of G if any two vertices in C are adjacent
to each other. The *clique number* $\omega(G)$ is the number of vertices in a maximum
clique of G. The CLIQUE problem tests if a graph contains a clique of size at
least k for some given integer $k \geq 0$. For a positive integer k, a k-*coloring* of G
is a mapping $c : V \to \{1, 2, \ldots, k\}$ such that $c(u) \neq c(v)$ whenever $uv \in E$. The
chromatic number $\chi(G)$ is the smallest integer k for which G has a k-coloring.
The COLORING problem tests if a graph has a k-coloring for some given integer k.
If k is fixed, that is, not part of the input, then we write k-COLORING instead.

A graph $G = (V, E)$ is a *split graph* if G has a *split partition*, which is a
partition of its vertex set into a clique K and an independent set I. A graph
is *cobipartite* if it is the complement of a *bipartite* (2-colorable) graph. A graph
is *chordal* if it has no induced cycles on more than three vertices. A graph is
perfect if the chromatic number of every induced subgraph equals the size of a
largest clique in that subgraph. Let C_n, P_n and K_n denote the n-vertex cycle,
path and clique, respectively. Let $K_{n,m}$ denote the complete bipartite graph with
partition classes of size m and n, respectively. The *cobanner*, *bull* and *butterfly*
are displayed in Fig. 1. We finish this section by stating some known results.

Lemma 1 ([14]). CLIQUE *is* NP-*complete for the following classes:* (C_5, P_5)-*free
graphs,* $K_{1,3}$-*free graphs, cobanner-free graphs and* $(bull, P_5)$-*free graphs.*

Lemma 2 ([10]). *Let H be a graph. For the class of H-free graphs,* COLORING
is polynomial-time solvable if H is an induced subgraph of P_4 or of $P_1 + P_3$ and
NP-*complete otherwise.*

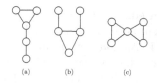

Fig. 1. (a) Cobanner. (b) Bull. (c) Butterfly.

Lemma 3 ([11]). 3-COLORING *is* NP-*complete for the class of K_3-free graphs.*

Lemma 4 ([7]). 1-CONTRACTION BLOCKER(ω) *is* NP-*complete for graphs with clique number* 3.

Lemma 5 ([7]). *For* $\pi \in \{\chi, \omega\}$, *both problems* CONTRACTION BLOCKER(π) *and* DELETION BLOCKER(π) *can be solved in polynomial time for* P_4-*free graphs, but are* NP-*compete on split graphs.*

3   Hardness Conditions and Results for Perfect Graphs

In this section we give some results that we need for the proofs of our main results in later sections. In the proof of Lemma 4 [7] it is readily seen that the graph obtained in the reduction as input graph for 1-CONTRACTION BLOCKER(ω) is in fact ($K_4, \overline{2P_1 + P_2}$, butterfly)-free. This gives us the following result.

Lemma 6 ([7]). 1-CONTRACTION BLOCKER(ω) *is* NP-*complete for the class of* ($K_4, \overline{2P_1 + P_2}, butterfly$)-*free graphs.*

Let $\mathcal{G}$ be a graph class closed under adding a vertex-disjoint copy of the same graph or of a complete graph. We call such a graph class *clique-proof*. The following result establishes a close relation between COLORING (resp. CLIQUE) and 1-CONTRACTION BLOCKER(χ) (resp. 1-CONTRACTION BLOCKER(ω)).

Theorem 1. *Let* $\mathcal{G}$ *be a clique-proof graph class. Then the following two statements hold:*

(i) if COLORING *is* NP-*complete for* $\mathcal{G}$, *then so is* 1-CONTRACTION BLOCKER(χ).

(ii) if CLIQUE *is* NP-*complete for* $\mathcal{G}$, *then so is* 1-CONTRACTION BLOCKER(ω).

Proof. We only give the proof for COLORING and 1-CONTRACTION BLOCKER(χ), as the proof for CLIQUE and 1-CONTRACTION BLOCKER(ω) can be obtained by the same arguments. Let $\mathcal{G}$ be a graph class that is clique-proof. From a given graph $G \in \mathcal{G}$ and integer $\ell \geq 1$ we construct the graph $G' = 2G + K_{\ell+1}$. Note that $G' \in \mathcal{G}$ by definition and that $\chi(G') = \max\{\chi(G), \ell + 1\}$. We claim that G is ℓ-colorable if and only if G' can be 1-contracted into a graph G^* with $\chi(G^*) \leq \chi(G') - 1$. First suppose that G is ℓ-colorable.

Then, in G', we contract an edge of $K_{\ell+1}$ in order to obtain a graph G^* that is ℓ-colorable. Conversely, suppose that G' can be 1-contracted into a graph G^* with $\chi(G^*) \leq \chi(G') - 1$. As contracting an edge in a copy of G does not lower the chromatic number, the contracted edge must be in $K_{\ell+1}$. Then, as $\chi(G^*) \leq \chi(G') - 1$, this implies that $\chi(G') = \ell + 1$ and $\chi(G^*) = \ell$. Hence, as $\chi(G^*) = \max\{\chi(G), \ell\}$, we conclude that $\chi(G) \leq \ell$.   □

Our next result is on cobipartite graphs (we omit its proof).

Theorem 2. *For $\pi \in \{\chi, \omega\}$,* CONTRACTION BLOCKER(π) *is* NP-*complete for cobipartite graphs.*

As cobipartite graphs are $3P_1$-free, we immediately obtain the following.

Corollary 1. *For $\pi \in \{\chi, \omega\}$,* CONTRACTION BLOCKER(π) *is* NP-*complete for $3P_1$-free graphs.*

We will continue with some further results on subclasses of perfect graphs. We need a known lemma.

Lemma 7 ([7]). *Let $G = (V, E)$ be a C_4-free graph and let $v_1v_2 \in E$. Let $G|v_1v_2$ be the graph obtained after the contraction of v_1v_2 and let v_{12} be the new vertex replacing v_1 and v_2. Then every maximal clique K in $G|v_1v_2$ containing v_{12} corresponds to a maximal clique K' in G and vice versa, such that*

(a) either $|K| = |K'|$ and $K \setminus \{v_{12}\} = K' \setminus \{v_1\}$;
(b) or $|K| = |K'|$ and $K \setminus \{v_{12}\} = K' \setminus \{v_2\}$;
(c) or $|K| = |K'| - 1$ and $K \setminus \{v_{12}\} = K' \setminus \{v_1, v_2\}$.

Moreover, every maximal clique in $G|v_1v_2$ not containing v_{12} is a maximal clique in G and vice versa.

Theorem 3. *For $\pi \in \{\chi, \omega\}$, 1-*CONTRACTION BLOCKER(π) *is* NP-*complete for chordal graphs.*

Proof. Since chordal graphs are perfect and closed under taking edge contractions, we may assume without loss of generality that $\pi = \omega$. Let $G = (V, E)$ be a graph that together with an integer k forms an instance of VERTEX COVER, which is the problem of deciding whether a graph G has a *vertex cover* of size at most k, that is, a subset S of vertices of size at most k such that each edge is incident with at least one vertex of S. VERTEX COVER is a well-known NP-complete problem (see [9]).

From G we construct a chordal graph G' as follows. We introduce a new vertex y not in G. We represent each edge e of G by a clique K_e in G' of size $|V|$ so that $K_e \cap K_f = \emptyset$ whenever $e \neq f$. We represent each vertex v of G by a vertex in G' that we also denote by v. Then we let the vertex set of G' be $V \cup \bigcup_{e \in E} K_e \cup \{y\}$. We add an edge between every vertex in K_e and a vertex $v \in V$ if and only if v is incident with e in G. In G' we let the vertices of V form a clique. Finally, we add all edges between y and any vertex in $V \cup \bigcup_{e \in E} K_e$.

Note that the resulting graph G' is indeed chordal. Also note that $\omega(G') = |V|+3$ (every maximum clique consists of y, the vertices of a clique K_e and their two neighbours in V).

We claim that G has a vertex cover of size at most k if and only if G' can be k-contracted to a graph H with $\omega(H) \leq \omega(G') - 1$. First suppose that G has a vertex cover U of size at most k. For each vertex $v \in U$, we contract the corresponding vertex v in G' to y. As $|U| \leq k$, this means that we k-contracted G' into a graph H. Since U is a vertex cover, we obtain $\omega(H) \leq |V|+2 = \omega(G')-1$.

Now suppose that G' can be k-contracted to a graph H with $\omega(H) \leq \omega(G')-1$. Let S be a corresponding sequence of edge contractions (so $|S| \leq k$ holds). By Lemma 7 and the fact that chordal graphs are closed under taking edge contractions, we find that no contraction in S results in a new maximum clique. Hence, as we need to reduce the size of each maximum clique $K_{uv} \cup \{u, v, y\}$ by at least 1, we may assume without loss of generality that each contraction in S concerns an edge with both its end-vertices in $V \cup \{y\}$. We construct a set U as follows. If S contains the contraction of an edge uy we select u. If S contains the contraction of an edge uv, we select one of u, v arbitrarily. Because each maximum clique $K_{uv} \cup \{u, v, y\}$ must be reduced, we find that $U \subseteq V$ is a vertex cover. By construction, $|U| \leq k$. This completes the proof.   □

Similar arguments as in the above proof can be readily used to show the following.

Theorem 4. *For $\pi \in \{\chi, \omega\}$, 1-DELETION BLOCKER(π) is NP-complete for chordal graphs.*

We will finish this section with a result on C_4-free perfect graphs.

Theorem 5. *For $\pi \in \{\chi, \omega\}$, 1-CONTRACTION BLOCKER(π) is NP-complete for the class of C_4-free perfect graphs.*

Proof. Let $\pi = \omega$, or equivalently, $\pi = \chi$. We adapt the construction used in the proof of Lemma 4 by doing as follows for each edge e of the graph G in this proof. First we subdivide e. This gives us two new edges e_1 and e_2. We introduce two new non-adjacent vertices u_e and v_e and make them adjacent to both end-vertices of e_1. Denote the resulting graph by G^*. Notice that we do not create any induced C_4 this way. Hence G^* is C_4-free. The vertices of the original graph together with the subdivision vertices form a bipartite graph on top of which we placed a number of triangles. Hence, G^* contains no induced hole of odd size and no induced antihole of odd size, where a *hole* is an induced cycle on at least five vertices and an *antihole* is the complement of a hole. Then, by the Strong Perfect Graph Theorem [5], G^* is perfect as well.

We increase the allowed number of edge contractions accordingly and observe that, because of the presence of the vertices u_e and v_e for each edge e, we are always forced to contract the edge e_1, which gives us back the original construction extended with a number of pendant edges (which do not play a role). Note that we have left the class of C_4-free perfect graphs after contracting away the triangles, but this is allowed.   □

4 Contraction Blocker in H-Free Graphs

In this section, we will consider both problems CONTRACTION BLOCKER(π) and d-CONTRACTION BLOCKER(π) for $\pi \in \{\omega, \chi\}$ and present our classification results for H-free graphs. We start with $\pi = \chi$ and H being a connected graph. In this case, we obtain a complete dichotomy for both problems CONTRACTION BLOCKER(χ) and d-CONTRACTION BLOCKER(χ) concerning their computational complexity.[1]

Theorem 6. *Let H be a connected graph. If H is an induced subgraph of P_4 then* CONTRACTION BLOCKER(χ) *is polynomial-time solvable for H-free graphs. Otherwise even* 1-CONTRACTION BLOCKER(χ) *is* NP-*hard for H-free graphs.*

Proof. Let H be a connected graph. If H is an induced subgraph of P_4, then we use Lemma 5. Now suppose that H is not an induced subgraph of P_4. Then COLORING is NP-complete for H-free graphs by Lemma 2. If H is not a clique, then the class of H-free graphs is clique-proof. Hence, we can use Theorem 1. So suppose H is a clique. It suffices to show NP-completeness for $H = K_3$. We reduce from 3-COLORING restricted to K_3-free graphs. This problem is NP-complete by Lemma 3. Let G be a K_3-free graph representing an instance of 3-COLORING. We obtain an instance of 1-CONTRACTION BLOCKER(χ) as follows. Take two copies of G and the 4-chromatic Grötzsch graph F (see [17], p. 184). Call the resulting graph G', i.e. $G' = 2G + F$. We claim that G is 3-colorable if and only if it is possible to contract precisely one edge of G' so that the new graph G^* has chromatic number $\chi(G') - 1$. We prove this claim via similar arguments as used in the proof of Theorem 1. □

For the case when H is a general graph (not necessarily connected), we obtain a complete dichotomy for CONTRACTION BLOCKER(χ).

Theorem 7. *Let H be a graph. If H is an induced subgraph of P_4 then* CONTRACTION BLOCKER(χ) *is polynomial-time solvable for H-free graphs, otherwise it is* NP-*hard for H-free graphs.*

Proof. If H is connected then we use Theorem 6. Suppose H is disconnected. If H contains a component that is not an induced subgraph of P_4 then we use Theorem 6 again. Assume that each connected component of H is an induced subgraph of P_4. If $2P_2 \subseteq_i H$ or $3P_1 \subseteq_i H$ then we use Lemma 5 and the fact that split graphs are $(2P_2, C_4, C_5)$-free (see [8]) or Corollary 1, respectively. Hence, $H \in \{2P_1, P_2 + P_1\}$, so $H \subseteq_i P_4$ and we can use again Theorem 6. □

Completing the classification of the computational complexity of d-CONTRACTION BLOCKER(χ) for general graphs H (not necessarily connected) is still open.

We now consider the case $\pi = \omega$. Also in this case we obtain a complete dichotomy when H is connected.

---

[1] We can modify the gadgets for proving NP-completeness for the case $d = 1$ in a straightforward way to obtain NP-completeness for every constant $d \geq 2$. A similar remark holds for other theorems. Details will be given in the journal version.

Theorem 8. *Let H be a connected graph. If H is an induced subgraph of P_4 or of $\overline{P_1 + P_3}$ then* CONTRACTION BLOCKER(ω) *is polynomial-time solvable for H-free graphs. Otherwise* 1-CONTRACTION BLOCKER(ω) *is* NP-*hard for H-free graphs.*

Proof. Let H be a connected graph. If H contains an induced C_4, use Theorem 5. If H has an induced K_4, $\overline{2P_1 + P_2}$ or butterfly, use Lemma 6. If H contains an induced $K_{1,3}$, C_5, P_5, bull or cobanner, use Lemma 1 with Theorem 1. So we may assume that H is $(C_4, C_5, P_5, K_{1,3}, K_4, \overline{2P_1 + P_2}$, bull, butterfly, cobanner)-free.

We claim that H is an induced subgraph of P_4 or of $\overline{P_1 + P_3}$. For contradiction, assume that $H \not\subseteq_i P_4$ and $H \not\subseteq_i \overline{P_1 + P_3}$. First suppose that H contains no cycle. Then, as H is connected, H is a tree. Because H is $K_{1,3}$-free, H is a path. Our assumption that H is not an induced subgraph of P_4 or of $\overline{P_1 + P_3}$ implies that H contains an induced P_5, which is not possible as H is P_5-free.

Now suppose that H contains a cycle C. Then C must have exactly three vertices, because H is (C_4, C_5, P_5)-free. As H is not an induced subgraph of $\overline{P_1 + P_3}$, we find that H contains at least one vertex x not on C. As H is connected, we may assume that x has a neighbour on C. Because H is $(\overline{2P_1 + P_2}, K_4)$-free, x has exactly one neighbour on C. Let v be this neighbour. Hence, H contains an induced $\overline{P_1 + P_3}$ (consisting of x, v and the other two vertices of C). As H is not an induced subgraph of $\overline{P_1 + P_3}$ and H is connected, it follows that H contains a vertex $y \notin V(C) \cup \{x\}$ that is adjacent to a vertex on C or to x.

First suppose y is adjacent to a vertex of C. Then, as H is $(\overline{2P_1 + P_2}, K_4)$-free, y has exactly one neighbour u in C. If $u = v$ then H either contains an induced claw (if x and y are non-adjacent) or an induced butterfly (if x and y are adjacent). Since, by our assumption, this is not possible, it follows that $u \neq v$. Then, because H is bull-free, we deduce that x and y are adjacent. However, then the vertices, u, v, x, y form an induced C_4, which is not possible as H is C_4-free. We conclude that y is not adjacent to a vertex of C, so y must be adjacent to x only. But then H contains an induced cobanner, a contradiction. Hence, H is an induced subgraph of P_4 or of $\overline{P_1 + P_3}$ as we claimed.

If H is an induced subgraph of P_4 then we use Lemma 5. If H is an induced subgraph of $\overline{P_1 + P_3}$, then we know from [12] that either G is K_3-free or G is complete multipartite. In the first case one must contract all the edges of an H-free graph in order to decrease its clique number. Hence CONTRACTION BLOCKER(ω) is polynomial-time solvable for K_3-free graphs. In the second case H is P_4-free, so we can use Lemma 5 again. $\qquad\square$

For general graphs H, we have one open case for CONTRACTION BLOCKER(ω) (while for d-CONTRACTION BLOCKER(ω) there are many more open cases).

Theorem 9. *Let $H \neq K_3 + P_1$ be a graph. If H is an induced subgraph of P_4 or of $\overline{P_1 + P_3}$ then* CONTRACTION BLOCKER(ω) *is polynomial-time solvable for H-free graphs, otherwise it is* NP-*hard for H-free graphs.*

Proof. If H is connected, use Theorem 8. Suppose H is disconnected. If H contains a component that is not an induced subgraph of P_4 or $\overline{P_1 + P_3}$ then we use

Theorem 8 again. Assume that each component of H is an induced subgraph of P_4 or $\overline{P_1 + P_3}$. If $2P_2 \subseteq_i H$ or $3P_1 \subseteq_i H$ then we use Lemma 5 or Corollary 1, respectively. Hence, $H \in \{2P_1, P_2 + P_1, K_3 + P_1\}$. In the first two cases $H \subseteq_i P_4$ and thus we can use Theorem 8, whereas we excluded the last case.   □

5   Deletion Blocker in H-Free Graphs

We adapt the proof of Theorem 1 to present relations between COLORING and 1-DELETION BLOCKER(χ) and between CLIQUE and 1-DELETION BLOCKER(ω).

Theorem 10. *Let $\mathcal{G}$ be a clique-proof graph class. Then the following two statements hold:*

(i) if COLORING is NP-complete for $\mathcal{G}$, then so is 1-DELETION BLOCKER(χ).
(ii) if CLIQUE is NP-complete for $\mathcal{G}$, then so is 1-DELETION BLOCKER(ω).

We notice a relation between 1-DELETION BLOCKER(ω) and VERTEX COVER.

Lemma 8. *Let G be a triangle-free graph containing at least one edge and let $k \geq 1$ be an integer. Then (G, k) is a yes-instance for 1-DELETION BLOCKER(ω) if and only if (G, k) is a yes-instance for VERTEX COVER.*

Proof. Let $G = (V, E)$ be a triangle-free graph with $|E| \geq 1$. Thus, $\omega(G) = 2$. Let $k \geq 1$ be an integer. First suppose that (G, k) is a yes-instance for VERTEX COVER and let V' be a solution, i.e. for every edge $e \in E$, there exists a vertex $v \in V'$ such that v is an endvertex of e. It follows that by deleting all vertices in V', we obtain a graph G' containing no edges and hence $\omega(G') \leq 1$. We conclude that (G, k) is a yes-instance for 1-DELETION BLOCKER(ω). Conversely, suppose that (G, k) is a yes-instance for 1-DELETION BLOCKER(ω) and let V' be a solution, i.e. the graph obtained form G by deleting the vertices in V' satisfies $\omega(G') \leq 1$. But this implies that G' contains no edges and thus V' is a vertex cover of size at most k. So (G, k) is a yes-instance for VERTEX COVER.   □

Corollary 2. 1-DELETION BLOCKER(ω) *is NP-complete for the class of (C_3, C_4)-free graphs.*

Proof. This follows immediately from Lemma 8 and the fact that VERTEX COVER is NP-complete for (C_3, C_4)-free graphs (see [14]).   □

We are now ready to prove the first main result of this section.

Theorem 11. *Let H be a connected graph. If H is an induced subgraph of P_4, then DELETION BLOCKER(ω) is polynomial-time solvable on H-free graphs. Otherwise 1-DELETION BLOCKER(ω) is NP-hard for H-free graphs.*

Proof. If H contains a cycle C_r, $r \in \{3, 4\}$, we use Corollary 2. If H contains a cycle C_r, ≥ 5, we use Lemma 1 combined with Theorem 10. Hence, we may assume now that H is a tree. If H contains an induced $K_{3,1}$, we use Lemma 1 combined with Theorem 10. Thus, H is a path. If this path has length at most 4, we use Lemma 5. Otherwise, we use Lemma 1 combined with Theorem 10. This completes the proof.   □

If H is disconnected, finding such a dichotomy is open. In particular, the cases when $H \in \{2P_2, 3P_1\}$ are unknown. Moreover, in contrast to the CONTRACTION BLOCKER(ω) problem, DELETION BLOCKER(ω) is polynomial-time solvable on cobipartite graphs [6], which form a subclass of $3P_1$-free graphs. We now focus on $\pi = \chi$. The proof of Theorem 6 can easily be adapted to get the following.

Theorem 12. *Let H be a connected graph. If H is an induced subgraph of P_4, then DELETION BLOCKER(χ) is polynomial-time solvable on H-free graphs. Otherwise, 1-DELETION BLOCKER(χ) is NP-hard for the class of H-free graphs.*

If H is disconnected, it seems much harder to get a dichotomy even when d is part of the input. In contrast to the case of ω, we can prove that DELETION BLOCKER(χ) is polynomial-time solvable for $3P_1$-free graphs.

Theorem 13. DELETION BLOCKER(χ) *can be solved in polynomial time for the class of $3P_1$-free graphs.*

Proof. Let $G = (V, E)$ be a $3P_1$-free graph with $|V| = n$ and let $k \geq 1$ be an integer. Consider an instance (G, k, d) of DELETION BLOCKER(χ). We proceed as follows. First consider an optimal coloring of G, which can be obtained in polynomial time [10]. Since G is $3P_1$-free, the size of each color class is at most 2. Also the number of color classes of size 1 is the same for every optimal coloring of G. Let ℓ be this number. Hence, there are $\frac{n-\ell}{2}$ color classes of size 2 and $\chi(G) = \ell + \frac{n-\ell}{2}$. Now (G, k, d) is a yes-instance if and only if we can obtain a graph G' from G by deleting at most k vertices such that $\chi(G') \leq \chi(G) - d = \ell + \frac{n-\ell}{2} - d$. Since G' is also $3P_1$-free, the color classes in any optimal coloring of G' have size at most 2 and thus, G' contains at most $2(\ell + \frac{n-\ell}{2} - d) = n + \ell - 2d$ vertices. In other words, we need to delete at least $2d - \ell$ vertices from G in order to get such a graph G'. So (G, k, d) is clearly a no-instance if $k < 2d - \ell$. Next we will show that if $k \geq 2d - \ell$, then (G, k, d) is a yes-instance and this will complete the proof. If $d \leq \ell$, we delete d vertices representing color classes of size 1. If $d > \ell$, we delete the ℓ vertices representing the color classes of size 1 and $2(d - \ell)$ vertices of $d - \ell$ color classes of size 2. This way, we clearly obtain a graph G' whose chromatic number is exactly $\chi(G) - d$.   □

6   Conclusions

We considered the problems $(d\text{-})$CONTRACTION BLOCKER(π) and $(d\text{-})$DELETION BLOCKER(π), where $\pi \in \{\chi, \omega\}$. We mainly focused on H-free graphs and analyzed the computational complexity of these problems. We obtained a complete dichotomy for both problems and both when d is fixed and when d is part of the input, if H is a connected graph. If H is an arbitrary graph that is not necessarily connected, further research is needed: What is the complexity of the problems d-CONTRACTION BLOCKER(χ) and d-CONTRACTION BLOCKER(ω) for H-free graphs when H is disconnected? What is the complexity of CONTRACTION BLOCKER(ω) for $(K_3 + P_1)$-free graphs? What are the complexities of

Table 1. Results for subclasses of perfect graphs closed under edge contraction (apart from the classes of bipartite and perfect graphs), where NP-c stands for NP-complete and P for polynomial-time solvable; results marked with a * correspond to results of this paper; the unmarked results for perfect graphs follow directly from other results.

	CONTRACTION BLOCKER(π)		DELETION BLOCKER(π)	
Class	$\pi = \alpha$	$\pi = \omega(= \chi)$	$\pi = \alpha$	$\pi = \omega(= \chi)$
Bipartite	?	P (trivial)	P [6]	P*
Cobipartite	$d = 1$: NP-c [7]	NP-c*; d fixed: P [7]	P*	P [6]
Chordal	?	$d = 1$: NP-c*	?	$d = 1$: NP-c*
Interval	?	P [7]	?	P [7]
Split	NP-c; d fixed: P [7]	NP-c; d fixed: P [7]	NP-c; d fixed: P [6]	NP-c; d fixed: P [6]
Cograph	P [7]	P [7]	P [7]	P [7]
C_4-free Perfect	?	$d = 1$: NP-c*	?	?
Perfect	$d = 1$: NP-c	$d = 1$: NP-c	NP-c; d fixed: ?	$d = 1$: NP-c

(d-)DELETION BLOCKER(χ) and (d-)DELETION BLOCKER(ω) for H-free graphs when H is disconnected? In particular, what is the complexity of d-DELETION BLOCKER(ω) for $2P_2$-free graphs and $3P_1$-free graphs?

Besides considering the parameters χ and ω, we may of course choose any other graph parameter π, such as $\pi = \alpha$, where α is the independence number (the size of a largest independent set in a graph). Note that d-DELETION BLOCKER(ω) in a graph G is equivalent to d-DELETION BLOCKER(α) in its complement $\overline{G}$. Studying the complexity of d-CONTRACTION BLOCKER(α) and d-DELETION BLOCKER(α) for H-free graphs is left as future research.

In addition to our results on H-free graphs, we also obtained some new results for subclasses of perfect graphs. We used these as auxiliary results for our classifications but also in order to continue a line of research started in [7]. Table 1 gives an overview of the known results and the new results of this paper for such classes of graphs. Notice that $\chi = \omega$ holds by definition of a perfect graph. In the table we also added results for CONTRACTION BLOCKER(α) and DELETION BLOCKER(α), since these problems have been studied in [6,7] and since some of our new results immediately imply corresponding results for the case $\pi = \alpha$. In particular, the polynomial-time solvability of d-DELETION BLOCKER(ω) for bipartite graphs (and therefore d-DELETION BLOCKER(α) in cobipartite graphs) follows from Corollary 2 and the fact that VERTEX COVER is polynomial-time solvable in bipartite graphs. The proof that shows that CONTRACTION BLOCKER(ω) is polynomial-time solvable for interval graphs can easily be adapted to show that DELETION BLOCKER(ω) is polynomial-time solvable for interval graphs.

As can be seen from Table 1 there are several open cases (marked by "?"). Some of these open cases form challenging open problems related to interval and chordal graphs, namely what is the complexity of CONTRACTION BLOCKER(α) and d-CONTRACTION BLOCKER(α) for interval graphs and for chordal graphs? What are the complexities of the problems DELETION BLOCKER(α) and d-DELETION BLOCKER(α) for interval graphs and for chordal graphs?

References

1. Bazgan, C., Bentz, C., Picouleau, C., Ries, B.: Blockers for the stability number and the chromatic number. Graphs Comb. **31**, 73–90 (2015)
2. Bazgan, C., Toubaline, S., Tuza, Z.: Complexity of most vital nodes for independent set in graphs related to tree structures. In: Iliopoulos, C.S., Smyth, W.F. (eds.) IWOCA 2010. LNCS, vol. 6460, pp. 154–166. Springer, Heidelberg (2011)
3. Bazgan, C., Toubaline, S., Tuza, Z.: The most vital nodes with respect to independent set and vertex cover. Discrete Appl. Math. **159**, 1933–1946 (2011)
4. Bentz, C., Costa, M.-C., de Werra, D., Picouleau, C., Ries, B.: Weighted Transversals and blockers for some optimization problems in graphs. In: Progress in Combinatorial Optimization. ISTE-WILEY (2012)
5. Chudnovsky, M., Robertson, N., Seymour, P.D., Thomas, R.: The strong perfect graph theorem. Ann. Math. **164**, 51–229 (2006)
6. Costa, M.-C., de Werra, D., Picouleau, C.: Minimum d-blockers and d-transversals in graphs. J. Comb. Optim. **22**, 857–872 (2011)
7. Diner, Ö.Y., Paulusma, D., Picouleau, C., Ries, B.: Contraction blockers for graphs with forbidden induced paths. In: Paschos, V.T., Widmayer, P. (eds.) CIAC 2015. LNCS, vol. 9079, pp. 194–207. Springer, Heidelberg (2015)
8. Földes, S., Hammer, P.L.: Split graphs. Congressus Numerantium **19**, 311–315 (1977). 8th South-Eastern Conference on Combinatorics, Graph Theory and Computing
9. Garey, M.R., Johnson, D.S.: Computers and Intractability: A Guide to the Theory of NP-Completeness. Freeman, New York (1979)
10. Král', D., Kratochvíl, J., Tuza, Z., Woeginger, G.J.: Complexity of coloring graphs without forbidden induced subgraphs. In: Brandstädt, A., Le, V.B. (eds.) WG 2001. LNCS, vol. 2204, pp. 254–262. Springer, Heidelberg (2001)
11. Maffray, F., Preissmann, M.: On the NP-completeness of the k-colorability problem for triangle-free graphs. Discrete Math. **162**, 313–317 (1996)
12. Olariu, S.: Paw-free graphs. Inf. Process. Lett. **28**, 53–54 (1988)
13. Pajouh, F.M., Boginski, V., Pasiliao, E.L.: Minimum vertex blocker clique problem. Networks **64**, 48–64 (2014)
14. Poljak, S.: A note on the stable sets and coloring of graphs. Comment. Math. Univ. Carolin. **15**, 307–309 (1974)
15. Ries, B., Bentz, C., Picouleau, C., de Werra, D., Costa, M.-C., Zenklusen, R.: Blockers and transversals in some subclasses of bipartite graphs: when caterpillars are dancing on a grid. Discrete Math. **310**, 132–146 (2010)
16. Toubaline, S.: Détermination des éléments les plus vitaux pour des problèmes de graphes. Ph.D. Thesis, Université Paris-Dauphine (2010)
17. West, D.B.: Introduction to Graph Theory. Prentice-Hall, Upper Saddle River (1996)

The Parity Hamiltonian Cycle Problem in Directed Graphs

Hiroshi Nishiyama[✉], Yukiko Yamauchi, Shuji Kijima,
and Masafumi Yamashita

Graduate School of Information Science and Electrical Engineering,
Kyushu University, Fukuoka, Japan
{hiroshi.nishiyama,yamauchi,kijima,mak}@inf.kyushu-u.ac.jp

Abstract. This paper investigates a variant of the Hamiltonian cycle, the *parity Hamiltonian cycle* (*PHC*) problem: a PHC in a *directed* graph is a closed walk (possibly using an arc more than once) which visits every vertex odd number of times. Nishiyama et al. (2015) investigated the *undirected* version of the PHC problem, and gave a simple characterization that a connected undirected graph has a PHC if and only if it has even order or it is non-bipartite. This paper gives a complete characterization when a directed graph has a PHC, and shows that the PHC problem in a directed graph is solved in polynomial time. The characterization, unlike with the undirected case, is described by a linear system over GF(2).

Keywords: Hamiltonian cycle · *T*-joins · Linear system over GF(2)

1 Introduction

It is said that the graph theory has its origin in the seven bridges of Königsberg settled by Leonhard Euler [2]. An *Eulerian* cycle, named after him in modern terminology, is a cycle which uses every edge exactly once, and it is now well-known that a connected undirected graph has an Eulerian cycle if and only if every vertex has an even degree. A *Hamiltonian* cycle (HC), a similar but completely different notion, is a cycle which visits every vertex exactly once. In contrast to the clear characterization of an Eulerian graph, the question if a given graph has a Hamiltonian cycle is a celebrated NP-complete problem due to Karp [11]. The HC problem is widely interested in computer science or mathematics, and has been approached with several variants or related problems. The traveling salesman problem (TSP) in a graph, which is NP-hard since the HC problem is so, is regarded as a relaxed version of the HC problem, in which the condition of visiting number on each vertex is relaxed to more than once. Another example may be a two-factor (in cubic graphs), which relaxes the condition of the connectivity of an HC, but a two-factor must contain each vertex exactly once (cf. [3,4,8,9]).

It could be a natural idea for the HC problem to modify the condition on the visiting number keeping the connectivity condition. The *parity Hamiltonian cycle*

© Springer International Publishing Switzerland 2016
R. Cerulli et al. (Eds.): ISCO 2016, LNCS 9849, pp. 50–58, 2016.
DOI: 10.1007/978-3-319-45587-7_5

(*PHC*) problem, which this paper is involved in, is a variant of the Hamiltonian cycle problem: a PHC is a closed walk (possibly using each edge more than once) which visits every vertex an odd number of times. Note that the PHC problem allows natural variations, directed or undirected, cycle or path, so does the HC. Brigham et al. [5] showed that any connected undirected graph has a parity Hamiltonian path or cycle, by giving an algorithm based on the depth first search. Thirty years later, Nishiyama et al. [14] investigated the PHC problem in *undirected* graphs, and gave a complete characterization that a connected undirected graph has a PHC if and only if it has an even order or it is non-bipartite. They also showed that any graph satisfying the condition admits a PHC which uses each edge at most four times, by presenting an algorithm to find a PHC using T-joins. On the other hand, the PHC problem becomes NP-complete if each edge is restricted to be used in a PHC at most three times.

This paper investigates a directed version of the PHC problem. We give two characterizations when a directed graph admits a PHC. The characterizations, unlike with the undirected case, are described by linear systems over GF(2). Our characterizations directly imply that the PHC problem in a directed graph is solved in polynomial time. We then give a faster algorithm to recognize if a directed graph has a PHC, which runs in linear time without (explicitly) solving the linear system over GF(2). In the linear time algorithm, T-joins play a key role. We also discuss a problem extended to GF(p), in Sect. 4.

Notice that the condition that an HC visits each vertex $1 \in \mathbb{R}$ times is replaced by $1 \in$ GF(2) times in a PHC. Modification of the field is found in group-labeled graphs or nowhere-zero flows [10, 12]. It was recently shown that the extension complexity of the TSP is exponential [6, 7, 17], while it is an interesting question if the PHC problem has an efficient (extended) formulation over GF(2).

2   Definitions and Notations

This section introduces definitions and notations. A *directed graph* (*digraph* for short) $D = (V, A)$ is given by a vertex set V and an arc set A (sometimes we use $V(D)$ and $A(D)$ to clarify the graph which we are focusing on). Let $\delta^+(v)$ (resp. $\delta^-(v)$) for $v \in V$ denote the set of *outgoing* (resp. *incoming*) arcs; that is, arcs that leave v (resp. enter v). The sizes $|\delta^+(v)|$ and $|\delta^-(v)|$ are called the *out-degree* and the *in-degree* of v, respectively.

A *directed walk* is a sequence of vertices and arcs $v_0 a_1 \cdots a_\ell v_\ell$, where $a_i = (v_{i-1}, v_i) \in A$ for each i ($1 \leq i \leq \ell$). A directed walk is *closed* if $v_\ell = v_0$. A *directed path* is a directed walk which contains each vertex at most once except the start vertex v_0 and the end vertex v_ℓ. A directed closed path is called a *directed cycle*. A digraph D is *strongly connected* if there exists a directed path from u to v for any pair of vertices $u, v \in V(D)$. For convenience, we often represent a directed closed walk by an integer vector $\tilde{x} \in \mathbb{Z}_{\geq 0}^A$, in which $\tilde{x}(a)$ denotes the number of occurrences of arc a in the closed walk.

A *cycle basis* of D is a set of directed cycles $\{C_1, C_2, \ldots, C_k\}$ which satisfies conditions: (i) Their incidence vectors $c_1, c_2, \ldots, c_k \in \{0, 1\}^A$ are linearly independent over $GF(2)$, and (ii) the incidence vector of every cycle, including those are not directed, can be represented as a linear combination of $c_1, c_2, \ldots, c_k$. We call each cycle C_i a *fundamental cycle*. It is known that the size k of cycle basis is equal to $|A(D)| - |V(D)| + 1$ [1], and a cycle basis of a digraph can be found in linear time [1,16].

Parity Hamiltonian Cycle Problem. A *parity Hamiltonian cycle* (PHC for short) of a digraph D is a directed closed walk in which each vertex appears odd number of times except the starting vertex. In other words, a PHC is a connected closed walk which satisfies a parity condition: $\sum_{a \in \delta^+(v)} \tilde{x}(a) \equiv \sum_{a \in \delta^-(v)} \tilde{x}(a) \equiv 1 \pmod{2}$ for each $v \in V$. Note that a PHC may use each arc more than once, unlike with HC's. The *parity Hamiltonian cycle problem* is a decision problem to decide whether an input graph has a PHC. Note that only strongly connected digraphs have PHC's, thus we assume that the input digraph is strongly connected in what follows.

3   Main Results

In this section we explain our main result. In Sect. 3.1 we show a characterization of digraphs which have PHC's, and in Sect. 3.2 we give an algorithm for the PHC problem and one for finding a PHC, and discuss their time complexities.

3.1   Characterization

To state our main result, we define two matrices M and Q. For a digraph D, let $M^+ = [m_{va}^+]$ and $M^- = [m_{va}^-] \in \{0, 1\}^{|V| \times |A|}$ be matrices respectively defined by

$$m_{va}^+ = \begin{cases} 1 \text{ if } a \in \delta^+(v), \\ 0 \text{ otherwise,} \end{cases} \text{ and } \quad m_{va}^- = \begin{cases} 1 \text{ if } a \in \delta^-(v), \\ 0 \text{ otherwise.} \end{cases}$$

Thus the parity condition of a PHC is written as $M^+ \tilde{x} \equiv M^- \tilde{x} \equiv 1 \pmod{2}$. We define a matrix M over $\{0, 1\}^{2|V| \times |A|}$ by

$$M = \begin{bmatrix} M^+ \\ M^- \end{bmatrix}.$$

Let $\{C_1, C_2, \ldots, C_k\}$ $(k = |A| - |V| + 1)$ be a cycle basis of D and let $c_1, c_1, \ldots, c_k \in \{0, 1\}^A$ their incidence vectors. Let $R = [c_1, c_1, \ldots, c_k]$. We define a matrix $Q \in \{0, 1\}^{|V| \times k}$ by

$$Q = M^+ R. \tag{1}$$

Remark that $Q = M^- R$ holds, since for each i, the column vector $q_i \in \{0, 1\}^V$ of Q is a vector such that $q_i(v) = 1$ if and only if C_i contains $v \in V$.

Now we are ready to state our main result.

Theorem 1. *The following three conditions are equivalent:*

(a) *A strongly connected digraph $D = (V, A)$ has a PHC,*
(b) $Mx \equiv 1 \pmod 2$ *has a solution* $x \in \{0, 1\}^A$,
(c) $Q\beta \equiv 1 \pmod 2$ *has a solution* $\beta \in \{0, 1\}^k$,

where $k = |A| - |V| + 1$ and 1 denotes the all 1 vector.

Proof. The proofs of (c) $\Rightarrow$ (a) and (a) $\Rightarrow$ (b) are easy, while the other way (b) $\Rightarrow$ (a) and (a) $\Rightarrow$ (c), as well as (b) $\Rightarrow$ (c) directly are not trivial. First we show (a) $\Leftrightarrow$ (b), then we show (a) $\Leftrightarrow$ (c).

(a) $\Rightarrow$ (b). Let $\tilde{x} \in \mathbb{Z}_{\geq 0}^A$ be a vector in which $\tilde{x}(a)$ denotes the number of uses of $a \in A$ in a PHC. By the parity condition of PHC, we have $M^+\tilde{x} \equiv M^-\tilde{x} \equiv 1$ (mod 2). Then let $x \in \{0, 1\}^A$ be defined by $x \equiv \tilde{x} \pmod 2$, we have $M^+x \equiv M^-x \equiv 1 \pmod 2$, and thus $Mx \equiv 1 \pmod 2$.

(b) $\Rightarrow$ (a). Suppose that $x \in \{0, 1\}^A$ is a solution of $Mx \equiv 1 \pmod 2$, then we explain how to construct a PHC. Remark that a graph indicated by x satisfies the parity condition of the visiting number on each vertex, but may not satisfy the Eulerian condition, meaning that $\sum_{a \in \delta^+(v)} x(a) = \sum_{a \in \delta^-(v)} x(a)$ may not hold for some vertex v, and connectivity.

First, we construct $x' \in \mathbb{Z}_{\geq 0}^A$ satisfying both of the parity condition $Mx' \equiv 1$ (mod 2) and the Eulerian condition $\sum_{a \in \delta^+(v)} x'(a) = \sum_{a \in \delta^-(v)} x'(a)$ for each $v \in V$. Let $\phi(v) = \sum_{a \in \delta^+(v)} x(a) - \sum_{a \in \delta^-(v)} x(a)$ for each $v \in V$, denoting the difference between out-degree and in-degree of v in x. Then x is Eulerian if and only if $\phi(v) = 0$ for all v. Notice that $\sum_{v \in V} \phi(v) = 0$ holds since the total of out-degrees is equal to the total of in-degrees. We also remark that $\phi(v)$ is even for each $v \in V$, since $Mx \equiv 1 \pmod 2$ implies that both of out-degree $(\sum_{a \in \delta^+(v)} x(a))$ and in-degree $(\sum_{a \in \delta^-(v)} x(a))$ are odd. Then we apply the following Procedure 1 to x:

Procedure 1

1. Find $u, v \in V$ such that $\phi(u) < 0$ and $\phi(v) > 0$.
2. Find a directed path P from u to v (P always exists since D is strongly connected).
3. $x(a) := x(a) + 2$ for each $a \in A(P)$.

Procedure 1 preserves the parity condition $Mx \equiv 1 \pmod 2$, and decreases the value of $\sum_{v \in V} |\phi(v)|$ (by four). By recursively applying Procedure 1 until $\sum_{v \in V} |\phi(v)|$ is zero, we obtain a desired closed walk x'.

If x' suggests a connected walk, we obtain a PHC. Suppose that x' is not connected. Then we apply the following Procedure 2 to x':

Procedure 2

1. Find $u, v \in V$ which are in distinct connected components.
2. Find directed paths P from u to v and P' from v to u.
3. $x'(a) := x'(a) + 2$ for each $a \in A(P) \cup A(P')$.

Procedure 2 preserves the parity condition $M\boldsymbol{x} \equiv \mathbf{1}$ (mod 2) and the Eulerian condition, and decreases the number of connected components. By recursively applying Procedure 2, we obtain a connected walk, which is in fact a PHC.

(c) $\Rightarrow$ (a). We construct a PHC from the solution $\boldsymbol{\beta} \in \{0,1\}^k$. Let

$$\alpha_i = \begin{cases} 1 & \text{if } \beta_i = 1, \\ 2 & \text{if } \beta_i = 0, \end{cases} \tag{2}$$

and set $\tilde{\boldsymbol{x}} = R\boldsymbol{\alpha}$. Notice that $\tilde{\boldsymbol{x}}$ indicates a closed walk since it is a sum of cycles. We claim that the closed walk indicated by $\tilde{\boldsymbol{x}}$, say γ, is a PHC. The walk γ is connected since γ uses all edges of D at least once, and D is strongly connected. Then, we have

$$M^+\tilde{\boldsymbol{x}} = M^+R\boldsymbol{\alpha} = Q\boldsymbol{\alpha} \equiv Q\boldsymbol{\beta} \equiv \mathbf{1} \pmod{2},$$

where the second last congruence comes from (2) and the last congruence follows from the assumption that $\boldsymbol{\beta}$ is a solution of $Q\boldsymbol{\beta} \equiv \mathbf{1}$ (mod 2). Hence, γ satisfies the parity condition, and thus γ is a PHC.

(a) $\Rightarrow$ (c). To show the necessity we show the following lemma.

Lemma 2. *Let γ be any closed walk of D, and let $\tilde{\boldsymbol{x}} \in \mathbb{Z}_{\geq 0}^A$ be a vector in which $\tilde{x}(a)$ denotes the number of uses of $a \in A$ in γ. Then $R\boldsymbol{\beta} \equiv \tilde{\boldsymbol{x}}$ (mod 2) has a solution $\boldsymbol{\beta} \in \{0,1\}^k$.*

Proof. Since γ is an Eulerian walk in a multi-digraph, meaning that γ consists of simple cycles, $\tilde{\boldsymbol{x}}$ is represented by

$$\tilde{\boldsymbol{x}} = \sum_{j=1}^{\ell} \alpha_j \gamma_j, \tag{3}$$

with appropriate positive integer ℓ, where each α_j is a nonnegative integer and each $\gamma_j \in \{0,1\}^A$ is the incidence vector of a directed cycle of D. Remark that each γ_j is represented by a linear combination of incidence vectors of fundamental cycles $\boldsymbol{c}_1, \ldots, \boldsymbol{c}_k$, such that $\gamma_j \equiv \sum_{i=1}^k \beta'_{ij}\boldsymbol{c}_i$ (mod 2) for some 0-1 coefficients β'_{ij} for each j. Let $\beta_i \in \{0,1\}$ be defined by $\beta_i \equiv \sum_{j=1}^{\ell} \beta'_{ij}\alpha_j$ (mod 2), then

$$\tilde{\boldsymbol{x}} \equiv \sum_{j=1}^{\ell} \alpha_j \sum_{i=1}^k \beta'_{ij}\boldsymbol{c}_i \equiv \sum_{i=1}^k \beta_i \boldsymbol{c}_i \pmod{2}$$

holds. Notice that $\sum_{i=1}^k \beta_i \boldsymbol{c}_i = R\boldsymbol{\beta}$, then we obtain the claim. □

Suppose that γ is a PHC of D, and that $\tilde{\boldsymbol{x}} \in \mathbb{Z}_{\geq 0}^A$ is a vector in which $\tilde{x}(a)$ denotes the number of uses of $a \in A$ in γ. Since a PHC is a closed walk, Lemma 2 implies that there is a vector $\boldsymbol{\beta} \in \{0,1\}^k$ such that $\tilde{\boldsymbol{x}} \equiv R\boldsymbol{\beta}$ (mod 2). Then

$$Q\boldsymbol{\beta} = M^+R\boldsymbol{\beta} \equiv M^+\tilde{\boldsymbol{x}} \equiv \mathbf{1} \pmod{2},$$

where the last congruence comes from the fact that $\tilde{\boldsymbol{x}}$ indicates a PHC. □

3.2   Recognition in Linear Time

By Theorem 1, we can decide whether or not a given directed graph D has a PHC in polynomial time, by solving a linear system $Mx \equiv 1 \pmod 2$ or $Q\beta \equiv 1 \pmod 2$, which costs $\mathcal{O}(|V||A|^2)$ time. This section improves the time complexity for the query to $\mathcal{O}(|A|)$.

Given an undirected graph $G = (V, E)$ and $T \subseteq V$, a T-join of G is an edge set J such that every vertex in T has odd degree and every other vertex has even degree in the subgraph induced by J. There exists a T-join of G if and only if $|C \cap T|$ is even in each component C of G [15]. Also, a T-join of any undirected graph can be found in linear time [13].

For a digraph $D = (V, A)$, let $BG(D)$ be an undirected bipartite graph $(V^+, V^-; E)$, where V^+ and V^- are the copies of V and $E = \{u^+v^- \mid (u, v) \in A\}$. One can easily see that the map is bijective. Observe that M of D coincides with the incidence matrix of $BG(D)$, and hence $|E| = |A|$.

Lemma 3. $Mx \equiv 1 \pmod 2$ *has a solution if and only if* $BG(D) = (V^+, V^-; E)$ *has a* $(V^+ \cup V^-)$-*join.*

Proof. Let F be any subset of E, and let $x_F \in \{0, 1\}^E$ be its incidence vector. Since M is the incidence matrix of $BG(D)$, the v-th entry of the vector Mx_F, $(Mx_F)_v$, denotes the degree of v in x_F. Let $x \in \{0, 1\}^E$ be a solution of $Mx \equiv 1 \pmod 2$. Then x indicates a subgraph of $BG(D)$ in which every vertex has odd degree, which is a $(V^+ \cup V^-)$-join of $BG(D)$. Conversely, if $x_F \in \{0, 1\}^E$ is the incidence vector of a $(V^+ \cup V^-)$-join F, x_F satisfies $Mx_F \equiv 1 \pmod 2$.   □

Since $BG(D)$ and a T-join are computed in linear time, we see the following.

Theorem 4. *The PHC problem in digraphs is solved in linear time.*   □

Finally we remark the time complexity to find a PHC of a given directed graph D. The proof of Lemma 3 implies that we can obtain a solution $x \in \{0, 1\}^A$ of $Mx \equiv 1 \pmod 2$ by finding a $(V^+ \cup V^-)$-join of $BG(D)$. Once we obtain a solution x, we can construct a PHC according to the proof of Theorem 3.1 for (b) $\Rightarrow$ (a). The algorithm is summarized in Algorithm 3.1.

It takes $\mathcal{O}(|A|)$ time in line 1. In line 2, we repeatedly find paths, each path is found in $\mathcal{O}(|A|)$ time and repeated $\mathcal{O}(|A|)$ time thus $\mathcal{O}(|A|^2)$ time in total. In line 3, we repeatedly find pairs of paths, each is done in $\mathcal{O}(|A|)$ and repeated $\mathcal{O}(|V|)$ time, thus $\mathcal{O}(|V||A|)$ time in total. Consequently the time complexity of Algorithm 3.1 is $\mathcal{O}(|A|^2)$.

4   Extension to GF(p)

This section is concerned with the following problem, generalization of the PHC problem: Given a digraph D and an integer p and an integer vector $r \in \{0, 1, \ldots, p - 1\}^A$, decide if there exists a closed walk which visits each

---

Algorithm 3.1 Finding a PHC in a digraph.

1: Find a $(V^+ \cup V^-)$-join J of $BG(D)$ and $\boldsymbol{x} \leftarrow \chi_J$
2: Repeat Procedure 1 until $\boldsymbol{x}$ satisfies the Eulerian condition
3: Repeat Procedure 2 until $\boldsymbol{x}$ becomes connected
4: return $\boldsymbol{x}$

---

vertex v $r(v)$ times modulo p. In other words, the problem asks to find a connected closed walk that satisfies the condition $M^+\tilde{\boldsymbol{x}} \equiv M^-\tilde{\boldsymbol{x}} \equiv \boldsymbol{r} \pmod{p}$, where $\tilde{\boldsymbol{x}} \in \mathbb{Z}_{\geq 0}^A$ is a vector in which $\tilde{x}(a)$ denotes the number of uses of arc a in the closed walk. One can see that this is the PHC problem when $p = 2$ and $\boldsymbol{r} = \boldsymbol{1}$. We give a characterization similar to Theorem 1 (b).

Theorem 5. *A strongly connected digraph D has a connected closed walk which satisfies $M^+\tilde{\boldsymbol{x}} \equiv M^-\tilde{\boldsymbol{x}} \equiv \boldsymbol{r} \pmod{p}$ if and only if $M^+\boldsymbol{x} \equiv M^-\boldsymbol{x} \equiv \boldsymbol{r} \pmod{p}$ has a solution $\boldsymbol{x} \in \{0, \ldots, p-1\}^A$.*

Proof. The proof is similar to (a) $\Leftrightarrow$ (b) of Theorem 1.

Necessity. Let $\tilde{\boldsymbol{x}} \in \mathbb{Z}_{\geq 0}^A$ be a vector in which $\tilde{x}(a)$ denotes the number of uses of $a \in A$ in a connected closed walk which satisfies $M^+\tilde{\boldsymbol{x}} \equiv M^-\tilde{\boldsymbol{x}} \equiv \boldsymbol{r}$ $\pmod{p}$. Then let $\boldsymbol{x} \in \{0, \ldots, p-1\}^A$ be defined by $\boldsymbol{x} \equiv \tilde{\boldsymbol{x}} \pmod{p}$, we have $M^+\boldsymbol{x} \equiv M^-\boldsymbol{x} \equiv \boldsymbol{r} \pmod{p}$.

Sufficiency. Suppose that $\boldsymbol{x} \in \{0, \ldots, p-1\}^A$ is a solution of $M^+\boldsymbol{x} \equiv M^-\boldsymbol{x} \equiv \boldsymbol{r}$ $\pmod{p}$, then we explain how to construct a closed walk which satisfies $M^+\tilde{\boldsymbol{x}} \equiv M^-\tilde{\boldsymbol{x}} \equiv \boldsymbol{r} \pmod{p}$. Remark that a graph indicated by $\boldsymbol{x}$ may not satisfy the Eulerian condition, meaning that $\sum_{a \in \delta^+(v)} x(a) = \sum_{a \in \delta^-(v)} x(a)$ may not hold for some vertex v.

First, we construct $\boldsymbol{x}' \in \mathbb{Z}_{\geq 0}^A$ satisfying the condition $M^+\boldsymbol{x}' \equiv M^-\boldsymbol{x}' \equiv \boldsymbol{r}$ $\pmod{p}$ and the Eulerian condition $\sum_{a \in \delta^+(v)} x'(a) = \sum_{a \in \delta^-(v)} x'(a)$ for each $v \in V$. Let $\phi(v) = \sum_{a \in \delta^+(v)} x(a) - \sum_{a \in \delta^-(v)} x(a)$ for each $v \in V$, denoting the difference between out-degree and in-degree of v in $\boldsymbol{x}$. Then $\boldsymbol{x}$ is Eulerian if and only if $\phi(v) = 0$ for all v. Notice that $\sum_{v \in V} \phi(v) = 0$ holds since the total of out-degrees is equal to the total of in-degrees. We also remark that $\phi(v)$ is a multiple of p for each $v \in V$, since $M^+\boldsymbol{x} \equiv M^-\boldsymbol{x} \equiv \boldsymbol{r} \pmod{p}$ implies that both of out-degree ($\sum_{a \in \delta^+(v)} x(a)$) and in-degree ($\sum_{a \in \delta^-(v)} x(a)$) are $r(v)$ modulo p. Then we apply the following Procedure 1' to $\boldsymbol{x}$:

Procedure 1'

1. Find $u, v \in V$ such that $\phi(u) < 0$ and $\phi(v) > 0$.
2. Find a directed path P from u to v (P always exists since D is strongly connected).
3. $x(a) := x(a) + p$ for each $a \in A(P)$.

Procedure 1' preserves the condition $M^+\boldsymbol{x} \equiv M^-\boldsymbol{x} \equiv \boldsymbol{r} \pmod{p}$, and decreases the value of $\sum_{v \in V} |\phi(v)|$ (by $2p$). By recursively applying Procedure 1' until $\sum_{v \in V} |\phi(v)|$ is zero, we obtain a desired closed walk $\boldsymbol{x}'$.

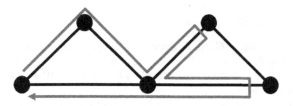

Fig. 1. An undirected graph which has a PHC but does not have a PHC orientation. The arrow indicates a PHC.

If x' suggests a connected walk, we obtain a closed walk which satisfies $M^+\tilde{x} \equiv M^-\tilde{x} \equiv r \pmod{p}$. Suppose that x' is not connected. Then we apply the following Procedure 2' to x':

Procedure 2'

1. Find $u, v \in V$ which are in distinct connected components.
2. Find directed paths P from u to v and P' from v to u.
3. $x'(a) := x'(a) + p$ for each $a \in A(P) \cup A(P')$.

Procedure 2' preserves the condition $M^+x \equiv M^-x \equiv r \pmod{p}$ and the Eulerian condition, and decreases the number of connected components. By recursively applying Procedure 2', we obtain a connected closed walk which satisfies $M^+\tilde{x} \equiv M^-\tilde{x} \equiv r \pmod{p}$. $\qquad\square$

If p is prime or power of a prime, the linear system $M^+x \equiv M^-x \equiv r \pmod{p}$ is solved over $\mathrm{GF}(p)$, and we obtain a desired closed walk in polynomial time. Otherwise $\mathrm{GF}(p)$ is not a field, and we need an extra observation to solve the equation efficiently.

5   Concluding Remarks

This paper gave two characterizations when a directed graph has a PHC. We have also shown that the characterization by M is generalized to problems over $\mathrm{GF}(p)$.

The PHC orientation problem is a problem to decide if a given undirected graph has an orientation which has a PHC. Figure 1 shows an example of an undirected graph which has a PHC, but does not admit a PHC orientation. It is open if the PHC orientation problem is solved in polynomial time. Another interesting question is if a PHC in a directed graph has an efficient (extended) formulation. Notice that minimizing the length of a PHC is NP-hard, since a PHC with length n is exactly a Hamiltonian cycle. A further connection between PHC and HC is a future work.

Acknowledgements. This work is partly supported by JSPS KAKENHI Grant Number 15K15938, 25700002, 15H02666, and Grant-in-Aid for Scientific Research on Innovative Areas MEXT Japan "Exploring the Limits of Computation (ELC)."

References

1. Bang-Jensen, J., Gutin, G.Z.: Digraphs: Theory, Algorithms and Applications. Springer, London (2008)
2. Biggs, N., Lloyd, E., Wilson, R.: Graph Theory 1736–1936. Oxford University Press, Oxford (1986)
3. Boyd, S., Iwata, S., Takazawa, K.: Finding 2-factors closer to TSP tours in cubic graphs. SIAM J. Discrete Math. **27**, 918–939 (2013)
4. Boyd, S., Sitters, R., van der Ster, S., Stougie, L.: TSP on cubic and subcubic graphs. In: Günlük, O., Woeginger, G.J. (eds.) IPCO 2011. LNCS, vol. 6655, pp. 65–77. Springer, Heidelberg (2011)
5. Brigham, R.C., Dutton, R.D., Chinn, P.Z., Harary, F.: Realization of parity visits in walking a graph. Coll. Math. J. **16**, 280–282 (1985)
6. Fiorini, S., Massar, S., Pokutta, S., Tiwary, H.R., de Wolf, R.: Exponential lower bounds for polytopes in combinatorial optimization. J. ACM **62**, 17 (2015)
7. Fiorini, S., Massar, S., Pokutta, S., Tiwary, H.R., de Wolf, R.: Linear vs. semidefinite extended formulations: exponential separation and strong lower bounds. In: Proceedings of STOC 2012, pp. 95–106 (2012)
8. Hartvigsen, D.: Extensions of Matching Theory, Ph.D. thesis, Carnegie Mellon University, Pittsburgh, PA (1984)
9. Hartvigsen, D., Li, Y.: Maximum cardinality simple 2-matchings in subcubic graphs. SIAM J. Optim. **21**, 1027–1045 (2011)
10. Joglekar, M., Shah, N., Diwan, A.A.: Balanced group-labeled graphs. Discrete Math. **312**, 1542–1549 (2012)
11. Karp, R.M.: Reducibility among combinatorial problems. In: Miller, R.E., Thatcher, J.W., Bohlinger, J.D. (eds.) Complexity of Computer Computations. The IBM Research Symposia Series, pp. 85–103. Springer, New York (1972)
12. Kochol, M.: Polynomials associated with nowhere-zero flows. J. Comb. Theor. B **84**, 260–269 (2002)
13. Korte, B., Vygen, J.: Combinatorial Optimization: Theory and Algorithms, 5th edn. Springer, Heidelberg (2012)
14. Nishiyama, H., Kobayashi, Y., Yamauchi, Y., Kijima, S., Yamashita, M.: The parity Hamiltonian cycle problem. arXiv.org e-Print archive abs/1501.06323
15. Schrijver, A.: Combinatorial Optimization. Springer, Heidelberg (2003)
16. Seymour, P., Thomassen, C.: Characterization of even directed graphs. J. Comb. Theor. B **42**, 36–45 (1987)
17. Yannakakis, M.: Expressing combinatorial optimization problems by linear programs. J. Comput. Syst. Sci. **43**, 441–466 (1991)

Lovász-Schrijver PSD-Operator on Claw-Free Graphs

Silvia Bianchi[1], Mariana Escalante[1], Graciela Nasini[1],
and Annegret Wagler[2($\boxtimes$)]

[1] FCEIA, Universidad Nacional de Rosario, Rosario, Argentina
{sbianchi,mariana,nasini}@fceia.unr.edu.ar
[2] LIMOS (UMR 6158 CNRS), University Blaise Pascal, Clermont-Ferrand, France
wagler@isima.fr

Abstract. The subject of this work is the study of the Lovász-Schrijver PSD-operator LS_+ applied to the edge relaxation ESTAB(G) of the stable set polytope STAB(G) of a graph G. We are interested in the problem of characterizing the graphs G for which STAB(G) is achieved in one iteration of the LS_+-operator, called LS_+-perfect graphs, and to find a polyhedral relaxation of STAB(G) that coincides with LS_+(ESTAB(G)) and STAB(G) if and only if G is LS_+-perfect. An according conjecture has been recently formulated (LS_+-Perfect Graph Conjecture); here we verify it for the well-studied class of claw-free graphs.

1 Introduction

The context of this work is the study of the stable set polytope, some of its linear and semi-definite relaxations, and graph classes for which certain relaxations are tight. Our focus lies on those graphs where a single application of the Lovász-Schrijver positive semi-definite operator introduced in [22] to the edge relaxation yields the stable set polytope.

The *stable set polytope* STAB(G) of a graph $G = (V, E)$ is defined as the convex hull of the incidence vectors of all stable sets of G (in a stable set all nodes are mutually nonadjacent). Two canonical relaxations of STAB(G) are the *edge constraint stable set polytope*

$$\text{ESTAB}(G) = \{\mathbf{x} \in [0,1]^V : x_i + x_j \leq 1, ij \in E\},$$

and the *clique constraint stable set polytope*

$$\text{QSTAB}(G) = \{\mathbf{x} \in [0,1]^V : x(Q) = \sum_{i \in Q} x_i \leq 1, \ Q \subseteq V \text{ clique}\}$$

(in a clique all nodes are mutually adjacent, hence a clique and a stable set share at most one node). We have STAB(G) $\subseteq$ QSTAB(G) $\subseteq$ ESTAB(G) for any

This work was supported by an ECOS-MINCyT cooperation (A12E01), a MATH-AmSud cooperation (PACK-COVER), PID-CONICET 0277, PICT-ANPCyT 0586.

© Springer International Publishing Switzerland 2016
R. Cerulli et al. (Eds.): ISCO 2016, LNCS 9849, pp. 59–70, 2016.
DOI: 10.1007/978-3-319-45587-7_6

graph, where $STAB(G)$ equals $ESTAB(G)$ for bipartite graphs, and $QSTAB(G)$ for perfect graphs only [5].

According to a famous characterization achieved by Chudnovsky et al. [3], perfect graphs are precisely the graphs without chordless cycles C_{2k+1} with $k \geq 2$, termed *odd holes*, or their complements, the *odd antiholes* $\overline{C}_{2k+1}$. This shows that odd holes and odd antiholes are the only *minimally imperfect graphs*.

Perfect graphs turned out to be an interesting and important class with a rich structure and a nice algorithmic behavior [18]. However, solving the stable set problem for a perfect graph G by maximizing over $QSTAB(G)$ does not work directly [17], but only via a detour involving a geometric representation of graphs [21] and a semi-definite relaxation $TH(G)$ introduced in [18].

For some $N \in \mathbf{Z}_+$, an orthonormal representation of a graph $G = (V, E)$ is a sequence $(\mathbf{u_i} : i \in V)$ of $|V|$ unit-length vectors $\mathbf{u_i} \in \mathbf{R}^N$, such that $\mathbf{u_i}^T \mathbf{u_j} = 0$ for all $ij \notin E$. For any orthonormal representation of G and any additional unit-length vector $\mathbf{c} \in \mathbf{R}^N$, the orthonormal representation constraint is $\sum_{i \in V} (\mathbf{c}^T \mathbf{u_i})^2 x_i \leq 1$. $TH(G)$ denotes the convex set of all vectors $\mathbf{x} \in \mathbf{R}_+^{|V|}$ satisfying all orthonormal representation constraints for G. For any graph G,

$$STAB(G) \subseteq TH(G) \subseteq QSTAB(G)$$

holds and approximating a linear objective function over $TH(G)$ can be done with arbitrary precision in polynomial time [18]. Most notably, the same authors proved a beautiful characterization of perfect graphs:

$$G \text{ is perfect} \Leftrightarrow TH(G) = STAB(G) \Leftrightarrow TH(G) = QSTAB(G). \tag{1}$$

For all imperfect graphs, $STAB(G)$ does not coincide with any of the above relaxations. It is, thus, natural to study further relaxations and to combinatorially characterize those graphs where $STAB(G)$ equals one of them.

Linear relaxations and related graphs. A natural generalization of the clique constraints are rank constraints $\mathbf{x}(G') = \sum_{i \in G'} x_i \leq \alpha(G')$ associated with arbitrary induced subgraphs $G' \subseteq G$. By the choice of the right hand side $\alpha(G')$, denoting the size of a largest stable set in G', rank constraints are valid for $STAB(G)$. A graph G is called *rank-perfect* by [32] if and only if $STAB(G)$ is described by rank constraints only.

By definition, rank-perfect graphs include all perfect graphs. By restricting the facet set to rank constraints associated with certain subgraphs, several well-known graph classes are defined, e.g., *near-perfect graphs* [29] where only rank constraints associated with cliques and the whole graph are allowed, or *t-perfect* [5] and *h-perfect graphs* [18] where rank constraints associated with edges, triangles and odd holes resp. cliques of arbitrary size and odd holes suffice.

As common generalization of perfect, t-perfect, and h-perfect graphs, the class of *a-perfect graphs* was introduced in [33] as graphs G where $STAB(G)$ is given by rank constraints associated with antiwebs. An *antiweb* A_n^k is a graph with n nodes $0, \ldots, n-1$ and edges ij if and only if $k \leq |i - j| \leq n - k$ and $i \neq j$.

Antiwebs include all complete graphs $K_n = A_n^1$, odd holes $C_{2k+1} = A_{2k+1}^k$, and their complements $\overline{C}_{2k+1} = A_{2k+1}^2$. Antiwebs are a-perfect by [33].

A more general type of inequalities is obtained from complete joins of antiwebs, called *joined antiweb constraints*

$$\sum_{i \leq k} \frac{1}{\alpha(A_i)} x(A_i) + x(Q) \leq 1,$$

associated with the join of some antiwebs $A_1, \ldots, A_k$ and a clique Q. We denote the linear relaxation of STAB(G) obtained by all joined antiweb constraints by ASTAB*(G). By construction, we see that

$$\text{STAB}(G) \subseteq \text{ASTAB}^*(G) \subseteq \text{QSTAB}(G) \subseteq \text{ESTAB}(G).$$

In [6], a graph G is called *joined a-perfect* if and only if STAB(G) coincides with ASTAB*(G). Besides all a-perfect graphs, further examples of joined a-perfect graphs are *near-bipartite graphs* (where the non-neighbors of every node induce a bipartite graph) due to [30].

A semi-definite relaxation and LS_+-perfect graphs. In the early nineties, Lovász and Schrijver introduced the PSD-operator LS_+ (called N_+ in [22]) which, applied to ESTAB(G), generates a positive semi-definite relaxation of STAB(G) stronger than TH(G) (see Sect. 2.1 for details). In order to simplify the notation we write $LS_+(G) = LS_+(\text{ESTAB}(G))$. In [22] it is shown that

$$\text{STAB}(G) \subseteq LS_+(G) \subseteq \text{ASTAB}^*(G).$$

As in the case of perfect graphs, the stable set problem can be solved in polynomial time for the class of graphs for which $LS_+(G) = \text{STAB}(G)$. These graphs are called LS_+-*perfect*, and all other graphs LS_+-*imperfect* (note that they are also called N_+-(im)perfect, see e.g. [1]). In [1], the authors look for a characterization of LS_+-perfect graphs similar to the characterization (1) for perfect graphs. More precisely, they intend to find an appropriate polyhedral relaxation $P(G)$ of STAB(G) such that

$$G \text{ is } LS_+\text{-perfect} \Leftrightarrow LS_+(G) = \text{STAB}(G) \Leftrightarrow LS_+(G) = P(G). \qquad (2)$$

A conjecture has been recently proposed in [2], which can be equivalently reformulated as follows [12]:

Conjecture 1 (LS_+-Perfect Graph Conjecture). A graph G is LS_+-perfect if and only if $LS_+(G) = \text{ASTAB}^*(G)$.

The results of Lovász and Schrijver [22] prove that joined a-perfect graphs are LS_+-perfect, thus, the conjecture states that LS_+-perfect graphs coincide with joined a-perfect graphs and that ASTAB*(G) is the studied polyhedral relaxation of STAB(G) playing the role of $P(G)$ in (2).

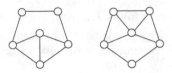

Fig. 1. The graphs G_{LT} (on the left) and G_{EMN} (on the right).

In addition, every subgraph of an LS_+-perfect graph is also LS_+-perfect. This motivates the definition of *minimally LS_+-imperfect graphs* as these LS_+-imperfect graphs whose proper induced subgraphs are all LS_+-perfect. The two smallest such graphs, found by [10,20], are depicted in Fig. 1.

Conjecture 1 has been already verified for near-perfect graphs by [1], for *fs-perfect graphs* (where the only facet-defining subgraphs are cliques and the graph itself) by [2], for *webs* (the complements $W_n^k = \overline{A}_n^k$ of antiwebs) by [11] and for *line graphs* (obtained by turning adjacent edges of a root graph into adjacent nodes of the line graph) by [12], see Sect. 2.1 for details.

The LS_+-Perfect Graph Conjecture for claw-free graphs. The aim of this contribution is to verify Conjecture 1 for a well-studied graph class containing all webs, all line graphs and the complements of near-bipartite graphs: the class of *claw-free graphs* (i.e., the graphs not containing the complete join of a single node and a stable set of size three).

Claw-free graphs attracted much attention due to their seemingly asymmetric behavior w.r.t. the stable set problem. On the one hand, the first combinatorial algorithms to solve the problem in polynomial time for claw-free graphs [23,28] date back to 1980. Therefore, the polynomial equivalence of optimization and separation due to [18] implies that it is possible to optimize over the stable set polytope of a claw-free graph in polynomial time. On the other hand, the problem of characterizing the stable set polytope of claw-free graphs in terms of an explicit description by means of a facet-defining system, originally posed in [18], was open for several decades. This motivated the study of claw-free graphs and its different subclasses, that finally answered this long-standing problem only recently (see Sect. 2.2 for details).

To verify the conjecture for claw-free graphs, we need not only to rely on structural results and complete facet-descriptions of their stable set polytope, but also to ensure that all facet-inducing subgraphs different from cliques, antiwebs or their complete joins are LS_+-imperfect. A graph G is said to be *facet-defining* if $STAB(G)$ has a full-support facet.

The paper is organized as follows: In Sect. 2, we present the State-of-the-Art on LS_+-perfect graphs (including families of LS_+-imperfect graphs needed for the subsequent proofs) and on claw-free graphs, their relevant subclasses and the results concerning the facet-description of their stable set polytopes from the literature. In Sect. 4, we verify, relying on the previously presented results, Conjecture 1 for the studied subclasses of claw-free graphs. As a conclusion, we obtain as our main result:

Theorem 1. *The LS_+-Perfect Graph Conjecture is true for all claw-free graphs.*

We close with some further remarks and an outlook to future lines of research.

2   State-of-the-Art

2.1   About LS_+-Perfect Graphs

We denote by $e_0, e_1, \ldots, e_n$ the vectors of the canonical basis of $\mathbf{R}^{n+1}$ (where the first coordinate is indexed zero), $\mathbf{1}$ the vector with all components equal to 1 and S_+^n the convex cone of $(n \times n)$ symmetric and positive semi-definite matrices with real entries. Given a convex set K in $[0,1]^n$, let

$$\operatorname{cone}(K) = \left\{ \begin{pmatrix} x_0 \\ \mathbf{x} \end{pmatrix} \in \mathbf{R}^{n+1} : \mathbf{x} = x_0 \mathbf{y}; \ \mathbf{y} \in K \right\}.$$

Then, we define the polyhedral set

$$\begin{aligned}
M_+(K) = \{ Y \in S_+^{n+1} : \ & Y e_0 = \operatorname{diag}(Y), \\
& Y e_i \in \operatorname{cone}(K), \\
& Y(e_0 - e_i) \in \operatorname{cone}(K), \ i = 1, \ldots, n \},
\end{aligned}$$

where $\operatorname{diag}(Y)$ denotes the vector whose i-th entry is Y_{ii}, for every $i = 0, \ldots, n$. Projecting this lifting back to the space $\mathbf{R}^n$ results in

$$LS_+(K) = \left\{ \mathbf{x} \in [0,1]^n : \begin{pmatrix} 1 \\ \mathbf{x} \end{pmatrix} = Y e_0, \text{ for some } Y \in M_+(K) \right\}.$$

In [22], Lovász and Schrijver proved that $LS_+(K)$ is a relaxation of the convex hull of integer solutions in K and that $LS_+^n(K) = \operatorname{conv}(K \cap \{0,1\}^n)$, where $LS_+^0(K) = K$ and $LS_+^k(K) = LS_+(LS_+^{k-1}(K))$ for every $k \geq 1$.

In this work we focus on the behavior of a single application of the LS_+-operator to the edge relaxation of the stable set polytope of a graph.

Recall that G_{LT} and G_{EMN} are the smallest LS_+-imperfect graphs. Further LS_+-imperfect graphs can be obtained by applying operations preserving LS_+-imperfection.

In [20], the *stretching* of a node v is introduced as follows: Partition its neighborhood $N(v)$ into two nonempty, disjoint sets A_1 and A_2 (so $A_1 \cup A_2 = N(v)$, and $A_1 \cap A_2 = \emptyset$). A stretching of v is obtained by replacing v by two adjacent nodes v_1 and v_2, joining v_i with every node in A_i for $i \in \{1,2\}$, and subdividing the edge $v_1 v_2$ by one node w. In [20] it is shown:

Theorem 2 ([20]). *The stretching of a node preserves LS_+-imperfection.*

Hence, all stretchings of G_{LT} and G_{EMN} are LS_+-imperfect, see Fig. 2 for some examples.

In [12], the authors characterized LS_+-perfect line graphs by showing that the only minimally LS_+-imperfect line graphs are stretchings of G_{LT} and G_{EMN} and occur as subgraphs in all facet-defining line graphs different from cliques and odd holes:

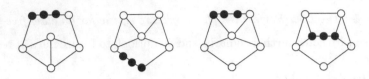

Fig. 2. Some node stretchings (v_1, w, v_2 in black) of G_{LT} and G_{EMN}.

Theorem 3 ([12]). *A facet-defining line graph G is N_+-perfect if and only if G is a clique or an odd hole.*

The proof relies on a result of Edmonds & Pulleyblank [8] who showed that a line graph $L(H)$ is facet-defining if and only if H is a 2-connected hypomatchable graph (that is, for all nodes v of H, $H-v$ admits a perfect matching). Such graphs H have an ear decomposition $H_0, H_1, \ldots, H_k = H$ where H_0 is an odd hole and H_i is obtained from H_{i-1} by adding an odd path (ear) between distinct nodes of H_{i-1}. In [12], it is shown that the line graph $L(H_1)$ of any ear decomposition is a node stretching of G_{LT} and G_{EMN} and, thus, LS_+-imperfect by [20].

Again, using stretchings of G_{LT} and G_{EMN} and exhibiting one more minimally LS_+-imperfect graph, namely the web W_{10}^2, LS_+-perfect webs are characterized in [11] as follows:

Theorem 4 ([11]). *A web is LS_+-perfect if and only if it is perfect or minimally imperfect.*

The proof shows that all imperfect not minimally imperfect webs with stability number 2 contain G_{EMN} and all webs W_n^2 different from W_7^2, W_{10}^2, some stretching of G_{LT}. Furthermore, all other webs contain some LS_+-imperfect $W_{n'}^2$ and are, thus, also LS_+-imperfect.

Finally, in [1], there is another family of LS_+-imperfect graphs presented that will play a central role in some subsequent proofs:

Theorem 5 ([1]). *Let G be a graph with $\alpha(G) = 2$ such that $G - v$ is an odd antihole for some node v. G is LS_+-perfect if and only if v is completely joined to $G - v$.*

2.2   About Claw-Free Graphs

In several respects, claw-free graphs are generalizations of line graphs. An intermediate class between line graphs and claw-free graphs form *quasi-line graphs*, where the neighborhood of any node can be partitioned into two cliques (i.e., quasi-line graphs are the complements of near-bipartite graphs).

For this class, it turned out that so-called clique family inequalities suffice to describe the stable set polytope. Given a graph G, a family $\mathcal{F}$ of cliques and an integer $p < n = |\mathcal{F}|$, the clique family inequality $(\mathcal{F}, p)$ is the following valid inequality for STAB(G)

$$(p - r) \sum_{i \in W} x_i + (p - r - 1) \sum_{i \in W_o} x_i \leq (p - r) \left\lfloor \frac{n}{p} \right\rfloor \tag{3}$$

where $r = n \, mod \, p$ and W (resp. W_o) is the set of nodes contained in at least p (resp. exactly $p - 1$) cliques of $\mathcal{F}$.

This generalizes the results of Edmods [7] and Edmonds & Pulleyblank [8] that $STAB(L(H))$ is described by clique constraints and rank constraints

$$x(L(H')) \leq \frac{1}{2}(|V(H')| - 1) \tag{4}$$

associated with the line graphs of 2-connected hypomatchable induced subgraphs $H' \subseteq H$. Note that the rank constraints of type (4) are special clique family inequalities. Chudnovsky and Seymour [4] extended this result to a superclass of line graphs. They showed that a connected quasi-line graph G is either a fuzzy circular interval graph or $STAB(G)$ is given by clique constraints and rank constraints of type (4).

Let $\mathcal{C}$ be a circle, $\mathcal{I}$ a collection of intervals in $\mathcal{C}$ without proper containments and common endpoints, and V a multiset of points in $\mathcal{C}$. The *fuzzy circular interval graph* $G(V, \mathcal{I})$ has node set V and two nodes are adjacent if both belong to one interval $I \in \mathcal{I}$, where edges between different endpoints of the same interval may be omitted.

Semi-line graphs are either line graphs or quasi-line graphs without a representation as a fuzzy circular interval graph. Due to [4], semi-line graphs are rank-perfect with line graphs as only facet-defining subgraphs.

Eisenbrand et al. [9] proved that clique family inequalities suffice to describe the stable set polytope of fuzzy circular interval graphs. Stauffer [31] verified a conjecture of [25] that every facet-defining clique family inequality of a fuzzy circular interval graph G is associated with a web in G.

All these results together complete the picture for quasi-line graphs. However, there arc claw-free graphs which are not quasi-line; the 5-wheel is the smallest such graph and has stability number 2. Due to Cook (see [30]), all facets for graphs G with $\alpha(G) = 2$ are *clique-neighborhood constraints*

$$2x(Q) + x(N'(Q)) \leq 2 \tag{5}$$

where $Q \subseteq G$ is a clique and $N'(Q) = \{v \in V(G) : Q \subseteq N(v)\}$. Therefore all non-trivial facets in this case are 1, 2-valued.

This is not the case for graphs G with $\alpha(G) = 3$. In fact, all the known difficult facets of claw-free graphs occur in this class. Some non-rank facets with up to five different non-zero coefficients are presented in [16,19]. All of these facets turned out to be so-called *co-spanning 1-forest constraints* due to [26], where it is also shown that it is possible to build a claw-free graph with stability number three inducing a co-spanning 1-forest facet with b different left hand side coefficients, for every positive integer b.

The problem of characterizing $STAB(G)$ when G is a connected claw-free but not quasi-line graph with $\alpha(G) \geq 4$ was studied by Galluccio et al.: In a series of results [13–15], it is shown that if such a graph G does not contain a clique cutset, then 1,2-valued constraints suffice to describe $STAB(G)$. Here, besides 5-wheels, different rank and non-rank facet-defining inequalities of the geared graph G shown in Fig. 3 play a central role.

In addition, graphs of this type can be decomposed into strips. A *strip*
(G, a, b) is a (not necessarily connected) graph with two designated simplicial
nodes a and b (a node is *simplicial* if its neighborhood is a clique). A claw-free
strip containing a 5-wheel as induced subgraph is a *5-wheel strip*. Given two
node-disjoint strips (G_1, a_1, b_1) and (G_2, a_2, b_2), their *composition* is the union
of $G_1 \setminus \{a_1, b_1\}$ and $G_2 \setminus \{a_2, b_2\}$ together with all edges between $N_{G_1}(a_1)$ and
$N_{G_2}(a_2)$, and between $N_{G_1}(b_1)$ and $N_{G_2}(b_2)$ [4].

As shown in [24], this composition operation can be generalized to more than
two strips: Every claw-free but not quasi-line graph G with $\alpha(G) \geq 4$ admits
a decomposition into strips, where at most one strip is quasi-line and all the
remaining ones are 5-wheel strips having stability number at most 3. There are
only three "basic" types of 5-wheel strips (see Fig. 3) which can be extended by
adding nodes belonging to the neighborhood of the 5-wheels (see [24] for details).

Note that a claw-free but not quasi-line graph G with $\alpha(G) \geq 4$ containing
a clique cutset may have a facet-inducing subgraph G' with $\alpha(G') = 3$ (inside a
5-wheel strip of type 3), see [27] for examples.

Taking all these results together into account gives the complete list of facets
needed to describe the stable set polytope of claw-free graphs.

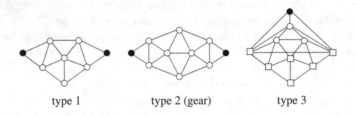

| type 1 | type 2 (gear) | type 3 |

Fig. 3. The three types of basic 5-wheel strips.

3 LS_+-Perfect Graph Conjecture for Claw-Free Graphs

In this section, we verify the LS_+-Perfect Graph Conjecture for all relevant
subclasses of claw-free graphs.

Graphs with $\alpha(G) = 2$ play a crucial role. Relying on the behavior of the stable
set polytope under taking complete joins [5] and the result on LS_+-(im)perfect
graphs G with $\alpha(G) = 2$ (Theorem 5), we can prove:

Theorem 6. *All facet-inducing LS_+-perfect graphs G with $\alpha(G) = 2$ are odd
antiholes or complete joins of odd antihole(s) and a (possibly empty) clique.*

This shows that all facet-inducing LS_+-perfect graphs G with $\alpha(G) = 2$ are
joined a-perfect, and we conclude:

Corollary 1. *The LS_+-Perfect Graph Conjecture is true for graphs with stability number 2.*

Quasi-line graphs partition into the two subclasses of semi-line graphs and fuzzy circular interval graphs.

Chudnovsky and Seymour [4] proved that the stable set polytope of a semi-line graph is given by rank constraints associated with cliques and the line graphs of 2-connected hypomatchable graphs. Together with the result from [12] (presented in Theorem 3), we directly conclude that the LS_+-Perfect Graph Conjecture holds for semi-line graphs.

Based on the results of Eisenbrand et al. [9] and Stauffer [31], combined with the characterization of LS_+-imperfect webs from [11] (Theorem 4), we are able to show:

Theorem 7. *All facet-inducing LS_+-perfect fuzzy circular interval graphs are cliques, odd holes or odd antiholes.*

As a consequence, every LS_+-perfect fuzzy circular interval graph is a-perfect. This verifies the LS_+-Perfect Graph Conjecture for fuzzy circular interval graphs.

Since the class of quasi-line graphs partitions into semi-line graphs and fuzzy circular interval graphs, we obtain as direct consequence:

Corollary 2. *The LS_+-Perfect Graph Conjecture is true for quasi-line graphs.*

Claw-free graphs that are not quasi-line are distinguished according to their stability number.

Relying on the behavior of the stable set polytope under clique identification [5] and the result on LS_+-(im)perfect graphs from Theorem 5, we can prove:

Theorem 8. *Every facet-defining claw-free not quasi-line graph G with $\alpha(G) = 3$ is LS_+-imperfect.*

Hence, the only facet-defining subgraphs G' of LS_+-perfect claw-free not quasi-line graphs G with $\alpha(G) = 3$ have $\alpha(G') = 2$ and are, by Theorem 6, cliques, odd antiholes or their complete joins. We conclude that LS_+-perfect facet-defining claw-free not quasi-line graphs G with $\alpha(G) = 3$ are joined a-perfect and, thus, the LS_+-Perfect Graph Conjecture is true for this class.

To treat the case of claw-free not quasi-line graphs G with $\alpha(G) \geq 4$, we rely on the decomposition of such graphs into strips, where at most one strip is quasi-line and all the remaining ones are 5-wheel strips [24]. By noting that 5-wheel strips of type 3 contain G_{LT} and exhibiting LS_+-imperfect line graphs in the other two cases, we are able to show:

Theorem 9. *Every facet defining claw-free not quasi-line graph G with $\alpha(G) \geq 4$ is LS_+-imperfect.*

This together with Theorem 8 shows that the only facet-defining subgraphs G' of LS_+-perfect claw-free not quasi-line graphs G with $\alpha(G) \geq 4$ have $\alpha(G') = 2$ and are, by Theorem 6, cliques, odd antiholes or their complete joins. Thus, every LS_+-perfect claw-free not quasi-line graph G with $\alpha(G) \geq 4$ is joined a-perfect and, thus, the LS_+-Perfect Graph Conjecture holds true for this class.

Combining Corollary 1 with the above results shows that all LS_+-perfect claw-free but not quasi-line graphs are joined a-perfect and we obtain:

Corollary 3. *The LS_+-Perfect Graph Conjecture is true for all claw-free graphs that are not quasi-line.*

Finally, we obtain our main result as direct consequence of Corollaries 2 and 3: The LS_+-Perfect Graph Conjecture is true for all claw-free graphs.

4   Conclusion and Future Research

The context of this work was the study of LS_+-perfect graphs, i.e., graphs where a single application of the Lovász-Schrijver PSD-operator LS_+ to the edge relaxation yields the stable set polytope. Hereby, we are particularly interested in finding an appropriate polyhedral relaxation $P(G)$ of STAB(G) that coincides with $LS_+(G)$ and STAB(G) if and only if G is LS_+-perfect. An according conjecture has been recently formulated (LS_+-Perfect Graph Conjecture); here we verified it for the well-studied class of claw-free graphs (Theorem 1).

Note further that, besides verifying the LS_+-Perfect Graph Conjecture for claw-free graphs, we obtained the complete list of all minimally LS_+-imperfect claw-free graphs. In fact, the results in [1, 11, 12] imply that the following graphs are minimally LS_+-imperfect:

- graphs G with $\alpha(G) = 2$ such that $G - v$ is an odd antihole for some node v, not completely joined to $G - v$,
- the web W_{10}^2,
- LS_+-imperfect line graphs (which are all node stretchings of G_{LT} or G_{EMN}).

Our results from Sect. 3 on facet-defining LS_+-perfect claw-free graphs imply that they are the only minimally LS_+-imperfect claw-free graphs.

Finally, the subject of the present work has parallels to the well-developed research area of perfect graph theory also in terms of polynomial time computability. In fact, it has the potential of reaching even stronger results due the following reasons. Recall that calculating the value

$$\eta_+(G) = \max \mathbf{1}^T x, x \in LS_+(G)$$

can be done in polynomial time for every graph G by [22]. Thus, the stable set problem can be solved in polynomial time for a strict superset of perfect graphs, the LS_+-perfect graphs, by $\alpha(G) = \eta_+(G)$. Hence, our future lines of research include to find

- new families of graphs where the conjecture holds (e.g., by characterizing the minimally LS_+-imperfect graphs within the class),
- new subclasses of LS_+-perfect or joined a-perfect graphs,
- classes of graphs G where STAB(G) and $LS_+(G)$ are "close enough" to have $\alpha(G) = \lfloor \eta_+(G) \rfloor$.

In particular, the class of graphs G with $\alpha(G) = \lfloor \eta_+(G) \rfloor$ can be expected to be large since $LS_+(G)$ is a much stronger relaxation of STAB(G) than TH(G). In all cases, the stable set problem could be solved in polynomial time in these graph classes by optimizing over $LS_+(G)$. Finally, note that $LS_+(P(G))$ with

$$\text{STAB}(G) \subseteq P(G) \subseteq \text{ESTAB}(G)$$

clearly gives an even stronger relaxation of STAB(G) than $LS_+(G)$. However, already optimizing over $LS_+(\text{QSTAB}(G))$ cannot be done in polynomial time anymore for all graphs G by [22]. Hence, in view of the polynomial time solvability of the stable set problem, LS_+-perfect graphs or their generalizations satisfying $\alpha(G) = \lfloor \eta_+(G) \rfloor$ are the most promising cases in this context.

References

1. Bianchi, S., Escalante, M., Nasini, G., Tunçel, L.: Near-perfect graphs with polyhedral $N_+(G)$. Electron. Notes Discrete Math. **37**, 393–398 (2011)
2. Bianchi, S., Escalante, M., Nasini, G., Tunçel, L.: Lovász-Schrijver PSD-operator and a superclass of near-perfect graphs. Electron. Notes Discrete Math. **44**, 339–344 (2013)
3. Chudnovsky, M., Robertson, N., Seymour, P., Thomas, R.: The strong perfect graph theorem. Ann. Math. **164**, 51–229 (2006)
4. Chudnovsky, M., Seymour, P.: The structure of claw-free graph (unpliblished manuscript) (2004)
5. Chvátal, V.: On certain polytopes associated with graphs. J. Comb. Theory (B) **18**, 138–154 (1975)
6. Coulonges, S., Pêcher, A., Wagler, A.: Characterizing and bounding the imperfection ratio for some classes of graphs. Math. Program. A **118**, 37–46 (2009)
7. Edmonds, J.R.: Maximum matching and a polyhedron with (0,1) vertices. J. Res. Nat. Bur. Stand. **69B**, 125–130 (1965)
8. Edmonds, J.R., Pulleyblank, W.R.: Facets of I-matching polyhedra. In: Berge, C., Chuadhuri, D.R. (eds.) Hypergraph Seminar. LNM, pp. 214–242. Springer, Heidelberg (1974)
9. Eisenbrand, F., Oriolo, G., Stauffer, G., Ventura, P.: The stable set polytope of quasi-line graphs. Combinatorica **28**, 45–67 (2008)
10. Escalante, M., Montelar, M.S., Nasini, G.: Minimal N_+-rank graphs: progress on Lipták and Tunçel's conjecture. Oper. Res. Lett. **34**, 639–646 (2006)
11. Escalante, M., Nasini, G.: Lovász and Schrijver N_+-relaxation on web graphs. In: Fouilhoux, P., Gouveia, L.E.N., Mahjoub, A.R., Paschos, V.T. (eds.) ISCO 2014. LNCS, vol. 8596, pp. 221–229. Springer, Heidelberg (2014)
12. Escalante, M., Nasini, G., Wagler, A.: Characterizing N_+-perfect line graphs. Int. Trans. Oper. Res. (2016, to appear)
13. Galluccio, A., Gentile, C., Ventura, P.: Gear composition and the stable set polytope. Oper. Res. Lett. **36**, 419–423 (2008)
14. Galluccio, A., Gentile, C., Ventura, P.: The stable set polytope of claw-free graphs with stability number at least four. I. Fuzzy antihat graphs are W-perfect. J. Comb. Theory Ser. B **107**, 92–122 (2014)
15. Galluccio, A., Gentile, C., Ventura, P.: The stable set polytope of claw-free graphs with stability number at least four. II. Striped graphs are G-perfect. J. Comb. Theory Ser. B **108**, 1–28 (2014)

16. Giles, R., Trotter, L.E.: On stable set polyhedra for K1,3 -free graphs. J. Comb. Theory **31**, 313–326 (1981)
17. Grötschel, M., Lovász, L., Schrijver, A.: The ellipsoid method and its consequences in combinatorial optimization. Combinatorica **1**, 169–197 (1981)
18. Grötschel, M., Lovász, L., Schrijver, A.: Geometric Algorithms and Combinatorial Optimization. Springer, New York (1988)
19. Liebling, T.M., Oriolo, G., Spille, B., Stauffer, G.: On non-rank facets of the stable set polytope of claw-free graphs and circulant graphs. Math. Methods Oper. Res. **59**(1), 25–35 (2004)
20. Lipták, L., Tunçel, L.: Stable set problem and the lift-and-project ranks of graphs. Math. Program. A **98**, 319–353 (2003)
21. Lovász, L.: On the Shannon capacity of a graph. IEEE Trans. Inf. Theory **25**, 1–7 (1979)
22. Lovász, L., Schrijver, A.: Cones of matrices and set-functions and 0–1 optimization. SIAM J. Optim. **1**, 166–190 (1991)
23. Minty, G.: On maximal independent sets of vertices in claw-free graphs. J. Comb. Theory **28**, 284–304 (1980)
24. Oriolo, G., Pietropaoli, U., Stauffer, G.: A new algorithm for the maximum weighted stable set problem in claw-free graphs. In: Lodi, A., Panconesi, A., Rinaldi, G. (eds.) IPCO 2008. LNCS, vol. 5035, pp. 77–96. Springer, Heidelberg (2008)
25. Pêcher, A., Wagler, A.: Almost all webs are not rank-perfect. Math. Program. B **105**, 311–328 (2006)
26. Pêcher, A., Wagler, A.: On facets of stable set polytopes of claw-free graphs with stability number three. Discrete Math. **310**, 493–498 (2010)
27. Pietropaoli, U., Wagler, A.: Some results towards the description of the stable set polytope of claw-free graphs. In: Proceedings of ALIO/EURO Workshop on Applied Combinatorial Optimization, Buenos Aires (2008)
28. Sbihi, N.: Algorithme de recherche d'un stable de cardinalité maximum dans un graphe sans étoile. Discrete Math. **29**, 53–76 (1980)
29. Shepherd, F.B.: Near-perfect matrices. Math. Program. **64**, 295–323 (1994)
30. Shepherd, F.B.: Applying Lehman's theorem to packing problems. Math. Program. **71**, 353–367 (1995)
31. Stauffer, G.: On the stable set polytope of claw-free graphs. Ph.D. thesis, Swiss Institute of Technology in Lausanne (2005)
32. Wagler, A.: Critical edges in perfect graphs. Ph.D. thesis, TU Berlin and Cuvillier Verlag Göttingen (2000)
33. Wagler, A.: Antiwebs are rank-perfect. 4OR **2**, 149–152 (2004)

Benders Decomposition for Capacitated Network Design

Sara Mattia[✉]

Istituto di Analisi dei Sistemi ed Informatica,
Consiglio Nazionale Delle Ricerche, Rome, Italy
sara.mattia@iasi.cnr.it

Abstract. Given a capacitated network, we consider the problem of choosing the edges to be activated to ensure the routing of a set of traffic demands. Both splittable and unsplittable flows are investigated. We present polyhedral results and develop a branch-and-cut algorithm based on a Benders decomposition approach to solve the problem.

1 Introduction

In traditional network design problems the aim is to find capacities for the edges to route a set of demands between node pairs (commodities). However, many real-life networks are already capacitated and the decision is to select the edges to be activated to ensure the routing of the commodities. This is the case, for example, when the edges can be switched on and off to save energy [2]. We refer to this problem as the Edge Activation (EA) problem. We denote by EAU the problem with unsplittable flows (commodities must be routed on a single path) and by EAS the one with unrestricted (splittable) flows. The capacity installation version of the problem, also known as the Network Loading (NL) problem, has received a lot of attention in the literature and many variants of the problem have been considered. Instead, the EA problem is somehow less studied and there are very few attempts to use a Benders-like approach. Unsplittable flows are considered in [2,5,8], while papers [10,12,13] deal with splittable flows. In [2] a problem with edge activation and survivability requirements is considered. Working and backup paths for the commodities must be provided and the edges used only by backup paths are switched off to save energy. In [5] the authors discuss a problem where multiple capacity modules are available on the edges and they can be activated separately, but the commodities cannot be split, not even on modules of the same edge. They present valid inequalities obtained from the polyhedron of the single edge problem and develop a branch-and-cut algorithm. In [8] a Benders decomposition approach to solve a connectivity problem with survivability restrictions and hop constraints is discussed. Lagrangian relaxations are investigated in [10] and a branch-and-cut-and-price approach is presented in [12]. In [13] a Lagrangian heuristic to be embedded into a branch-and-bound scheme is proposed.

In this paper we present a Benders-like reformulation for both the EAU and the EAS problem. We identify valid and facet-defining inequalities and propose a

© Springer International Publishing Switzerland 2016
R. Cerulli et al. (Eds.): ISCO 2016, LNCS 9849, pp. 71–80, 2016.
DOI: 10.1007/978-3-319-45587-7_7

branch-and-cut algorithm to solve the problem. Projecting out the routing variables is a popular approach for problems with splittable flows [4,6,7,11,15,19], but not for unsplittable ones. To the best of our knowledge, the only recent papers considering Benders decomposition for unsplittable flows are [2,8]. However, both papers differ from what we do here. In [2] the authors do not use a pure capacity formulation, as only a part of the flow variables is projected out. In the problem addressed in [8] only connectivity must be ensured or, equivalently, the amount of each demand is one. This has some consequences that do not hold in our case, making our problem more difficult to solve. We show that, differently from the NL problem, the convex-hull of integer feasible solutions of the Benders formulation is not completely described by inequalities having a metric left-hand-side, non even for splittable flows. On the other hand, some classes of inequalities that are known to be facets for the NL problem, remain facets for EAS, but not necessarily for EAU. We concentrate on inequalities deriving from partitions of the node set and provide conditions for facets of a problem corresponding to a partition (p-node problem, see Sect. 3) to be extended to facets of the original problem.

The paper is structured as follows. In Sect. 2 we present formulations for the EAU and the EAS problem. In Sect. 3 we investigate the polyhedral properties of the projected formulation of EAU and EAS, providing valid and facet-defining inequalities. In Sect. 4 we discuss the results of a preliminary computational testing. In Sect. 5 conclusions are presented.

2   Models

Let $G(V, E)$ be an undirected graph and let K be the set of commodities. Each commodity $k \in K$ has a source node s_k, a destination node t_k and an amount d_k. Each edge $e \in E$ can be activated at cost $c_e > 0$. We assume that the capacity U is the same for all the edges and that $U > d_k$ for all $k \in K$. Both the EAS and the EAU problem are NP-hard, as they include the Steiner tree problem. Set $N_i = \{j \in V : (i,j) \in E\}$ denotes the neighborhood of node i. The EAU problem can be formulated in a compact way (constraints and variables are polynomial in the input size) as follows.

$$\text{FU}\quad \min \sum_{e \in E} c_e x_e$$

$$\sum_{j \in N_i} (f_{ij}^k - f_{ji}^k) = \begin{cases} 1 & i = s_k \\ -1 & i = t_k \\ 0 & \text{otherwise} \end{cases} \quad i \in V, k \in K \tag{1}$$

$$\sum_{k \in K} d_k (f_{ij}^k + f_{ji}^k) \leq U x_e \qquad e \in E \tag{2}$$

$$\mathbf{x} \in \{0, 1\}^{|E|}, \mathbf{f} \in \{0, 1\}^{2 \times |E| \times |K|}$$

Binary variable x_e represents the activation of edge $e \in E$. Binary variable f_{ij}^k (f_{ji}^k) takes value one if commodity k is routed on edge e from i to j (from j to i)

and zero otherwise. Constraints (1) ensure that the commodities are routed and constraints (2) impose that the edge capacities are not exceeded. Let FS be the flow formulation for the EAS problem obtained by relaxing the integrality requirements on $\mathbf{f}$.

Now we derive the non-compact projected formulations using a Benders-like approach. Let us consider the EAS problem. A vector $\mathbf{0} \leq \bar{\mathbf{x}} \leq \mathbf{1}$ corresponds to a feasible set of active edges if the following problem admits a solution $(\beta, \mathbf{f})$ with $\beta \geq 0$.

$$\text{FSfix} \quad \max \beta$$

$$(\pi_i^k) \quad \sum_{j \in N_i} (f_{ij}^k - f_{ji}^k) = -d_{s_k i} \quad i \in V, k \in K \tag{3}$$

$$(\mu_e) \quad \sum_{k \in K} (f_{ij}^k + f_{ji}^k) \leq U \bar{x}_e - \beta \quad e \in E \tag{4}$$

$$\mathbf{f} \in \mathbb{R}_+^{2 \times |E| \times |K|}$$

If not so, by duality we get the feasibility conditions (5), leading to the capacity formulation below, where variables π and μ are the dual variables corresponding to constraints (3) and (4) respectively.

$$\text{PS} \quad \min \sum_{e \in E} c_e x_e$$

$$\sum_{e \in E} \mu_e x_e \geq \sum_{k \in K} \pi_{t_k}^k \frac{d_k}{U} \quad \mu \geq 0 \tag{5}$$

$$\mathbf{x} \in \{0, 1\}^{|E|}$$

For the NL problem inequalities (5) are also known as *metric* inequalities [14,23]. The right-hand-side can be strengthened obtaining the so-called *tight metric* (TM) inequalities [4]. This is done by replacing the right-hand-side by the optimal value R_μ, obtained when μ is used as objective function of the problem. Here we call inequalities (5) *Benders* (BE) inequalities.

When the flows are unsplittable, PS is not a formulation of the problem anymore, not even for a 2-node problem (see Sect. 3). One option is to replace the BE inequalities by their TM version (R_μ must be computed considering unsplittable solutions). However, such a formulation would be of little practical use, as separating the TM inequalities is very hard in practice, as it requires the solution of bilevel programs, even for splittable flows [18]. Another possibility is to use *combinatorial Benders*-like (CB) cuts [9], similarly to what done in [8].

$$\text{PU} \quad \min \sum_{e \in E} c_e x_e$$

$$\sum_{e \in E_y^0} x_e \geq 1 \ \mathbf{y} \in F \tag{6}$$

$$\mathbf{x} \in \{0, 1\}^{|E|}$$

F is the set of the activation vectors $\mathbf{y}$ not allowing an unsplittable routing and $E_{\mathbf{y}}^0 = \{e \in E : y_e = 0\}$.

Lemma 1. PU *is a formulation of the EAU problem.*

3   Facets and Valid Inequalities

We denote by $\mathcal{FS}$, $\mathcal{FU}$, $\mathcal{PS}$ and $\mathcal{PU}$ respectively the convex-hull of the binary feasible solutions of FS, FU, PS and PU. We now study $\mathcal{PS}$ and $\mathcal{PU}$. Any inequality that is valid for $\mathcal{PS}$ ($\mathcal{FS}$) is valid for $\mathcal{PU}$ ($\mathcal{FU}$) and any inequality that is valid for $\mathcal{PS}$ ($\mathcal{PU}$) is valid for $\mathcal{FS}$ ($\mathcal{FU}$). An edge e is a *bridge* if its removal makes the problem infeasible. We also say that $T \subseteq E$ is a *bridge set* if the simultaneous removal of the edges in T makes the problem infeasible. Without loss of generality, we assume that E contains no bridges. If so, the following holds.

Lemma 2. $\mathcal{PS}$ *(PU) is full-dimensional.*

Let us now consider non-negativity constraints and upper-bound inequalities.

$$x_e \geq 0 \quad e \in E \tag{7}$$
$$x_e \leq 1 \quad e \in E \tag{8}$$

Theorem 1. *The following results hold:*

1. *inequalities (7) are facet-defining for $\mathcal{PS}$ (PU) if no $\{e, h\}$ is a bridge set, for any $h \in E \setminus \{e\}$;*
2. *inequalities (8) are facet-defining for $\mathcal{PS}$ (PU).*

When $\mathbf{x}$ is an unrestricted general integer instead of a binary vector, $\mathcal{PS}$ and $\mathcal{PU}$ are completely described by (7) and TM inequalities [4,20,21]. In our case this is no longer true.

Example 1. Let $G(V, E)$ be the complete undirected graph on five nodes. Let the demands be $d_{12} = d_{13} = d_{23} = d_{45} = 1$, $d_{ij} = 0$ otherwise and assume that $U = 1$. Inequality $x_{12} + x_{13} + x_{14} + x_{23} + x_{24} \geq 2$ is a facet of $\mathcal{PS}$, but it is neither a bound inequality nor a TM. The same happens for unsplittable flows.

However, we can prove that no inequality with negative coefficients can be a facet but inequalities (8).

Theorem 2. *Let $\mathbf{a}^T \mathbf{x} \geq b$ be a valid inequality for $\mathcal{PS}$ (PU). If $\exists e : a_e < 0$ then either it is the upper bound inequality (8) or it is not a facet of $\mathcal{PS}$ (PU).*

We note that any TM inequality is a binary knapsack, therefore all the inequalities that are valid for the knapsack polyhedron corresponding to a given TM are valid for $\mathcal{PS}$ and $\mathcal{PU}$. A popular class of valid inequalities for the NL problem includes inequalities derived from partitions of the node set [3,16,22]. Here we give conditions for them to provide facets for the EA problem. Consider a partition $P = \{V_1 : \ldots V_p\}$ of the node set (*p-partition*). Let E_I be the set of edges having endpoints in the same set of the partition and let $E_C = E \setminus E_I$. Let K_I be the commodities having source and destination in the same set and let $K_C = K \setminus K_I$. Denote by $G_P(P, E_P)$ the graph having a node for any set of the partition and an edge for any original edge $e \in E_C$. Denote by K_P the demand set having a commodity for any original commodity in K_C. Hence, each commodity (edge) in K_C (E_C) corresponds to one commodity (edge) of K_P (E_P) and there may exist parallel commodities (edges). Given a *p-partition* P, let the corresponding *p-node* EAS (EAU) problem be the problem on graph G_P with demands K_P. Let $\mathbf{x}$ be a feasible solution of the *p-node* problem, let the complementary problem associated with the *p-node* problem and with $\mathbf{x}$ be the problem on graph $G_{\mathbf{x}}^c(V, E_{\mathbf{x}}^c)$ with demand set K. Edge set $E_{\mathbf{x}}^c$ contains the edges in E_I, while the ones in E_C are either already active (if $x_e = 1$) or removed (if $x_e = 0$).

Definition 1. *A partition* $P = \{V_1 : \ldots V_p\}$ *is shrinkable if, for any feasible solution* $\mathbf{x}$ *of the p-node problem, e is not a bridge for the complementary problem corresponding to* $\mathbf{x}$ *for any* $e \in E_I$.

Let $\mathcal{PS}_p$ ($\mathcal{PU}_p$) be the convex-hull of the binary feasible solutions of the projected formulation of the *p-node* EAS (EAU) problem corresponding to partition P. Let $\mathbf{a}^T \mathbf{x} \geq b$ be a valid inequality for $\mathcal{PS}_p$ ($\mathcal{PU}_p$), we call inequality $\sum_{e \in E_C} a_{t_e} x_e \geq b$ the *extended* inequality derived from $\mathbf{a}^T \mathbf{x} \geq b$, where t_e is the original edge corresponding to $e \in E_p$. $\mathcal{PS}_p$ ($\mathcal{PU}_p$) and $\mathcal{PS}$ ($\mathcal{PU}$) are related in the following way.

Theorem 3. *The following results hold:*

1. *if* $\mathbf{a}^T \mathbf{x} \geq b$ *is valid for* $\mathcal{PS}_p$ ($\mathcal{PU}_p$), *then the extended inequality is valid for* $\mathcal{PS}$ ($\mathcal{PU}$);
2. *if* P *is shrinkable and each* V_i *is connected, then for any facet* $\mathbf{a}^T \mathbf{x} \geq b$ *of* $\mathcal{PS}_p$ ($\mathcal{PU}_p$), *the extended inequality is a facet of* $\mathcal{PS}$ ($\mathcal{PU}$).

Let a cut $\{S : V - S\}$ be a 2-partition of the nodes and let $\delta(S)$ and $K(S)$ be the edges and the demands separated by the cut. Let $d(S) = \sum_{k \in K(S)} d_k / U$ and $D(S) = \lceil d(S) \rceil$. Inequalities (9) are known as *cutset* (CS) inequalities.

$$\sum_{e \in \delta(S)} x_e \geq D(S) \quad S \subseteq V \tag{9}$$

The CS inequalities define facets of $\mathcal{PS}_2$.

Theorem 4. *The following results hold:*

1. *the CS inequalities are valid for* $\mathcal{PS}_2$;
2. *they are facet-defining for* $\mathcal{PS}_2$ *if and only if* $|E| > D(S) > 0$.

It follows from Theorem 3 that the extended CS inequality is valid and facet-defining for $\mathcal{PS}$, under some conditions. The same does not hold for $\mathcal{PU}$.

Example 2. Consider a graph with two nodes and four parallel edges with $U = 10$. Suppose to have three parallel commodities, each of them with amount 6. Let $\{\{1\} : \{2\}\}$ be the unique non trivial cut. The right-hand-side of the corresponding CS inequality is 2, but there is no feasible unsplittable solution using less than three active edges.

Solving the 2-node EAS problem is trivially polynomial, as it is solving the 2-node NL problem, both for splittable and for unsplittable flows. On the contrary, solving the 2-node EAU problem is hard, as it corresponds to a bin-packing problem, which consists of minimizing the number of bins needed to pack a set of objects of different sizes.

Theorem 5. *The 2-node EAU problem is NP-hard.*

Then, we can strengthen the CS inequalities using the same technique adopted for the bin-packing problem [17]. The aim is to replace value $D(S)$ by a better lower bound on the optimal value. $D(S)$ is close to the optimum when $d_k << U$ for all k. An alternative bound can be computed as follows. Let $0 \le \alpha \le U/2$, $K^1(S) = \{k \in K(S) : d_k > U - \alpha\}$, $K^2(S) = \{k \in K(S) : U/2 < d_k \le U - \alpha\}$, $K^3 = \{\alpha \le d_k \le U/2\}$ and let $b(S)$ be the value computed as follow.

$$b(S) = |K^1(S)| + |K^2(S)| + \max\left\{0, \left\lceil \frac{\sum_{k \in K^3(S)} d_k - (|K^2(S)|U - \sum_{k \in K^2(S)} d_k)}{U} \right\rceil\right\}$$

Finally, let $B(S) = \max_{\alpha \ge 0}\{b(S)\}$. The strengthened cutset (SCS) inequality (10) is valid for $\mathcal{PU}$ and dominates the CS inequality, since $B(S) \ge D(S)$.

$$\sum_{e \in \delta(S)} x_e \ge B(S) \quad S \subseteq V \tag{10}$$

Although the worst case performance of lower bound $B(S)$ is better than the one of $D(S)$, which is $1/2$, it is still $2/3$ [17]. Therefore, SCS are not necessarily facets too.

Consider now a 3-partition $P = \{V_1 : V_2 : V_3\}$ of the node set. Let $D(V_i)$ be the right-hand-side of the CS inequality corresponding to cut $\{V_i : V \setminus V_i\}$ for $i = 1, \ldots, 3$. Let E_{ij} be the set of edges going from V_i to V_j and let $E = E_{12} \cup E_{13} \cup E_{23}$. Let $D(3P) = \left\lceil \frac{D(V_1) + D(V_2) + D(V_3)}{2} \right\rceil$. The 3-partition (3P) inequality below is clearly valid for $\mathcal{PS}$ and $\mathcal{PU}$, although not necessarily facet defining for $\mathcal{PU}$, even if values $D(V_i)$ are replaced by $B(V_i)$.

$$\sum_{e \in E_{12}} x_e + \sum_{e \in E_{13}} x_e + \sum_{e \in E_{23}} x_e \ge D(3P) \quad P = \{V_1 : V_2 : V_3\} \tag{11}$$

4   Preliminary Computational Testing

We now discuss the computational performance of the mentioned formulations and inequalities, based on preliminary experiments. We generated instances considering two U values (u_1 and u_5 with $u_5 > u_1$), two node sizes $(15, 20)$, three edge connectivity values (20 %, 40 %, 60 %) and three demand density values (30 %, 60 %, 90 %). For each configuration (number of nodes, edges, demands and capacity) three instances are generated, for a total number of 108 instances. Network topologies are generated using the GT-ITM tool [1]. We added randomly generated edges for the nodes having degree less than two. The demands are randomly generated. The computation has been done on a 4× Intel Core i5@3.20 GHz using single thread CPLEX 12.6 with a time limit of one hour.

We implemented and tested six different algorithms: flow formulation for the splittable and the unsplittable problem solved by cplex with default settings (FDS and FDU); flow formulation for the splittable and the unsplittable problem solved by adding cutting planes (FCS and FCU); Benders formulation for the splittable and the unsplittable problem (FBS and FBU). In all the approaches with generation of cuts we use CS inequalities and 3P inequalities. For FBS and FBU we also use Benders cuts and, for BU only, combinatorial Benders cuts. For the flow formulation we also add variable upper bound inequalities (VA) and energy cover inequalities (ECOV) as in [2].

$$\text{VA} \quad f_{ij}^k + f_{ji}^k \leq x_e \qquad e = (i,j) \in E, k \in K$$

$$\text{ECOV} \quad \sum_{k \in C}(f_{ij}^k + f_{ji}^k) \leq (|C| - 1)x_e \quad C \subseteq K : \sum_{k \in C} d_k > U, e = (i,j) \in E$$

We add one cut at a time, checking the inequalities in the following order: VA (if the case), CS, 3P, BE (if the case), ECOV (if the case), CB (if the case and if the current solution is integer). Cuts are added only at the root node, but for feasibility cuts BE and CB, which are used also in the other nodes. For FBS and FBU CPLEX cuts and heuristics are turned off, as they proved to be of limited utility. Single node CS inequalities are added to the initial formulation for all the cutting plane approaches. VA inequalities are polynomially many and they are separated by lookup. CS and 3P inequalities are separated heuristically, by randomly choosing an initial partition and then applying a local search. BE inequalities are separated solving the dual of FSfix (for aggregated commodities). ECOV are separated as in [2]. CE inequalities are first separated heuristically and then exactly using formulation FU for fixed $\mathbf{x}$. The heuristic separation works as follows. Given the current solution $\bar{x}$, we build an auxiliary graph $G_k(E_{\bar{x}}^1)$ including only the edges that take value one in the current solution. If there is no s_k-t_k path in $G(E_{\bar{x}}^1)$ for at least one commodity, then the solution is not feasible.

Tables 1, 2, 3, and 4 report number of instances solved depending on the number of nodes, edges, demands and capacity size respectively. Figures 1, 2, 3, and 4 show how the average computing time changes according to the considered parameters. Computing time is set to the time limit for the unsolved instances. For all the approaches the computing time increases when the considered control parameters increase and, on average, the Benders formulation is faster than the other two approaches, although some approaches can solve more instances. Instead, when the capacity increases the problem becomes easier to solve. For large capacities, FB seems to have a great advantage on the other approaches (Table 4), both for

Table 1. Solved (nodes)

	splittable		unspslittable	
	n15	n20	n15	n20
FD	54	44	42	13
FC	54	45	42	10
FB	54	43	40	30

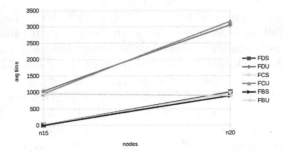

Fig. 1. Average times (nodes)

Table 2. Solved (edges)

	splittable			unspslittable		
	e20	e40	e60	e20	e40	e60
FD	36	34	28	28	16	11
FC	36	35	28	26	16	10
FB	36	33	28	33	22	15

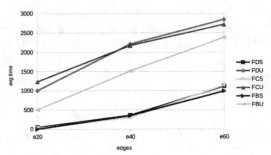

Fig. 2. Average times (edges)

Table 3. Solved (demands)

	splittable			unspslittable		
	d30	d60	d90	d30	d60	d90
FD	34	33	31	26	16	13
FC	35	33	31	24	15	13
FB	34	32	31	26	23	21

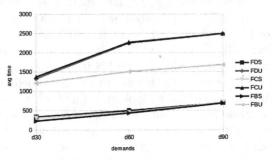

Fig. 3. Average times (demands)

Table 4. Solved (capacity)

	splittable		unspslittable	
	u1	u5	u1	u5
FD	52	46	21	34
FC	52	47	18	34
FB	43	54	19	51

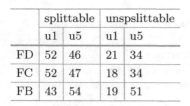
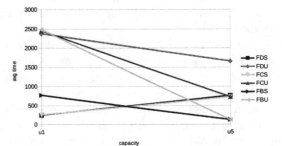

Fig. 4. Average times (capacity)

splittable and for unsplittable flows. For unsplittable flows, the behavior of the Benders formulation is less influenced by the control parameters than the other approaches. In fact, the number of solved instances for FD and FC immediately decreases if the number of nodes, edges or demands increases, whereas it is not so for FB (Tables 1, 2, and 3). The same does not happen for splittable flows, where all the algorithms seem to show a behavior that is less dependent from the control parameters. This is possibly do to the fact that the considered instances are more difficult for unsplittable than for splittable flows and then larger instances should also be tested.

5   Conclusions

We considered the problem of choosing the edges to be activated in a capacitated network to guarantee the routing of a set of commodities. Two routing policies are investigated: splittable (unrestricted routing) and unsplittable (single path routing). We studied both from the theoretical and from the computational point of view a Benders-like formulation of the problem. We presented polyhedral results showing that, contrary to what happens for the capacity allocation version of the problem, the Benders-like formulation is not completely defined by inequalities having metric left-hand-side coefficients. We focused on inequalities generated by subproblems obtained by partitioning the node set. We gave a condition stating when a facet of the p-node problem can be extended to a facet of the original problem. We also proved that for splittable flows the well-known cutset inequalities remain facets for the splittable problem, whereas this is not true for unsplittable flows. Indeed, contrary to what happen for the NL problem, even the 2-node unsplittable EA problem is NP-hard, as it is equivalent to a bin-packing problem. A preliminary testing showed that, on average, the Benders formulation is faster than the complete formulation. Moreover, for unsplittable flows it is less affected by an increasing of the number of edge, nodes and commodities.

References

1. GT-ITM: Georgia Tech Internetwork Topology Models. http://www.cc.gatech.edu/fac/Ellen.Zegura/gt-itm/
2. Addis, B., Carello, G., Mattia, S.: Energy-aware survivable networks. In: Proceedings of INOC 2015. ENDM, vol. 52, pp. 133–140 (2016)
3. Agarwal, Y.: K-partition-based facets of the network design problem. Networks **47**(3), 123–139 (2006)
4. Avella, P., Mattia, S., Sassano, A.: Metric inequalities and the network loading problem. Disc. Opt. **4**, 103–114 (2007)
5. Benhamichea, A., Mahjoub, R., Perrot, N., Uchoa, E.: Unsplittable non-additive capacitated network design using set functions polyhedra. Comp. Oper. Res. **66**, 105–115 (2016)
6. Bienstock, D., Chopra, S., Günlük, O., Tsai, C.Y.: Minimum cost capacity installation for multicommodity network flows. Math. Prog. Ser. B **81**, 177–199 (1998)

7. Bienstock, D., Mattia, S.: Using mixed-integer programming to solve power grid blackout problems. Disc. Opt. **4**, 115–141 (2007)
8. Botton, Q., Fortz, B., Gouveia, L., Poss, M.: Benders decomposition for the hop-constrained survivable network design problem. INFORMS J. Comput. **25**, 13–26 (2013)
9. Codato, G., Fischetti, M.: Combinatorial Benders' cuts for mixed-integer linear programming. Oper. Res. **54**(4), 756–766 (2006)
10. Crainic, T., Frangioni, A., Gendron, B.: Bundle-based relaxation methods for multicommodity capacitated fiwed charge network design. Disc. Appl. Math. **112**, 73–99 (2001)
11. Fortz, B., Poss, M.: An improved benders decomposition applied to a multi-layer network design problem. Oper. Res. Lett. **37**(5), 777–795 (2009)
12. Gendron, B., Larose, M.: Branch-and-price-and-cut for large-scale multicommodity capacitated fixed-charge network design. EURO J. Comput. Optim. **2**, 55–75 (2014)
13. Holmberg, K., Yuan, D.: A lagrangian heuristic based branch-and-bound approach for the capacitated network design problem. Oper. Res. **48**, 461–481 (2000)
14. Iri, M.: On an extension of the max-flow min-cut theorem to multicommodity flows. J. Oper. Res. Soc. Jpn. **13**, 129–135 (1971)
15. Lee, C., Lee, K., Park, S.: Benders decomposition approach for the robust network design problem with flow bifurcations. Networks **62**(1), 1–16 (2013)
16. Magnanti, T., Mirchandani, P., Vachani, R.: The convex hull of two core capacitated network design problems. Math. Prog. **60**, 233–250 (1993)
17. Martello, S., Toth, P.: Lower bounds and reduction procedures for the bin packing problem. Disc. Appl. Math. **28**, 59–70 (1990)
18. Mattia, S.: Separating tight metric inequalities by bilevel programming. Oper. Res. Lett. **40**(6), 568–572 (2012)
19. Mattia, S.: Solving survivable two-layer network design problems by metric inequalities. Comput. Optim. Appl. **51**(2), 809–834 (2012)
20. Mattia, S.: A polyhedral study of the capacity formulation of the multilayer network design problem. Networks **62**(1), 17–26 (2013)
21. Mattia, S.: The robust network loading problem with dynamic routing. Comput. Optim. Appl. **54**(3), 619–643 (2013)
22. Mattia, S.: The cut property under demand uncertainty. Networks **66**(2), 159–168 (2015)
23. Onaga, K., Kakusho, O.: On feasibility conditions of multicommodity flows in network. IEEE Trans. Circ. Theor. **18**(4), 425–429 (1971)

Modelling and Solving the Joint Order Batching and Picker Routing Problem in Inventories

Cristiano Arbex Valle[1(✉)], John E. Beasley[2], and Alexandre Salles da Cunha[1]

[1] Departamento de Ciência da Computação, Universidade Federal de Minas Gerais,
Belo Horizonte, Brazil
{arbex,acunha}@dcc.ufmg.br
[2] Mathematical Sciences, Brunel University, Uxbridge, UK
john.beasley@brunel.ac.uk

Abstract. In this work we investigate the problem of order batching and picker routing in inventories. These are labour and capital intensive problems, often responsible for a substantial share of warehouse operating costs. In particular, we consider the case of online grocery shopping in which orders may be composed of dozens of items. To the best of our knowledge, no exact algorithms have been proposed for this problem. We therefore introduce three integer programming formulations for the joint problem of batching and routing, one of them involving exponentially many constraints to enforce connectivity requirements and two compact formulations based on network flows. For the former we implement a branch-and-cut algorithm which separates connectivity constraints. We built a test instance generator, partially based on publicly-available real world data, in order to compare empirically the three formulations.

Keywords: Order batching · Picker routing · Inventory management · Integer programming

1 Introduction

Warehouses require intensive product handling operations, amongst those order picking is known to be the most labour and machine intensive. Its cost is estimated to be as much as 55 % [15] of total warehouse operating expenses, as online shopping has grown in popularity in recent years we also believe that this figure is likely increasing. Order picking is defined as being *the process of retrieving products from storage in response to specific customer requests.*

Material handling activities can be differentiated as parts-to-picker systems, in which automated units deliver the items to stationary pickers, and picker-to-parts systems, in which pickers walk/ride through the warehouse collecting requested items. With respect to the latter, three planning problems can be

Cristiano Arbex Valle is funded by CPNq grant 401367/2014-2. Alexandre Salles da Cunha is partially funded by FAPEMIG grant CEX - PPM-00187-15 and CNPq grants 303677/2015-5, 471464/2013-9 and 200493/2014-0.

© Springer International Publishing Switzerland 2016
R. Cerulli et al. (Eds.): ISCO 2016, LNCS 9849, pp. 81–97, 2016.
DOI: 10.1007/978-3-319-45587-7_8

distinguished: assignment of products to locations, grouping of customer orders into batches, routing of order pickers [13]. This paper deals jointly with the last two activities, which are critical to the efficiency of warehouse operations.

In particular, we consider the case of online grocery shopping, where orders can be composed of dozens of items. As practical aspects of this problem we highlight (i) the heterogeneity of products (which can be of various shapes, sizes or expiration dates) and (ii) the fact that the picking can be performed in warehouses that are either closed or open to the public (such as supermarkets); due to such features pickers generally walk/ride the warehouse, collecting products manually instead of relying on automated systems.

In this paper we formulate and solve the Joint Order Batching and Picker Routing Problem (JOBPRP). The task is to find minimum-cost closed walks, not necessarily Hamiltonian, where each picker visits all locations required to pick all products from their assigned orders. Locations may be visited more than once if necessary. To the best of our knowledge, no exact methods have been proposed to solve JOBPRP. We introduce three integer programming formulations for JOBPRP, each one being the basis for a different branch-and-bound algorithm. One of the formulations is a new directed model that involves exponentially many constraints to enforce connectivity requirements for closed walks. The other two are compact formulations based on network flows. We introduce a Branch-and-cut algorithm that relies on the non-compact model and put the compact formulations into the CPLEX branch-and-bound solver. We also introduce a JOBPRP test instance generator based on publicly available real-world data. In particular, we consider the special case where orders may be composed of dozens of items; in such situations mixing orders in the same basket or splitting an order among different pickers is generally avoided to reduce order processing errors.

The remainder of this paper is organised as follows. Section 2 presents a description and a literature review of the problem. Section 3 introduces JOBPRP as a graph optimisation problem and Sect. 4 presents the three integer programming formulations aforementioned. Section 5 briefly discusses some implementation details. Finally, Sect. 6 empirically compares the three formulations and, in Sect. 7, we present some concluding remarks and future research directions.

2   Problem Description and Literature Review

We consider a warehouse to be composed of vertical aisles which contain slots on both sides, each slot holds one type of product; slots can also be stacked vertically in shelves. We assume that pickers move in the centre of an aisle and that products on both sides can be reached by the picker. A warehouse may also contain cross-aisles. Every warehouse has one cross-aisle in the top and another in the bottom, it may also contain extra cross-aisles which divide aisles into subaisles.

A typical warehouse layout with three cross-aisles and two shelves, together with its sparse graph representation, can be seen in Fig. 1. White vertices represent locations from where the picker reaches products, i.e. from the top left

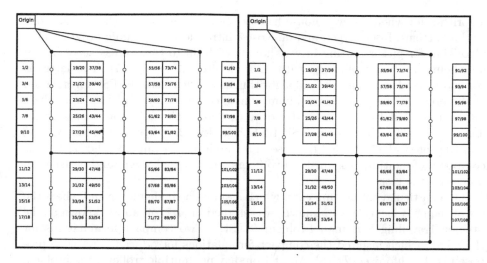

Fig. 1. Example of warehouse / graph layout   **Fig. 2.** Warehouse with reduced graph

white vertex the picker reaches products in slots 1, 2, 19 and 20. Black vertices represent "artificial" locations which connect aisles and cross-aisles, while the "Origin" indicates the starting and return point for pickers. Often, however, not all locations need to be visited as the joint set of orders does not contain products in all locations. It is therefore possible to reduce the graph that depicts the warehouse by eliminating such locations, as it can be seen in Fig. 2, where, for example, no products in slots 3, 4, 21 and 22 are needed.

Pickers use trolleys that accommodate a limited number of baskets. The number of baskets necessary to carry each order is assumed to be known. In fact, in this work, we consider that all baskets carry a fixed number of items, irrespective of their shapes and sizes. Baskets needed to carry a batch must not exceed trolley capacity. Following general practice in supermarkets, orders from different customers cannot be put together in the same basket and a single order cannot be split between different trolleys. Mixing and dividing orders could reduce picking time, however this benefit is often offset by necessary post-processing and a higher risk of errors.

There is a vast literature on the problems of order batching and picking. An extensive survey relating to different picking strategies was presented by de Koster *et al.* [15], where order picking is shown to be the most critical activity for productivity growth in the sector. Wave picking is an example of an alternative manual picking strategy, where the warehouse is divided in zones, each zone being assigned to one or more pickers; however in this strategy post-processing of orders is necessary. In this paper we deal with batch picking, where a picker collects products for one or more orders.

Many algorithms have been proposed for the problem of routing a single picker. An evaluation of several heuristics is given in [26], it also includes an exact algorithm for when the warehouse has two cross-aisles (one in the top and one in the bottom). Roodbergen and de Koster introduced a dynamic programming algorithm for picker routing in warehouses with up to three cross-aisles in [28] and heuristics for warehouses with multiple cross-aisles in [27]. The latter is compared to a branch-and-bound algorithm based on a classical TSP formulation of the problem, which assumes a hamiltonian circuit in a complete graph where all vertices must be visited exactly once (no artificial vertices are considered). Theys et al. [29] adapted the TSP LKH heuristic [20] for single-picker routing and Dekker et al. [6] studied the case of a Dutch retail company whose layout is composed of multiple cross aisles, two floors and different origin and destination points for the pickers.

One particular interesting case, which assumes a single picker and where the set of required locations are known, was studied in [4,7,24] and is known as the Steiner Travelling Salesman Problem (STSP). Two compact formulations for the STSP, which inspired the compact formulations introduced here, were proposed by Letchford et al. [18]. Apart from having multiple trolleys, the problem proposed in this work differs from the STSP since the assignment of required vertices to each route is not defined a priori.

Several heuristics have also been proposed for batching and routing multiple pickers, very often routing distances are estimated using single-picker heuristics during the solution of the batching problem. An extensive survey of batching methods can be found in [11].

For warehouses with two cross-aisles, a VNS heuristic was proposed by Albareda-Sambola et al. [1] and Tabu Search and Hill Climbing heuristics were introduced by Henn and Wäscher [13], the latter methods were adapted in [12] to a problem that also considers order sequencing. Azadnia et al. [2] proposed heuristics that solve sequencing and batching first, then routing in a second stage. For warehouses with three cross-aisles, Matusiak et al. [22] proposed a simulated annealing algorithm which includes precedence constraints (for instance, a heavier box has to be at the bottom of a container), in their method the routing of candidate batches is solved with the A*-algorithm [10].

For general warehouses, several batching methods are presented by de Koster et al. [16], where routing distances are estimated using single-picker methods. Gibson and Sharp [8] introduced batching heuristics based on two and four-dimensional space-filling curves. A genetic algorithm that jointly considers both routing and batching was proposed by Hsu et al. [14]. An integer programming formulation for batching is shown in [17], where routing distances are estimated. The problem is solved with a heuristic based on fuzzy logic. An association-based clustering approach to batching considering customer demands (instead of routing distance) is presented by Chen et al. [3].

As far as we are aware of, no exact methods have been proposed for the joint batching and routing problem. Also, the specific case of grocery shopping, where orders are generally composed of dozens items, has not been explicitly dealt with in any of the previous works here cited.

3   The Joint Problem of Order Batching and Picker Routing: A Graph Optimisation Problem

Let $\mathcal{P}$ be the set of products whose storage slot in the warehouse are known and $\mathcal{L}$ be the set of locations in the warehouse from where a picker can collect products, i.e. the middle of an aisle containing products on both sides, in different shelves, and from where the picker can reach those products. Each location $L \in \mathcal{L}$ contains a subset $P_L \subset \mathcal{P}$ of products. A warehouse graph representation includes a vertex for every location, represented by the white vertices in Fig. 1.

Let O be the set of orders that must be collected, each order $o \in O$ contains a subset of products $P_o \subseteq \mathcal{P}$. Accordingly, for each order $o \in O$, the subset of locations $L_o \subseteq \mathcal{L}$ contains all products in P_o - it is possible that multiple products of the same order are in a single location. $L(O) = \bigcup_{o \in O} L_o$ represents the set of locations that contains all products that need to be picked in all orders. Also let $d_{\ell m} \geq 0$ be the distance between locations $\ell, m \in \mathcal{L}$, symmetric so that $d_{\ell m} = d_{m\ell}$.

In practice, very often $L(O) \neq \mathcal{L}$, i.e. there are locations for which none of their products are present in any of the outstanding orders. In such cases, the graph that represents a warehouse can be reduced by eliminating all vertices that represent locations in $\mathcal{L} \setminus L(O)$. An example of such reduced graph can be seen in Fig. 2, where several vertices were removed as discussed previously. If v is eliminated, and there are arcs $(u, v), (v, u), (w, v), (v, w)$, we create arcs $(u, w), (w, u)$ where $d_{uw} = d_{wu} = d_{uv} + d_{vw} = d_{wv} + d_{vu}$.

Let $\mathcal{T}$ denote the set of available pickers (or trolleys), $T = |\mathcal{T}|$, B be the number of baskets that a trolley can carry and b_o be the known number of baskets needed to carry order $o \in O$. We assume that a basket will only contain products from a single order, even if it is partially empty, i.e., it is not possible to put products of different orders in the same basket. Finally, let s be the origin point from where the trolleys depart (and to where they must return), and accordingly let $d_{s\ell} \geq 0$ be the distance between location ℓ and s.

To define JOBPRP, we introduce a directed and connected graph $D = (V, A)$. The set of vertices V is given by the union of s (the origin point), a set $V(O)$ containing a vertex for every location $\ell \in L(O)$ and a set V_A of "artificial locations". These artificial locations are placed in corners between aisles and cross-aisles, and do not contain products that must be picked. We also define sets V_o containing a vertex for every location $\ell \in L_o$, $\bigcup_{o \in O} V_o = V(O)$. Thus, $V = \{s\} \cup V(O) \cup V_A$ and $|V| = 1 + |V(O)| + |V_A|$. In Fig. 2, the origin represents s, the 17 white vertices represent $V(O)$ and the 9 black vertices represent V_A.

Given a vertex $i \in V(O)$, define $\ell(i) : V(O) \rightarrow L(O)$ as the location to which the vertex refers. Arcs in set A connect two neighbour vertices of V. For instance, arc (i, j) connects location $\ell(i)$ to location $\ell(j)$. $d_{\ell(i),\ell(j)} \geq 0$ represents the distance between vertices i and j. For readability we write d_{ij} instead of $d_{\ell(i),\ell(j)}$.

A solution to JOBPRP in D is a collection of $T^* \leq T$ closed walks, one for each trolley. Each closed walk $t = 1, \ldots, T^*$ is associated with a subgraph $(V_t \cup \{s\}, A_t)$ of D. Each walk starts at s, traverses a set $A_t \subseteq A$ of selected arcs and returns to s. For each $t = 1, \ldots, T^*$, let $O_t \subseteq O$ denote the subset of orders collected during walk t and, for $i \in V$, let $O^i \subseteq O$ be the set of orders which must pick a product in vertex i. For $i \in V_A \cup \{s\}$, $O^i = \emptyset$.

The following requirements must be met by solutions to JOBPRP:

- Capacity constraints impose that, for every $t = 1, \ldots, T^*$, $\sum_{o \in O_t} b_o \leq B$.
- Indivisibility of the products P_o in each order $o \in O_t$ collected by each trolley imposes that, for every $o \in O_t$, there exists at least one walk t where if $i \in V_t : \ell(i) \in L_o$, then every $j : \ell(j) \in L_o$ must also be visited by t, i.e., $j \in V_t$.
- Finally, in order to guarantee that the T^* walks collect all orders, we must impose that $\cup_{t=1}^{T^*} O_t = O$. We assume that there are at least as many orders as there are trolleys available.

JOBPRP is then the problem of finding $T^* \leq T$ closed walks, meeting the requirements outlined above, in order to minimize the total length $\sum_{t=1}^{T^*} \sum_{(i,j) \in A_t} d_{ij}$.

In real applications, D is very likely a sparse graph, since vertices in $V(O)$ typically have only two neighbours while those in V_A do not connect to more than four other vertices. Because of the types of graphs resulting from typical applications, a feasible trolley walk for JOBPRP may be forced to not only repeat vertices but also visit vertices from non-collected orders. For instance, this could occur if a product from a non-collected order happens to be between two distinct products in a collected order. However, under reasonable assumptions, we can prove that the optimal solution to JOBPRP is composed of walks that may repeat vertices, but will not repeat arcs. An explicit proof can be found in the appendix of [19], for a closely related problem.

Lemma 1. *Let us assume that D is such that for every arc (i, j) there is a corresponding arc (j, i) and that $d_{ij} = d_{ji} \geq 0$. Then, in the optimal solution T^* to JOBPRP, every walk $t \in T^*$ visits each arc in A_t exactly once.*

4   Integer Programming Formulations for the JOBPRP

Based on Lemma 1, we introduce three integer programming formulations for the JOBPRP. These formulations allow vertices to be visited multiple times, but enforce that each arc cannot be traversed more than once. The first model uses exponentially many constraints to enforce connectivity of the closed walks. The other two are based on single and multiple commodity network flows. For all of them, let $\delta^-(W) = \{(i,j) \in A : i \notin W, j \in W\}$, $\delta^+(W) = \{(i,j) \in A : i \in W, j \notin W\}$ and $A(W) = \{(i,j) \in A : i \in W, j \in W\}$.

4.1   A Formulation Based on Exponentially Many Constraints

The first formulation uses binary decision variables z_{ot} to indicate whether ($z_{ot} = 1$) or not ($z_{ot} = 0$) trolley t picks order $o \in O$, x_{tij} to indicate whether ($x_{tij} = 1$) or not

($x_{tij} = 0$) arc $(i, j) \in A$ is traversed by trolley t, y_{ti} to indicate whether ($y_{ti} = 1$) or not ($y_{ti} = 0$) vertex $i \in V \setminus \{s\}$ is visited by trolley t. In addition, the model also uses variables $g_{ti} \in \mathbb{Z}_+$ to indicate the outdegree of vertex $i \in V$ in the closed walk for trolley t.

JOBPRP can be stated as the following Integer Program:

$$\min \left\{ \sum_{t \in T} \sum_{(i,j) \in A} d_{ij} x_{tij} : (z, x, y, g) \in P_g \cap (\mathbb{B}^{|O|T}, \mathbb{B}^{|A|T}, \mathbb{B}^{(|V|-1)T}, \mathbb{Z}^{T|V|}) \right\}, \tag{1}$$

where polytope P_g is given by:

$$\sum_{o \in O} b_o z_{ot} \leq B, \qquad \forall t \in T \tag{2}$$

$$\sum_{t \in T} z_{ot} = 1, \qquad \forall o \in O \tag{3}$$

$$\sum_{(i,j) \in \delta^+(i)} x_{tij} \geq z_{ot}, \qquad \forall o \in O, t \in T, i : \ell(i) \in L_o, \tag{4}$$

$$\sum_{(i,j) \in \delta^+(i)} x_{tij} = \sum_{(j,i) \in \delta^-(i)} x_{tji}, \qquad \forall i \in V, t \in T \tag{5}$$

$$\sum_{(s,j) \in \delta^+(s)} x_{tsj} \geq z_{ot}, \qquad \forall t \in T, o \in O \tag{6}$$

$$\sum_{(i,j) \in \delta^+(i)} x_{tij} = g_{ti}, \qquad \forall i \in V, t \in T \tag{7}$$

$$y_{ti} \geq x_{tij}, \qquad \forall (i, j) \in A, t \in T \tag{8}$$

$$\sum_{j \in W} g_{tj} \geq y_{ti} + \sum_{(j,k) \in A(W)} x_{tjk}, \quad \forall i \in W, W \subseteq V \setminus \{s\}, |W| > 1, t \in T \tag{9}$$

$$0 \leq x_{tij} \leq 1, \qquad \forall (i, j) \in A, t \in T \tag{10}$$

$$z_{ot} \geq 0, \qquad \forall o \in O, t \in T \tag{11}$$

$$g_{ti} \geq 0, \qquad \forall i \in V, t \in T \tag{12}$$

$$y_{ti} \leq 1, \qquad \forall i \in V, t \in T \tag{13}$$

Constraints (2) guarantee that the number of baskets in a trolley does not exceed its available capacity while constraints (3) ensure that each order will be collected by precisely one trolley. Constraints (4) enforce that if an order is assigned to a trolley, then every vertex that stores a product of this order will be visited by the trolley at least once. Constraints (5) make sure that for every arc that reaches a vertex, there is one that leaves it. Constraints (6) ensure that if a trolley picks any order, then it necessarily departs from the origin. Constraints (7) and (8) define the outdegree and the y variables for each vertex. Note that if the maximum outdegree of each vertex was not allowed to be greater than one, but instead, if $g_{ti} \in \{0, 1\}$, we would have $y_{ti} = g_{ti}$, and constraints (9) would change to the generalized subtour breaking constraints $\sum_{(j,k) \in A(W)} x_{tjk} \leq \sum_{j \in W \setminus \{i\}} y_{tj}$ [21]. Constraints (9) do allow subtours found in closed walks as long as at least one vertex in the cycle has an outdegree of 2.

4.2   Compact Formulations

The JOBPRP formulations we discuss in this section make use of network flows in order to model the trolleys' closed walks. The first model assigns one commodity for each order $o \in O$ and makes $|L_o|$ units of that commodity available at the origin s. Whenever a trolley t implements an order o, $|L_o|$ units of that commodity are shipped from s and one unit of that commodity must be retained by each vertex $i \in V(O) : \ell(i) \in L_o$. This introduces real valued flow variables f_{ij}^{ot}, to indicate the amount of commodity from order o passing through arc (i, j) in trolley t. The formulation is

$$\min \left\{ \sum_{t \in \mathcal{T}} \sum_{(i,j) \in A} d_{ij} x_{tij} : (z, x, f) \in P_f \cap (\mathbb{B}^{|O|T}, \mathbb{B}^{|A|T}, \mathbb{R}^{T|A||O|}) \right\}, \quad (14)$$

where polyhedral region P_f is the intersection of (2)–(6), (10)–(11) and:

$$\sum_{(j,i) \in \delta^-(i)} f_{ji}^{ot} - \sum_{(i,j) \in \delta^+(i)} f_{ij}^{ot} = z_{ot}, \quad \forall i \in V(O), o \in O^i, t \in \mathcal{T} \quad (15)$$

$$\sum_{(j,i) \in \delta^-(i)} f_{ji}^{ot} - \sum_{(i,j) \in \delta^+(i)} f_{ij}^{ot} = 0, \quad \forall i \in V \setminus \{s\}, o \in O \setminus O^i, t \in \mathcal{T} \quad (16)$$

$$0 \leq f_{ij}^{ot} \leq |L_o| x_{tij}, \quad \forall (i,j) \in A, o \in O, t \in \mathcal{T} \quad (17)$$

Constraints (15) and (16) impose flow balance constraints. According to (15), whenever an order o has products to be picked in location $\ell(i)$ and trolley t collects that order ($z_{ot} = 1$), vertex i must keep one unit of commodity ot. On the other hand, constraints (16) enforce that vertices must not retain any commodity whose order is not assigned to their locations.

A multi-commodity network flow formulation for JOBPRP can be obtained by defining flow variables h_{jk}^{oti} to indicate if commodity oti passes through arc $(j, k) \in A$. Whenever $z_{ot} = 1, t \in \mathcal{T}, o \in O$, one unit of commodity $oti, i : \ell(i) \in L_o$ is shipped from s to vertex i. The formulation is

$$\min \left\{ \sum_{t \in \mathcal{T}} \sum_{(i,j) \in A} d_{ij} x_{tij} : (z, x, h) \in P_h \cap (\mathbb{B}^{|O|T}, \mathbb{B}^{|A|T}, \mathbb{R}^{T|A||O||V(O)|}) \right\}, \quad (18)$$

where polytope P_h is given by (2)–(6), (10)–(11) and:

$$\sum_{(j,i)\in\delta^-(i)} h_{ji}^{oti} - \sum_{(i,j)\in\delta^+(i)} h_{ij}^{oti} = z_{ot}, \quad \forall o \in O, t \in T, i \in V_o, \quad (19)$$

$$\sum_{(k,j)\in\delta^-(j)} h_{kj}^{oti} - \sum_{(j,k)\in\delta^+(j)} h_{jk}^{oti} = 0, \quad \forall o \in O, \forall j \in V \setminus \{s,i\}, t \in T, i \in V_o, \quad (20)$$

$$\sum_{(j,s)\in\delta^-(s)} h_{js}^{oti} - \sum_{(s,j)\in\delta^+(s)} h_{sj}^{oti} = -z_{ot}, o \in O, t \in T, \forall i \in V_o, \quad (21)$$

$$0 \leq h_{jk}^{oti} \leq x_{tjk}, \quad \forall o \in O, t \in T, i \in V_o, (j,k) \in A \quad (22)$$

Constraints (19) ensure that commodity oti is retained by vertex i if and only if order o is picked by trolley t. Constraints (20) enforce flow balance and constraints (21) make sure that one unit of commodity oti leaves the origin.

4.3   Symmetry Breaking Constraints

Due to the artificial indexation of trolleys, formulations P_g, P_f and P_h suffer significantly from symmetry. That means that identical order to walk assignments lead to different vector representations, when we simply change the indices of the trolleys to which the closed walks are assigned. Branch-and-bound algorithms based on symmetric formulations tend to perform poorly, since they enumerate search regions that essentially lead to the same solutions. As an attempt to overcome symmetry for JOBPRP, we add the following constraints to formulations P_g, P_f and P_h:

$$\sum_{t=1}^{o} z_{ot} \geq 1, \quad o = 1, \dots, T \quad (23)$$

Here we ensure that the first order, $o = 1$, will be picked by the trolley assigned to the first index $t = 1$. The second order, $o = 2$, is either picked by either trolley 1 or 2, and so forth.

5   Implementation Details

In this section, we highlight some implementation details regarding the algorithm based on the formulations outlined above.

Branch-and-cut. We employ a Branch-and-cut algorithm [25] which separates inequalities (9). Let $\overline{g}_{ti}, \overline{y}_{ti}$ and $\overline{x}_{tij}$ be the values taken by the corresponding variables g_{ti}, y_{ti} and x_{tij} in an optimal solution to polytope P_g. For every $t \in T^*$, let $\overline{V}^t \subseteq V \setminus s$ be the set of vertices where $\overline{y}_{ti} > 0$ and $A(\overline{V}^t) \subseteq A$ be the set of arcs with both ends in $\overline{V}^t$ where $\overline{x}_{tij} > 0$. The problem consists of finding a subset of vertices $\overline{W} \subseteq \overline{V}^t$ and a vertex $i \in \overline{W}$ for which a constraint of type (9) is violated.

This problem can be polynomially solved by finding the minimum cut (maximum flow) on a network given by the graph $\overline{D}^t = (\overline{V}^t, A(\overline{V}^t))$ with capacities $\overline{x}_{tij} \forall (i,j) \in A(\overline{V}^t)$. Given any arbitrary $\overline{W} \subset \overline{V}^t$, constraints (7) ensure that $\sum_{j \in \overline{W}} \overline{g}_{tj} - \sum_{(j,k) \in A(\overline{W})} \overline{x}_{tjk} = \sum_{(j,\ell) \in \delta^+(\overline{W})} \overline{x}_{tj\ell}$, which is exactly the value of the cut that separates $\overline{W}$ from $\overline{V}^t \setminus \overline{W}$.

To solve the minimum cut/maximum flow problem, we employ the push-relabel algorithm first introduced by Goldberg and Tarjan [9]. Since the problem has to be solved separately for every $t \in T^*$, a violated cut for t is temporarily kept in a cut pool and checked for violation in subsequent iterations for other trolleys.

Heuristic. We implemented a constructive heuristic to provide an initial solution for all three formulations. Batching is computed via the time savings heuristic as explained in Sect. 2.2.1 of [16], but the savings matrix is not recalculated as orders are clustered.

For the estimation of partial route distances, we employed picker routing heuristics introduced in [27], which are specific for warehouses with multiple cross-aisles. Partial routes are computed by running each of the following heuristics: S-shape, Largest gap, Combined and Combined+. The best solution among these four different methods is taken as the input value in the savings matrix.

Solver Tuning. We employed CPLEX 12.6.0 [5] as the the solver for all three formulations. Branching priority is given to z_{ot} variables and probing level is set to 1. For the compact formulations, the remaining parameters are set to default values, including enabled presolve and the automatic generation of solver-separated cuts.

For formulation P_g, in which we explicitly separate constraints (9), we disable the solver presolve (including dual and nonlinear reductions). Most solver cuts are also disabled, except for Lift-and-project, Zero-half, Mixed Integer Rounding (MIR), Gomory and Cover cuts. All the other settings are left as default. Parallel processing is disabled for P_g and set to the solver default value for P_f and P_h.

6   Computational Experience

We conducted a series of experiments to empirically compare the algorithms based on the three formulations introduced in Sect. 4. The generation of test problems is explained below.

6.1   Test Problems

In order to simulate a realistic supermarket environment, we make use of the publicly available Foodmart database [23]. The database is composed of 2 years worth of anonymised customer purchases for a chain of supermarkets. There are a total of 1560 distinct products, separated in product classes containing 4 different category levels. It also contains approximately 270000 orders for the period 1997–1998, each composed of a customer id, a list of distinct products purchased, the number of items for each distinct product and the purchase date.

Warehouse. No information about warehouse layouts and product placement exists in Foodmart, therefore we built a warehouse layout generator to simulate both. The generator is based on a previous version developed and kindly provided by Dr. Birger Raa [29]. The generator creates warehouses that must be able to hold a minimum predetermined number of distinct products given a (fixed) number of aisles, cross-aisles and shelves. Arbitrary lengths are also given, in metres, for aisles and cross-aisles width, as well as rack depth and slot width. The distance from the origin to its closest artificial vertex (the black vertex in the top left corner of Fig. 1) is also given.

The generator computes the number of slots a shelf in an aisle side must have in order to hold at least the required number of products, while keeping the number of empty slots to a minimum. As an example, the warehouse in Fig. 1 was computed for a minimum of 104 products, while having 3 aisles, 3 cross-aisles and 2 shelves. Each shelf in each aisle side must have at least 9 slots, so that the total number of individual products in the warehouse is 108 (four slots would be empty in this example). The generator also computes the position of cross-aisles such that aisles are divided in subaisles as equal (in terms of number of slots) as possible.

The placement of products in slots is done by sorting all products from the highest category level to the lowest, and placing them in consecutive slots, so that similar products are close to each other.

A single warehouse layout is used for all test sets created for this work. It contains 8 aisles, 3 cross-aisles and 3 shelves. Each shelf in each aisle side holds 33 slots, so this warehouse can store 1584 distinct products (enough for all 1560 products in the Foodmart database). The distance from the origin to the first artificial vertex is 4 m, the aisle and cross-aisle widths are 3 m, and both the slot width and rack depth are 1 m.

Orders. We observed in the Foodmart database that orders are generally very small (the vast majority containing up to only 4 or 5 distinct products). On the other hand, online orders (which inspired this problem) may be composed of dozens of items.

To produce larger orders, we combined different Foodmart orders into a single one. For every customer, all of their purchases made in the first Δ days are combined into a single order. The combined order may contain not only more distinct products, but also a higher quantity of items of a single product.

A test instance is taken as the O orders with the highest number of distinct products. If $O = 5$, the 5 largest combined orders make up the test set; if $O = 6$, we take the same orders as in $O = 5$ plus the sixth largest combined order. We created several test instances for $\Delta = \{5, 10, 20\}$ and $O = \{5, \ldots, 30\}$. Table 1 shows the number of distinct products, the total number of items and the number of required baskets for the largest 30 orders (for each value of Δ). We consider that each basket may carry up to 40 items.

Capacity of Baskets and Number of Trolleys. With regard to parameter values we set $B = 8$, each basket holding a maximum of 40 items, irrespective of their sizes or weights. For every test instance, we define the number of trolleys $T = \left\lceil \frac{\sum_{o \in O} b_o}{B} + 0.2 \right\rceil$. Finding the exact minimum T required to service all orders is an optimisation problem on its own, we however do not tackle this problem in this work. Not all trolleys need to be used as the solution may leave some trolleys idle.

Table 1. Largest combined orders for $\Delta = \{5, 10, 20\}$

Order index	$\Delta = 5$			$\Delta = 10$			$\Delta = 20$		
	Products	Items	Baskets	Products	Items	Baskets	Products	Items	Baskets
1	18	63	2	23	79	2	23	79	2
2	11	39	1	23	66	2	23	66	2
3	11	39	1	12	34	1	16	45	2
4	11	35	1	11	40	1	15	42	2
5	8	28	1	11	39	1	14	50	2
6	8	28	1	11	39	1	14	49	2
7	7	28	1	11	36	1	14	40	1
8	7	27	1	11	35	1	13	42	2
9	7	27	1	11	31	1	13	41	2
10	7	24	1	10	38	1	13	40	1
11	7	24	1	10	36	1	13	39	1
12	7	23	1	10	32	1	12	48	2
13	7	23	1	10	30	1	12	45	2
14	7	22	1	10	29	1	12	43	2
15	7	21	1	9	28	1	12	41	2
16	7	20	1	9	26	1	12	39	1
17	7	20	1	8	27	1	12	36	1
18	7	20	1	8	25	1	12	34	1
19	7	20	1	8	22	1	12	34	1
20	7	19	1	7	26	1	12	33	1
21	7	19	1	7	23	1	11	43	2
22	7	19	1	7	23	1	11	40	1
23	7	18	1	7	22	1	11	39	1
24	7	18	1	7	20	1	11	39	1
25	7	10	1	7	20	1	11	38	1
26	6	21	1	7	20	1	11	38	1
27	6	19	1	7	19	1	11	37	1
28	6	17	1	7	18	1	11	36	1
29	6	14	1	7	17	1	11	33	1
30	6	9	1	7	10	1	11	31	1

Table 2. Instances features

O	$\Delta = 5$					$\Delta = 10$					$\Delta = 20$				
	$\sum_{o \in O} b_o$	T	T^*	$\|V\|$	$\|A\|$	$\sum_{o \in O} b_o$	T	T^*	$\|V\|$	$\|A\|$	$\sum_{o \in O} b_o$	T	T^*	$\|V\|$	$\|A\|$
5	6	1	1	76	192	7	2	1	90	220	10	2	2	104	248
6	7	2	1	82	204	8	2	1	101	242	11	2	2	114	268
7	8	2	1	89	218	9	2	2	109	258	13	2	2	123	286
8	9	2	2	95	230	10	2	2	114	268	14	2	2	133	306
9	10	2	2	101	242	11	2	2	123	286	16	3	2	141	322
10	11	2	2	106	252	12	2	2	131	302	17	3	3	147	334
11	12	2	2	111	262	13	2	2	140	320	19	3	3	149	338
12	13	2	2	116	272	14	2	2	145	330	20	3	3	153	346
13	14	2	2	121	282	15	3	2	150	340	21	3	3	157	354
14	15	3	2	128	296	16	3	2	156	352	23	4	3	162	364
15	16	3	2	132	304	17	3	3	160	360	24	4	3	169	378
20	21	3	3	153	346	22	3	3	179	398	32	5	4	196	432
25	26	4	4	174	388	27	4	4	195	430	38	5	5	211	462
30	31	5	4	186	412	32	5	4	207	454	43	6	6	228	496

Details about test instances are shown in Table 2. For each value of $\Delta = \{5, 10, 20\}$, there are five columns: The total number of baskets needed to carry all orders (labelled as $\sum_{o \in O} b_o$), the number of available trolleys T, the actual number of trolleys T^* required to service all orders and the number of vertices $|V|$ and edges $|A|$ of each respective graph D. For higher values of Δ, both the total number of products and the number of distinct products increase, this is reflected in the higher number of baskets required and the larger number of vertices and arcs for higher values of Δ.

6.2   Computational Results

In this section, we present computational results for our three formulations. We used an Intel Xeon with 24 cores @ 2.40 GHz with 32 GB of RAM and Linux as the operating system. The code was written in C++ and CPLEX 12.6.0 [5] was used as the mixed-integer solver. The heuristic described in Sect. 5 is used to warm start the three algorithms with valid JOBPRP upper bounds. A maximum time limit of 6 CPU hours was imposed to each algorithm and instance.

Table 3 compares the three algorithms for a selection of instances from our test set. For each algorithm, we include six columns. **T(s)** denotes the total elapsed time, in seconds, **UB** and **LB** respectively represent the best upper and lower bounds obtained at the end of the search, when either the instance was solved to proven optimality or when the time limit has hit. **GAP** is defined as $100(UB - LB)/UB$. **FLB** is the lower bound obtained at the end of the first node of the branch-and-bound search tree and **NS** stands for the total number of nodes investigated during the search.

Table 3. Comparison of algorithms based on formulations P_g, P_f and P_h. A symbol "–" for **T(s)** entries indicates that the instance was not solved when the time limit was hit.

Δ	O	Branch-and-cut algorithm based on P_g						Branch-and-bound algorithm based on P_f						Branch-and-bound algorithm based on P_h					
		T(s)	UB	GAP	LB	FLB	NS	T(s)	UB	GAP	LB	FLB	NS	T(s)	UB	GAP	LB	FLB	NS
5	5	8.2	348.6	–	–	345.6	132	**4.3**	348.6	–	–	328.7	1136	21.7	348.6	–	–	345.6	61
	6	10.7	364.8	–	–	364.8	4	**9.8**	364.8	–	–	353.7	456	23.5	364.8	–	–	364.8	1
	7	**18.2**	374.8	–	–	372.8	176	47.1	374.8	–	–	357.3	2333	105.1	374.8	–	–	372.8	182
	8	168.2	503.8	–	–	403.7	1195	**116.5**	503.8	–	–	400.9	6882	331.7	503.8	–	–	409.9	198
	9	474.1	539.6	–	–	411.9	3376	**463.4**	539.6	–	–	390.6	24782	858.3	539.6	–	–	421.4	759
	10	1517.5	581.4	–	–	433.2	6940	**849.3**	581.4	–	–	434.1	45617	941.5	581.4	–	–	451.1	129
	11	9816.6	613.5	–	–	442.4	58893	3931.0	613.5	–	–	427.6	297642	**2853.6**	613.5	–	–	460.1	1168
	12	–	621.8	1.6	611.7	450.3	76162	**2958.1**	621.4	–	–	456.2	166718	7519.4	621.4	–	–	472.4	5600
	13	14618.4	623.4	–	–	461.0	50388	**3428.0**	623.4	–	–	447.9	185231	3907.0	623.4	–	–	478.5	223
	14	–	647.5	10.1	581.9	467.7	14572	–	**639.3**	5.9	601.4	445.3	175800	–	**639.3**	**0.1**	638.5	495.0	5067
	15	–	657.3	10.6	587.9	474.1	16981	–	661.5	9.6	597.9	450.0	184100	**18578.3**	653.4	–	–	500.2	1135
	20	–	**930.5**	**29.7**	654.5	519.2	14001	–	932.8	33.6	619.0	478.8	91400	–	985.3	42.5	566.9	552.0	101
	25	–	1252.4	**50.5**	619.9	539.8	5072	–	**1193.9**	52.3	569.4	494.5	12553	–	1252.4	100.0	0.0	–	0
	30	–	**1387.4**	**56.0**	610.8	548.0	2820	–	1509.3	64.2	541.0	519.0	946	–	1509.3	100.0	0.0	–	0
10	5	14.4	371.1	–	–	369.1	4	**9.2**	371.1	–	–	360.9	413	27.4	371.1	–	–	371.1	1
	6	**19.4**	377.1	–	–	374.7	178	34.9	377.1	–	–	365.6	1187	197.3	377.1	–	–	374.7	401
	7	**228.0**	549.8	–	–	428.3	651	514.6	549.8	–	–	405.6	39097	650.7	549.8	–	–	424.7	175
	8	**551.9**	584.2	–	–	438.3	3555	707.8	584.2	–	–	395.6	73183	723.6	584.2	–	–	436.0	235
	9	10472.3	637.4	–	–	447.2	70742	**4934.4**	637.4	–	–	423.2	429781	–	637.4	0.4	634.8	464.1	11372
	10	–	661.8	1.6	650.9	468.6	89314	**13810.6**	661.8	–	–	428.2	1092715	–	661.9	0.9	655.8	480.2	4348
	11	–	709.8	7.8	654.7	486.0	57096	–	701.8	3.7	676.0	443.8	923700	–	**699.8**	**2.0**	685.7	504.7	3233
	12	–	721.7	7.3	669.1	493.6	58310	–	707.8	1.2	699.0	445.5	1219600	**20067.4**	707.7	–	–	509.2	2514
	13	–	**725.8**	11.2	644.6	503.3	29495	–	745.4	16.7	620.8	461.0	328300	–	738.2	**10.9**	657.6	522.2	285
	14	–	746.2	13.9	642.1	511.0	26700	–	**735.8**	17.4	608.2	463.1	240200	–	743.2	**11.1**	660.6	530.4	149
	15	–	**930.5**	31.6	636.9	525.6	8474	–	940.0	**31.1**	647.5	474.8	191000	–	1045.3	42.6	599.5	556.3	13
	20	–	1055.3	**33.0**	707.0	563.0	12300	–	**1042.4**	36.9	657.4	507.4	74214	–	1113.3	42.7	637.5	603.7	14
	25	–	1300.4	**47.8**	678.7	593.4	4623	–	**1294.0**	54.8	585.3	532.6	5814	–	1300.4	100.0	0.0	–	0
	30	–	1590.1	**58.2**	664.9	609.2	3483	–	**1589.5**	62.5	595.6	551.1	645	–	1615.5	100.0	0.0	–	0
20	5	**273.1**	573.8	–	–	443.6	3147	498.3	573.8	–	–	427.9	62001	512.5	573.8	–	–	445.9	720
	6	3015.6	656.2	–	–	479.6	36421	**1742.6**	656.2	–	–	450.7	229530	2460.8	656.2	–	–	487.2	1608
	7	5316.9	689.8	–	–	499.5	50250	6847.5	689.8	–	–	478.1	806672	**3220.9**	689.8	–	–	515.4	1240
	8	14848.3	697.8	–	–	505.2	116551	**7080.1**	697.8	–	–	457.0	772937	9247.2	697.8	–	–	504.1	4129
	9	–	**727.7**	4.7	693.9	505.1	67558	–	735.4	9.7	664.1	471.1	743300	–	**727.7**	**0.1**	726.9	527.0	3292
	10	–	952.5	**19.9**	763.3	538.8	21163	–	944.0	22.4	732.8	491.7	568000	–	**927.0**	21.6	726.5	557.3	343
	11	–	1037.3	27.1	755.8	546.3	18031	–	**1017.0**	**23.6**	777.2	502.3	1467500	–	1022.6	29.9	716.8	580.3	177
	12	–	1051.3	27.0	767.4	558.0	19900	–	**1017.0**	**23.1**	781.7	501.3	761400	–	1048.8	31.4	719.4	585.3	190
	13	–	1049.0	**26.8**	768.1	559.9	20947	–	**1032.9**	27.1	753.0	512.4	409700	–	1075.3	39.0	656.3	601.0	94
	14	–	**1064.9**	**32.8**	715.7	563.8	10004	–	**1064.9**	35.0	691.9	525.5	79300	–	1078.9	43.1	613.7	613.7	1
	15	–	1126.7	36.1	720.2	560.1	7542	–	**1071.0**	**35.7**	688.6	525.7	70900	–	1300.0	100.0	0.0	–	0
	20	–	1428.4	**48.8**	730.8	614.7	3605	–	**1426.1**	53.6	661.9	565.5	7011	–	1428.4	100.0	0.0	–	0
	25	–	1775.4	**58.5**	736.5	635.6	3019	–	**1757.5**	63.1	647.6	578.6	3202	–	1777.5	100.0	0.0	–	0
	30	–	2076.0	**64.7**	733.2	654.0	1409	–	2076.0	69.2	638.3	589.1	36	–	2076.0	100.0	0.0	–	0

The branch-and-bound algorithms based on formulations P_h and P_f managed to solve 19 out of 42 instances to proven optimality. The Branch-and-cut algorithm based on P_g solved only 17 instances, within the imposed time limit. For the 21 instances that were solved to optimality by one of the methods introduced here, the best elapsed times are highlighted in bold. Out of these 21 cases, the branch-and-bound algorithm based on P_f had the best CPU times in 12 instances. However, two instances could only be solved by the branch-and-bound method based on P_h.

For the other 21 unsolved instances, we highlight, in bold, the best upper bound and the smallest duality gap yet to be closed, when the time limit was hit. Once again, the branch-and-bound algorithm based on P_f is shown to be a competitive choice, as it provided, at the end of the time limit, the best upper bound in 13 out of these 21 instances. As the number of orders grows, however, the branch-and-cut algorithm based on P_g consistently obtained lower gaps and higher final lower bounds. For $\Delta = 20$, for instance, it obtained the strongest final lower bounds for every instance where $O \geq 13$.

The strongest root node lower bounds were obtained by formulation P_h. As one could expect, the evaluation of multi-commodity lower bounds become very expensive as the size of the instances grows, so that these bounds could not be evaluated within the time limit for 8 cases in our test set. The root node lower bounds obtained by the algorithm that relies on P_g is often weaker than P_h counterparts. Despite that, the Branch-and-cut algorithm based on P_g is more scalable, although for large instances, its final optimality gaps are high (64.7 % for $\Delta = 20$ and $O = 30$).

All three formulations benefited from contraints (23). Although not reported here due to space constraints, FLBs were on average 3.4 % lower when these constraints were not included (for all formulations).

In summary, the algorithms based on formulations P_f and P_h tend to perform better than that based on P_g for small and medium sized instances. A possible reason is that commercial solvers such as CPLEX are very competitive due to presolve reduction and built-in general-purpose cuts generation. Important CPLEX features had to be disabled when our callback separation routine was turned on, for the Branch-and-cut algorithm based on P_g. We have observed, however, that P_g, having less variables than P_f and P_h, is more scalable as instances get larger.

7   Conclusions and Future Work

In this work we investigated the joint order batching and picker routing problem (JOBPRP) in inventories. In particular, we considered the case of online grocery shopping where orders may be composed of dozens of items. According to our literature review, no exact methods have been proposed for jointly solving both batching and routing as a single problem.

We introduced three formulations and branch-and-bound algorithms for the JOBPRP. The first model involves exponentially many constraints to enforce connectivity and is solved via a branch-and-cut algorithm. The other two are compact formulations based on network flows, the first considering a single commodity and the second considering multiple commodities. For all three formulations, we proposed symmetry breaking constraints and we implemented a con structive heuristic to warm start the three branch-and-bound algorithms.

We also developed an instance generator that creates warehouse layouts with multiple aisles, cross-aisles and shelves. We make use of publicly available real-world supermarket data to generate orders and to place products in the warehouse. We generated instances with a varying number of orders of varying sizes.

We empirically compared the formulations and algorithms and concluded that the formulations based on network flows tend to perform better for small and medium sized instances. We believe this is in part due to the competitiveness of CPLEX, the mixed-integer solver chosen. However, as the number of orders grows, the formulation with an exponential number of connectivity constraints is more scalable as, within the time limit, stronger lower bounds are obtained.

In future we plan to conduct a theoretical study of the strength of each formulation and to characterize new valid inequalities that could be added to reinforce their linear relaxation bounds. On the practical side, we also intend to develop meta-heuristics to obtain stronger upper bounds, in the expectation of solving (or nearly solving) instances of industrial scale in competitive computational times.

References

1. Albareda-Sambola, M., Alonso-Ayuso, A., Molina, E., de Blas, C.S.: Variable neighborhood search for order batching in a warehouse. Asia Pac. J. Oper. Res. (APJOR) **26**(05), 655–683 (2009)
2. Azadnia, A.H., Taheri, S., Ghadimi, P., Saman, M.Z.M., Wong, K.Y.: Order batching in warehouses by minimizing total tardiness: a hybrid approach of weighted association rule mining and genetic algorithms. Sci. World J. **2013**(1), 1–13 (2013)
3. Chen, M.C., Wu, H.P.: An association-based clustering approach to order batching considering customer demand patterns. Omega **33**(4), 333–343 (2004)
4. Cornuéjols, G., Fonlupt, J., Naddef, D.: The travelling salesman problem on a graph and some related integer polyhedra. Math. Program. **33**(1), 1–27 (1985)
5. CPLEX Optimizer: IBM (2016). http://www-01.ibm.com/software/integration/optimization/cplex-optimizer/. Accessed 8 Feb 2016
6. Dekker, R., de Koster, M.B.M., Roodbergen, K.J., van Kalleveen, H.: Improving order-picking response time at Ankor's warehouse. Interfaces **34**(4), 303–313 (2004)
7. Fleischmann, B.: A cutting-plane procedure for the traveling salesman problem on a road network. Eur. J. Oper. Res. **21**(3), 307–317 (1985)
8. Gibson, D.R., Sharp, G.P.: Order batching procedures. Eur. J. Oper. Res. **58**(1), 57–67 (1992)
9. Goldberg, A.V., Tarjan, R.E.: A new approach to the maximum flow problem. J. ACM **35**(4), 921–940 (1988)
10. Hart, P., Nilsson, N., Raphael, B.: A formal basis for the heuristic determination of minimum cost paths. IEEE Trans. Syst. Sci. Cybern. **4**(2), 100–107 (1968)
11. Henn, S., Koch, S., Wäscher, G.: Order batching in order picking warehouses: a survery of solution approaches. In: Manzini, R. (ed.) Warehousing in the Global Supply Chain, pp. 105–137. Springer, London (2012). Chap. 6
12. Henn, S., Schmid, V.: Metaheuristics for order batching and sequencing in manual order picking systems. Comput. Ind. Eng. **66**(2), 338–351 (2013)
13. Henn, S., Wäscher, G.: Tabu search heuristics for the order batching problem in manual order picking systems. Eur. J. Oper. Res. **222**(3), 484–494 (2012)
14. Hsu, C.M., Chen, K.Y., Chen, M.C.: Batching orders in warehouses by minimizing travel distance with genetic algorithms. Comput. Ind. **56**(2), 169–178 (2004)
15. de Koster, R., Le-Duc, T., Roodbergen, K.J.: Design and control of warehouse order picking: a literature review. Eur. J. Oper. Res. **182**(2), 481–501 (2007)
16. de Koster, R., van der Poort, E.S., Wolters, M.: Efficient orderbatching methods in warehouses. Int. J. Prod. Res. **37**(7), 1479–1504 (1999)

17. Lam, C.H., Choy, K., Ho, G., Lee, C.: An order-picking operations system for managing the batching activities in a warehouse. Int. J. Syst. Sci. **45**(6), 1283–1295 (2014)
18. Letchford, A.N., Nasiri, S.D., Theis, D.O.: Compact formulations of the Steiner traveling salesman problem and related problems. Eur. J. Oper. Res. **228**(1), 83–92 (2013)
19. Letchford, A.N., Oukil, A.: Exploiting sparsity in pricing routines for the capacitated arc routing problem. Comput. Oper. Res. **36**(7), 2320–2327 (2009)
20. Lin, S., Kernighan, B.W.: An effective heuristic algorithm for the traveling-salesman problem. Oper. Res. **21**(2), 498–516 (1973)
21. Lucena, A.: Steiner problem in graphs: Lagrangean relaxation and cutting planes. COAL Bull. **21**(1), 2–8 (1992)
22. Matusiak, M., de Koster, R., Kroon, L., Saarinen, J.: A fast simulated annealing method for batching precedence-constrained customer orders in a warehouse. Eur. J. Oper. Res. **236**(3), 968–977 (2014)
23. MySQL Foodmart Database (2008). http://pentaho.dlpage.phi-integration.com/mondrian/mysql-foodmart-database. Accessed 31 Jan 2016
24. Orloff, C.S.: A fundamental problem in vehicle routing. Networks **4**(1), 35–64 (1974)
25. Padberg, M., Rinaldi, G.: A branch-and-cut algorithm for resolution of large scale of symmetric traveling salesman problem. SIAM Rev. **33**(1), 60–100 (1991)
26. Petersen II, C.G.: An evaluation of order picking routing policies. Int. J. Oper. Prod. Manage. **17**(11), 1098–1111 (1997)
27. Roodbergen, K.J., de Koster, R.: Routing methods for warehouses with multiple cross aisles. Int. J. Prod. Res. **39**(9), 1865–1883 (2001)
28. Roodbergen, K.J., de Koster, R.: Routing order pickers in a warehouse with a middle aisle. Eur. J. Oper. Res. **133**(1), 32–43 (2001)
29. Theys, C., Bräysy, O., Dullaert, W., Raa, B.: Using a TSP heuristic for routing order pickers in warehouses. Eur. J. Oper. Res. **200**(3), 755–763 (2010)

Uniqueness of Equilibria in Atomic Splittable Polymatroid Congestion Games

Tobias Harks[1] and Veerle Timmermans[2(✉)]

[1] Institute of Mathematics, University of Augsburg, 86135 Augsburg, Germany
tobias.harks@math.uni-augsburg.de
[2] Department of Quantitative Economics, Maastricht University,
6200 MD Maastricht, The Netherlands
v.timmermans@maastrichtuniversity.nl

Abstract. We study uniqueness of Nash equilibria in atomic splittable congestion games and derive a uniqueness result based on polymatroid theory: when the strategy space of every player is a *bidirectional flow polymatroid*, then equilibria are unique. Bidirectional flow polymatroids are introduced as a subclass of polymatroids possessing certain exchange properties. We show that important cases such as base orderable matroids can be recovered as a special case of bidirectional flow polymatroids. On the other hand we show that matroidal set systems are in some sense necessary to guarantee uniqueness of equilibria: for every atomic splittable congestion game with at least three players and non-matroidal set systems per player, there is an isomorphic game having multiple equilibria. Our results leave a gap between base orderable matroids and general matroids for which we do not know whether equilibria are unique.

1 Introduction

We revisit the issue of uniqueness of equilibria in *atomic splittable congestion games*. In this class of games there is a finite set of resources E, a finite set of players N, and each player $i \in N$ is associated with a weight $d_i \geq 0$ and a collection of allowable subsets of resources $\mathcal{S}_i \subseteq 2^E$. A strategy for player i is a (possibly fractional) distribution $\boldsymbol{x}_i \in \mathbb{R}_+^{|\mathcal{S}_i|}$ of the weight over the allowable subsets $\mathcal{S}_i$. Thus, we can compactly represent the strategy space of every player $i \in N$ by the following polytope

$$P_i := \{\boldsymbol{x}_i \in \mathbb{R}_+^{|\mathcal{S}_i|} : \sum_{S \in \mathcal{S}_i} x_S = d_i\}. \tag{1}$$

We denote by $\boldsymbol{x} = (\boldsymbol{x}_i)_{i \in N}$ the overall strategy profile. The induced load under $\boldsymbol{x}_i$ at resource e is defined as $x_{i,e} := \sum_{S \in \mathcal{S}_i : e \in S} x_S$ and the total load on e is then given as $x_e := \sum_{i \in N} x_{i,e}$. Resources have player-specific cost functions $c_{i,e} : \mathbb{R}_+ \to \mathbb{R}_+$ which are assumed to be non-negative, increasing, differentiable and convex. The total cost of player i in strategy distribution $\boldsymbol{x}$ is defined as

$$\pi_i(\boldsymbol{x}) = \sum_{e \in E} x_{i,e}\, c_{i,e}(x_e).$$

© Springer International Publishing Switzerland 2016
R. Cerulli et al. (Eds.): ISCO 2016, LNCS 9849, pp. 98–109, 2016.
DOI: 10.1007/978-3-319-45587-7_9

Each player wants to minimize the total cost on the used resources and a Nash equilibrium is a strategy profile x from which no player can unilaterally deviate and reduce its total cost. Using that the strategy space is compact and cost functions are increasing and convex Kakutanis' fixed point theorem implies the existence of a Nash equilibrium.

Example 1. A well-known special case of the above formulation arises when the resources E correspond to edges of a graph $G = (V, E)$ and the allowable subsets S_i correspond to the set of s_i-t_i-paths for some $(s_i, t_i) \in V \times V$. In this case, we speak of an atomic splittable *network* congestion game.

1.1 Uniqueness of Equilibria

Uniqueness of equilibria is fundamental to predict the outcome of distributed resource allocation: if there are multiple equilibria it is not clear upfront which equilibrium will be selected by the players. An intriguing question in the field of atomic splittable congestion games is the possible non-uniqueness of equilibria. Multiple equilibria x, y exist whenever there exists a player i and resource e such that $x_{i,e} \neq y_{i,e}$. A variant on this question is whether or not there exist multiple equilibria such that there exists at least one resource e for which $x_e \neq y_e$. We call this variant "uniqueness up to induced load on the resources".

For non-atomic players and network congestion games on directed graphs, Milchtaich [18] proved that Nash equilibria are not unique when cost functions are player-specific. Uniqueness is only guaranteed if the underlying graph is two terminal s-t-*nearly-parallel*. Richman and Shimkin [22] extended this result to hold for atomic splittable network games. Bhaskar et al. [5] looked at uniqueness up to induced load on the resources. They proved that even when all players experience the same cost on a resource, there can exist multiple equilibria. They further proved that for two players, the Nash equilibrium is unique if and only if the underlying undirected graph is generalized series-parallel. For multiple players of two types (players are of the same type if they have the same weight and share the same origin-destination pair), there is a unique equilibrium if and only if the underlying undirected graph is s-t-series-parallel. For more than two types of players, there is a unique equilibrium if and only if the underlying undirected graph is generalized nearly-parallel.

1.2 Our Results and Outline of the Paper

In this paper we study the uniqueness of equilibria for general set systems $(S_i)_{i \in N}$. Interesting combinatorial structures of the S_i's beyond paths may be trees, forests, Steiner trees or tours in a directed or undirected graph, or bases of matroids.

As our main result we give a sufficient condition for uniqueness based on the theory of polymatroids. We show that if the strategy space of every player is a polymatroid base polytope satisfying a special exchange property – we term this class of polymatroids *bidirectional flow polymatroids* – the equilibria are unique.[1]

---

[1] The formal definition of bidirectional flow polymatroids appears in Definition 1.

We demonstrate that bidirectional flow polymatroids are quite general, as they contain *base-orderable matroids, gammoids, transversal and laminar matroids*. For an overview of special cases that follow from our main result, we refer to the full version of this paper [14].

The uniqueness result is stated in Sect. 4. In Sect. 5 we show that base-orderable matroids are a special case of bidirectional flow polymatroids. Definitions of polymatroid congestion games and bidirectional flow polymatroids are introduced in Sects. 2 and 3, respectively. In Sect. 6 we complement our uniqueness result by showing the following. Consider a game with at least three players for which the set systems S_i of all players $i \in N$ are *not* bases of a matroid. Then there exists a game with strategy spaces $\phi(S_i)$ isomorphic to S_i which admits multiple equilibria. Here, the term *isomorphic* means that there is no a priori description on how the individual strategy spaces of players interweave in the ground set of resources. Our results leave a gap between general matroids and base orderable matroids for which we do not know whether or not equilibria are unique.

In Sect. 7 we consider uniqueness of equilibria if the set systems S_i correspond to paths in an undirected graph. The instance used for showing multiplicity of equilibria of non-matroid games can be seen as a 3-player game played on an undirected 3-vertex cycle graph. From this we can derive a new characterization of uniqueness of equilibria in undirected graphs. If we assume at least three players and if we do not specify beforehand which vertices of the graph serve as sources or sinks, an undirected graph induces unique equilibria if and only if the graph has no cycle of length at least 3.

1.3 Further Related Work

Atomic splittable (network) congestion games have been first proposed by Orda et al. [20] and Altman et al. [3] in the context of modeling routing in communication networks. Other applications include traffic and freight networks (cf. Cominetti et al. [8]) and scheduling (cf. Huang [16]). Haurie and Marcotte [15] showed that classical nonatomic congestion games (cf. Beckmann et al. [4] and Wardrop [26]) can be modeled as atomic splittable congestion games by constructing a sequence of games and taking the limit with respect to the number of players. It follows that atomic splittable congestion games are strictly more general than their nonatomic counterpart. Cominetti et al. [8], Harks [10] and Roughgarden and Schoppmann [23] studied the price of anarchy in atomic splittable congestion games.

Integral polymatroid congestion games were introduced in Harks, Klimm and Peis [11] and later they were studied from an optimization perspective in Harks, Oosterwijk and Vredeveld [12]. Polymatroid theory was recently used in the context of nonatomic congestion games, where it is shown that matroid set systems are immune to the Braess paradox, see Fujishige et al. [9].

2   Polymatroid Congestion Games

In polymatroid congestion games we assume that the strategy space for every player corresponds to a *polymatroid base polytope*.

In order to define polymatroids we first have to introduce submodular functions. A function $\rho : 2^E \rightarrow \mathbb{R}$ is called submodular if $\rho(U) + \rho(V) \geq \rho(U \cup V) + \rho(U \cap V)$ for all $U, V \subseteq E$. It is called monotone if $\rho(U) \leq \rho(V)$ for all $U \subseteq V$, and normalized if $\rho(\emptyset) = 0$. Given a submodular, monotone and normalized function ρ, the pair (E, ρ) is called a *polymatroid*. The associated polymatroid base polytope is defined as:

$$P_\rho := \left\{ \boldsymbol{x} \in \mathbb{R}_+^E \mid x(U) \leq \rho(U) \ \forall U \subseteq E, \ x(E) = \rho(E) \right\},$$

where $x(U) := \sum_{e \in U} x_e$ for all $U \subseteq E$.

In a polymatroid congestion game, we associate with every player i a player-specific polymatroid (E, ρ_i) and assume that the strategy space of player i is defined by the (player-specific) polymatroid base polytope P_{ρ_i}. From now on, when we mention a polymatroid congestion game, we mean a weighted atomic splittable polymatroid congestion game. A special case of polymatroid congestion games are the matroid congestion games:

Example 2 (Matroid Congestion Games). Consider an *atomic splittable matroid congestion game*, where for every $i \in N$ the allowable subsets are the base set $\mathcal{B}_i$ of a matroid $\mathcal{M}_i = (E, \mathcal{I}_i)$. The rank function $rk_i : 2^E \rightarrow \mathbb{R}$ of matroid $\mathcal{M}_i$ is defined as: $rk_i(S) := \max\{|U| \mid U \subseteq S \text{ and } U \in \mathcal{I}_i\}$ for all $S \subseteq E$, and is submodular, monotone and normalized [21]. Moreover, the characteristic vectors of the bases in $\mathcal{B}_i$ are exactly the vertices of the polymatroid base polytope P_{rk_i}. It follows that the polytope $P_i := \{\boldsymbol{x} \in \mathbb{R}_+^{|\mathcal{B}_i|} \mid \sum_{B \in \mathcal{B}_i} x_B = d_i\}$ corresponds to strategy distributions that lead to load vectors in the following polytope:

$$P_{d_i \cdot rk_i} = \left\{ \boldsymbol{x}_i \in \mathbb{R}_+^E \mid x_i(U) \leq d_i \cdot rk_i(U) \ \forall U \subseteq E, x_i(E) = d_i \cdot rk_i(E) \right\}.$$

Hence matroid congestion games are a special case of polymatroid congestion games. Two practical examples of polymatroid congestion games can be found in the full version of this paper [14].

3   Bidirectional Flow Polymatroids

We provide a sufficient condition for a class of polymatroid congestion games to have a unique Nash equilibrium. We prove that if the strategy space of every player is the base polytope of a *bidirectional flow polymatroid*, Nash equilibria are unique. In order to define the class of bidirectional flow polymatroids we first discuss some basic properties of polymatroids. We start with a generalization of the strong exchange property for matroids. Let $\chi_e \in \mathbb{Z}^{|E|}$ be the characteristic vector with $\chi_e(e) = 1$, and $\chi_e(e') = 0$ for all $e' \neq e$.

Lemma 1 (Strong exchange property polymatroids (Murota [19])). *Let P_ρ be a polymatroid base polytope defined on (E, ρ). Let $x, y \in P_\rho$ and suppose $x_e > y_e$ for some $e \in E$. Then there exists an $e' \in E$ with $x_{e'} < y_{e'}$ and an $\epsilon > 0$ such that:*

$$x + \epsilon(\chi_{e'} - \chi_e) \in P_\rho \text{ and } y + \epsilon(\chi_e - \chi_{e'}) \in P_\rho.$$

This exchange property will play an important role in the definition of bidirectional flow polymatroids. Given a strategy x in the base polytope of polymatroid (E, ρ), we are interested in the exchanges that can be made between x_e and $x_{e'}$ for some resources in $e, e' \in E$. For any $x, y \in P_\rho$ define the capacitated graph $D(x, y)$ on vertices E. An edge (e, e') exists if there is an $\epsilon > 0$ such that $x + \epsilon(\chi'_e - \chi_e) \in P_\rho$ and $y + \epsilon(\chi_e - \chi'_e) \in P_\rho$. For edge (e, e') we define capacity $\hat{c}_{x,y}(e, e')$ as follows:

$$\hat{c}_{x,y}(e, e') := \max\{\alpha | x + \alpha(\chi_{e'} - \chi_e) \in P_\rho \text{ and } y + \alpha(\chi_e - \chi_{e'}) \in P_\rho\}.$$

A *bidirectional flow* is a flow in $D(x, y)$ where every resource e with $x_e > y_e$ has supply of $x_e - y_e$ and every resource e with $x_e < y_e$ has a demand of $y_e - x_e$. Such a flow might not exist. In that case we say that x and y are *conflicting strategies*. We are now ready to define the class of *bidirectional flow polymatroids*:

Definition 1 (Bidirectional flow polymatroid). *A polymatroid (E, ρ) is called a* bidirectional flow polymatroid *if for every pair of vectors x, y in base polytope P_ρ, there exists a bidirectional flow in $D(x, y)$.*

Not all polymatroids are bidirectional flow polymatroids, like the graphic matroid on the K_4. Note that there does exist a flow if capacities are defined as $\hat{c}_{x,y}(e, e') := \max\{\alpha | x + \alpha(\chi_{e'} - \chi_e) \in P_\rho\}$ instead. This is proven by Wallacher and Zimmermann under the name of the *strong difference theorem* [25, Theorem 7], and a shorter proof can be found in the full version of this paper [14].

4   A Uniqueness Result

In this section we prove that when the strategy space of every player is the base polytope of a bidirectional flow polymatroid, equilibria are unique. We denote the *marginal cost* of player i on resource $e \in E$ by $\mu_{i,e}(x) = c_{i,e}(x_e) + x_{i,e} c'_{i,e}(x_e)$.

An equilibrium condition for polymatroid congestion games, a result that follows from [10, Lemma 1], is as follows:

Lemma 2. *Let x be a Nash equilibrium in a polymatroid congestion game. If $x_{i,e} > 0$, then for all $e' \in E$ for which there is an $\epsilon > 0$ such that $x_i + \epsilon(\chi_{e'} - \chi_e) \in P_{\rho_i}$, we have $\mu_{i,e}(x) \leq \mu_{i,e'}(x)$.*

In the rest of this section we will prove the following theorem:

Theorem 1. *If for a polymatroid congestion game, the strategy space for every player is the base polytope of a bidirectional flow polymatroid, then the equilibria of this game are unique in the sense that $x_{i,e} = y_{i,e}$ for all $i \in N$ and $e \in E$.*

From now on we assume $\boldsymbol{x} = (\boldsymbol{x}_i)_{i \in N}$ and $\boldsymbol{y} = (\boldsymbol{y}_i)_{i \in N}$ are strategy profiles, where strategies $\boldsymbol{x}_i$ and $\boldsymbol{y}_i$ are taken from the base polytope P_{ρ_i} of a player-specific bidirectional flow polymatroid. Before we prove Theorem 1, we first introduce some new notation. We define $E^+ = \{e \in E | x_e > y_e\}$ and $E^- = \{e \in E | x_e < y_e\}$ as the sets of *globally* overloaded and underloaded resources. Define $E^= = \{e \in E | x_e = y_e\}$ as the set of resources on which the total load does not change. In the same way we define player-specific sets of *locally* underloaded and overloaded resources $E^{i,+} = \{e \in E | x_{i,e} > y_{i,e}\}$ and $E^{i,-} = \{e \in E | x_{i,e} < y_{i,e}\}$. We also introduce player set $N_>^+ = \{i \in N | \sum_{e \in E^+}(x_{i,e} - y_{i,e}) > 0\}$. We can distinguish between two cases. Either $E = E^=$, thus $x_e = y_e$ for all resources $e \in E$, or $E \neq E^=$, which implies that E^+ and E^- are non-empty. Note that in this last case $N_>^+ \neq \emptyset$.

For each player i we create a graph $G(\boldsymbol{x}_i, \boldsymbol{y}_i)$ from graph $D(\boldsymbol{x}_i, \boldsymbol{y}_i)$ by adding a super-source s_i and a super-sink t_i to $D(\boldsymbol{x}_i, \boldsymbol{y}_i)$. We add edges from s_i to $e \in E^{i,+}$ with capacity $x_{i,e} - y_{i,e}$ and edges from $e \in E^{i,-}$ to t_i with capacity $y_{i,e} - x_{i,e}$. Graph $G(\boldsymbol{x}_i, \boldsymbol{y}_i)$ is visualized in Fig. 1.

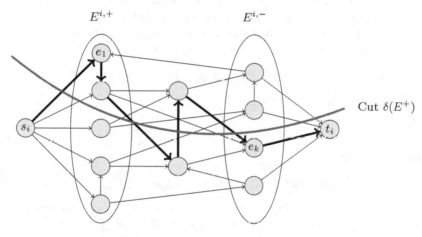

Fig. 1. Visualization of graph $G(\boldsymbol{x}_i, \boldsymbol{y}_i)$ and cut $\delta(E^+)$ used in the proof of Lemma 3.

Recall that strategies $\boldsymbol{x}_i$ and $\boldsymbol{y}_i$ are both chosen from the base polytope of a bidirectional flow polymatroid. Therefore there exists a flow f_i in $D(\boldsymbol{x}_i, \boldsymbol{y}_i)$ where every resource $e \in E^{i,+}$ has a supply of $x_{i,e} - y_{i,e}$ and $e \in E^{i,-}$ a demand of $y_{i,e} - x_{i,e}$. Using f_i we define a $s_i - t_i$ flow f_i' in $G(\boldsymbol{x}_i, \boldsymbol{y}_i)$ as follows:

$$f_i'(e, e') = \begin{cases} x_{i,e} - y_{i,e}, & \text{if } e = s_i \text{ and } e' \in E^{i,+}, \\ y_{i,e} - x_{i,e}, & \text{if } e \in E^{i,-} \text{ and } e' = t_i, \\ f_i(e, e'), & \text{otherwise.} \end{cases} \qquad (2)$$

Lemma 3. *For $x \neq y$, there exists a player i and a path $(s_i, e_1, \ldots, e_k, t_i)$ in $G(\boldsymbol{x}_i, \boldsymbol{y}_i)$ such that $e_1 \in E^{i,+} \cap (E^+ \cup E^=)$ and $e_k \in E^{i,-} \cap (E^- \cup E^=)$.*

Proof. If $E \neq E^=$, then $N_>^+ \neq \emptyset$, and we pick a player $i \in N_>^+$. Flow f_i' can be decomposed into flow carrying s_i-t_i paths, and we will show that there exists a path in this path decomposition that goes from s_i to a vertex $e_1 \in E^{i,+} \cap E^+$, and, after visiting possibly other vertices, finally goes through a vertex $e_k \in E^{i,-} \cap E^-$ to t_i. To see this consider the cut $\delta(E^+)$, following notation by Schrijver [24], as visualized in Fig. 1. Recall that $i \in N_>^+$, hence, $\sum_{e \in E^+}(x_{i,e} - y_{i,e}) > 0$. Thus, in f_i' more load enters E^+ from s_i, than leaves E^+ to t_i. This implies that in the flow decomposition of f_i' there must be a path that goes from s_i to a vertex $e_1 \in E^{i,+} \cap E^+$, crosses cut $\delta(E^+)$ an odd number of times to a vertex $e_k \in E^{i,-} \cap (E^- \cup E^=)$ before ending in t_i. As this is a flow-carrying path in f_i', it exists in $G(\boldsymbol{x}_i, \boldsymbol{y}_i)$.

If $E = E^=$, pick any player i for which there exists a resource e with $x_{i,e} \neq y_{i,e}$ and look at the path decomposition of f_i'. Every path $(s_i, e_1, \ldots, e_k, t_i)$ in this decomposition is a path such that $e_1 \in E^{i,+}$ and $e_k \in E^{i,-}$. As $E = E^=$, it also holds that $e_1 \in E^{i,+} \cap E^=$ and $e_k \in E^{i,-} \cap E^=$. As this is a flow-carrying path in f_i', it exists in $G(\boldsymbol{x}_i, \boldsymbol{y}_i)$.

Proof (Theorem 1). Assume $\boldsymbol{x}$ and $\boldsymbol{y}$ are both Nash equilibria. Using Lemma 3 we find a path $(s_i, e_1, \ldots, e_k, t_i)$ in $G(\boldsymbol{x}_i, \boldsymbol{y}_i)$ such that $e_1 \in E^{i,+} \cap (E^+ \cup E^=)$ and $e_k \in E^{i,-} \cap (E^- \cup E^=)$. Since every edge (e_j, e_{j+1}) exists in $G(\boldsymbol{x}_i, \boldsymbol{y}_i)$, for all $j \in \{1, \ldots, k-1\}$ we get:

$$\boldsymbol{x_i} + \epsilon(\chi_{e_{j+1}} - \chi_{e_j}) \in P_{\rho_i} \text{ and } \boldsymbol{y_i} + \epsilon(\chi_{e_j} - \chi_{e_{j+1}}) \in P_{\rho_i}.$$

Using Lemma 2 we obtain for $\boldsymbol{x}$:

$$\mu_{i,e_1}(\boldsymbol{x}) \leq \mu_{i,e_2}(\boldsymbol{x}) \leq \cdots \leq \mu_{i,e_k}(\boldsymbol{x}), \tag{3}$$

and similarly for $\boldsymbol{y}$:

$$\mu_{i,e_k}(\boldsymbol{y}) \leq \mu_{i,e_{k-1}}(\boldsymbol{y}) \leq \cdots \leq \mu_{i,e_1}(\boldsymbol{y}). \tag{4}$$

Recall that $\mu_{i,e}(\boldsymbol{x}) = c_{i,e}(x_e) + x_{i,e} c_{i,e}'(x_e)$. As $e_1 \in E^{i,+}$, we have that $x_{i,e_1} > y_{i,e_1}$. Because c_{i,e_1} is strictly increasing and $e_1 \in (E^+ \cup E^=)$ we get $c_{i,e_1}(x_{e_1}) \geq c_{i,e_1}(y_{e_1})$ and $c_{i,e_1}'(x_{e_1}) > 0$ using $x_{e_1} \geq x_{i,e_1} > 0$. Moreover, since c_{i,e_1} is convex, the slope of c_{i,e_1} is non-decreasing and, hence, $c_{i,e_1}'(x_{e_1}) \geq c_{i,e_1}'(y_{e_1})$. Putting things together, we get

$$\mu_{i,e_1}(\boldsymbol{y}) < \mu_{i,e_1}(\boldsymbol{x}). \tag{5}$$

Similarly, as $e_k \in E^{i,-} \cap (E^- \cup E^=)$, we have:

$$\mu_{i,e_k}(\boldsymbol{x}) \leq \mu_{i,e_k}(\boldsymbol{y}). \tag{6}$$

Combining (3), (4), (5) and (6), we have:

$$\mu_{i,e_k}(\boldsymbol{x}) \leq \mu_{i,e_k}(\boldsymbol{y}) \leq \mu_{i,e_1}(\boldsymbol{y}) < \mu_{i,e_1}(\boldsymbol{x}) \leq \mu_{i,e_k}(\boldsymbol{x}).$$

This is a contradiction and therefore either strategy $\boldsymbol{x_i}$ or $\boldsymbol{y_i}$ is not a Nash equilibrium for player i. □

5   Applications

In this section we demonstrate that bidirectional flow polymatroids are general enough to allow for meaningful applications. As described in Example 2, matroid congestion games belong to polymatroid congestion games. A subclass of matroids are *base orderable* matroids introduced by Brualdi [6] and Brualdi and Scrimger [7].

Definition 2 (Base orderable matroid). *A matroid $\mathcal{M} = (E, \mathcal{I})$ is called base orderable if for every pair of bases (B, B') there exists a bijective function $g_{B,B'} : B \to B'$ such that for all $e \in B$ both $B - e + g_{B,B'}(e) \in \mathcal{B}$ and $B' + e - g_{B,B'}(e) \in \mathcal{B}$.*

We prove that polymatroids defined by the rank function of a base orderable matroid belong to the class of bidirectional flow polymatroids. Therefore, all matroid congestion games for which the player-specific matroids are base orderable have unique equilibria.

Theorem 2. *Let rk be the rank function of a base orderable matroid (E, rk). Then, for any $d \geq 0$, the polymatroid $(E, d \cdot rk)$ is a bidirectional flow polymatroid.*

Proof. Polytope P_i in Example 2 describes exactly how some player-specific weight d_i can be divided over different bases in $\mathcal{B}_i$ to obtain a feasible strategy $\boldsymbol{x}_i \in P_{d_i \cdot rk_i}$. In this proof we use the same polytope structures, but remove the player specific index i. Thus polytope P describes how weight d can be divided over bases in $\mathcal{B}$ to obtain a feasible strategy $\boldsymbol{x} \in P_{d \cdot rk}$. We call vector $\boldsymbol{x}' \in P$ a *base decomposition* of $\boldsymbol{x}$ if it satisfies $x_e = \sum_{B \in \mathcal{B}; e \in B} x'_B$ for all $e \in E$. Given two vectors $\boldsymbol{x}, \boldsymbol{y} \in P_{d \cdot rk}$, we look at the differences between two base decompositions $\boldsymbol{x}', \boldsymbol{y}' \in P$. We introduce sets $\mathcal{B}^+, \mathcal{B}^- \subset \mathcal{B}$ that will contain respectively the *overloaded* and *underloaded bases*: $\mathcal{B}^+ = \{B \in \mathcal{B} | x'_B > y'_B\}$ and $\mathcal{B}^- = \{B \in \mathcal{B} | x'_B < y'_B\}$.

Using these sets we create the complete directed bipartite graph $D_B(\boldsymbol{x}, \boldsymbol{y})$ on vertices $(\mathcal{B}^+, \mathcal{B}^-)$, where bases $B \in \mathcal{B}^+$ have a supply $x'_B - y'_B$ and bases $B \in \mathcal{B}^-$ have a demand $y'_B - x'_B$. As the total supply equals the total demand, there exists a transshipment t from strategies $B \in \mathcal{B}^+$ to strategies $B' \in \mathcal{B}^-$, such that, when carried out, we obtain $\boldsymbol{y}'$ from $\boldsymbol{x}'$. We denote by $t_{(B,B')}$ the amount of load transshipped from $B \in \mathcal{B}^+$ to $B' \in \mathcal{B}^-$.

In the remainder of the proof, we use transshipment t to construct a flow f in graph $D(\boldsymbol{x}, \boldsymbol{y})$. As the polymatroid is defined by the rank function of a base orderable matroid, for every pair of bases (B, B') there exists a bijective function $g_{B,B'} : B \to B'$ such that both $B - e + g_{B,B'}(e) \in \mathcal{B}$ and $B' + e - g_{B,B'}(e) \in \mathcal{B}$ for all $e \in B$. Note that when $e \in B \cap B'$, $g_{B,B'}(e) = e$. Define

$$\mathcal{B}^2_{e,e'} := \left\{ (B, B') \in \mathcal{B}^+ \times \mathcal{B}^- | e \in B, e' \in B' \text{ and } g_{B,B}(e) = e' \right\}.$$

Then we define flow f as: $f_{(e,e')} = \sum_{(B,B')\in\mathcal{B}^2_{e,e'}} t_{B,B'}$ for all $(e,e') \in E \times E$. Flow f does satisfy all demands and supplies in $D(x,y)$ as f is created from base decompositions x', y' for strategy profiles x and y. If we define:

$$x'' := x' + \sum_{(B,B')\in\mathcal{B}^2_{e,e'}} t_{B,B'} \cdot \left(\chi_{B-e+g_{B,B'}(e)} - \chi_B\right) \in P,$$

then x'' is a base decomposition of strategy $x + f_{(e,e')}(\chi_{e'} - \chi_e)$, and thus $f_{(e,e')} \leq \hat{c}_{x,y}(e,e')$. Therefore f is a bidirectional flow between x and y.   □

For graphic matroids, the generalized series-parallel graph is the maximal graph structure that allows for a bidirectional flow between every pair of strategies (See the full version of this paper for the proof of this statement [14]).

6   Non-matroid Set Systems

We now derive necessary conditions on a given set system $(\mathcal{S}_i)_{i\in N}$ so that any atomic splittable congestion game based on $(\mathcal{S}_i)_{i\in N}$ admits unique equilibria. We show that the *matroid property* is a necessary condition on the players' strategy spaces that guarantees uniqueness of equilibria *without* taking into account how the strategy spaces of different players interweave.[2] To state this property mathematically precisely, we introduce the notion of *embeddings* of $\mathcal{S}_i$ in E. An embedding is a map $\tau := (\tau_i)_{i\in N}$, where every $\tau_i : E_i \to E$ is an injective map from $E_i := \cup_{S\in\mathcal{S}_i} S$ to E. For $X \subseteq E_i$, we denote $\tau_i(X) := \{\tau_i(e) \mid e \in X\}$. Mapping τ_i induces an isomorphism $\phi_{\tau_i} : \mathcal{S}_i \to \mathcal{S}'_i$ with $S \mapsto \tau_i(S)$ and $\mathcal{S}'_i := \{\tau_i(S) \mid S \in \mathcal{S}_i\}$. Isomorphism $\phi_\tau = (\phi_{\tau_i})_{i\in N}$ induces the isomorphic strategy space $\phi_\tau(\mathcal{S}) = (\phi_{\tau_i}(\mathcal{S}_i))_{i\in N}$.

Definition 3. *A family of set systems $\mathcal{S}_i \subseteq 2^{E_i}$, for $i \in N$ is said to have the* strong uniqueness property *if for all embeddings τ, the induced game with isomorphic strategy space $\phi_\tau(\mathcal{S})$ has unique Nash equilibria.*

Since for bases of matroids any embedding τ_i with isomorphism ϕ_{τ_i} has the property that $\phi_{\tau_i}(\mathcal{S}_i)$ is again a collection of bases of a matroid, we obtain the following immediate consequence of Theorem 1.

Corollary 1. *If $(\mathcal{S}_i)_{i\in N}$ consists of bases of a base-orderable matroid $M_i = (E, \mathcal{I}_i)$, $i \in N$, then $(\mathcal{S}_i)_{i\in N}$ possess the strong uniqueness property.*

For obtaining necessary conditions we need a certain property of non-matroids stated in the following Lemma. Its proof can be derived from the proof of Lemma 5.1 in [13], or the proof of Lemma 16 in [2].

---

[2] The term "interweaving" has been introduced by Ackermann et al. [1,2].

Lemma 4. *If $S_i \subseteq 2^{E_i}$ with $S_i \neq \emptyset$ is a non-matroid, then there exist $X, Y \in S_i$ and $\{a, b, c\} \subseteq X \Delta Y := (X \setminus Y) \cup (Y \setminus X)$ such that for each set $Z \in S_i$ with $Z \subseteq X \cup Y$, either $a \in Z$ or $\{b, c\} \subseteq Z$.*

Theorem 3. *Let $|N| \geq 3$ and assume that for all $i \in N$, S_i is a non-matroid set system. Then, $(S_i)_{i \in N}$ does not have the strong uniqueness property.*

Proof. We will show that there are embeddings $\tau_i : E_i \to E$, $i \in N$, such that the isomorphic strategy space $\phi_\tau(S) = (\phi_{\tau_1}(S_1), \ldots, \phi_{\tau_n}(S_n))$ admits a game with multiple equilibria. We can assume w.l.o.g. that each set system S_i forms an anti-chain (in the sense that $X \in S_i, X \subset Y$ implies $Y \notin S_i$) since cost functions are non-negative and strictly increasing. Let us call a non-empty set system $S_i \subseteq 2^{E_i}$ a *non-matroid* if S_i is an anti-chain and $(E_i, \{X \subseteq S : S \in S_i\})$ is not a matroid.

Let $\tilde{E} = \bigcup_{i \in N} \tau_i(E_i)$ denote the set of all resources under the embeddings $\tau_i, i \in N$. The costs on all resources in $\tilde{E} \setminus (\tau_1(E_1) \cup \tau_2(E_2) \cup \tau_3(E_3))$ are set to zero. Also, the demands of all players d_i with $i \in N \setminus \{1, 2, 3\}$ are set to zero. This way, the game is basically determined by the players $1, 2, 3$. We set the demands $d_1 = d_2 = d_3 = 1$.

Let us choose two sets X, Y in S_1 and $\{a, b, c\} \subseteq X \cup Y$ as described in Lemma 4. Let $e := \tau_1(a), f := \tau_1(b)$ and $g := \tau_1(c)$. We set the costs of all resources in $\tau_1(E_1) \setminus (\tau_1(X) \cup \tau_1(Y))$ to some very large cost M (large enough so that player 1 would never use any of these resources). The cost on all resources in $(\tau_1(X) \cup \tau_1(Y)) \setminus \{e, f, g\}$ is set to zero. This way, player 1 always chooses a strategy $\tau_1(Z) \subseteq \tau_1(X) \cup \tau_1(Y)$ which, by Lemma 4, either contains e, or it contains both f and g. We apply the same construction for player 2 and 3, only changing the role of e to act as f and g, respectively. Note that the so-constructed game is essentially isomorphic to the routing game illustrated in Fig. 2 if we interpret resource e as arc (s_1, t_1), resource f as arc (s_2, t_2), and resource g as arc (s_3, t_3).

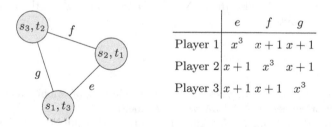

	e	f	g
Player 1	x^3	$x + 1$	$x + 1$
Player 2	$x + 1$	x^3	$x + 1$
Player 3	$x + 1$	$x + 1$	x^3

Fig. 2. Example that admits multiple Nash equilibria

Every player has two possible paths: the direct path using only one edge, or the indirect path using two edges. Using the cost functions in Fig. 2, we show that the game where everyone puts all their weight on the direct path is a Nash equilibrium, as is the game where everybody puts their weight on the indirect path.

If all players put their weight on the direct route, then player 1 cannot deviate to decrease it's costs, as: $c_{1,e}(1) + c'_{1,e}(1) \cdot 1 = 1 + 3 \leq 2 + 2 = c_{1,f}(1) + c_{1,g}(1)$. On the other hand, when all players put their weight on the indirect direct route, player 1 can also not deviate, as: $c_{1,f}(2) + c'_{1,f}(2) \cdot 1 + c_{1,g}(2) + c'_{1,g}(2) \cdot 1 = 3 + 1 + 3 + 1 \leq 8 = c_{1,e}(2)$. The same inequalities hold for player 2 and 3. And therefore everyone playing the direct route, or everyone playing the indirect route both results in a Nash equilibrium. □

7   A Characterization for Undirected Graphs

In Sect. 6 we proved that non-matroid set systems in general do not have the strong uniqueness property when there are at least three players, by constructing embeddings τ_i that lead to the counter example in Fig. 2. This example also gives new insights in uniqueness of equilibria in network congestion games. In the following, we give a characterization of graphs that guarantee uniqueness of Nash equilibria even when player-specific cost functions are allowed.

Definition 4. *An undirected (multi)graph G is said to have the* uniqueness *property if for any atomic splittable network congestion game on $G = (V, E)$, equilibria are unique.*

Note that in the above definition, we do not specify how source- and sink vertices are distributed in V. We obtain the following result which is related to Theorem 3 of Meunier and Pradeau [17], where a similar result is given for non-atomic congestion games with player-specific cost functions. The proof can be found in the full version of this paper [14].

Theorem 4. *An undirected graph has the uniqueness property if and only if G has no cycle of length 3 or more.*

Acknowledgements. We thank Umang Bhaskar, Britta Peis and Satoru Fujishige for fruitful discussions. We also thank Neil Olver for pointing out the connection to base orderable matroids.

References

1. Ackermann, H., Röglin, H., Vöcking, B.: On the impact of combinatorial structure on congestion games. J. ACM **55**(6), 1–22 (2008)
2. Ackermann, H., Röglin, H., Vöcking, B.: Pure Nash equilibria in player-specific and weighted congestion games. Theoret. Comput. Sci. **410**(17), 1552–1563 (2009)
3. Altman, E., Basar, T., Jimnez, T., Shimkin, N.: Competitive routing in networks with polynomial costs. IEEE Trans. Automat. Control **47**, 92–96 (2002)
4. Beckmann, M., McGuire, C., Winsten, C.: Studies in the Economics and Transportation. Yale University Press, New Haven (1956)

5. Bhaskar, U., Fleischer, L., Hoy, D., Huang, C.: Equilibria of atomic flow games are not unique. Math. Oper. Res. **41**(1), 92–96 (2015)
6. Brualdi, R.: Exchange systems, matchings, and transversals. J. Comb. Theor. **5**(3), 244–257 (1968)
7. Brualdi, R., Scrimger, E.: Comments on bases in dependence structures. Bull. Aust. Math. Soc. **1**, 161–167 (1969)
8. Cominetti, R., Correa, J.R., Stier-Moses, N.E.: The impact of oligopolistic competition in networks. Oper. Res. **57**(6), 1421–1437 (2009)
9. Fujishige, S., Goemans, M., Harks, T., Peis, B.: Matroids are immune to Braess paradox (2015). http://arxiv.org/abs/1504.07545F
10. Harks, T.: Stackelberg strategies and collusion in network games with splittable flow. Theor. Comput. Syst. **48**, 781–802 (2011)
11. Harks, T., Klimm, M., Peis, B.: Resource competition on integral polymatroids. In: Liu, T.-Y., Qi, Q., Ye, Y. (eds.) WINE 2014. LNCS, vol. 8877, pp. 189–202. Springer, Heidelberg (2014)
12. Harks, T., Oosterwijk, T., Vredeveld, T.: A logarithmic approximation for polymatroid matroid congestion games. Unpublished manuscript (2014)
13. Harks, T., Peis, B.: Resource buying games. Algorithmica **70**(3), 493–512 (2014)
14. Harks, T., Timmermans, V.: Uniqueness of equilibria in atomic splittable polymatroid congestion games (2015). arXiv, abs/1512.01375
15. Haurie, A., Marcotte, P.: On the relationship between Nash-Cournot and Wardrop equilibria. Networks **15**, 295–308 (1985)
16. Huang, C.-C.: Collusion in atomic splittable routing games. In: Aceto, L., Henzinger, M., Sgall, J. (eds.) ICALP 2011, Part II. LNCS, vol. 6756, pp. 564–575. Springer, Heidelberg (2011)
17. Meunier, F., Pradeau, T.: The uniqueness property for networks with several origin-destination pairs. Eur. J. Oper. Res. **237**(1), 245–256 (2012)
18. Milchtaich, I.: Topological conditions for uniqueness of equilibrium in networks. Math. Oper. Res. **30**(1), 225–244 (2005)
19. Murota, K.: Discrete Convex Analysis. SIAM, Philadelphia (2003)
20. Orda, A., Rom, R., Shimkin, N.: Competitive routing in multi-user communication networks. IEEE/ACM Trans. Netw. **1**, 510–521 (1993)
21. Pym, S., Perfect, H.: Submodular function and independence structures. J. Math. Anal. Appl. **30**(1), 1–31 (1970)
22. Richman, O., Shimkin, N.: Topological uniqueness of the Nash equilibrium for selfish routing with atomic users. Math. Oper. Res. **32**(1), 215–232 (2007)
23. Roughgarden, T., Schoppmann, F.: Local smoothness and the price of anarchy in splittable congestion games. J. Econom. Theor. **156**, 317–342 (2015). Computer Science and Economic Theory
24. Schrijver, A.: Combinatorial Optimization: Polyhedra and Efficiency. Springer, Heidelberg (2003)
25. Wallacher, C., Zimmermann, T.: A polynomial cycle canceling algorithm for submodular flows. Math. Program. **86**(1), 1–15 (1999)
26. Wardrop, J.: Some theoretical aspects of road traffic research. Proc. Inst. Civil Eng. **1**(Pt. II), 325–378 (1952)

A Coordinate Ascent Method for Solving Semidefinite Relaxations of Non-convex Quadratic Integer Programs

Christoph Buchheim[1], Maribel Montenegro[1(✉)], and Angelika Wiegele[2]

[1] Fakultät für Mathematik, Technische Universität Dortmund, Dortmund, Germany
christoph.buchheim@tu-dortmund.de, maribel.montenegro@math.tu-dortmund.de
[2] Department of Mathematics, Alpen-Adria-Universität Klagenfurt,
Klagenfurt, Austria
angelika.wiegele@aau.at

Abstract. We present a coordinate ascent method for a class of semidefinite programming problems that arise in non-convex quadratic integer optimization. These semidefinite programs are characterized by a small total number of active constraints and by low-rank constraint matrices. We exploit this special structure by solving the dual problem, using a barrier method in combination with a coordinate-wise exact line search. The main ingredient of our algorithm is the computationally cheap update at each iteration and an easy computation of the exact step size. Compared to interior point methods, our approach is much faster in obtaining strong dual bounds. Moreover, no explicit separation and reoptimization is necessary even if the set of primal constraints is large, since in our dual approach this is covered by implicitly considering all primal constraints when selecting the next coordinate.

Keywords: Semidefinite programming · Non-convex quadratic integer optimization · Coordinate descent method

1 Introduction

The importance of Mixed-Integer Quadratic Programming (MIQP) lies in both theory and practice of mathematical optimization. On one hand, a wide range of problems arising in practical applications can be formulated as MIQP. On the other hand, it is the most natural generalization of Mixed-Integer Linear Programming (MILP). However, it is well known that MIQP is NP-hard, as it contains MILP as a special case. Moreover, contrarily to what happens in MILP, the hardness of MIQP is not resolved by relaxing the integrality requirement on the variables: while convex quadratic problems can be solved in polynomial time by either the ellipsoid method [6] or interior point methods [5,9], the general problem of minimizing a non-convex quadratic function over a box is NP-hard, even if only one eigenvalue of the Hessian is negative [8].

© Springer International Publishing Switzerland 2016
R. Cerulli et al. (Eds.): ISCO 2016, LNCS 9849, pp. 110–122, 2016.
DOI: 10.1007/978-3-319-45587-7_10

Buchheim and Wiegele [2] proposed the use of semidefinite relaxations and a specialized branching scheme (Q-MIST) for solving unconstrained non-convex quadratic minimization problems where the variable domains are arbitrary closed subsets of $\mathbb{R}$. Their work is a generalization of the well-known semidefinite programming approach to the maximum cut problem or, equivalently, to unconstrained quadratic minimization over variables in the domain $\{-1, 1\}$. Q-MIST needs to solve a semidefinite program (SDP) at each node of the branch-and-bound tree, which can be done using any standard SDP solver. In [2], an interior point method was used for this task, namely the CSDP library [1]. It is well-known that interior point algorithms are theoretically efficient to solve SDPs, they are able to solve small to medium size problems with high accuracy, but they are memory and time consuming for large scale instances.

A related approach to solve the same kind of non-convex quadratic problems was presented by Dong [3]. A convex quadratic relaxation is produced by means of a cutting surface procedure, based on multiple diagonal perturbations. The separation problem is formulated as a semidefinite problem and is solved by coordinate-wise optimization methods. More precisely, the author defines a barrier problem and solves it using coordinate descent methods with exact line search. Due to the particular structure of the problem, the descent direction and the step length can be computed by closed formulae, and fast updates are possible using the Sherman-Morrison-Woodbury formula (we will refer to it just as Woodbury formula in the following). Computational results show that this approach produces lower bounds as strong as the ones provided by Q-MIST and it runs much faster for instances of large size.

In this paper, we adapt and generalize the coordinate-wise approach of [3] in order to solve the dual of the SDP relaxation arising in the Q-MIST approach. In our setting, it is still true that an exact coordinate-wise line search can be performed efficiently by using a closed-form expression, based on the Woodbury formula. Essentially, each iteration of the algorithm involves the update of one coordinate of the vector of dual variables and the computation of an inverse of a matrix that changes by a rank-two constraint matrix when changing the value of the dual variable. Altogether, our approach fully exploits the specific structure of our problem, namely a small total number of (active) constraints and low-rank constraint matrices of the semidefinite relaxation. Furthermore, in our model the set of dual variables can be very large, so that the selection of the best coordinate requires more care than in [3]. However, our new approach is much more efficient than the corresponding separation approach for the primal problem described in [2].

2   Preliminaries

We consider non-convex quadratic mixed-integer optimization problems of the form

$$
\begin{aligned}
\min \quad & x^\top \hat{Q} x + \hat{l}^\top x + \hat{c} \\
\text{s.t.} \quad & x \in D_1 \times \cdots \times D_n ,
\end{aligned}
\tag{1}
$$

where $\hat{Q} \in \mathbb{R}^{n \times n}$ is symmetric but not necessarily positive semidefinite, $\hat{l} \in \mathbb{R}^n$, $\hat{c} \in \mathbb{R}$, and $D_i = \{l_i, \ldots, u_i\} \subseteq \mathbb{Z}$ is finite for all $i = 1, \ldots, n$. Buchheim and Wiegele [2] have studied the more general case where each D_i is an arbitrary closed subset of $\mathbb{R}$. The authors have implemented a branch-and-bound approach called Q-MIST, it mainly consists in reformulating Problem (1) as a semidefinite optimization problem and solving a relaxation of the transformed problem within a branch-and-bound framework. In this section, first we describe how to obtain a semidefinite relaxation of Problem (1), then we formulate it in a matrix form and compute the dual problem.

2.1 Semidefinite Relaxation

Semidefinite relaxations for quadratic optimization problems can already be found in an early paper of Lovász in 1979 [7], but it was not until the work of Goemans and Williamson in 1995 [4] that they started to catch interest. The basic idea is as follows: given any vector $x \in \mathbb{R}^n$, the matrix $xx^\top \in \mathbb{R}^{n \times n}$ is rank-one, symmetric and positive semidefinite. In particular, also the augmented matrix

$$\ell(x) := \begin{pmatrix} 1 \\ x \end{pmatrix} \begin{pmatrix} 1 \\ x \end{pmatrix}^\top = \begin{pmatrix} 1 & x^\top \\ x & xx^\top \end{pmatrix} \in \mathbb{R}^{(n+1) \times (n+1)}$$

is positive semidefinite. This well-known fact leads to semidefinite reformulations of various quadratic problems. Defining a matrix

$$Q := \begin{pmatrix} \hat{c} & \frac{1}{2}\hat{l}^\top \\ \frac{1}{2}\hat{l} & \hat{Q} \end{pmatrix},$$

Problem (1) can be rewritten as

$$\min \quad \langle Q, X \rangle$$
$$\text{s.t.} \quad X \in \ell(D_1 \times \cdots \times D_n),$$

so that it remains to investigate the set $\ell(D_1 \times \cdots \times D_n)$. The following result was proven in [2].

Theorem 1. *Let $X \in \mathbb{R}^{(n+1) \times (n+1)}$ be symmetric. Then $X \in \ell(D_1 \times \cdots \times D_n)$ if and only if*

(a) $(x_{i0}, x_{ii}) \in P(D_i) := \text{conv}\{(u, u^2) \mid u \in D_i\}$ for all $i = 1, \ldots, n$,
(b) $x_{00} = 1$,
(c) $\text{rank}(X) = 1$, and
(d) $X \succeq 0$.

We derive that the following optimization problem is a convex relaxation of (1), obtained by dropping the rank-one constraint of Theorem 1 (c):

$$\min \quad \langle Q, X \rangle$$
$$\text{s.t.} \quad (x_{i0}, x_{ii}) \in P(D_i) \quad \forall i = 1, \ldots n \tag{2}$$
$$x_{00} = 1$$
$$X \succeq 0$$

This is an SDP, since the constraints $(x_{i0}, x_{ii}) \in P(D_i)$ can be replaced by a set of linear constraints, as discussed in the next section.

2.2   Matrix Formulation

In the case of finite D_i considered here, the set $P(D_i)$ is a polytope in $\mathbb{R}^2$ with $|D_i|$ many extreme points. It can thus be described equivalently by a set of $|D_i|$ linear inequalities.

Lemma 1. *For $D_i = \{l_i, \ldots, u_i\}$, the polytope $P(D_i)$ is completely described by lower bounding facets $-x_{ii} + (j + (j+1))x_{0i} \le j(j+1)$ for $j = l_i, l_i+1, \ldots, u_i-1$ and one upper bounding facet $x_{ii} - (l_i + u_i)x_{0i} \le -l_i u_i$.*

Exploiting $x_{00} = 1$, we may rewrite the polyhedral description of $P(D_i)$ presented in the previous lemma as

$$(1 - j(j+1))x_{00} - x_{ii} + (j + (j+1))x_{0i} \le 1, \ j = l_i, l_i+1, \ldots, u_i - 1$$
$$(1 + l_i u_i)x_{00} + x_{ii} - (l_i + u_i)x_{0i} \le 1 .$$

We write the resulting inequalities in matrix form as $\langle A_{ij}, X \rangle \le 1$. To keep analogy with the facets, the index ij represents the inequalities corresponding to lower bounding facets if $j = l_i, l_i + 1, \ldots, u_i - 1$ whereas $j = u_i$ corresponds to the upper facet; see Fig. 1.

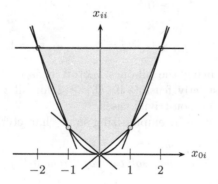

Fig. 1. The polytope $P(\{-2, -1, 0, 1, 2\})$. Lower bounding facets are indexed, from left to right, by $j = -2, -1, 0, 1$, the upper bounding facet is indexed by 2.

Moreover, we write the constraint $x_{00} = 1$ in matrix form as $\langle A_0, X \rangle = 1$, where $A_0 := e_0 e_0^\top$. In summary, Problem (2) can now be stated as

$$\begin{aligned}
\min \quad & \langle Q, X \rangle \\
\text{s.t.} \quad & \langle A_0, X \rangle = 1 \\
& \langle A_{ij}, X \rangle \le 1 \quad \forall j = l_i, \ldots, u_i, \forall i = 1, \ldots, n \\
& X \succeq 0.
\end{aligned} \qquad (3)$$

The following simple observation is crucial for our algorithm presented in the following section.

Lemma 2. *The constraint matrix A_0 has rank one. All constraint matrices A_{ij} have rank one or two. The rank of A_{ij} is one if and only if $j = u_i$ and $u_i - l_i = 2$.*

2.3   Dual Problem

In order to derive the dual problem of (3), we define

$$\mathcal{A}(X) := \begin{pmatrix} \langle A_0, X \rangle \\ \langle A_{ij}, X \rangle_{j \in \{l_i, \dots, u_i\}, i \in \{1, \dots, n\}} \end{pmatrix}$$

and associate a dual variable $y_0 \in \mathbb{R}$ with the constraint $\langle A_0, X \rangle = 1$ as well as dual variables $y_{ij} \leq 0$ with $\langle A_{ij}, X \rangle \leq 1$, for $j \in \{l_i, \dots, u_i\}$ and $i \in \{1, \dots, n\}$. We then define $y \in \mathbb{R}^{m+1}$ as

$$y := \begin{pmatrix} y_0 \\ (y_{ij})_{j \in \{l_i, \dots, u_i\}, i \in \{1, \dots, n\}} \end{pmatrix}.$$

The dual semidefinite program of Problem (3) is

$$\begin{aligned} \max \quad & \langle b, y \rangle \\ \text{s.t.} \quad & Q - \mathcal{A}^\top y \succeq 0 \\ & y_0 \in \mathbb{R} \\ & y_{ij} \leq 0 \quad \forall j = l_i, \dots, u_i, \forall i = 1, \dots, n, \end{aligned} \tag{4}$$

the vector $b \in \mathbb{R}^{m+1}$ being the all-ones vector. It is easy to verify that the primal problem (3) is strictly feasible if $|D_i| \geq 2$ for all $i = 1, \dots, n$, so that strong duality holds in all non-trivial cases.

We conclude this section by emphasizing some characteristics of any feasible solution of Problem (3).

Lemma 3. *Let X^* be a feasible solution of Problem (3). For $i \in \{1, \dots, n\}$, consider the active set*

$$\mathscr{A}_i = \{j \in \{l_i, \dots, u_i\} \mid \langle A_{ij}, X^* \rangle = 1\}$$

corresponding to variable i. Then

(i) for all $i \in \{1, \dots, n\}$, $|\mathscr{A}_i| \leq 2$, and
(ii) if $|\mathscr{A}_i| = 2$, then $x_{ii}^* = {x_{0i}^*}^2$ and $x_{i0}^* \in D_i$.

Proof. The polytope $P(D_i)$ is two-dimensional with non-degenerate vertices. Due to the way the inequalities $\langle A_{ij}, X \rangle \leq 1$ are defined it is impossible to have more than two inequalities intersecting at one point. Therefore, a given point $(x_{ii}, x_{i0}) \in P(D_i)$ satisfies zero, one, or two inequalities with equality. In the last case, we have $x_{ii} = x_{i0}^2$ by construction, which implies $x_{i0} \in D_i$.   □

For the dual problem (4), Lemma 3 (i) means that at most $2n+1$ out of the $m+1$ variables can be non-zero in an optimal solution. Clearly, such a small number of non-zero variables is beneficial in a coordinate-wise optimization method. Moreover, by Lemma 3 (ii), if two dual variables corresponding to the same primal variable are non-zero in an optimal dual solution, then this primal variable will obtain an integer feasible value in the optimal primal solution.

3 A Coordinate Ascent Method

We aim at solving the dual problem (4) by coordinate-wise optimization, in order to obtain fast lower bounds to be used inside the branch-and-bound framework Q-MIST. Our approach is motivated by an algorithm proposed by Dong [3]. The author formulates Problem (1) as a convex quadratically constrained problem, and devises a cutting surface procedure based on diagonal perturbations to construct convex relaxations. The separation problem turns out to be a semidefinite problem with convex non-smooth objective function, and it is solved by a primal barrier coordinate minimization algorithm with exact line search.

The dual Problem (4) has a similar structure to the semidefinite problem solved in [3], therefore similar ideas can be applied. Our SDP is more general however, it contains more general constraints with matrices of rank two (instead of one) and most of our variables are constrained to be non-positive. Another difference is that we deal with a very large number of constraints, out of which only a few are non-zero however. On the other hand, our objective function is linear, which is not true for the problem considered in [3].

As a first step, we introduce a penalty term modeling the semidefinite constraint $Q - \mathcal{A}^\top y \succeq 0$ of Problem (4) and obtain

$$\begin{aligned} \max \quad & f(y;\sigma) := \langle b, y \rangle + \sigma \log \det(Q - \mathcal{A}^\top y) \\ \text{s.t.} \quad & Q - \mathcal{A}^\top y \succ 0 \\ & y_0 \in \mathbb{R} \\ & y_{ij} \leq 0 \quad \forall j = l_i, \dots, u_i, \forall i = 1, \dots, n \end{aligned} \tag{5}$$

for $\sigma > 0$. The gradient of the objective function of Problem (5) is

$$\nabla_y f(y;\sigma) = b - \sigma \mathcal{A}((Q - \mathcal{A}^\top y)^{-1}).$$

For the following, we denote $W := (Q - \mathcal{A}^\top y)^{-1}$, so that

$$\nabla_y f(y;\sigma) = b - \sigma \mathcal{A}(W). \tag{6}$$

We will see later that, using the Woodbury formula, the matrix W can be updated quickly when changing the value of a dual variable, which is crucial for the performance of the algorithm proposed. We begin by describing a general algorithm to solve (5) in a coordinate maximization manner. In the following, we explain each step of this algorithm in detail.

Outline of a barrier coordinate ascent algorithm for Problem (4)

---

1 **Starting point:** choose any feasible solution y of (5);
2 **Direction:** choose a coordinate direction e_{ij};
3 **Step size:** using exact line search, determine the step length s;
4 **Move along chosen coordinate:** $y \leftarrow y + s e_{ij}$;
5 **Update** the matrix W accordingly;
6 **Decrease** the penalty parameter σ;
7 **Go to (2),** unless some stopping criterion is satisfied;

3.1   Definition of a Starting Point

If $Q \succ 0$, we can safely choose $y^{(0)} = 0$ as starting point. Otherwise, define $a \in \mathbb{R}^n$ by $a_i = (A_{iu_i})_{0i}$ for $i = 1, \ldots, n$. Moreover, define

$$\tilde{y} := \min\{\lambda_{min}(\hat{Q}) - 1, 0\},$$

$$y_0 := \hat{c} - \tilde{y} \sum_{i=1}^{n}(1 + l_i u_i) - 1 - (\tfrac{1}{2}\hat{l} - \tilde{y}a)^{\top}(\tfrac{1}{2}\hat{l} - \tilde{y}a),$$

and $y^{(0)} \in \mathbb{R}^{m+1}$ as

$$y^{(0)} := \begin{pmatrix} y_0 \\ (y_{ij})_{j \in \{l_i, \ldots, u_i\}, i \in \{1, \ldots, n\}} \end{pmatrix}, \quad y_{ij} = \begin{cases} \tilde{y}, & j = u_i, i = 1, \ldots, n \\ 0, & \text{otherwise.} \end{cases}$$

Then the following lemma can be proved.

Lemma 4. *The vector $y^{(0)}$ is feasible for* (5).

3.2   Choice of an Ascent Direction

We improve the objective function coordinate-wise: at each iteration k of the algorithm, we choose an ascent direction $e_{ij^{(k)}} \in \mathbb{R}^m$ where $ij^{(k)}$ is the coordinate of the gradient with maximum absolute value

$$ij^{(k)} := \arg\max_{ij}|\nabla_y f(y; \sigma)_{ij}| \, . \tag{7}$$

However, moving a coordinate ij to a positive direction is allowed only if $y_{ij} < 0$, so that the coordinate $ij^{(k)}$ in (7) has to be chosen among those satisfying

$$(\nabla_y f(y; \sigma)_{ij} > 0 \text{ and } y_{ij} < 0) \quad \text{or} \quad \nabla_y f(y; \sigma)_{ij} < 0 \, .$$

The entries of the gradient depend on the type of inequality. By (6), we have

$$\nabla_y f(y; \sigma)_{ij} = 1 - \sigma \langle W, A_{ij} \rangle.$$

The number of lower bounding facets for a single primal variable i is $u_i - l_i$, which is not polynomial in the input size from a theoretical point of view.

From a practical point of view, a large domain D_i may slow down the coordinate selection if all potential coordinates have to be evaluated explicitly.

However, the regular structure of the gradient entries corresponding to lower bounding facets for variable i allows to limit the search to at most three candidates per variable. To this end, we define the function

$$\varphi_i(j) := 1 - \sigma\langle W, A_{ij}\rangle = 1 - \sigma\big((1 - j(j+1))w_{00} + (2j+1)w_{0i} - w_{ii}\big)$$

and aim at finding a minimizer of $|\varphi|$ over $\{l_i, \ldots, u_i - 1\}$. As φ_i is a univariate quadratic function, we can restrict our search to at most three candidates, namely the bounds l_i and $u_i - 1$ and the rounded global minimizer of φ_i, if it belongs to $l_i, \ldots, u_i - 1$; the latter is

$$\left\lceil \frac{w_{0i}}{w_{00}} - \frac{1}{2} \right\rceil .$$

In summary, taking into account also the upper bounding facets and the coordinate zero, we need to test at most $4n + 1$ candidates in order to solve (7), independently of the bounds l_i and u_i.

3.3 Computation of the Step Size

We compute the step size $s^{(k)}$ by exact line search in the chosen direction. For this, we need to solve the following one-dimensional maximization problem

$$s^{(k)} = \arg\max_s\{f(y^{(k)} + se_{ij(k)}; \sigma) \mid Q - \mathcal{A}^\top(y^{(k)} + se_{ij(k)}) \succ 0, s \leq -y_{ij(k)}\} \quad (8)$$

unless the chosen coordinate is zero, in which case the upper bound on s is dropped. Note that $s \mapsto f(y^{(k)} + se_{ij(k)}; \sigma)$ is strictly concave on

$$\{s \in \mathbb{R} \mid Q - \mathcal{A}^\top(y^{(k)} + se_{ij(k)}) \succ 0\} .$$

By the first order optimality conditions, we thus need to find the unique $s^{(k)} \in \mathbb{R}$ satisfying the semidefinite constraint $Q - \mathcal{A}^\top(y^{(k)} + s^{(k)}e_{ij(k)}) \succ 0$ such that either

$$\nabla_s f(y^{(k)} + s^{(k)}e_{ij(k)}; \sigma) = 0 \quad \text{and} \quad y_{ij(k)} + s^{(k)} \leq 0$$

or

$$\nabla_s f(y^{(k)} + s^{(k)}e_{ij(k)}; \sigma) > 0 \quad \text{and} \quad s^{(k)} = -y_{ij(k)}^{(k)} .$$

In order to simplify the notation, we omit the superindex (k) in the following. From the definition,

$$f(y + se_{ij}; \sigma) = \langle b, y\rangle + s\langle b, e_{ij}\rangle + \sigma \log \det(Q - \mathcal{A}^\top y - s\mathcal{A}^\top(e_{ij}))$$
$$= \langle b, y\rangle + s + \sigma \log \det(W^{-1} - sA_{ij}).$$

Then, the gradient with respect to s is

$$\nabla_s f(y + se_{ij}; \sigma) = 1 - \sigma\langle A_{ij}, (W^{-1} - sA_{ij})^{-1}\rangle. \quad (9)$$

Now the crucial task is to compute the inverse of the matrix $W^{-1} - sA_{ij}$, which is of dimension $n + 1$. For this purpose, notice that W^{-1} is changed by a rank-one or rank-two matrix sA_{ij}; see Lemma 2. Therefore, we can compute both the inverse matrix $(W^{-1} - sA_{ij})^{-1}$ and the optimal step length by means of the Woodbury formula for the rank-one or rank-two update. In the latter case, the formula is quadratic in s and may thus yield two candidates for the optimal step length. In order to satisfy $Q - \mathcal{A}^\top(y + se_{ij}) \succ 0$, we then have to choose the smaller one of the candidates.

Finally, we have to point out that the zero coordinate can also be chosen as ascent direction, in that case the gradient is

$$\nabla_s f(y + se_0; \sigma) = 1 - \sigma\langle A_0, (W^{-1} - sA_0)^{-1}\rangle,$$

and the computation of the step size is analogous.

3.4   Algorithm Overview

Our approach to solve Problem (4) is summarized in Algorithm CD.

Algorithm CD: Barrier coordinate ascent algorithm for Problem (4)

Input: $Q \in \mathbb{R}^{(n+1)\times(n+1)}$
Output: A lower bound on the optimal value of Problem (3)
1 Use Lemma 4 to compute $y^{(0)}$ such that $Q - \mathcal{A}^\top y^{(0)} \succ 0$
2 Compute $W^{(0)} \leftarrow (Q - \mathcal{A}^\top y^{(0)})^{-1}$
3 **for** $k \leftarrow 0$ **until** *max-iterations* **do**
4   Choose a coordinate direction $e_{ij^{(k)}}$ as described in Sect. 3.2
5   Compute the step size $s^{(k)}$ as described in Sect. 3.3
6   Update $y^{(k+1)} \leftarrow y^{(k)} + s^{(k)} e_{ij^{(k)}}$
7   Update $W^{(k)}$ using the Woodbury formula
8   Update σ
9   Terminate if some stopping criterion is met
10 **return** $\langle b, y^{(k)}\rangle$

Before entering the main loop, the running time of Algorithm CD is dominated by the computation of the minimum eigenvalue of $\hat{Q}$ needed to compute $y^{(0)}$ and by the computation of the inverse matrix of $Q - \mathcal{A}^\top y^{(0)}$. Both can be done in $O(n^3)$ time. Each iteration of the algorithm can be performed in $O(n^2)$. Indeed, as discussed in Sect. 3.2, we need to consider $O(n)$ candidates for the coordinate selection, so that this task can be performed in $O(n^2)$ time. For calculating the step size and updating the matrix $W^{(k)}$, we also need $O(n^2)$ time using the Woodbury formula.

Notice that the algorithm produces a feasible solution $y^{(k)}$ of Problem (4) at every iteration and hence a valid lower bound $\langle b, y^{(k)}\rangle$ for Problem (3). In particular, when used within a branch-and-bound algorithm, this means that Algorithm CD can be stopped as soon as $\langle b, y^{(k)}\rangle$ exceeds a known upper bound for Problem (3). Otherwise, the algorithm can be stopped after a fixed number of iterations or when other criteria show that only a small further improvement of the bound can be expected.

The choice of an appropriate termination rule however is closely related to the update of σ performed in Step 8. The aim is to find a good balance between the convergence for fixed σ and the decrease of σ. In our implementation, we use the following rule: whenever the entry of the gradient corresponding to the chosen coordinate has an absolute value below 0.01, we multiply σ by 0.25. As soon as σ falls below 10^{-5}, we fix it to this value.

3.5   Two-Dimensional Update

In Algorithm CD, we change only one coordinate in each iteration, as this allows to update the matrix $W^{(k)}$ in $O(n^2)$ time using the Woodbury formula. This was due to the fact that all constraint matrices in the primal SDP (3) have rank at most two. However, taking into account the special structure of the constraint matrix A_0, one can see that every linear combination of any constraint matrix A_{ij} with A_0 still has rank at most two. In other words, we can simultaneously update the dual variables y_0 and y_{ij} and still recompute $W^{(k)}$ in $O(n^2)$ time.

In order to improve the convergence of Algorithm CD, we choose a coordinate ij as explained in Sect. 3.2 and then perform an exact plane-search in the two-dimensional space corresponding to the directions e_0 and e_{ij}, i.e., we solve the bivariate problem

$$\arg\max_{(s_0,s)} \left\{ f(y + s_0 e_0 + s e_{ij}; \sigma) \mid Q - \mathcal{A}^\top(y + s_0 e_0 + s e_{ij}) \succ 0, s \le -y_{ij} \right\}, \quad (10)$$

where we again omit the superscript (k) for sake of readibilty. Similar to the one-dimensional case in (8), due to strict concavity of $(s_0, s) \mapsto f(y + s_0 e_0 + s e_{ij}; \sigma)$ over $\{(s_0, s) \in \mathbb{R}^2 \mid Q - \mathcal{A}^\top(y + s_0 e_0 + s e_{ij}) \succ 0\}$, solving (10) is equivalent to finding the unique pair $(s_0, s) \in \mathbb{R}^2$ such that

$$\nabla_{s_0} f(y + s_0 e_0 + s e_{ij}; \sigma) = 0$$

and either

$$\nabla_s f(y + s_0 e_0 + s e_{ij}; \sigma) = 0 \quad \text{and} \quad y_{ij} + s \le 0$$

or

$$\nabla_s f(y + s_0 e_0 + s e_{ij}; \sigma) > 0 \quad \text{and} \quad s = -y_{ij}.$$

To determine (s_0, s), it thus suffices to set both gradients to zero and solve the resulting two-dimensional system of equations. If it turns out that $y_{ij} + s > 0$, we fix $s := -y_{ij}$ and recompute s_0 by solving

$$\nabla_{s_0} f(y + s_0 e_0 + s e_{ij}; \sigma) = 0.$$

Proceeding as before, we have

$$f(y + s_0 e_0 + s e_{ij}; \sigma) = \langle b, y \rangle + s_0 + s + \sigma \log \det(W^{-1} - s_0 A_0 - s A_{ij}),$$

and the gradients with respect to s_0 and s are

$$\nabla_{s_0} f(y + s_0 e_0 + s e_{ij}; \sigma) = 1 - \sigma \langle A_0, (W^{-1} - s_0 A_0 - s A_{ij})^{-1} \rangle$$
$$\nabla_s f(y + s_0 e_0 + s e_{ij}; \sigma) = 1 - \sigma \langle A_{ij}, (W^{-1} - s_0 A_0 - s A_{ij})^{-1} \rangle .$$

The matrix $s_0 A_0 + s A_{ij}$ is of rank two; replacing $(W^{-1} - s_0 A_0 - s A_{ij})^{-1}$ by the Woodbury formula and setting the gradients to zero, we obtain a system of two quadratic equations. Using these ideas, a slightly different version of Algorithm CD is obtained by changing Steps 5 and 6 adequately, which we call Algorithm CD2D.

4   Experiments

For our experiments, we generate random instances in the same way as proposed in [2]: the objective matrix is $\hat{Q} = \sum_{i=1}^n \mu_i v_i v_i^\top$, where the n numbers μ_i are chosen as follows: for a given value of $p \in [0, 100]$, the first $pn/100$ μ_i's are generated uniformly from $[-1, 0]$ and the remaining ones from $[0, 1]$. Additionally, we generate n vectors of dimension n, with entries uniformly at random from $[-1, 1]$, and orthonormalize them to obtain the vectors v_i. The parameter p represents the percentage of negative eigenvalues, so that $\hat{Q}$ is positive semidefinite for $p = 0$, negative semidefinite for $p = 100$ and indefinite for any other value $p \in (0, 100)$. The entries of the vector $\hat{l}$ are generated uniformly at random from $[-1, 1]$, and $\hat{c} = 0$. In this paper, we restrict our evaluation to ternary instances, i.e., instances with $D_i = \{-1, 0, 1\}$.

We evaluate the performance of both Algorithms CD and CD2D in the root node of the branch-and-bound tree and compare them with CSDP, the SDP solver used in [2]. Our experiments were performed on an Intel Xeon processor running at 2.5 GHz. Algorithms CD and CD2D were implemented in C++, using routines from the LAPACK package only in the initial phase for computing a starting point and the inverse matrix $W^{(0)}$.

The main motivation to consider a fast coordinate ascent method was to obtain quick and good lower bounds for the quadratic integer problem (1). We are thus interested in the improvement of the lower bound over time. In Fig. 2, we plotted the lower bounds obtained by CSDP and by the Algorithms CD and CD2D in the root node for two ternary instances of size $n = 100$, for the two values $p = 0$ and $p = 100$. Notice that we use a log scale for the y-axis.

From Fig. 2, we see that Algorithm CD2D clearly dominates both other approaches: the lower bound it produces exceeds the other bounds until all approaches come close to the optimum of (2). This is true in particular for the instance with $p = 100$. Even Algorithm CD is stronger than CSDP in the beginning, but then CSDP takes over. Note that the computation of the root bound for the instance shown in Fig. 2(a) involves one re-optimization due to separation. For this reason, the lower bound given by CSDP has to restart with a very weak value.

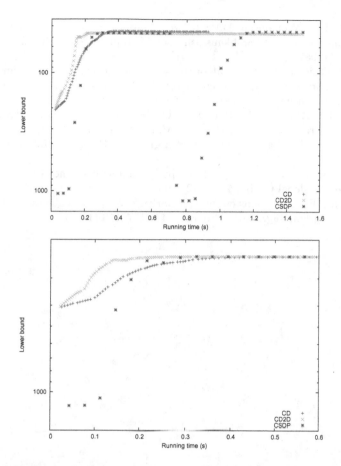

Fig. 2. Comparison of the lower bounds in the root node obtained by Q-MIST with CSDP, CD and CD2D; for $p = 0$ (top) and $p = 100$ (bottom)

As a next step, we will integrate the Algorithm CD2D into the branch-and-bound framework of Q-MIST. We are confident that this will improve the running times of Q-MIST significantly when choosing the stopping criteria carefully. This is left as future work.

References

1. Borchers, B.: CSDP, a C library for semidefinite programming. Optim. Methods Softw. **11**(1–4), 613–623 (1999)
2. Buchheim, C., Wiegele, A.: Semidefinite relaxations for non-convex quadratic mixed-integer programming. Math. Program. **141**(1–2), 435–452 (2013)
3. Dong, H.: Relaxing nonconvex quadratic functions by multiple adaptive diagonal perturbations. SIAM J. Optim. (accepted for publication)

4. Goemans, M.X., Williamson, D.P.: Improved approximation algorithms for maximum cut and satisfiability problems using semidefinite programming. J. ACM **42**(6), 1115–1145 (1995)
5. Kapoor, S., Vaidya, P.M.: Fast algorithms for convex quadratic programming and multicommodity flows. In: Proceedings of the 18th Annual ACM Symposium on Theory of Computing, pp. 147–159 (1986)
6. Kozlov, M.K., Tarasov, S.P., Hačijan, L.G.: The polynomial solvability of convex quadratic programming. USSR Comput. Math. Math. Phys. **20**(5), 223–228 (1980)
7. Lovász, L.: On the Shannon capacity of a graph. IEEE Trans. Inf. Theory **25**(1), 1–7 (1979)
8. Pardalos, P.M., Vavasis, S.A.: Quadratic programming with one negative eigenvalue is NP-hard. J. Global Optim. **1**, 15–22 (1991)
9. Ye, Y., Tse, E.: An extension of Karmarkar's projective algorithm for convex quadratic programming. Math. Program. **44**, 157–179 (1989)

MIP Formulations for a Rich Real-World Lot-Sizing Problem with Setup Carryover

Filippo Focacci[1], Fabio Furini[2(✉)], Virginie Gabrel[2], Daniel Godard[1], and Xueying Shen[1,2]

[1] DecisionBrain, 18 rue Yves Toudic, 75010 Paris, France
{ffocaci,daniel.godard,sylvia.shen}@decisionbrain.com
[2] PSL, CNRS, LAMSADE UMR 7243, Université Paris-Dauphine, 75775 Paris Cedex 16, France
{fabio.furini,virginie.gabrel}@dauphine.fr

Abstract. A rich lot-sizing problem is studied in this manuscript which comes from a real-world application. Our new lot-sizing problem combines several features, i.e., parallel machines, production time windows, backlogging, lost sale and setup carryover. Three mixed integer programming formulations are proposed. We theoretically and computationally compare these different formulations, testing them on real-world and randomly generated instances. Our study is the first step for efficiently tackling and solving this challenging real-world lot-sizing problem.

Keywords: Lot-sizing · Setup carryover · Mixed integer programming · Computational tests

1 Introduction

The *Lot-sizing Problem* (LSP) aims to plan the production in order to satisfy customer demands and to minimize operational costs. A number of different LSP variants have been studied in the literature developing *Mixed Integer Programming* (MIP) formulations. We refer the interested readers to [11] for a complete survey on the topic. In this manuscript, we study a rich real-world LSP variant presented in detail in the following.

Case Study. The problem is the core of a consulting project with an apparel company developed by DecisionBrain. The company runs multiple plants over Asia and provides services for many international brands. The production has to be planned in order to satisfy customer demands given limited resources. To produce one piece of clothes, a sequence of steps are required such as cutting, embroidery, sewing and washing. We concentrate on the bottleneck of the production process, i.e., the sewing process. This is the only step executed by workers instead of machines. A product can be executed by any production line consisting of a team of workers. When a production line switches production

© Springer International Publishing Switzerland 2016
R. Cerulli et al. (Eds.): ISCO 2016, LNCS 9849, pp. 123–134, 2016.
DOI: 10.1007/978-3-319-45587-7_11

from one product to another, a setup adjustment occurs which consumes capacities and generates costs. In this problem, each demand consists of a quantity of one product to be produced between its release date and due date. The release date is due to material availability and the production for this demand can only start from this date. Before the due date, the demand can be satisfied at any time without extra cost. After the due date, the demand can still be satisfied causing an additional tardiness cost. Two levels of tardiness costs are considered: the first is due to a transport cost (an air transport instead of a maritime one necessary to meet the due date), while the second is due to a discount offered to the customer for compensation. Moreover, the unsatisfaction of demands is also allowed with a very high penalty cost. The goal of the company is to plan its production in order to minimize the cost. Considering that the production lines are nearly fully loaded, an efficient production planning is crucial to increase company competitiveness. This project focuses on medium-term planning decisions, in which the planning horizon is divided into a set of consecutive time buckets, such as weeks. The medium-term planning decides the production quantities in each time bucket for each production line. This problem can be modeled as a capacitated lot-sizing problem with setup carryover, parallel machines, production time windows, backlogging and lost sale.

Main Problem Features. Any product can be produced on any production line, called machine for brevity in the following. The assigned machine must be tuned at a state according to the product, which is called a *setup state* for the product. When the machine switches from one setup state to another within a time bucket, a setup occurs which causes a setup capacity and a setup cost. The last setup state of a time bucket is carried over to the beginning of the next time bucket, which is known as setup carryover [4]. Therefore, if the last setup state of $t-1$ is for product i, then producing i at the beginning of t does not cause any setup cost or setup capacity. Setting up the machines for the production of a product is only allowed once for each time bucket. The LSP with setup carryover has been first studied in [4], in which a MIP model has been proposed and a fix-and-relax heuristic algorithm has been developed. Different MIP formulations have been also proposed, see [6–8,13,14].

Another major difficulties of our problem comes from *parallel machines*. Based on the application, we consider uniform parallel machines, in which the machines have independent capacities and the consumed capacity for each product only depends on the product itself. The introduction of parallel machines may lead to a large amount of symmetric solutions, therefore it increases the difficulties of the problem. In our application, there are up to 29 parallel machines with different capacities. In [15] two branch and bound algorithms are developed to solve the LSP with parallel machines. Also the Lagrangian relaxation is a widely used technique to address this problem [5,16].

In the LSP with *Production Time Window* (PTW), the production for each demand can only happen between its release time and due date. The release time represents several situations such as raw material available dates. The LSP with

PTW has been first addressed by [3]. Polyhedral studies and exact algorithms have been developed in [2,3,9,17].

Backlogging and *Lost sale* (BL) (see [12,18]) are also considered in our problem, which means that the demands can be satisfied later than their due dates or even not satisfied incurring in a penalty cost. These are quite common features in real-world applications when the production capacity is insufficient to satisfy all the demands. The LSP with PTW and BL has been studied in [1], for the case of single product and an uncapacitated machine.

Problem Definition. The input parameters of our problem are:

- $\mathcal{T} = \{1, 2, \ldots, T\}$: set of time buckets.
- $\mathcal{M} = \{1, 2, \ldots, M\}$: set of machines.
- $\mathcal{N} = \{1, 2, \ldots, N\}$: set of products.
- $\mathcal{D} = \{1, 2, \ldots, D\}$: set of demands.
- cap_{rt}: capacity of machine r in time bucket t ($r \in \mathcal{M}$, $t \in \mathcal{T}$).
- cap_i: capacity required by unitary production of product i ($i \in \mathcal{N}$).
- st_{ir}: setup capacity for product i on machine r ($i \in \mathcal{N}$, $r \in \mathcal{M}$).
- sc_{ir}: setup cost for product i on machine r ($i \in \mathcal{N}$, $r \in \mathcal{M}$).
- $p_d \in \mathcal{N}$: the required product of the demand d ($d \in \mathcal{D}$).
- q_d: quantity of product p_d required by demand d ($d \in \mathcal{D}$).
- b_d: release date of demand d ($d \in \mathcal{D}$).
- e_d^1: first due date of demand d ($d \in \mathcal{D}$). No extra cost in interval $[b_d, e_d^1)$.
- e_d^2: second due date of demand d ($d \in \mathcal{D}$).
- tc_d^1: unitary extra cost for demand d satisfied at or after e_d^1 ($d \in \mathcal{D}$).
- tc_d^2: unitary extra cost for demand d satisfied at or after e_d^2 ($d \in \mathcal{D}$).
- lc_d: unitary cost for unsatisfied demand d ($d \in \mathcal{D}$, $lc_d > tc_d^1 + tc_d^2$).

The problem is to decide for each machine $r \in \mathcal{M}$ and for each time bucket $t \in \mathcal{T}$, how much to produce of each product $i \in \mathcal{N}$. The objective is to minimize the total cost including lost sale cost, tardiness cost and setup cost without exceeding the machine capacities cap_{rt} for each machine r and time bucket t ($r \in \mathcal{M}$, $t \in \mathcal{T}$). Note that the production is used to satisfy demands directly so there are no inventory costs.

To the best of our knowledge, it is the first time that such a rich LSP is studied combining many of the important real-world features.

2 MIP Formulations

Different formulations concerning setup carryover have been proposed in the literature. In this section, we extend 3 MIP formulations for the LSP defined above. These formulations mainly differ in the way the setup carryover is modelled. They all share a common part concerning the Lost Sale costs, the Tardiness cost and material flow conservation constraints.

For each product $i \in \mathcal{N}$, each machine $r \in \mathcal{M}$, each time bucket $t \in \mathcal{T}$ and each demand $d \in \mathcal{D}$, we introduce the following decision variables:

– $x_{irt} \in \mathbb{R}^+$: the production quantity of product i on machine r during time t.
– $y_{dt} \in [0, q_d]$: the satisfied quantity of demand d in time bucket $t \geq b_d$.
– $y_d \in [0, q_d]$: the unsatisfied quantity of demand d.

The lost Lost Sale and Tardiness cost are defined as follows:

$$LostSaleCost = \sum_{d \in \mathcal{D}} lc_d y_d ,$$ (1)

$$TardinessCost = \sum_{d \in \mathcal{D}, t \in \mathcal{T}: t \geq e_d^1} tc_d^1 y_{dt} + \sum_{d \in \mathcal{D}, t \in \mathcal{T}: t \geq e_d^2} tc_d^2 y_{dt} ,$$ (2)

The constraints modelling the demand satisfaction can be written as follows:

$$\sum_{r \in \mathcal{M}} x_{irt} = \sum_{d \in \mathcal{D}: p_d = i} y_{dt} \qquad\qquad i \in \mathcal{N}, t \in \mathcal{T}$$ (3)

$$\sum_{t \in \mathcal{T}} y_{dt} + y_d = q_d \qquad\qquad d \in \mathcal{D}$$ (4)

$$0 \leq x_{irt} \qquad\qquad i \in \mathcal{N}, r \in \mathcal{M}, t \in \mathcal{T}$$ (5)

$$0 \leq y_{dt}, y_d \leq q_d \qquad\qquad d \in \mathcal{D}, t \geq b_d$$ (6)

The material flow conservation is formulated as (3). Constraints (4) guarantee that for each demand, the summation of the satisfied amount and the unsatisfied amount equals to the required quantity. Finally, due to the release date, y_{dt} is only defined from the release time b_d for each demand d in (6).

If no setup is considered, the problem is a linear programming problem, which can be solved in polynomial time. Therefore, the complexity mainly comes from the setup. The machine capacity is both consumed by production capacity $ProdCap(r, t)$ and setup capacity. The capacity consumption due to production can be expressed as follows:

$$ProdCap(r, t) = \sum_{i \in \mathcal{N}} cap_i x_{irt} \quad r \in \mathcal{M}, t \in \mathcal{T} .$$ (7)

In the next sections, we present 3 different ways to model the setup cost and setup capacity consumption.

2.1   Formulation 1

In Haase [8], a MIP formulation for the LSP on a single machine with setup carryover has been introduced. We adapt this formulation to our problem and introduce setup variables for each product $i \in \mathcal{N}$, each machine $r \in \mathcal{M}$ and each time bucket $t \in \mathcal{T}$ as follows:

– $v_{rt} \in [0, 1]$, $v_{rt} > 0$ indicates if more than one product is produced in time bucket t and for machine r.
– $z_{irt} \in \{0, 1\}$ equals to 1 if a setup state for product i on machine r exists in time bucket t and 0 otherwise.

- $z_{irt}^c \in \{0,1\}$ equals to 1 if the setup state for product i is carried over from time bucket $t-1$ to time bucket t on machine r and 0 otherwise.

Then the first formulation $(Form1)$ is formally given as follows $(\tilde{\mathcal{T}} = \mathcal{T} \setminus \{1\})$:

$$\min \quad (1) + (2) + \sum_{i \in \mathcal{N}, r \in \mathcal{M}, t \in \mathcal{T}} sc_{ir}(z_{irt} - z_{irt}^c) \tag{8}$$

$$s.t. \quad (3) - (6)$$

$$ProdCap(r,t) + \sum_{i \in \mathcal{N}} st_{ir}(z_{irt} - z_{irt}^c) \leq cap_{rt} \qquad r \in \mathcal{M}, t \in \mathcal{T} \tag{9}$$

$$x_{irt} \leq \Theta_{irt} z_{irt} \qquad i \in \mathcal{N}, r \in \mathcal{M}, t \in \mathcal{T} \tag{10}$$

$$\sum_{i \in \mathcal{N}} z_{irt}^c \leq 1 \qquad r \in \mathcal{M}, t \in \mathcal{T} \tag{11}$$

$$z_{irt}^c \leq z_{ir,t-1} \qquad i \in \mathcal{N}, r \in \mathcal{M}, t \in \tilde{\mathcal{T}} \tag{12}$$

$$z_{irt}^c \leq z_{irt} \qquad i \in \mathcal{N}, r \in \mathcal{M}, t \in \mathcal{T} \tag{13}$$

$$z_{irt}^c + z_{ir,t-1}^c + v_{r,t-1} \leq 2 \qquad i \in \mathcal{N}, r \in \mathcal{M}, t \in \tilde{\mathcal{T}} \tag{14}$$

$$N v_{rt} - \sum_{i \in \mathcal{N}} z_{irt} + 1 \geq 0 \qquad r \in \mathcal{M}, t \in \mathcal{T} \tag{15}$$

$$z_{irt}^c, z_{irt} \in \{0,1\} \qquad i \in \mathcal{N}, r \in \mathcal{M}, t \in \mathcal{T} \tag{16}$$

$$v_{rt} \in [0,1] \qquad r \in \mathcal{M}, t \in \mathcal{T} \tag{17}$$

where Θ_{irt} is a large enough constant that never unnecessarily limits the production x_{irt}.

According to the definition of setup carryover, a setup cost has to be paid when there is a setup ($z_{irt} = 1$) which is not carried over from the last time bucket ($z_{irt}^c = 0$). Constraints (9) ensure that the capacity is not exceeded on each machine in each time bucket, where the setup capacity consumption is formulated similarly to the setup cost. Constraints (10) link the production and the setup since a positive production of i on r at t requires a setup state for i on r at t. For the setup carryover, there is at most one setup state to be carried over to the next time bucket, which is guaranteed by constraints (11). A setup state of i on r carried over from $t-1$ to t implies that this state is included in both $t-1$ (12) and t (13). If there is more than one setup state in one time bucket, i.e., $v_{rt} > 0$, the initial setup state and the last setup state are necessarily different. This is formulated as constraints (14). Finally, to fulfill the definition of variable v_{rt}, we have the constraints (15) ($N = |\mathcal{N}|$, i.e., the number of different products).

2.2   Formulation 2

In Sox et al. [13], several MIP formulations for the LSP on single machine with setup carryover are presented. We adapt one of them to our problem introducing the following setup variables for each product $i \in \mathcal{N}$, each machine $r \in \mathcal{M}$ and each time bucket $t \in \mathcal{T}$:

– $z_{irt}^0 \in \{0,1\}$ equals to 1 if the initial setup state is for product i on machine r in time bucket t, implying that the final setup state for $t-1$ on r is for product i.
– $z_{irt}^+ \in \{0,1\}$ equals to 1 if there is a state switch for product i on machine r in time bucket t.

Then the second formulation ($Form2$) is formally given as follows:

$$\min \quad (1) + (2) + \sum_{i \in \mathcal{N}, r \in \mathcal{M}, t \in \mathcal{T}} sc_{ir} z_{irt}^+ \tag{18}$$

$$s.t. \quad (3) - (6)$$

$$ProdCap(r,t) + \sum_{i \in \mathcal{N}} st_{ir} z_{irt}^+ \leq cap_{rt} \qquad\qquad r \in \mathcal{M}, t \in \mathcal{T} \tag{19}$$

$$x_{irt} \leq \Theta_{irt}(z_{irt}^0 + z_{irt}^+) \qquad\qquad i \in \mathcal{N}, r \in \mathcal{M}, t \in \mathcal{T} \tag{20}$$

$$\sum_{i \in \mathcal{N}} z_{irt}^0 = 1 \qquad\qquad r \in \mathcal{M}, t \in \mathcal{T} \tag{21}$$

$$z_{irt}^0 \leq z_{ir,t-1}^0 + z_{ir,t-1}^+ \qquad\qquad i \in \mathcal{N}, r \in \mathcal{M}, t \in \tilde{\mathcal{T}} \tag{22}$$

$$z_{jr,t-1}^+ \leq 2 - z_{ir,t-1}^0 - z_{irt}^0 \qquad\qquad i, j \neq i \in \mathcal{N}, r \in \mathcal{M}, t \in \tilde{\mathcal{T}} \tag{23}$$

$$z_{irt}^0, z_{irt}^+ \in \{0,1\} \qquad\qquad i \in \mathcal{N}, r \in \mathcal{M}, t \in \mathcal{T} \tag{24}$$

The total setup cost is formulated as $\sum_{i \in \mathcal{N}, r \in \mathcal{M}, t \in \mathcal{T}} sc_{ir} z_{irt}^+$ since the setup cost has to be paid only when there is a setup switch ($z_{irt}^+ = 1$). Constraints (19) ensure that the total used capacity does not exceed the available capacity, where the setup capacity consumption is formulated similarly to the setup cost. Constraints (20) link the setup and production since a positive production of i on r at t requires a setup state for i on r at t, which is either from an initial setup state or a setup switch. There is a unique initial setup state for each time bucket on each machine, which is established by constraints (21). Also, the initial setup state must be one of the setup states in the previous time bucket (22). However, constraints (23) ensure that, on machine r during time bucket t, no setup switch is possible when the initial setup state and the last setup state of t (i.e., the initial setup state of the next time bucket $t+1$) are both for the same product.

2.3   Formulation 3

In Suerie et al. [14], a MIP formulation for the LSP on single or multi-level with setup carryover is presented. This formulation is similar to the $Form2$. In addition to the previously defined binary variables $z_{irt}^0 \in \{0,1\}$ and $z_{irt}^+ \in \{0,1\}$ for each product $i \in \mathcal{N}$, for each machine $r \in \mathcal{M}$, each time bucket $t \in \mathcal{T}$, additional variables $w_{rt} \in \{0,1\}$ are introduced. w_{rt} equals to 1 if only one product is produced on r in t, and 0 otherwise. Then the third formulation ($Form3$) is formally given as follows:

$$\min \quad (1) + (2) + \sum_{i \in \mathcal{N}, r \in \mathcal{M}, t \in \mathcal{T}} sc_{ir} z^+_{irt} \tag{25}$$

$$s.t. \quad (3) - (6), (19) - (22), (24)$$

$$z^0_{irt} + z^0_{ir,t-1} \le 1 + w_{r,t-1} \qquad i \in \mathcal{N}, r \in \mathcal{M}, t \in \tilde{\mathcal{T}} \tag{26}$$

$$z^+_{irt} + w_{rt} \le 1 \qquad i \in \mathcal{N}, r \in \mathcal{M}, t \in \mathcal{T} \tag{27}$$

$$0 \le w_{rt} \le 1 \qquad \forall r \in \mathcal{M}, t \in \mathcal{T} \tag{28}$$

Constraints (26) implies that on machine r, the initial setup states at $t - 1$ and t have to be different when more than one product is produced during $t - 1$ ($w_{r,t-1} = 0$). Constraints (27) ensures that $w_{rt} = 0$ when there are more than one setup state during time bucket t. Thanks to the other constraints of the formulation, it is not necessary to impose integrality contraints for the w_{rt} variables. Due to constraints (21) and (22), we have $z^0_{ir,t+1} = z^0_{irt} = 1$, which forces $w_{rt} = 1$.

3 Theoretical Comparison

In this section, we compare the previously introduced MIP formulations strength in terms of lower bounds given by the Linear Programming (LP) relaxation.

Let $(\tilde{x}, \tilde{y}, \tilde{z}^c, \tilde{z}, \tilde{v})$, $(\bar{x}, \bar{y}, \bar{z}^0, \bar{z}^+)$ and $(\dot{x}, \dot{y}, \dot{z}^0, \dot{z}^+, \dot{w})$ be optimal solutions of the LP relaxation of $Form1$, $Form2$ and $Form3$ respectively, while $f_1(\tilde{x}, \tilde{y}, \tilde{z}^c, \tilde{z}, \tilde{v})$, $f_2(\bar{x}, \bar{y}, \bar{z}^0, \bar{z}^+)$ and $f_3(\dot{x}, \dot{y}, \dot{z}^0, \dot{z}^+, \dot{w})$ are the corresponding optimal objective function values. Then the following theorem holds:

Theorem 1

$$f_2(\bar{x}, \bar{y}, \bar{z}^0, \bar{z}^+) = f_3(\dot{x}, \dot{y}, \dot{z}^0, \dot{z}^+, \dot{w}) \ge f_1(\tilde{x}, \tilde{y}, \tilde{z}^c, \tilde{z}, \tilde{v}) \ .$$

This theorem shows that the LP relaxation of $Form2$ and $Form3$ provide equivalently better lower bounds than $Form1$. The proof of the theorem requires a number of technicalities and for reason of space we decided to omit the proof.

A comparison of the formulation size is summarized in the Table 1. They all have the same number of binary variables, while $Form2$ has less continuous variables than the other two. The number of constraints increases in the order $Form3$, $Form1$ and $Form2$.

Table 1. Formulation size comparison

Form	Number of variables	# Binary	Number of constraints
Form1	$3NMT + MT + DT + D$	$2NMT$	$NT + MT + 3NMT - 2NM + \Pi$
Form2	$3NMT + DT + D$	$2NMT$	$MT + NM(T-1) + \frac{N(N-1)}{2}M(T-1) + \Pi$
Form3	$3NMT + MT + DT + D$	$2NMT$	$MT + 2NMT - NM + \Pi$

$\Pi := NMT + NT + MT + D$

4   Computational Comparison

In this section, we test the performances of the three MIP formulations, by comparing the computational difficulties to find optimal solutions using CPLEX 12.6 as a MIP solver.

The testbed consists of 3 real-world instances and 810 randomly generated instances. The 3 real-world instances have the following features: $T = 27, 36, 30$, $M = 3, 28, 29$, $N = 3, 18, 1$, $D = 313, 1188, 595$, and $R = 0.99, 0.14, 0.33$, where R is a indicator representing the capacity usage defined as $R = \frac{\sum_{d \in \mathcal{D}} cap_{p_d} q_d}{\sum_{r \in \mathcal{M}, t \in \mathcal{T}} cap_{rt}}$. Although R does not consider the setup capacity, it is an indicator of the machine loads. For the randomly generated instances, we consider the following parameters: $T \in \{4, 9, 13\}$, $M \in \{1, 5, 10\}$, $N \in \{4, 8, 12\}$, $D \in \{50, 100, 200\}$ and $R \approx \{0.75, 0.9\}$. Assuming that one time bucket corresponds to one week, we choose the number of time buckets from one month ($T = 4$) to a season ($T = 13$). The other parameters are chosen to generate instances with different levels of difficulty. For each possible combination of the parameters, we generate five different instances to limit bias. All the experiments run on one core of an Intel Core i7-4790 2.50 GHz 3.60, with 16 GB shared memory, under the Linux Ubuntu 12.4 operating system.

In the Table 2, we present the computational results using CPLEX to solve the MIP and LP models on the benchmark instances setting a time limit of 10 min. In the table, the computing time is expressed in seconds. The results are presented for the randomly generated instances in the upper part of the table and for the 3 real-world instances in the bottom part. For the randomly generated instances, we give the average results for each value of the parameter T, M, D, N and R. Averages values over all the randomly generated instances are reported in the Row TOT/AVG. The Column #Opt reports the total number of optimally solved instances for each formulation. In the first two columns, we present the parameters and their values. In Column LPGap, we measure the quality of the LP relaxation by showing the average LPGap, which is calculated as $\frac{BestMip - LPVal}{BestMip}$, where $bestMip$ is the best known MIP solution for the instance and $LPVal$ is the optimal LP relaxation objective value. On all instances except two, the three formulations are characterized by the same $LPVal$. In the remaining two cases the difference is very small, thus we only report the LPGap once in the table. The value of Θ_{irt} has been set to $\Theta_{irt} = \min\left\{\frac{cap_{rt}}{cap_i}, \sum_{d:p_d=i, t \geq b_d} q_d\right\}$. In Column #Opt and MipTime, we report the number of instances solved to prove optimality within the time limit of 10 min and the average computing time. In Column #Nodes and Gap, we report the number of explored nodes and, for the instances not solved to optimality within the time limit, we report the exit gap. This gap represents the relative difference between the primal and the dual bounds computed by the CPLEX at the time limit. For the real instances, we report the objective function values returned by the solver in Column MipObj.

As far as the computing time necessary to calculate the LP relaxation is concerned, the average values over all the instances are 0.37 s, 0.74 s and 0.30 s for $Form1$, $Form2$ and $Form3$ respectively. Thanks to the shortest average

Table 2. Formulation comparison

	LPGap	Form1				Form2				Form3			
		#Opt	MipTime	#Nodes	ExitGap	#Opt	MipTime	#Nodes	ExitGap	#Opt	MipTime	#Nodes	ExitGap
T 4	30.6	165	32.5	85970	11.8	175	35.2	49088	11.9	178	36.2	55347	11.4
9	49.8	123	68.7	41168	29.1	128	61.2	26469	29.0	131	65.3	31207	29.0
13	58.4	91	112.3	35839	38.9	108	102.3	20961	36.5	107	96.6	25667	38.3
M 1	65.3	212	48.0	18748	12.7	217	55.1	12988	12.6	218	53.4	13563	14.2
5	47.1	97	67.5	50892	35.9	101	45.7	30784	36.2	105	58.3	37983	35.9
10	26.4	70	104.4	93338	27.2	93	91.1	52746	25.3	93	81.4	60674	26.4
D 50	46.3	151	66.0	63995	26.3	156	53.4	35211	26.7	159	57.8	43104	25.9
100	47.0	126	56.5	47988	30.1	135	59.5	32839	29.6	137	56.6	37585	30.0
200	45.5	102	68.1	50994	29.9	120	72.4	28469	27.5	120	70.0	31532	29.8
N 4	35.0	231	46.5	58790	11.5	258	47.0	25497	9.7	258	43.8	22985	13.3
8	48.5	87	94.8	64440	25.3	94	86.7	46402	25.4	94	85.7	55293	24.9
12	55.3	61	82.7	39746	33.0	59	80.9	24619	33.6	64	93.4	33943	32.9
R 0.75	48.3	211	59.5	41299	31.7	226	50.2	23728	30.3	230	51.9	29214	31.5
0.90	44.2	168	68.3	67353	26.7	185	74.1	40617	26.0	186	72.1	45600	26.6
TOT/AVG	46.3	379	63.4	54326	28.9	411	60.9	32173	27.9	416	60.9	37407	28.7

Instance	LPGap	MipObj	MipTime	#Nodes	ExitGap	MipObj	MipTime	#Nodes	ExitGap	MipObj	MipTime	#Nodes	ExitGap
R1	14.2	6.55e5	12.2	1827	0.0	6.55e5	2.6	434	0.0	6.55e5	3.1	404	0.0
R2	94.6	2.24e6	TL	2	97.4	2.96e7	TL	0	99.8	1.09e6	TL	452	94.6
R3	0.0	1.07e4	0.1	0	0.0	1.07e4	0.0	0	0.0	1.07e4	0.0	0	0.0

computing time for the LP relaxation, $Form3$ is able to explore more nodes within the given time limit. As expected, according to the Column #Opt, the table shows that as the problem size increases, the instances become more difficult to solve. The parameter N, which has an impact on the number of binary variables, affects the most the solvability of the instances. Take for example $Form2$, when N increases from 4 to 12, the number of instances decreases from 258 to 59, whereas when D increases from 50 to 200, the number is only reduced by 36. The number of time buckets T has a smaller impact on the computing time. We can observe that $Form1$ explores a higher number of nodes, this is probably due to the fact that it struggles to find good quality integer solutions. On the other hand, $Form2$ and $Form3$ explore almost the same number of nodes for the randomly generated instances, while the $Form3$ explores more nodes for the 3 real-world instances. Regarding the number of randomly generated instances solved to be proven optimality, $Form3$ solves 416 instances and $Form2$ solves 411 instances, while $Form1$ solves only 379. A similar behavior can be also observed for the exit gap. For the real-world instances, we observe similar results, i.e., the $Form3$ shows the best performance. Hence, according to the computational experiments, $Form3$ shows the best overall computational performance. This is due to the fact that it has the least number of constraints with the same number of binary variables compared to the other two formulations, and its LP relaxation can be solved faster.

We also compare the performances of the three formulations using the Performance Profile [10] in Fig. 1. Let $t_{p,s}$, $o_{p,s}$ be the computing time and the objective function value of instance $p \in P$, where P is the set of instances, given by formulation $s \in S = \{Form1, Form2, Form3\}$. Let b_p equal to $\min_{s \in S}\{o_{p,s}\}$. The performance ratio $\rho_s(\tau)$ of a given $\tau \geq 1$ is defined as follows:

$$\rho_s(\tau) = \frac{1}{|P|} \sum_{p \in P, r_{p,s} \leq \tau} 1 \,, \quad r_{p,s} = \begin{cases} \frac{t_{p,s}}{\min_{s \in S}\{t_{p,s}\}} & \text{if } |\frac{o_{p,s} - b_p}{b_p}| \leq \delta \,, \\ \rho_M & \text{otherwise} \,, \end{cases}$$

where ρ_M is a large enough number such that $\rho_M > r_{p,s}$ for $p \in P$ and $s \in S$. When $\delta = 0$, the ratio $\rho_s(\tau)$ represents the percentage of instances for which a given formulation s returns the best known solutions given time $\tau \min_{s \in S}\{t_{p,s}$ for each instance. Due to the numerical precision of CPLEX, we set $\delta = 10^{-6}$. In the Fig. 1, the horizontal axis represents the "Time Factor τ" using the logarithmic scale. The vertical axis represents the performance ratio $\rho_s(\tau)$. When $\tau = 1$, the performance ratio gives the percentage of instances that are solved fastest and best by each formulation. By "best" we mean computing the best known objective function value. We observe that $Form3$ obtains the best known solution in shortest computing time for approximately 42 % instances, 36 % for $Form2$ and 30 % for $Form1$. Moreover, when τ increase, the performance ratio tends to the percentage of instances that are solved best by each formulation within the time limit. $Form3$ obtains the best solutions for approximately 78 % instances. Therefore, we can conclude that $Form3$ gives the overall best performance.

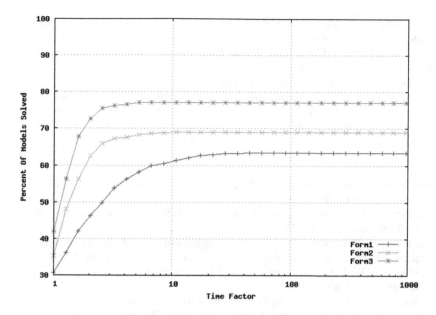

Fig. 1. Performance profile

5 Conclusion

We study a new variant of the LSP, which is based on a real-world application. The problem combines, for the first time, several classical features of the LSP such as setup carryover and PTW. We present and compare three different MIP formulations of the problem. We prove that one of the formulations is weaker since it may provide worse LP relaxation bounds. A set of instances are randomly generated and extensive computational experiments are conducted to compare these formulations. The results show that one of the formulation gives the overall best performance on both real-world instances and randomly generated instances. A library of instances is available online, and we hope that this can stimulate further research on this very challenging rich real-world LSP (http://decisionbrain.com/ISCO2016).

References

1. Absi, N., Kedad-Sidhoum, S., Dauzère-Pérès, S.: Uncapacitated lot-sizing problem with production time windows, early productions, backlogs and lost sales. Int. J. Prod. Res. **49**(9), 2551–2566 (2011)
2. Brahimi, N., Dauzère-Pérès, S., Wolsey, L.A.: Polyhedral and lagrangian approaches for lot sizing with production time windows and setup times. Comput. Oper. Res. **37**(1), 182–188 (2010)
3. Dauzère-Pérès, S., Brahimi, N., Najid, N., Nordli, A.: The single-item lot sizing problem with time windows. Technical report 02/4/AUTO, École des Mines de Nantes, France (2002)

4. Dillenberger, C., Escudero, L.F., Wollensak, A., Zhang, W.: On solving a large-scale resource allocation problem in production planning. In: Fandel, G., Gulledge, T., Jones, A. (eds.) Operations Research in Production Planning and Control, pp. 105–119. Springer, Heidelberg, Berlin (1993)
5. Fiorotto, D.J., Araujo, S.A.D.: Reformulation and a Lagrangian heuristic for lot sizing problem on parallel machines. Ann. Oper. Res. **217**(1), 213–231 (2014)
6. Gopalakrishnan, M.: A modified framework for modelling set-up carryover in the capacitated lotsizing problem. Int. J. Prod. Res. **38**(14), 3421–3424 (2000)
7. Gören, H.G., Tunal, S.: Solving the capacitated lot sizing problem with setup carryover using a new sequential hybrid approach. Appl. Intell. **42**(4), 805–816 (2015)
8. Haase, K.: Lotsizing and Scheduling for Production Planning. LNEM, vol. 408. Springer, Berlin (1994)
9. Hwang, H.: Dynamic lot-sizing model with production time windows. Nav. Res. Log. **54**(6), 692–701 (2007)
10. Mittelmann, H.D., Pruessner, A.: A server for automated performance analysis of benchmarking data. Optim. Method Softw. **21**(1), 105–120 (2006)
11. Pochet, Y., Wolsey, L.A.: Production Planning by Mixed Integer Programming. Springer, New York (2006)
12. Sandbothe, R.A., Thompson, G.L.: A forward algorithm for the capacitated lot size model with stockouts. Oper. Res. **38**(3), 474–486 (1990)
13. Sox, C.R., Gao, Y.: The capacitated lot sizing problem with setup carry-over. IIE Trans. **31**(2), 173–181 (1999)
14. Suerie, C., Stadtler, H.: The capacitated lot-sizing problem with linked lot sizes. Manage. Sci. **49**(8), 1039–1054 (2003)
15. Toledo, F.M.B.: Dimensionamento de Lotes em Máquinas Paralelas. Tese (Doutorado), Faculdade de Engenharia Elétrica e Computação, Universidade Estadual de Campinas, Campinas (1998)
16. Toledo, F.M.B., Armentano, V.A.: A lagrangian-based heuristic for the capacitated lot-sizing problem in parallel machines. Eur. J. Oper. Res. **175**(2), 1070–1083 (2006)
17. Wolsey, L.A.: Lot-sizing with production and delivery time windows. Math. Program. **107**(3), 471–489 (2006)
18. Zangwill, W.I.: A backlogging model and a multi-echelon model of a dynamic economic lot size production system - a network approach. Manage. Sci. **15**(9), 506–527 (1969)

Towards an Accurate Solution of Wireless Network Design Problems

Fabio D'Andreagiovanni[1,2,3]($\boxtimes$) and Ambros M. Gleixner[1]

[1] Department of Mathematical Optimization, Zuse Institute Berlin (ZIB),
Takustr. 7, 14195 Berlin, Germany
{d.andreagiovanni,gleixner}@zib.de
[2] DFG Research Center MATHEON, Einstein Center for Mathematics (ECMath),
Straße des 17. Juni 136, 10623 Berlin, Germany
[3] Institute for System Analysis and Computer Science, National Research Council
of Italy (IASI-CNR), via dei Taurini 19, 00185 Rome, Italy

Abstract. The optimal design of wireless networks has been widely studied in the literature and many optimization models have been proposed over the years. However, most models directly include the signal-to-interference ratios representing service coverage conditions. This leads to mixed-integer linear programs with constraint matrices containing tiny coefficients that vary widely in their order of magnitude. These formulations are known to be challenging even for state-of-the-art solvers: the standard numerical precision supported by these solvers is usually not sufficient to reliably guarantee feasible solutions. Service coverage errors are thus commonly present. Though these numerical issues are known and become evident even for small-sized instances, just a very limited number of papers has tried to tackle them, by mainly investigating alternative non-compact formulations in which the sources of numerical instabilities are eliminated. In this work, we explore a new approach by investigating how recent advances in exact solution algorithms for linear and mixed-integer programs over the rational numbers can be applied to analyze and tackle the numerical difficulties arising in wireless network design models.

Keywords: Linear programming · Precise solutions · Network design · Wireless telecommunications systems

1 Introduction

In the last decade, the presence of wireless communications in our everyday life has greatly expanded and wireless networks have thus increased in number,

This work was partly conducted within the Research Campus Modal funded by the German Federal Ministry of Education and Research (BMBF, Grant no. 05M14ZAM). It was also partially supported by the *Einstein Center for Mathematics Berlin* through Project ROUAN (MI4) (ROUAN) and by BMBF through Project VINO (Grant no. 05M13ZAC) and Project *ROBUKOM* (Grant no. 05M10ZAA) [4].

© Springer International Publishing Switzerland 2016
R. Cerulli et al. (Eds.): ISCO 2016, LNCS 9849, pp. 135–147, 2016.
DOI: 10.1007/978-3-319-45587-7_12

size and complexity. In this context, the traditional design approach adopted by professionals, based on trial-and-error supported by simulation, has exhibited many limitations: this approach is not able to pursue an efficient exploitation of scarce and precious radio resources, such as frequencies and channel bandwidth, and the need for exact mathematical optimization approaches has increased.

The problem of designing a wireless network can be essentially described as that of configuring a set of transmitters in order to cover with a telecommunication service a set of receivers, while guaranteeing a minimum quality of service. Over the years, many optimization models have been proposed for designing wireless networks (see [9,12,20] for an introduction). However, most models have opted for so-called *natural formulations*, which directly include the formulas used to assess service coverage conditions. This leads to the definition of mixed-integer programs whose constraint matrices contain tiny coefficients that greatly vary in their order of magnitude. Furthermore, the natural formulations commonly include also the notorious big-M coefficients to represent disjunctive service coverage constraints. These formulations are known to be challenging even for state-of-the-art solvers. Additionally, the standard numerical precision supported by these solvers is usually not sufficient to reliably guarantee feasible solutions [23]. If returned solutions are verified in a post-optimization phase, it is thus common to find service coverage errors.

Though these numerical issues are known and can be found even in the case of instances of small size, it is interesting to note that just a very limited number of papers has tried to tackle them: the majority of these works rely on the definition of alternative non-compact formulations that reduce the numerical drawbacks of natural formulations (see the next section for a review of the main approaches). In contrast to these works, we propose here a new approach: we investigate how recent advances in exact solution algorithms for (integer) linear programs over the rational numbers can be applied to analyze and tackle the numerical difficulties arising in wireless network design.

Our main original contributions are in particular:

1. we present the first formal discussion about why even effective state-of-the-art solvers fail to correctly discriminate between feasible and infeasible solutions in wireless network design;
2. we assess, for the first time in literature, both formally and computationally the actual benefits coming from scaling the very small coefficients involved in natural formulations; coefficient scaling is a practice that is adopted by many professionals and scholars dealing with wireless network design, with the belief of eliminating numerical errors; we show that just adopting scaling is not sufficient to guarantee accurate feasibility of solutions returned by floating-point solvers;
3. we show how extended-precision solvers can be adopted to check the correctness of solutions returned by floating-point solvers and, if errors are present, to get correct valorization of the continuous variables of the problem.

Our computational experiments are made over a set of realistic instances defined in collaboration with a major European telecommunication company.

The remainder of this paper is organized as follows: in Sect. 2, we formally characterize the wireless network design problem and introduce the natural formulations; in Sect. 3, we discuss the question of accuracy in Mixed Integer Programming (MIP) solvers, addressing in particular the issues arising in wireless network design; in Sect. 4, we present our computational experiments over realistic network instances.

2   The Wireless Network Design Problem

For modeling purposes, a wireless network can be described as a set of transmitters T that provide a telecommunication service to a set of receivers R. Transmitters and receivers are characterized by a location and a number of radio-electrical parameters (e.g., power emission and transmission frequency). The *Wireless Network Design Problem* (WND) consists in establishing the location and suitable values for the parameters of the transmitters with the goal of optimizing an objective function that expresses the interest of the decision maker: common objectives are the maximization of a revenue function associated with wireless service coverage or, assuming a green-network perspective, the minimization of the total power emission of the network transmitters. For an exhaustive introduction to the WND, we refer the reader to [9,12,20].

Given a receiver $r \in R$ that we want to cover with service, we must choose a single transmitter $s \in S$, called *server*, that provides the telecommunication service to r. Once the server of a receiver is chosen, all the other transmitters are *interferers* and deteriorate the quality of service obtained by r from its server s.

From an analytical point of view, if we denote by p_t the power emission of a transmitter $t \in T$, a receiver $r \in R$ is considered covered with service (or briefly *served*) when the ratio of the service power to the sum of the interfering powers (*Signal-to-Interference Ratio - SIR*) is above a threshold $\delta > 0$, which depends on the desired quality of service [25]:

$$SIR_{rs}(p) = \frac{a_{rs(r)} \cdot p_{s(r)}}{N + \sum_{t \in T \setminus \{s(r)\}} a_{rt} \cdot p_t} \geqslant \delta. \tag{1}$$

In this inequality: (i) $s(r) \in T$ is the server of receiver r; (ii) the power $P_t(r)$ that r receives from a transmitter $t \in T$ is proportional to the emitted power p_t by a factor $a_{rt} \in [0, 1]$, i.e. $P_t(r) = a_{rt} \cdot p_t$. The factor a_{rt} is called *fading coefficient* and summarizes the reduction in power that a signal experiences while propagating from t to r [25]; (iii) in the denominator, we highlight the presence of the system noise $N > 0$ among the interfering signals.

By simple algebra operations, inequality (1) can be transformed into the following linear inequality, commonly called *SIR inequality*:

$$a_{rs(r)} \cdot p_{s(r)} - \delta \sum_{t \in T \setminus \{s(r)\}} a_{rt} \cdot p_t \geqslant \delta \cdot N. \tag{2}$$

Since service coverage assessment is a central element in the design of any wireless network, the SIR inequality constitutes the core of any optimization problem used in wireless network design. If we just focus attention on setting power emissions, we can define the so-called *Power Assignment Problem* (PAP), in which we want to fix the power emission of each transmitter in order to serve a set of receivers, while minimizing the sum of all power emissions. By introducing a non-negative decision variable $p_t \in [0, P_{\max}]$ to represent the feasible power emission range of a transmitter $t \in T$, the PAP can be easily formulated as the following pure Linear Program (LP):

$$\min \sum_{t \in T} p_t \qquad\qquad\qquad\qquad\qquad\qquad \text{(PAP)}$$

$$a_{rs(r)} \cdot p_{s(r)} - \delta \sum_{t \in T \setminus \{s(r)\}} a_{rt} \cdot p_t \geq \delta \cdot N \qquad\qquad \forall\, r \in R \qquad (3)$$

$$0 \leq p_t \leq P_{\max} \qquad\qquad\qquad\qquad \forall\, t \in T, \qquad (4)$$

where (3) are the SIR inequalities associated with receivers to be served.

In a hierarchy of WND problems (see [9,23] for details), the PAP constitutes a basic WND problem that lies at the core of virtually all more general WND problems. A particularly important generalization of the PAP is constituted by the *Scheduling and Power Assignment Problem (SPAP)* [9,12,22,23], where, besides the power emissions, it is also necessary to choose the assignment of a served receiver to a transmitter in the network that acts as server of the receiver. This can be easily modeled by introducing 0-1 service assignment variables, obtaining the following natural formulation:

$$\max \sum_{r \in R} \sum_{t \in T} \pi_t \cdot x_{rt} \qquad\qquad\qquad\qquad\qquad \text{(SPAP)}$$

$$a_{rs} \cdot p_s - \delta \sum_{t \in T \setminus \{s\}} a_{rt} \cdot p_t + M \cdot (1 - x_{rs}) \geq \delta \cdot N \quad \forall\, r \in R, s \in T \quad (5)$$

$$\sum_{t \in T} x_{rt} \leq 1 \qquad\qquad\qquad\qquad\qquad \forall\, r \in R \qquad (6)$$

$$0 \leq p_t \leq P_{\max} \qquad\qquad\qquad\qquad \forall\, t \in T \qquad (7)$$

$$x_{rt} \in \{0, 1\} \qquad\qquad\qquad\qquad \forall\, r \in R, t \in T , \quad (8)$$

which includes: (i) additional binary variables x_{rt} to represent that receiver r is served by transmitter t; (ii) modified SIR inequalities, defined for each possible server transmitter $s \in T$ of a receiver r, including large constant values $M > 0$ to activate/deactivate the corresponding SIR inequalities (as expressed by the constraint (6) each user may be served by at most one transmitter and thus at most one SIR inequality must be satisfied for each receiver); (iii) a modified objective function aiming at maximizing the revenue obtained from serving transmitters (every receiver grants a revenue $\pi_t > 0$).

Drawbacks of SIR-based Formulations. The natural (mixed-integer) linear programming formulations associated with the PAP and the SPAP and based on the direct inclusion of the SIR inequalities are widely adopted for the WND in different application contexts, such as DVB-T, (e.g., [22,23]), UMTS (e.g., [2]), WiMAX (e.g., [9,12]), and wireless mesh networks and jamming (e.g., [10,13]). In principle, such formulations can be solved by MIP solvers, but, as clearly pointed out in works like [9,12,20,23], in practice:

(1) the fading coefficients may vary in a wide range (e.g., in DVB-T instances, difference between coefficients may exceed 90 dB), leading to very ill-conditioned coefficient matrices that make the solution process *numerically unstable*;

(2) in the case of SPAP-like formulations, the big-M coefficients lead to extremely weak bounds that may greatly decrease the effectiveness of solvers implementing state-of-the-art versions of branch-and-bound techniques;

(3) the resulting coverage plans are often unreliable and may contain errors, i.e. SIR constraints recognized as satisfied by an MIP solver are actually violated.

Though these issues are known, it is interesting to note that just a limited number of works in the wide literature about WND has tried to tackle them and natural formulations are still widely used. We refer the reader to [9,20] for a review of works that have tried to tackle these drawbacks and we recall here some more relevant ones. One of the first works that has identified the presence and effects of numerical issues in WND is [22], where a GRASP algorithm is proposed to solve very large instances of the SPAP, arising in the design of DVB-T networks. Other exact solution approaches have aimed at eliminating the source of numerical instabilities (i.e., the fading and big-M coefficients) by considering non-compact formulations: in [5], a formulation based on cover inequalities is introduced for a maximum link activation problem; in [9,12], it is instead shown how using a *power-indexed* formulation, modeling power emissions by discrete power variables allows to define a peculiar family of generalized upper bound cover inequalities that provide (strong) formulations. In [9], it is also presented an alternative formulation based on binary expansion of variables, which can become strong in some relevant practical cases, thanks to the superincreasing property of the adopted expansion coefficients. In [11], it is proposed the definition of a non-compact formulation purely based on assignment variables that relates to a maximum feasible subsystem problem. Finally, in [8], the numerical instabilities are tackled by a genetic heuristic exploiting power discretization.

According to a widespread belief, numerical instabilities in WND may be eliminated by multiplying all the fading coefficients of the problem by a large power of 10 (typically 10^{12}). However, in our direct experience with real world instances of several wireless technologies (e.g., DVB-T [12], WiMAX [9,12]), this did neither improve the performance of the solver nor of the quality of solutions found, which were still subject to coverage errors.

3   Numerical Accuracy in Linear Programming Solvers

Wireless network design problems are not only combinatorially complex, but as was argued before, also numerically sensitive. State-of-the-art MIP solvers employ floating-point arithmetic, hence their arithmetic computations are subject to round-off errors. This makes it necessary to allow for small violations of the constraints, bounds, and integrality requirements when checking solutions for feasibility. To this end, MIP solvers typically use a combination of absolute and relative tolerance to define their understanding of feasibility. A linear inequality $\alpha^\mathsf{T} x \leqslant \alpha_0$ is considered as satisfied by a point x^* if

$$\frac{\alpha^\mathsf{T} x^* - \alpha_0}{\max\{|\alpha^\mathsf{T} x^*|, |\alpha_0|, 1\}} \leqslant \epsilon_{\text{feas}} \tag{9}$$

with a feasibility tolerance $\epsilon_{\text{feas}} > 0$.[1] If the activity $\alpha^\mathsf{T} x^*$ and right-hand side α_0 are below one in absolute value, an absolute violation of up to $\epsilon_{\text{feas}} > 0$ is allowed. Otherwise, a relative tolerance is applied and larger violations are accepted. Typically, ϵ_{feas} ranges between 10^{-6} and 10^{-9}.

Feasibility of SIR Inequalities. When employing floating-point arithmetic to optimize wireless network design problems containing SIR inequalities, care is required when enforcing and checking their feasibility. First, since the coefficients and right-hand side of the SIR inequality (2) are significantly below 10^{-9} in absolute value, the inequality (9) results in a very loose definition of feasibility. The allowed absolute violation may be larger than the actual right-hand side.

Second, though the original SIR inequality (1) is equivalent to its linear reformulation (2), if we check their violation with respect to numerical tolerances, they behave differently. Indeed, an (absolute) violation $\epsilon_{\text{linear}} = \delta N - (a_{rs} p_s - \delta \sum_{t \in T \setminus \{s\}} a_{rt} p_t)$ of (2) corresponds to a much larger violation of (1), since

$$\epsilon_{\text{SIR}} = \delta - \frac{a_{rs} p_s}{N + \sum_{t \in T \setminus \{s\}} a_{rt} p_t} = \frac{\epsilon_{\text{linear}}}{N + \sum_{t \in T \setminus \{s\}} a_{rt} p_t} \tag{10}$$

and the sum of noise and interference signals $N + \sum_{t \in T \setminus \{s\}} a_{rt} p_t$ typically has an order of 10^{-9} or smaller. In combination with the feasibility tolerances promised by standard MIP solvers ($\approx 10^{-9}$), this would at best guarantee violations in the order of 1 for the original problem formulation.

The Impact of Scaling. Internally, MIP solvers may apply scaling factors to rows and columns of the constraint matrix in order to improve the numerical stability. Primarily, this aims at improving the condition numbers of basis matrices during the solution of LPs. However, from (9) it becomes apparent that an external, a priori scaling of constraints made by the user can change the very definition of feasibility: if the activity and right-hand side are significantly below 1 in absolute value, then scaling up tightens the feasible region. Precisely, with a scaling factor $S > 1$, if $|S\alpha^\mathsf{T} x^*| < 1$ and $|S\alpha_0| < 1$, then

$$\frac{S(\alpha^\mathsf{T} x^* - \alpha_0)}{\max\{|S\alpha^\mathsf{T} x^*|, |S\alpha_0|, 1\}} \leqslant \epsilon_{\text{feas}} \Leftrightarrow \frac{(\alpha^\mathsf{T} x^* - \alpha_0)}{\max\{|\alpha^\mathsf{T} x^*|, |\alpha_0|, 1\}} \leqslant \frac{\epsilon_{\text{feas}}}{S}, \tag{11}$$

---

[1] This is the definition of feasibility used by the academic MINLP solver SCIP [1,26]. While we do not know for certain the numerical definitions used by closed-source commercial solvers, we think that they follow a similar practice.

and the absolute tolerance can be decreased by a factor of $1/S$. This can then be used to arrive at a sufficiently strict definition of feasibility for constraints with very small coefficients, such as the SIR inequalities (2).

Advances in Exact LP and MIP Solving. Although the floating-point numerics used in today's state-of-the-art MIP solvers yield reliable results for the majority of problems and applications, there are cases in which results of higher accuracy are desired or needed, such as verification problems, computer proofs, or simply numerically instable instances. In the following we will review recent advances in methods for solving LPs and MIPs exactly over the rational numbers.

Trivially, of course, one can obtain an exact solution algorithm by performing all computations in exact arithmetic. However, for all but a few instances of interest, this idea is not sufficiently performant. As a starting point, it has been observed that LP bases returned by floating-point solvers are often optimal for real world problems [14]. For example, [21] could compute optimal bases to all of the NETLIB LP instances using only floating-point LP solvers and subsequently certifying them in exact rational arithmetic.

Following these observations, Applegate et al. [3] developed a simplex-based general-purpose exact LP solver, QSopt_ex, which exploits this behavior to achieve fast computation times on average. If an optimal basis is not identified by the double-precision subroutines, more simplex pivots are performed using increased levels of precision until the exact rational solution is identified (see also [15]).

Recently, Gleixner et al. [16,17] have developed an iterative refinement procedure for solving LPs with high accuracy, by solving a sequence of closely related LPs in order to compute primal and dual correction terms. The procedure avoids rational LU factorizations and solves LP in extended precision and hence often computes solutions with only tiny violations faster than QSopt_ex. Although not an exact method in itself, it can be used to speed up QSopt_ex significantly.

Finally, exact LP solving is a crucial subroutine for solving MIPs exactly. Once a promising assignment for the integer variables has been found, an exact LP solver can be used to compute feasible values for the continuous variables or prove that this integer assignment does not admit a fully feasible solution vector.

The majority of LPs within a MIP solution process, however, is solved to bound the objective value of the optimal solution. Solving these exactly does provide safe dual bounds, but can result in a large slow-down. The key to obtain a faster exact MIP solver is to avoid exact LP solving by correcting the dual solution obtained from a floating-point LP solver, see [24]. Cook et al. [6,7] have followed this approach to develop an exact branch-and-bound algorithm available as an extension of the solver SCIP [26].

In the following section, we will investigate empirically how these tools can be applied to analyze and address the numerical difficulties encountered in solving wireless network design problems.

4   Computational Experiments

The goal of our experiments was twofold: first, in order to test whether MIP solvers can be reliably used as decision tools for wireless network design models as introduced in Sect. 2, we analyzed the accuracy of primal solutions returned by a state-of-the-art MIP solver; second, we investigated the practical applicability and performance of the exact solution methods described in the previous section.

Experimental Setup. Experimental setup. The experiments were conducted on a computer with a 64bit Intel Xeon E3-1290 v2 CPU (4 cores, 8 threads) at 3.7 GHz with 8 MB cache and 16 GB main memory. We ran all jobs separately to avoid random noise in the measured running time that might be caused by cache-misses if multiple processes share common resources. We used CPLEX 12.5.0.0 [19] (default, deterministic parallel with up to four threads), QSopt_ex 2.5.10 [3] with EGlib 2.6.10 and GMP 4.3.1 [18], and SoPlex 2.0 [27] with GMP 5.0.5 [18] (both single-thread).

Test Instances. We performed our experiments on realistic instances of a WiMAX network, defined in cooperation with a major European telecommunications company. The instances correspond to various scenarios of a single-frequency network adopting a single transmission scheme.[2] For each instance, we solved the corresponding SPAP model from Sect. 2. We considered ten instances with between 100 and 900 receivers ($|R|$) and between 8 and 45 transmitters ($|T|$). The maximum emission power $P_{\max}$ of each transmitter was set equal to 30 dBmW and the SIR threshold (δ) was between 8 dB and 11 dB.[3]

Accuracy of MIP Solutions. In our first experiment, we ran CPLEX with a time limit of one hour (because of the combinatorial complexity of the problems, only the smallest instances can be solved to optimality within this limit) and checked the feasibility of the best primal solution returned. Table 1 shows the results for the unscaled instances, Table 2 shows the results for the instances with the linearized SIR inequalities (2) multiplied by $S = 10^{12}$ as in Sect. 2.

The first two columns give the size of each instance, while the second two columns state the smallest and largest absolute value in the coefficients and right-hand sides of the SIR constraints. These values differ by up to 10^{12}, a first indicator of numerical instability. Column "obj." gives the objective value of the solution at the end of the solving process that we checked, i.e., the number of receivers served by one transmitter. We report both the maximum violation of the original SIR inequalities (1) in column "SIR viol." and their linearization (2). Both for scaled and unscaled models, the results show that they differ by a factor of up to 10^{12}. This demonstrates that the linearized SIR inequalities must be satisfied with a very tight tolerance if we want to guarantee a reasonably small tolerance, 10^{-6}, say, for the original problem statement.

---

[2] For more details on WiMAX networks, see [9].

[3] The smallest MIP has 808 variables, 900 constraints, and 8 000 nonzeros, the largest instance contains 32 436 variables, 33 300 constraints, and 1 231 200 nonzeros.

Table 1. A posteriori check and exact verification of binary assignments from floating-point MIP solutions for instances without scaling.

Instance				Post processing					Exact LP	
$|R|$	$|T|$	α_{min}	α_{max}	obj.	linear viol.	SIR viol.	served	unserved	stat.	time
100	8	$4 \cdot 10^{-17}$	$4 \cdot 10^{-8}$	41	$1.7 \cdot 10^{-10}$	12.6	13	28	∅	0.2
169	20	$1 \cdot 10^{-19}$	$3 \cdot 10^{-8}$	73	$6.4 \cdot 10^{-11}$	6.3	1	72	∅	20.3
225	20	$2 \cdot 10^{-19}$	$2 \cdot 10^{-8}$	176	$1.2 \cdot 10^{-10}$	6.3	5	171	∅	10.0
256	40	$4 \cdot 10^{-19}$	$3 \cdot 10^{-8}$	155	$9.0 \cdot 10^{-11}$	6.3	15	140	∅	103.0
400	25	$8 \cdot 10^{-20}$	$2 \cdot 10^{-8}$	373	$1.2 \cdot 10^{-10}$	6.3	7	366	∅	55.7
400	40	$8 \cdot 10^{-20}$	$2 \cdot 10^{-8}$	301	$9.0 \cdot 10^{-11}$	6.3	13	288	∅	233.7
441	45	$8 \cdot 10^{-20}$	$2 \cdot 10^{-8}$	312	$1.0 \cdot 10^{-10}$	6.3	15	297	∅	440.5
529	40	$8 \cdot 10^{-20}$	$2 \cdot 10^{-8}$	421	$9.0 \cdot 10^{-11}$	6.3	13	408	∅	337.1
625	25	$2 \cdot 10^{-17}$	$5 \cdot 10^{-5}$	280	$1.9 \cdot 10^{-9}$	6.0	225	55	∅	113.2
900	36	$2 \cdot 10^{-20}$	$9 \cdot 10^{-9}$	890	$7.7 \cdot 10^{-11}$	2.5	14	876	∅	660.1

Table 2. A posteriori check and exact verification of binary assignments from floating-point MIP solutions for instances scaled with 10^{12}.

Instance				Post processing					Exact LP	
$|R|$	$|T|$	α_{min}	α_{max}	obj.	linear viol.	SIR viol.	served	unserved	stat.	time
100	8	$4 \cdot 10^{-5}$	$4 \cdot 10^{5}$	28	$7.1 \cdot 10^{-17}$	$4.6 \cdot 10^{-6}$	24	4	✓	0.1
169	20	$1 \cdot 10^{-7}$	$3 \cdot 10^{5}$	44	$7.3 \cdot 10^{-17}$	$8.0 \cdot 10^{-6}$	43	1	✓	2.1
225	20	$3 \cdot 10^{-7}$	$2 \cdot 10^{5}$	42	$6.2 \cdot 10^{-17}$	$7.0 \cdot 10^{-6}$	38	4	✓	0.9
256	40	$4 \cdot 10^{-7}$	$3 \cdot 10^{5}$	72	$8.1 \cdot 10^{-17}$	$5.1 \cdot 10^{-6}$	62	10	✓	11.7
400	25	$8 \cdot 10^{-8}$	$2 \cdot 10^{5}$	77	$6.7 \cdot 10^{-17}$	$1.4 \cdot 10^{-5}$	71	6	✓	5.5
400	40	$8 \cdot 10^{-8}$	$2 \cdot 10^{5}$	95	$6.7 \cdot 10^{-17}$	$7.9 \cdot 10^{-6}$	85	10	✓	20.2
441	45	$8 \cdot 10^{-8}$	$2 \cdot 10^{5}$	101	$8.9 \cdot 10^{-16}$	$4.8 \cdot 10^{-5}$	89	12	✓	35.5
529	40	$8 \cdot 10^{-8}$	$2 \cdot 10^{5}$	101	$8.8 \cdot 10^{-15}$	$6.1 \cdot 10^{-4}$	96	5	✓	29.9
625	25	$8 \cdot 10^{-5}$	$5 \cdot 10^{7}$	417	$1.9 \cdot 10^{-14}$	$1.9 \cdot 10^{-3}$	415	2	✓	6.0
900	36	$2 \cdot 10^{-8}$	$9 \cdot 10^{4}$	202	$8.1 \cdot 10^{-18}$	$1.4 \cdot 10^{-6}$	200	2	✓	58.0

As it can be seen, the results for the unscaled models are significantly worse in this respect: although the violation of the linearized constraint looks quite small, the original SIR inequalities are strongly violated. As a result, these solutions cannot be implemented in practice.[4]

The column "served" states the number of receivers r served by a transmitter s for which the corresponding quantity $SIR_{rs}(p)$ is at least $\delta - 10^{-6}$.

---

[4] Although with this kind of unreliability, this does not matter anymore, note that the numerical difficulties during the solving process are also reflected in the lower objective values obtained by the unscaled models.

Table 3. Exact computation of the power vector via QSopt_ex versus iterative refinement via SoPlex to a tolerance of 10^{-25} for instances scaled with 10^{12}.

Instance					QSopt_ex		SoPlex		
$|R|$	$|T|$	α_{min}	α_{max}	obj.	stat.	time	max. viol.	time	rel. [%]
100	8	$4\cdot10^{-5}$	$4\cdot10^{5}$	28	✓	0.1	$3.7\cdot10^{-29}$	0.1	-0.0
169	20	$1\cdot10^{-7}$	$3\cdot10^{5}$	44	✓	2.1	$2.4\cdot10^{-39}$	1.0	-52.4
225	20	$3\cdot10^{-7}$	$2\cdot10^{5}$	42	✓	0.9	$2.2\cdot10^{-29}$	0.5	-44.4
256	40	$4\cdot10^{-7}$	$3\cdot10^{5}$	72	✓	11.7	$1.6\cdot10^{-36}$	2.5	-78.6
400	25	$8\cdot10^{-8}$	$2\cdot10^{5}$	77	✓	5.5	$1.8\cdot10^{-27}$	1.3	-76.4
400	40	$8\cdot10^{-8}$	$2\cdot10^{5}$	95	✓	20.2	$6.9\cdot10^{-40}$	4.4	-78.2
441	45	$8\cdot10^{-8}$	$2\cdot10^{5}$	101	✓	35.5	$3.1\cdot10^{-40}$	6.3	-82.2
529	40	$8\cdot10^{-8}$	$2\cdot10^{5}$	101	✓	29.9	$5.7\cdot10^{-27}$	4.9	-83.6
625	25	$8\cdot10^{-5}$	$5\cdot10^{7}$	417	✓	6.0	$3.0\cdot10^{-29}$	2.8	-53.3
900	36	$2\cdot10^{-8}$	$9\cdot10^{4}$	202	✓	58.0	$5.4\cdot10^{-40}$	11.7	-79.8

This gives the (cardinality of the) subset of receivers that can reliably be served by the power vector p of the MIP solution. It is evident these values are significantly below the claimed objective value of the MIP solution for the unscaled models. Although the situation is much better for the scaled models, also these exhibit a notable number of receivers that are incorrectly claimed to be served.

Exact Verification of Binary Assignments. As these first results show, the values of the binary variables in the MIP solutions are not supported by the power vector given by the continuous variables. In our second experiment, we tried to test whether the binary part of the solutions are correct in the sense that there exists a power vector p that satisfies these receiver-transmitter assignments. To this end, we fixed the binary variables to their value in the MIP solution and solved the remaining LP, effectively obtaining a PAP instance as defined in Sect. 2, exactly with QSopt_ex. Note that this is a pure feasibility problem.

For the unscaled models, all LPs turned out to be infeasible, as is indicated by the symbol "∅" in Table 1. On the contrary, the LPs obtained from the scaled models could all be verified as feasible. Hence the exact LP solver computed a power vector p to serve all receivers as claimed by the MIP solver.

Additionally, we can see that proving the infeasibility of the unscaled LPs took notably longer than proving the scaled LPs feasible. The reason is that in the first case, QSopt_ex always had to apply increased 128bit arithmetic, while for the scaled LPs, the basis information after initial double-precision solve turned out to be already exactly feasible.

Exact MIP Solving. We stress that the approach above only yields proven primal bounds on the optimal objective value. Because CPLEX uses floating-point LP bounds, it is unclear whether optimal solutions have been cut off. In order to further investigate this, we tried to apply the exact extension of the SCIP solver.

However, for all but the smallest instances, we could not get any results. For the instance with 225 transmitters and 20 receivers, the solving took over 20 h, 139 097 820 branch-and-bound nodes, and more than 7 GB peak memory usage. The result was 42 and thus confirmed the optimality of the solution found by CPLEX.

The slow performance is not really surprising, as the current implementation is a pure branch-and-bound algorithm and lacks many of the sophisticated features of today's state-of-the-art MIP solvers. Hence, this should not be taken as a proof that exact MIP solvers are in principle not applicable to this application.

Accurate Computation of the Power Vector. Arguably, computing the power vector exactly is more than necessary for the practical application, and the running times of QSopt_ex with almost one minute for the largest LP may become a bottleneck. However, in practice it suffices to compute a power vector that satisfies the original SIR inequalities (1) within a reasonably small tolerance. In our last experiment, we tested whether the idea of iterative refinement available in the SoPlex solver, can achieve this faster than an exact LP solver. We used an (absolute) tolerance of 10^{-25}, which for the scaled models suffices to guarantee a tolerance of the same order of magnitude for (1).

Table 3 shows the results: the actually reached maximum violation of the LP rows (as small as 10^{-40}), the solving time, and its relative difference to the running times of QSopt_ex. For all but the two instances that are solved within one second, SoPlex is at least twice as fast as QSopt_ex. Note, however, that the implementation of both solvers, in particular the simplex method, differs in many details, and so we cannot draw a reliable conclusion, let alone on such a limited test set. However, it suggests that iterative refinement may be more suited to the practical setting of certain applications.

5   Conclusion

This paper has tried to highlight a number of numerical issues that must be considered when solving MIP models for wireless network design. We demonstrated that the linearization of the crucial SIR inequalities in combination with the definition of feasibility used in floating-point solvers can lead to completely unreliable results and that an a priori scaling of the constraints can help, but it is not able to make the solutions completely reliable. We also showed that the current performance of exact MIP solvers is not sufficient to address the combinatorial difficulty of these models. On the positive side, we could show that recent advances in exact and accurate LP solving are of great help for computing reliable primal solutions. So far, we have applied these only as a post processing after the MIP solution process. Ideally, however, the accurate solution of LPs on the continuous variables should be integrated into the branch-and-bound process and used as a direct verification of the primal bound given by the incumbent solution. An important next step will be to extend the experiments to larger sets of instances including other types of wireless technologies, such as DVB-T.

References

1. Achterberg, T.: SCIP: solving constraint integer programs. Math. Prog. Comput. **1**(1), 1–41 (2009)
2. Amaldi, E., Capone, A., Malucelli, F., Mannino, C.: Optimization problems and models for planning cellular networks. In: Resende, M., Pardalos, P. (eds.) Handbook of Optimization in Telecommunication, pp. 917–939. Springer, Heidelberg (2006)
3. Applegate, D.L., Cook, W., Dash, S., Espinoza, D.G.: QSopt_ex (2007). http://www.dii.uchile.cl/~daespino/ESolver_doc/
4. Bauschert, T., Büsing, C., D'Andreagiovanni, F., Koster, A.M.C.A., Kutschka, M., Steglich, U.: Network planning under demand uncertainty with robust optimization. IEEE Commun. Mag. **52**, 178–185 (2014)
5. Capone, A., Chen, L., Gualandi, S., Yuan, D.: A new computational approach for maximum link activation in wireless networks under the sinr model. IEEE Trans. Wirel. Commun. **10**(5), 1368–1372 (2011)
6. Cook, W., Koch, T., Steffy, D.E., Wolter, K.: An exact rational mixed-integer programming solver. In: Günlük, O., Woeginger, G.J. (eds.) IPCO 2011. LNCS, vol. 6655, pp. 104–116. Springer, Heidelberg (2011)
7. Cook, W., Koch, T., Steffy, D.E., Wolter, K.: A hybrid branch-and-bound approach for exact rational mixed-integer programming. Math. Prog. Comp. **5**, 305–344 (2013)
8. D'Andreagiovanni, F.: On improving the capacity of solving large-scale wireless network design problems by genetic algorithms. In: Di Chio, C., et al. (eds.) EvoApplications 2011, Part II. LNCS, vol. 6625, pp. 11–20. Springer, Heidelberg (2011)
9. D'Andreagiovanni, F.: Pure 0–1 programming approaches to wireless network design. 4OR-Q. J. Oper. Res. **10**(2), 211–212 (2012). doi:10.1007/s10288-011-0162-z
10. D'Andreagiovanni, F.: Revisiting wireless network jamming by SIR-based considerations and multiband robust optimization. Optim. Lett. **9**(8), 1495–1510 (2015)
11. D'Andreagiovanni, F., Mannino, C., Sassano, A.: Negative cycle separation in wireless network design. In: Pahl, J., Reiners, T., Voß, S. (eds.) INOC 2011. LNCS, vol. 6701, pp. 51–56. Springer, Heidelberg (2011)
12. D'Andreagiovanni, F., Mannino, C., Sassano, A.: GUB covers and power indexed formulations for wireless network design. Manage. Sci. **59**(1), 142–156 (2013)
13. Dely, P., D'Andreagiovanni, F., Kassler, A.: Fair optimization of mesh-connected WLAN hotspots. Wirel. Commun. Mob. Com. **15**(5), 924–946 (2015)
14. Dhiflaoui, M., Funke, S., Kwappik, C., Mehlhorn, K., Seel, M., Schömer, E., Schulte, R., Weber, D.: Certifying and repairing solutions to large LPs: How good are LP-solvers? In: Proceedings of SODA 2003, pp. 255–256. SIAM (2003)
15. Espinoza, D.G.: On Linear Programming, Integer Programming and Cutting Planes. Ph.D. thesis, Georgia Institute of Technology (2006)
16. Gleixner, A.M., Steffy, D.E., Wolter, K.: Improving the accuracy of linear programming solvers with iterative refinement. In: Proceedings ISSAC 2012, Grenoble (2012)
17. Gleixner, A.M., Steffy, D.E., Wolter, K.: Iterative refinement for linear programming. INFORMS J. Comput. **28**(3), 449–464 (2016)
18. GNU Multiple Precision Arithmetic Library Version. http://gmplib.org/
19. CPLEX. https://www-01.ibm.com/software/commerce/optimization/cplex-optimizer

20. Kennington, J., Olinick, E., Rajan, D.: Wireless Network Design: Optimization Models and Solution Procedures. Springer, Heidelberg (2010)
21. Koch, T.: The final NETLIB-LP results. Oper. Res. Lett. **32**(2), 138–142 (2004)
22. Mannino, C., Rossi, F., Smriglio, S.: The network packing problem in terrestrial broadcasting. Oper. Res. **54**(6), 611–626 (2006)
23. Mannino, C., Rossi, F., Smriglio, S.: A unified view in planning broadcasting networks. DIS Technical report, vol. 8. Aracne Editrice, Roma (2007)
24. Neumaier, A., Shcherbina, O.: Safe bounds in linear and mixed-integer linear programming. Math. Program. **99**(2), 283–296 (2004)
25. Rappaport, T.S.: Wireless Communications: Principles and Practice. Prentice Hall, Upper Saddle River (2001)
26. SCIP: Solving Constraint Integer Programs. http://scip.zib.de
27. SoPlex. The Sequential object-oriented simPlex. http://soplex.zib.de/

Approximability and Exact Resolution of the Multidimensional Binary Vector Assignment Problem

Marin Bougeret, Guillerme Duvillié$^{(\boxtimes)}$, and Rodolphe Giroudeau

LIRMM, Université Montpellier 2, Montpellier, France
{marin.bougeret,guillerme.duvillie,rodolphe.giroudeau}@lirmm.fr

Abstract. In this paper we consider the multidimensional binary vector assignment problem. An input of this problem is defined by m disjoint multisets $V^1, V^2, \ldots, V^m$, each composed of n binary vectors of size p. An output is a set of n disjoint m-tuples of vectors, where each m-tuple is obtained by picking one vector from each multiset V^i. To each m-tuple we associate a p dimensional vector by applying the bit-wise AND operation on the m vectors of the tuple. The objective is to minimize the total number of zeros in these n vectors. We denote this problem by $\min \sum 0$, and the restriction of this problem where every vector has at most c zeros by $(\min \sum 0)_{\#0 \leq c}$. $(\min \sum 0)_{\#0 \leq 2}$ was only known to be **APX**-complete, even for $m = 3$ [5]. We show that, assuming the unique games conjecture, it is **NP**-hard to $(n - \varepsilon)$-approximate $(\min \sum 0)_{\#0 \leq 1}$ for any fixed n and ε. This result is tight as any solution is a n-approximation. We also prove without assuming UGC that $(\min \sum 0)_{\#0 \leq 1}$ is **APX**-complete even for $n = 2$, and we provide an example of $n - f(n, m)$-approximation algorithm for $\min \sum 0$. Finally, we show that $(\min \sum 0)_{\#0 \leq 1}$ is polynomial-time solvable for fixed m (which cannot be extended to $(\min \sum 0)_{\#0 \leq 2}$ according to [5]).

1 Introduction

1.1 Problem Definition

In this paper we consider the multidimensional binary vector assignment problem denoted by $\min \sum 0$. An input of this problem (see Fig. 1) is described by m multisets $V^1, \ldots, V^m$, each multiset V^i containing n binary p-dimensional vectors. For any $j \in [n]^1$, and any $i \in [m]$, the jth vector of multiset V^i is denoted v_j^i, and for any $k \in [p]$, the kth coordinate of v_j^i is denoted $v_j^i[k]$.

The objective of this problem is to create a set S of n stacks. A stack $s = (v_1^s, \ldots, v_m^s)$ is an $m - tuple$ of vectors such that $v_i^s \in V^i$, for any $i \in [m]$. Furthermore, S has to be such that every vector of the input appears in exactly one created stack.

---

^{1} Note that $[n]$ stands for $\{1, 2, \ldots, n\}$.

© Springer International Publishing Switzerland 2016
R. Cerulli et al. (Eds.): ISCO 2016, LNCS 9849, pp. 148–159, 2016.
DOI: 10.1007/978-3-319-45587-7_13

We now introduce the operator $\wedge$ which assigns to a pair of vectors (u, v) the vector given by $u \wedge v = (u[1] \wedge v[1], u[2] \wedge v[2], \ldots, u[p] \wedge v[p])$. We associate to each stack s a unique vector given by $v_s = \bigwedge_{i \in [m]} v_i^s$.

The cost of a vector v is defined as the number of zeros in it. More formally if v is p-dimensional, $c(v) = p - \sum_{k \in [p]} v[k]$. We extend this definition to a set of stacks $S = \{s_1, \ldots, s_n\}$ as follows: $c(S) = \sum_{s \in S} c(v_s)$.

The objective is then to find a set S of n disjoint stacks minimizing the total number of zeros. This leads us to the following definition of the problem:

Optimization Problem 1 $\min \sum 0$

Input m multisets of n p-dimensional binary vectors.

Output   A set S of n disjoint stacks minimizing $c(S)$.

Throughout this paper, we denote $(\min \sum 0)_{\#0 \leq c}$ the restriction of $\min \sum 0$ where the number of zeros per vector is upper bounded by c.

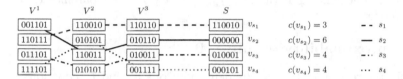

Fig. 1. Example of $\min \sum 0$ instance with $m = 3, n = 4, p = 6$ and of a feasible solution S of cost $c(S) = 17$.

1.2   Related Work

The dual version of the problem called $\max \sum 1$ (where the objective is to maximize the total number of 1 in the created stacks) has been introduced by Reda et al. in [8] as the "yield maximization problem in Wafer-to-Wafer 3-D Integration technology". They prove the **NP**-completeness of $\max \sum 1$ and provide heuristics without approximation guarantee. In [6] we proved that, even for $n = 2$, for any $\varepsilon > 0$, $\max \sum 1$ is $\mathcal{O}(m^{1-\varepsilon})$ and $\mathcal{O}(p^{1-\varepsilon})$ inapproximable unless $\mathbf{P} = \mathbf{NP}$. We also provide an ILP formulation proving that $\max \sum 1$ (and thus $\min \sum 0$) is $\mathbf{FPT}^2$ when parameterized by p.

We introduced $\min \sum 0$ in [4] where we provide in particular $\frac{4}{3}$-approximation algorithm for $m = 3$. In [5], authors focus on a generalization of $\min \sum 0$, called MULTI DIMENSIONAL VECTOR ASSIGNMENT, where vectors are not necessary binary vectors. They extend the approximation algorithm of [4] to get a $f(m)$-approximation algorithm for arbitrary m. They also prove the **APX**-completeness of the $(\min \sum 0)_{\#0 \leq 2}$ for $m = 3$. This result was the only known inapproximability result for $\min \sum 0$.

---

[2] *i.e.* admits an algorithm in $f(p)poly(|I|)$ for an arbitrary function f.

1.3   Contribution

In Sect. 2 we study the approximability of $\min \sum 0$. Our main result in this section is to prove that assuming UGC, it is **NP**-hard to $(n - \varepsilon)$-approximate $(\min \sum 0)_{\#0 \leq 1}$ (and thus $\min \sum 0$) for any fixed $n \geq 2$, $\forall \varepsilon > 0$. This result is tight as any solution is a n-approximation.

Notice that this improves the only existing negative result for $\min \sum 0$, which was the **APX**-hardness of [5] (implying only no-**PTAS**).

We also show how this reduction can be used to obtain the **APX**-hardness for $(\min \sum 0)_{\#0 \leq 1}$ for $n = 2$ unless $\mathbf{P} = \mathbf{NP}$, which is weaker negative result, but does not require UGC. We then give an example $n - f(n, m)$ approximation algorithm for the general problem $\min \sum 0$.

In Sect. 3, we consider the exact resolution of $\min \sum 0$. We focus on *sparse* instances, *i.e.* instances of $(\min \sum 0)_{\#0 \leq 1}$. Indeed, recall that authors of [5] show that $(\min \sum 0)_{\#0 \leq 2}$ is **APX**-complete even for $m = 3$, implying that $(\min \sum 0)_{\#0 \leq 2}$ cannot be polynomial-time solvable for fixed m unless $\mathbf{P} = \mathbf{NP}$. Thus, it is natural to ask if $(\min \sum 0)_{\#0 \leq 1}$ is polynomial-time solvable for fixed m. Section 3 is devoted to answer positively to this question. Notice that the question of determining if $(\min \sum 0)_{\#0 \leq 1}$ is **FPT** when parameterized by m remains open.

2   Approximability of $\min \sum 0$

We refer the reader to [1,7] for the definitions of Gap and L-reductions.

2.1   Inapproximability Results for $(\min \sum 0)_{\#0 \leq 1}$

From now we suppose that $\forall k \in [p]$, $\exists i$, $\exists j$ such that $v_j^i[k] = 0$. In other words, for any solution S and $\forall k$, there exists a stack s such that $v_s[k] = 0$. Otherwise, we simply remove such a coordinate from every vector of every set, and decrease p by one. Since this coordinate would be set to 1 in all the stacks of all solutions, such a preprocessing preserves approximation ratios and exact results.

In a first time, we define the following polynomial-time computable function f which associates an instance of $(\min \sum 0)_{\#0 \leq 1}$ to any k-uniform hypergraph, *i.e.* an hypergraph $G = (U, E)$ such that every hyperedges of E contains exactly k distinct elements of U.

Definition of f. We consider a k-uniform hypergraph $G = (U, E)$. We call f the polynomial-time computable function that creates an instance of $(\min \sum 0)_{\#0 \leq 1}$ from a G as follows.

1. We set $m = |E|$, $n = k$ and $p = |U|$.
2. For each hyperedge $e = \{u_1, u_2, \ldots, u_k\} \in E$, we create the set V^e containing k vectors $\{v_j^e, j \in [k]\}$, where for all $j \in [k]$, $v_j^e[u_j] = 0$ and $v_j^e[l] = 1$ for $l \neq u_j$. We say that a vector v **represents** $u \in U$ iff $v[u] = 0$ and $v[l \neq u] = 1$ (and thus vector v_j^e represents u_j).

An example of this construction is given in Fig. 2.

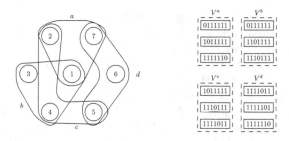

Fig. 2. Illustration of the reduction from an hypergraph $G = (U = \{1, 2, 3, 4, 5, 6, 7\}$, $E = \{\{1, 2, 7\}, \{1, 3, 4\}, \{2, 4, 5\}, \{5, 6, 7\}\})$ to an instance $(\min \sum 0)_{\#0 \leq 1}$

Negative Results Assuming UGC. We consider the following problem. Notice that what we call a vertex cover in a k-regular hypergraph $G = (U, E)$ is a set $U' \subseteq U$ such that for any hyperedge $e \in E$, $U' \cap e \neq \emptyset$.

Decision Problem 1 ALMOST Ek VERTEX COVER

Input   *We are given an integer $k \geq 2$, two arbitrary positive constants ε and δ and a k-uniform hypergraph $G = (U, E)$.*

Output   *Distinguish between the following cases:*

> **YES Case** *there exist k disjoint subsets $U^1, U^2, \ldots, U^k \subseteq U$, satisfying $|U^i| \geq \frac{1-\varepsilon}{k}|U|$ and such that every hyperedge contains at most one vertex from each U^i.*
> **NO Case** *every vertex cover has size at least $(1 - \delta)|U|$.*

It is shown in [2] that, assuming UGC, this problem is **NP**-complete.

Theorem 1. *For any fixed $n \geq 2$, for any constants $\varepsilon, \delta > 0$, there exists a $\frac{n-n\delta}{1+n\varepsilon}$-Gap reduction from ALMOST Ek VERTEX COVER to $(\min \sum 0)_{\#0 \leq 1}$. Consequently, under UGC, for any fixed n $(\min \sum 0)_{\#0 \leq 1}$ is **NP**-hard to approximate within a factor $(n - \varepsilon')$ for any $\varepsilon' > 0$.*

Proof. We consider an instance I of ALMOST Ek VERTEX COVER defined by two positive constants δ and ϵ, an integer k and a k-regular hypergraph $G = (U, E)$.

We use the function f previously defined to construct an instance $f(I)$ of $\min \sum 0$. Let us now prove that if I is a positive instance, $f(I)$ admits a solution S of cost $c(S) < (1 + n\varepsilon)|U|$, and otherwise any solution S of $f(I)$ has cost $c(S) \geq n(1 - \delta)|U|$.

NO Case. Let S be a solution of $f(I)$. Let us first remark that for any stack $s \in S$, the set $\{k : v_s[k] = 0\}$ defines a vertex cover in G. Indeed, s contains exactly one vector per set, and thus by construction s selects one vertex per hyperedge in G. Remark also that the cost of s is equal to the size of the corresponding vertex cover.

Now, suppose that I is a negative instance. Hence each vertex cover has a size at least equal to $(1 - \delta)|U|$, and any solution S of $f(I)$, composed of exactly n stacks, verifies $c(S) \geq n(1 - \delta)|U|$.

YES Case. If I is a positive instance, there exists k disjoint sets $U^1, U^2, \ldots, U^k \subseteq U$ such that $\forall i = 1, \ldots, k$, $|U^i| \geq \frac{1-\varepsilon}{k}|U|$ and such that every hyperedge contains at most one vertex from each U^i.

We introduce the subset $X = U \backslash \bigcup_{i=1}^k U^i$. By definition $\{U^1, U^2, \ldots, U^k, X\}$ is a partition of U and $X \leq \varepsilon|U|$. Furthermore, $U^i \cup X$ is a vertex cover $\forall i = 1, \ldots, k$. Indeed, each hyperedge $e \in E$ that contains no vertex of U^i, contains at least one vertex of X since e contains k vertices.

We now construct a solution S of $f(I)$. Our objective is to construct stacks $\{s_i\}$ such that for any i, the zeros of s_i are included in $U_i \cup X$ (i.e. $\{l : v_{s_i}[l] = 0\} \subseteq U_i \cup X$). For each $e = \{u_1, \ldots, u_k\} \in E$, we show how to assign exactly one vector of V^e to each stack $s_1, \ldots, s_k$. For all $i \in [k]$, if v_j^e represents a vertex u with $u \in U^i$, then we assign v_j^e to s_i. W.l.o.g., let $S'_e = \{s_1, \ldots, s_{k'}\}$ (for $k' \leq k$) be the set of stacks that received a vertex during this process. Notice that as every hyperedge contains at most one vertex from each U^i, we only assigned one vector to each stack of S'_e. After this, every unassigned vector $v \in V^e$ represents a vertex of X (otherwise, such a vector v would belong to a set U^i, $i \in k'$, a contradiction). We assign arbitrarily these vectors to the remaining stacks that are not in S'_e. As by construction $\forall i \in [k]$, $v_s i$ contains only vectors representing vertices from $U^i \cup X$, we get $c(s_i) \leq |U^i| + |X|$.

Thus, we obtain a feasible solution S of cost $c(S) = \sum_{i=1}^k c(s_i) \leq k|X| + \sum_{i=1}^k |U^i|$. As by definition we have $|X| + \sum_{i=1}^k |U^i| = |U|$, it follows that $c(S) \leq |U| + (k - 1)\varepsilon|U|$ and since $k = n$, $c(S) < |U|(1 + n\varepsilon)$.

If we define $a(n) = (1 + n\varepsilon)|U|$ and $r(n) = \frac{n(1-\delta)}{(1+n\varepsilon)}$, the previous reduction is a $r(n)$-Gap reduction. Furthermore, $\lim_{\delta,\varepsilon \to 0} r(n) = n$, thus it is **NP**-hard to approximate $(\min \sum 0)_{\#0 \leq 1}$ within a ratio $(n - \varepsilon')$ for any $\varepsilon' > 0$.

$\square$

Notice that, as a function of n, this inapproximability result is optimal. Indeed, we observe that any feasible solution S is an n-approximation as, for any instance I of $\min \sum 0^3$, $Opt(I) \geq p$ and for any solution S, $c(S) \leq pn$.

Negative Results Without Assuming UGC. Let us now study the negative results we can get when only assuming $\mathbf{P} \neq \mathbf{NP}$. Our objective is to prove that $(\min \sum 0)_{\#0 \leq 1}$ is **APX**-hard, even for $n = 2$. To do so, we present a reduction from ODD CYCLE TRANSVERSAL, which is defined as follows. Given an input graph $G = (U, E)$, the objective is to find an odd cycle transversal of minimum size, i.e. a subset $T \subseteq U$ of minimum size such that $G[U \backslash T]$ is bipartite.

---

[3] Recall that we assume $\forall k \in [p], \exists i, \exists j$ such that $v_j^i[k] = 0$.

For any integer $\gamma \geq 2$, we denote $\mathcal{G}_\gamma$ the class of graphs $G = (U, E)$ such that any optimal odd cycle transversal T has size $|T| \geq \frac{|U|}{\gamma}$. Given $\mathcal{G}$ a class of graphs, we denote $OCT_\mathcal{G}$ the ODD CYCLE TRANSVERSAL problem restricted to $\mathcal{G}$.

Lemma 1. *For any constant $\gamma \geq 2$, there exists an L-reduction from $OCT_{\mathcal{G}_\gamma}$ to $(\min \sum 0)_{\#0 \leq 1}$ with $n = 2$.*

Proof. Let us consider an integer γ, an instance I of $OCT_{\mathcal{G}_\gamma}$, defined by a graph $G = (V, E)$ such that $G \in \mathcal{G}_\gamma$. W.l.o.g., we can consider that G contains no isolated vertex.

Remark that any graph can be seen as a 2-uniform hypergraph. Thus, we use the function f previously defined to construct an instance $f(I)$ of $(\min \sum 0)_{\#0 \leq 1}$ such that $n = 2$. Since, G contains no isolated vertex, $f(I)$ contains no position k such that $\forall i \in [m]$, $\forall j \in [n]$, $v_j^i[k] = 1$.

Let us now prove that I admits an odd cycle transversal of size t if and only if $f(I)$ admits a solution of cost $p + t$.

$\Leftarrow$ We consider an instance $f(I)$ of $(\min \sum 0)_{\#0 \leq 1}$ with $n = 2$ admitting a solution $S = \{s_A, s_B\}$ with cost $c(S) = p + t$. Let us specify a function g which produces from S a solution $T = g(I, S)$ of $OCT_{\mathcal{G}_\gamma}$, i.e. a set of vertices of U such that $G[U \backslash T]$ is bipartite.

We define $T = \{u \in U : v_{s_A}[u] = v_{s_B}[u] = 0\}$, the set of coordinates equal to zero in both s_A and s_B. We also define $A = \{u \in V : v_{s_A}[u] = 0 \text{ and } v_{s_B}[u] = 1\}$ (resp. $B = \{u \in V : v_{s_B}[u] = 0 \text{ and } v_{s_A}[u] = 1\}$), the set of coordinates set to zero only in s_A (resp. s_B). Notice that $\{T, A, B\}$ is a partition of U.

Remark that A and B are independent sets. Indeed, suppose that $\exists \{u, v\} \in E$ such that $u, v \in A$. As $\{u, v\} \in E$ there exists a set $V^{(u,v)}$ containing a vector that represents u and another vector that represents v, and thus these vectors are assigned to different stacks. This leads to a contradiction. It follows that $G[U \backslash T]$ is bipartite and T is an odd cycle transversal.

Since $c(S) = |A| + |B| + 2|T| = p + |T| = p + t$, we get $|T| = t$.

$\Rightarrow$ We consider an instance I of $OCT_{\mathcal{G}_\gamma}$ and a solution T of size t. We now construct a solution $S = \{s_A, s_B\}$ of $f(I)$ from T.

By definition, $G[U \backslash T]$ is a bipartite graph, thus the vertices in $U \backslash T$ may be split into two disjoint independent sets A and B. For each edge $e \in E$, the following cases can occur:

- if $\exists u \in e$ such that $u \in A$, then the vector corresponding to u is assigned to s_A, and the vector corresponding to $e \setminus \{u\}$ is assigned to s_B (and the same rule holds by exchanging A and B)
- otherwise, u and $v \in T$, and we assign arbitrarily v_u^e to s_A and the other to s_B.

We claim that the stacks s_A and s_B describe a feasible solution S of cost at most $p + t$.

Since, for each set, only one vector is assigned to s_A and the other to s_B, the two stacks s_A and s_B are disjoint and contain exactly m vectors. S is therefore a feasible solution.

Remark that v_{s_A} (resp. v_{s_B}) contains only vectors v such that $v[k] = 0 \implies k \in A \cup T$ (resp. $k \in B \cup T$), and thus $c(v_A) \leq |A| + |T|$ (resp. $c(v_B) \leq |B| + |T|$). Hence $c(S) \leq |A| + |B| + 2|T| = p + t$.

Let us now prove that this reduction is an L-reduction.

1. By definition, any instance I of $OCT_{\mathcal{G}_\gamma}$ verifies $|Opt(I)| \geq |U|/\gamma$. Thus,

$$Opt(f(I)) \leq |U| + Opt(I) \leq (\gamma + 1)Opt(I)$$

2. We consider an arbitrary instance I of $OCT_{\mathcal{G}_\gamma}$, $f(I)$ the corresponding instance of $(\min \sum 0)_{\#0 \leq 1}$, S a solution of $f(I)$ and $T = g(I), S$ the corresponding solution of I.
   We proved $|T| - Opt(I) = c(S) - |U| - (Opt(f(I)) - |U|) = c(S) - Opt(f(I))$.

   Therefore, we get an L-reduction for $\alpha = \gamma + 1$ and $\beta = 1$.   □

Lemma 2. ([3]). *There exist a constant γ and $\mathcal{G} \subset \mathcal{G}_\gamma$ such that $OCT_{\mathcal{G}}$ is **APX**-hard.*

The following result is now immediate.

Theorem 2. $(\min \sum 0)_{\#0 \leq 1}$ *is **APX**-hard, even for $n = 2$.*

2.2 Approximation Algorithm for $\min \sum 0$

Let us now show an example of algorithm achieving a $n - f(n, m)$ ratio. Notice that the $(n - \epsilon)$ inapproximability result holds for fixed n and $\#0 = 1$, while the following algorithm is polynomial-time computable when n is part of the input and $\#0$ is arbitrary.

Proposition 1. *There is a polynomial-time $n - \frac{n-1}{n\rho(n,m)}$ approximation algorithm for $\min \sum 0$, where $\rho(n, m) > 1$ is the approximation ratio for independent set in graphs that are the union of m complete n-partite graphs.*

Proof. Let I be an instance of $\min \sum 0$. Let us now consider an optimal solution $S^* = \{s_1^*, \ldots, s_n^*\}$ of I. For any $i \in [n]$, let $Z_i^* = \{l \in [p] : v_{s_i^*}[l] = 0 \text{ and } v_{s_t^*}[l] = 1, \forall t \neq i\}$ be the set of coordinates equal to zero only in stack s_i^*. Let $\Delta = \sum_{i=1}^n |Z_i^*|$. Notice that we have $c(S^*) \geq \Delta + 2(p - \Delta)$, as for any coordinate l outside $\bigcup_i Z_i^*$, there are at least two stacks with a zero at coordinate l. W.l.o.g., let us suppose that Z_1^* is the largest set among $\{Z_i^*\}$, implying $|Z_1^*| \geq \frac{\Delta}{n}$.

Given a subset $Z \subset [p]$, we will construct a solution $S = \{s_1, \ldots, s_n\}$ such that for any $l \in Z$, $v_{s_1}[l] = 0$, and for any $i \neq 1$, $v_{s_i}[l] = 1$. Informally, the zero at coordinates Z will appear only in s_1, which behaves as a "trash" stack. The cost of such a solution is $c(S) \leq c(s_1) + \sum_{i=2}^n c(s_i) \leq p + (n - 1)(p - |Z|)$. Our objective is now to compute such a set Z, and to lower bound $|Z|$ according to $|Z_1^*|$.

Let us now define how we compute Z. Let $P = \{l \in [p] : \forall i \in [m], |\{j : v_j^i[l] = 0\}| \leq 1\}$ be the subset of coordinates that are never nullified in two different vectors of the same set. We will construct a simple undirected graph $G = (P, E)$, and thus it remains to define E. For vector v_j^i, let $Z_j^i = Z(v_j^i) \cap P$, where $Z(v) \subseteq [p]$ denotes the set of null coordinates of vector v. For any $i \in [m]$, we add to G the edges of the complete n-partite graph $G^i = (\{Z_1^i \times \cdots \times Z_n^i\})$ (*i.e.* for any $j_1, j_2, v_1 \in Z_{j_1}^i, v_2 \in Z_{j_2}^i$, we add edge $\{v_1, v_2\}$ to G). This concludes the description of G, which can be seen as the union of m complete n-partite graphs.

Let us now see the link between independent set in G and our problem. Let us first see why Z_1^* is a independent set in G. Recall that by definition of Z_1^*, for any $l \in Z_1^*$, $v_{s_1^*}[k] = 0$, but $v_{s_j^*}[k] = 1$, $j \geq 2$. Thus, it is immediate that $Z_1^* \subseteq P$. Moreover, assume by contradiction that there exists an edge in G between to vertices l_1 and l_2 of Z_1^*. This implies that there exists $i \in [m]$, j_1 and $j_2 \neq j_1$ such that $v_{j_1}^i[l_1] = 0$ and $v_{j_2}^i[l_2] = 0$. As by definition of Z_1^* we must have $v_{s_j}[k_1] = 1$ and $v_{s_j}[k_2] = 1$ for $j \geq 2$, this implies that s_1^* must contains both $v_{j_1}^i$ and $v_{j_2}^i$, a contradiction. Thus, we get $Opt(G) \geq |Z_1^*|$, where $Opt(G)$ is the size of a maximum independent set in G.

Now, let us check that for any independent set $Z \subseteq P$ in G, we can construct a solution $S = \{s_1, \ldots, s_n\}$ such that for any $l \in Z$, $v_{s_1}[l] = 0$, and for any $i \neq 1$, $v_{s_i}[l] = 1$. To construct such a solution, we have to prove that we can add in s_1 all the vectors v such that $\exists l \in Z$ such that $v[l] = 0$. However, this last statement is clearly true as for any $i \in [m]$, there is at most one vector v_j^i with $Z(v_j^i) \subseteq Z$.

Thus, any $\rho(n, m)$ approximation algorithm gives us a set Z with $|Z| \geq \frac{|Z_1^*|}{\rho(n,m)} \geq \frac{\Delta}{n\rho(n,m)}$, and we get a ratio of $\frac{p+(n-1)(p-\frac{\Delta}{n\rho(n,m)})}{2p-\Delta} \leq n - \frac{n-1}{n\rho(n,m)}$ for $\Delta = p$.

$\square$

Remark 1. We can get, for example, $\rho(n, m) = mn^{m-1}$ using the following algorithm. For any $i \in [m]$, let $G^i = (A_1^i, \ldots, A_n^i)$ be the i-th complete n-partite graph. W.l.o.g., suppose that A_1^1 is the largest set among $\{A_j^i\}$. Notice that $|A_1^1| \geq \frac{Opt}{m}$. The algorithm starts by setting $S_1 = A_1^1$ (S_1 may not be an independent set). Then, for any i from 2 to m, the algorithm set $S_i = S_{i-1} \setminus (\cup_{j \neq j_0} A_j^i)$, where $j_0 = \arg\max_j\{|S_{i-1} \cap A_j^i|\}$. Thus, for any i we have $|S_i| \geq \frac{|S_{i-1}|}{n}$, and S_i is an independent set when considering only edges from $\cup_{l=1}^i G^l$. Finally, we get an independent set of G of size $|S_m| \geq \frac{S_1}{n^{m-1}} \geq \frac{Opt}{mn^{m-1}}$.

3   Exact Resolution of Sparse Instances

The section is devoted to the exact resolution of $\min \sum 0$ for sparse instances where each vector has at most one zero ($\#0 \leq 1$). As we have seen in Sect. 2, $(\min \sum 0)_{\#0 \leq 1}$ remains **NP**-hard (even for $n = 2$). Thus it is natural to ask if $(\min \sum 0)_{\#0 \leq 1}$ is polynomial-time solvable for fixed m (for general n). This section is devoted to answer positively to this question. Notice that we cannot extend this result to a more general notion of sparsity as $(\min \sum 0)_{\#0 \leq 2}$ is

APX-complete for $m = 3$ [5]. However, the question if $(\min \sum 0)_{\#0 \leq 1}$ is fixed parameter tractable when parameterized by m is left open.

We first need some definitions, and refer the reader to Fig. 3 where an example is depicted.

Definition 1

- *For any* $l \in [p], i \in [m]$, *we define* $B^{(l,i)} = \{v_j^i : v_j^i[l] = 0\}$ *to be the set of vectors of set* i *that have their (unique) zero at position* l. *For the sake of homogeneous notation, we define* $B^{(p+1,i)} = \{v_j^i : v_j^i$ *is a 1 vector*$\}$. *Notice that the* $B^{(l,i)}$ *form a partition of all the vectors of the input, and thus an input of* $(\min \sum 0)_{\#0 \leq 1}$ *is completely characterized by the* $B^{(l,i)}$.
- *For any* $l \in [p+1]$, *the* **block** $B^l = \bigcup_{i \in [m]} B^{(l,i)}$.

Informally, the idea to solve $(\min \sum 0)_{\#0 \leq 1}$ in polynomial time for fixed m is to parse the input block after block using a dynamic programming algorithm. When arriving at block B^l we only need to remember for each $c \subseteq [m]$ the number x_c of "partial stacks" that have only one vector for each $V^i, i \in c$. Indeed, we do not need to remember what is "inside" these partial stacks as all the remaining vectors from $B^{l'}, l' \geq l$ cannot "match" (*i.e.* have their zero in the same position) the vectors in these partial stacks.

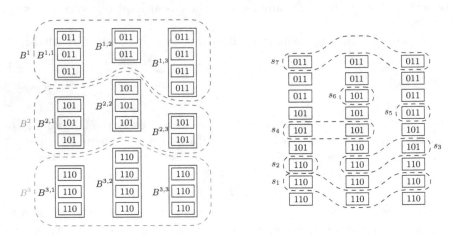

Fig. 3. Left: instance I of $(\min \sum 0)_{\#0 \leq 1}$ partitioned into blocks. Right: A profile $P = \{x_{\{0\}} = 2, x_{\{1\}} = 1, x_{\{2\}} = 1, x_{\{3\}} = 1, x_{\{1,2\}} = 1, x_{\{1,3\}} = 1, x_{\{2,3\}} = 1, x_{\{1,2,3\}} = 1\}$ encoding a set S of partial stacks of I containing two empty stacks. The support of s_7 is $sup(s_7) = \{1, 3\}$ and has cost $c(s_7) = 1$.

Definition 2

- *A* **partial stack** $s = \{v_{i_1}^s, \ldots, v_{i_k}^s\}$ *of* I *is such that* $\{i_x \in [m], x \in [k]\}$ *are pairwise disjoints, and for any* $x \in [k]$, $v_{i_x}^s \in V^{i_x}$. *The* **support** *of a partial stack* s *is* $sup(s) = \{i_x, x \in [k]\}$. *Notice that a stack* s *(i.e. non partial) has* $sup(s) = [m]$.

– *The cost is extended in the natural way: the cost of a partial stack $c(s) = c(\bigwedge_{x \in [k]} v_{i_x}^s)$ is the number of zeros of the bitwise AND of the vectors of s.*

We define the notion of profile as follows:

Definition 3. *A **profile** $P = \{x_c, c \subseteq [m]\}$ is a set of 2^m positive integers such that $\sum_{c \subseteq [m]} x_c = n$.*

In the following, a profile will be used to encode a set S of n partial stacks by keeping a record of their support. In other words, $x_c, c \subseteq [m]$ will denote the number of partial stacks in S of support c. This leads us to introduce the notion of reachable profile as follows:

Definition 4. *Given two profiles $P = \{x_c : c \subseteq [m]\}$ and $P' = \{x'_{c'} : c' \subseteq [m]\}$ and a set $S = \{s_1, \ldots, s_n\}$ of n partial stacks, P' is said reachable from P through S iff there exist n couples $(s_1, c_1), (s_2, c_2), \ldots, (s_n, c_n)$ such that:*

– *For each couple (s, c), $sup(s) \cap c = \emptyset$.*
– *For each $c \subseteq [m], |\{(s_j, c_j) : c_j = c, j = 1, \ldots, n\}| = x_c$. Intuitively, the configuration c appears in exactly x_c couples.*
– *For each $c' \subseteq [m], |\{(s_j, c_j) : sup(s_j) \cup c_j = c', j = 1, \ldots, n\}| = x'_{c'}$. Intuitively, there exist exactly $x'_{c'}$ couples that, when associated, create a partial of profile c'.*

Given two profiles P and P', P' is said reachable from P, if there exists a set S of n partial stacks such that P' is reachable from P through S.

Intuitively, a profile P' is reachable from P through S if every partial stack of the set encoded by P can be assigned to a unique partial stack from S to obtain a set of new partial stacks encoded by P'.

Remark that, given a set of partial stacks S only their profile is used to determine whether a profile is reachable or not. An example of a reachable profile is given on Fig. 4.

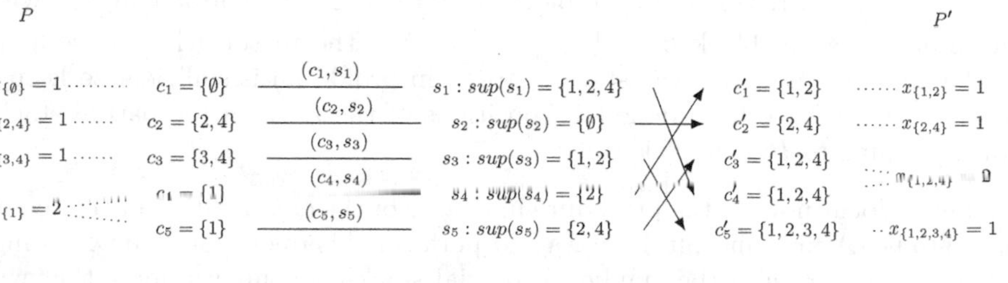

Fig. 4. Example of a profile $P' = \{x_{\{1,2\}} = 1, x_{\{2,4\}} = 1, x_{\{1,2,4\}} = 2, x_{\{1,2,3,4\}} = 1\}$ reachable from $P = \{x_{\{\emptyset\}} = 1, x_1 = 2, x_{\{2,4\}} = 1, x_{\{3,4\}} = 1\}$ through $S = \{s_1 : sup(s_1) = \{1, 2, 4\}, s_2 : sup(s_2) = \{\emptyset\}, s_3 : sup(s_3) = \{1, 2\}, s_4 : sup(s_4) = \{2\}, s_5 : sup(s_5) = \{2, 4\}\}$.

We introduce now the following problem Π. We then show that this problem can be used to solve $(\min\sum 0)_{\#0\leq 1}$ problem, and we present a dynamic programming algorithm that solves Π in polynomial time when m is fixed.

Optimization Problem 2 Π

Input (l, P) with $l \in [p+1]$, P a profile.

Output   A set of n partial stacks $S = \{s_1, s_2, \ldots, s_n\}$ such that S is a partition of $\mathcal{B} = \bigcup_{l' \geq l} B^{l'}$ and for every $c \subseteq [m]$, $|\{s \in S | sup(s) = [m] \setminus c\}| = x_c$ and such that $c(S) = \sum_{j=1}^{n} c(s_j)$ is minimum.

Remark that an instance I of $(\min\sum 0)_{\#0\leq 1}$ can be solved optimally by solving optimally the instance $I' = (1, P = \{x_\emptyset = n, x_c = 0, \forall c \neq \emptyset\})$ of Π. The optimal solution of I' is indeed a set of n partial disjoint stacks of support $[m]$ of minimum cost.

We are now ready to define the following dynamic programming algorithm that solves any instance (l, P) of Π by parsing the instance block after block and branching for each of these blocks on every reachable profile.

Function MinSumZeroDP(l, P)

> **if** $k == p+1$ **then**
> > **return** 0;
>
> **return** $\min(c(S') + \text{MinSumZeroDP}(l+1, P'))$, with P' reachable from P
> through S', where S' partition of B^l;

Note that this dynamic programming assumes the existence of a procedure that enumerates *efficiently* all the profiles P' that are reachable from P. The existence of such a procedure will be shown thereafter.

Lemma 3. *For any instance of Π (l, P), MinSumZeroDP$(l, P) = Opt(l, P)$.*

Proof. Lemma 3 is true as in a given block l, the algorithm tries every reachable profile, and the zeros of vectors in blocks $\mathcal{B} = \bigcup_{l' < l} B^{l'}$ cannot be matched with those of vectors in block $\mathcal{B}' = \bigcup_{l' \geq l} B^{l'}$. This is the reason why the support of the already created partial stacks (stored in profile P) is sufficient to keep a record of what have been done (the positions of the zeros in the partial stacks corresponding to P is not relevant). □

Let us focus now on the procedure in charge of the enumeration of the reachable profile. A first and intuitive way to perform this operation is by guessing, for all $c, c' \subseteq [m]$, $y_{c,c'}$ the number of partial stacks in configuration c that will be turned into configuration c' with vectors of current block B^l. For each such guess it is possible to greedily verify that each $y_{c,c'}$ can be satisfied with the vectors of the current block. As each of the $y_{c,c'}$ can take values from 0 to n and c and c' can be both enumerated in $\mathcal{O}^*(n^{2^m})$, the previous algorithm runs in $\mathcal{O}^*(n^{2^{2m}})$.

This complexity can be improved as follows. The idea is to enumerate every possible profile P' and to verify using another dynamic programming algorithm if such a P' is reachable from P. We define $Aux_{P'}(P, X)$, that verifies if P' is reachable from P by using all vectors of X. If $X = \emptyset$, then the algorithm returns whether P is equal to P' or not. Otherwise, we consider the first vector v of X (we fix any arbitrary order) for which a branching is done on every possible assignment of v. More formally, the algorithm returns $\bigvee_{c \subseteq [m], x_c > 0, c \cap sup(v) = \emptyset} Aux_{P'}(P_2 = \{x'_l\}, X \setminus \{v\})$, where $x'_l = x_l - 1$ if $l = c$, $x'_l = x_l + 1$ if $l = c \cup sup(v)$, and $x'_l = x_l$ otherwise.

Using Aux in MinSumZeroDP, we get the following theorem.

Theorem 3. $(\min \sum 0)_{\#0 \leq 1}$ *can be solved in* $\mathcal{O}^*(n^{2^{m+2}})$.

We compute the overall complexity as follows: for each of the pn^{2^m} possible values of the parameters of MinSumZeroDP, the algorithm tries the n^{2^m} profiles P', and run for each one $Aux_{P'}$ in $\mathcal{O}^*(n^{2^m}nm)$ (the first parameter of Aux can take n^{2^m} values, and the second nm as we just encode how many vectors left in X).

References

1. Ausiello, G., Paschos, V.T.: Reductions, completeness and the hardness of approximability. Eur. J. Oper. Res. **172**(3), 719–739 (2006)
2. Bansal, N., Khot, S.: Inapproximability of hypergraph vertex cover and applications to scheduling problems. In: Abramsky, S., Gavoille, C., Kirchner, C., Meyer auf der Heide, F., Spirakis, P.G. (eds.) ICALP 2010. LNCS, vol. 6198, pp. 250–261. Springer, Heidelberg (2010)
3. Bougeret, M., Duvillié, G., Giroudeau, R.: Approximability and exact resolution of the multidimensional binary vector assignment problem. Research report, Lirmm; Université de Montpellier, HAL id: lirmm-01310648, May 2016
4. Dokka, T., Bougeret, M., Boudet, V., Giroudeau, R., Spieksma, F.C.R.: Approximation algorithms for the wafer to wafer integration problem. In: Erlebach, T., Persiano, G. (eds.) WAOA 2012. LNCS, vol. 7846, pp. 286–297. Springer, Heidelberg (2013)
5. Dokka, T., Crama, Y., Spieksma, F.C.: Multi-dimensional vector assignment problems. Discrete Optim. **14**, 111–125 (2014)
6. Duvillié, G., Bougeret, M., Boudet, V., Dokka, T., Giroudeau, R.: On the complexity of wafer-to-wafer integration. In: Paschos, V.T., Widmayer, P. (eds.) CIAC 2015. LNCS, vol. 9079, pp. 208–220. Springer, Heidelberg (2015)
7. Papadimitriou, C., Yannakakis, M.: Optimization, approximation, and complexity classes. In: Proceedings of the Twentieth Annual ACM Symposium on Theory of Computing, pp. 229–234. ACM (1988)
8. Reda, S., Smith, G., Smith, L.: Maximizing the functional yield of wafer-to-wafer 3-D integration. IEEE Trans. Very Large Scale Integr. (VLSI) Syst. **17**(9), 1357–1362 (2009)

Towards a Polynomial Equivalence Between {k}-Packing Functions and k-Limited Packings in Graphs

Valeria Leoni[1,2]($\boxtimes$) and María Patricia Dobson[1]

[1] Depto. de Matemática, Facultad de Ciencias Exactas, Ingeniería y Agrimensura,
Universidad Nacional de Rosario, Rosario, Argentina
`valeoni@fceia.unr.edu.ar`
[2] CONICET, Rosario, Argentina

Abstract. Given a positive integer k, the {k}-*packing function problem*
({k}PF) is to find in a given graph G, a function f of maximum weight
that assigns a non-negative integer to the vertices of G in such a way
that the sum of $f(v)$ over each closed neighborhood is at most k. This
notion was recently introduced as a variation of the k-limited packing
problem (kLP) introduced in 2010, where the function was supposed to
assign a value in $\{0,1\}$. For all the graph classes explored up to now,
{k}PF and kLP have the same computational complexity. It is an open
problem to determine a graph class where one of them is NP-complete
and the other, polynomially solvable. In this work, we first prove that
{k}PF is NP-complete for bipartite graphs, as kLP is known to be. We
also obtain new graph classes where the complexity of these problems
would coincide.

Keywords: Computational complexity · $\mathcal{F}$-free graph · Bipartite graph

1 Basic Definitions and Preliminaries

All the graphs in this paper are simple, finite and undirected.

For a graph G, $V(G)$ and $E(G)$ denote respectively its vertex and edge sets.
For any $v \in V(G)$, $N_G[v]$ is the *closed neighborhood* of v in G. For a given
graph G and a function $f : V(G) \to \mathbb{R}$, we denote $f(A) = \sum_{v \in A} f(v)$, where
$A \subseteq V(G)$. The *weight* of f is $f(V(G))$.

A graph H is *bipartite* if $V(G)$ is the union of two disjoint (possibly empty)
independent sets called *partite sets* of G. Equivalently, bipartite graphs are
defined as odd-cycle-free graphs, i.e. graphs that have no induced odd-cycle.

A graph is *complete* if $E(G)$ contains all edges corresponding to any pair of
distinct vertices from $V(G)$. The complete graph on n vertices is denoted by K_n.

Given G_1 and G_2 two graphs, the *strong product* $G_1 \otimes G_2$ is defined on the
vertex set $V(G_1) \times V(G_2)$, where two vertices u_1v_1 and u_2v_2 are adjacent if
and only if $u_1 = u_2$ and $(v_1, v_2) \in E(G_2)$, or $v_1 = v_2$ and $(u_1, u_2) \in E(G_1)$, or
$(v_1, v_2) \in E(G_2)$ and $(u_1, u_2) \in E(G_1)$.

© Springer International Publishing Switzerland 2016
R. Cerulli et al. (Eds.): ISCO 2016, LNCS 9849, pp. 160–165, 2016.
DOI: 10.1007/978-3-319-45587-7_14

Given a graph G and a positive integer k, a set $B \subseteq V(G)$ is a k-*limited packing* in G if each closed neighborhood has at most k vertices of B [7]. Observe that a k-limited packing in G can be considered as a function $f : V(G) \to \{0,1\}$ such that $f(N_G[v]) \leq k$ for all $v \in V(G)$. The maximum possible weight of a k-limited packing in G is denoted by $L_k(G)$. When $k = 1$, a k-limited packing in G is a *2-packing* in G and $L_k(G)$ is the known *packing number* of G, $\rho(G)$.

This concept is a good model for many utility location problems in operations research, for example the problem of locating garbage dumps in a city. In most of them, the utilities—garbage dumps—are necessary but probably obnoxious. That is why it is of interest to place the maximum number of utilities in such a way that no more than a given number of them (k) is near to each agent in a given scenario.

In order to expand the set of utility location problems to be modeled, the concept of $\{k\}$-packing function of a graph was introduced in [4] as a variation of a k-limited packing. Recalling the problem of locating garbage dumps in a given city, if a graph G and a positive integer k model the scenario, when dealing with $\{k\}$-packing functions we are allowed to locate more than one garbage dump in any vertex of G subject to there are at most k garbage dumps in each closed neighborhood. Formally, given a graph G and a positive integer k, a $\{k\}$-*packing function* of G is a function $f : V(G) \to \mathbb{Z}_+^0$ such that for all $v \in V(G)$, $f(N_G[v]) \leq k$. The maximum possible weight of a $\{k\}$-packing function of G is denoted by $L_{\{k\}}(G)$ [4].

Fig. 1. A graph G with $L_3(G) = 4$ and $L_{\{3\}}(G) = 6$.

Since any k-limited packing in G can be seen as a $\{k\}$-packing function of G, it is clear to see that $L_k(G) \leq L_{\{k\}}(G)$. For K_3, $L_3(K_3) = L_{\{3\}}(K_3) = 3$. Nevertheless, for the graph in Fig. 1, these numbers do not coincide.

The above definitions induce the study—started in [2,4]—of the computational complexity of the following decision problems:

k **LIMITED PACKING**, fixed $k \in \mathbb{Z}_+$ (kLP) [2]

Instance: (G,l), where G is a graph and $l \in \mathbb{Z}_+$.
Question: Does G have a k-limited packing of size at least l?

$\{k\}$-**PACKING FUNCTION**, fixed $k \in \mathbb{Z}_+$ ($\{k\}$PF) [4]

Instance: (G,l), where G is a graph and $l \in \mathbb{Z}_+$.
Question: Does G have a $\{k\}$-packing function of weight at least l?

Table 1. "NP-c", "P" and "?" mean NP-complete, polynomial and open problem, resp.

Class	k**LP**	$\{k\}$**PF**
General graphs	NP-c [3]	NP-c [5,6]
Strongly chordal	P [2]	P [4]
Dually chordal	?	P [5,6]
Doubly chordal	?	P [6]
P_4-lite	P [3]	P [4]
P_4-tidy	P [3]	P [4]
Bounded tree-width	P [4]	P [4]
Bounded clique-width	P [4]	P [4]
Split	NP-c [3]	NP-c [6]
Chordal	NP-c [3]	NP-c [6]
Bipartite	NP-c [3]	?

Table 1 summarizes the already known results on the complexity of $\{k\}$PF in contrast with kLP, for fixed $k \in \mathbb{Z}_+$.

It is an open problem to determine a graph class where one of these problems is NP-complete and the other, polynomially solvable.

In Sect. 2 we prove that $\{k\}$PF is NP-complete on bipartite graphs, answering in this way one of the open questions in Table 1.

In Sect. 3 we obtain new graph classes where the complexity of kLP and $\{k\}$PF would coincide.

2 $\{k\}$-Packing Functions on Bipartite Graphs

As Table 1 shows, it is already known that kLP is NP-complete on bipartite graphs [3]. The proof is based on a reduction from a variation of the classical domination problem on a bipartite graph to kLP on a bipartite graph.

In this section we state that also $\{k\}$PF is NP-complete for bipartite graphs. In this case the proof consists in a reduction from $\{k\}$PF in a general graph to $\{k\}$PF in a bipartite graph.

We have:

Theorem 1. *For every fixed $k \in \mathbb{Z}_+$, $\{k\}$PF is NP-complete on bipartite graphs.*

Proof. Let $k \in \mathbb{Z}_+$ be fixed. It is already known that $\{k\}$PF is NP-complete for general graphs [5].

Let (G, l) be an instance of $\{k\}$PF. We build a bipartite graph B in the following way (see Fig. 2). Let

$$X = \{x_v : v \in V(G)\} \cup \{x\}, \quad Y = \{y_v : v \in V(G)\} \cup \{y\}$$

be the partite sets of B. Let also

$$E(B) = \bigcup_{v \in V(G)} \{(x_v, y_u) : u \in N_G[v]\} \cup \{(x, y') : y' \in Y\}.$$

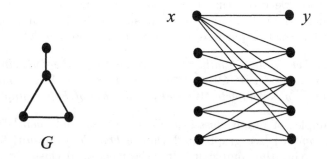

Fig. 2. Construction of B from a graph G in Theorem 1.

We will prove that

$$L_{\{k\}}(B) = L_{\{k\}}(G) + k.$$

On the one hand, let f be a $\{k\}$-packing function of G with weight $L_{\{k\}}(G)$. Consider the function $h : V(B) \to \mathbb{Z}_+^0$ defined as follows. For each $v \in V(G)$ let $h(x_v) = f(v)$ and $h(y_v) = 0$. Let also $h(x) = 0$ and $h(y) = k$. Notice that h is indeed a $\{k\}$-packing function of B with weight $L_{\{k\}}(G) + k$. Hence,

$$L_{\{k\}}(B) \geq L_{\{k\}}(G) + k.$$

On the other hand, let h be a $\{k\}$-packing function of B with weight $L_{\{k\}}(B)$. We can assume that h satisfies $h(x) = 0$ and $h(y_v) = 0$ for each $v \in V(G)$: if h does not satisfy these conditions, we can construct another $\{k\}$-packing function $\hat{h}$ of B of maximum weight, by defining $\hat{h}(x_v) = h(x_v)$ and $\hat{h}(y_v) = 0$ for each $v \in V(G)$, $\hat{h}(x) = 0$ and $\hat{h}(y) = \sum_{v \in V(G)} h(y_v) + h(x) + h(y)$. Now we construct a function $f : V(G) \to \mathbb{Z}_+^0$ by letting $f(v) = h(x_v)$ for each $v \in V(G)$. Clearly, f is a $\{k\}$-packing function of G with weight $L_{\{k\}}(B) - \hat{h}(y)$. Hence, $L_{\{k\}}(G) \geq L_{\{k\}}(B) - \hat{h}(y)$. Since $\hat{h}(y) = \sum_{v \in V(G)} h(y_v) + h(x) + h(y) = h(N_B[x]) \leq k$, it follows that

$$L_{\{k\}}(G) \geq L_{\{k\}}(B) - k.$$ $\square$

3   A General Result

Clearly, there are polynomial-time reductions from kLP ($\{k\}$PF) to $\{k\}$PF (kLP), since both problems are NP-complete in the general case. It is known a linear reduction from $\{k\}$PF to kLP that involves changes in the graph; more precisely, it is proved that $L_{\{k\}}(G) = L_k(G \otimes K_k)$, for every graph G and positive integer k [4]. This reduction is closed within some graph classes, for instance strongly chordal graphs and graphs with the parameter clique-width bounded by a constant. From these facts it is derived the polynomiality of $\{k\}$PF for strongly chordal graphs and graphs with the parameter clique-width bounded by a constant [4].

In this section we prove that the above reduction is closed within certain graph class defined by forbidden induced subgraphs, following the ideas of Theorem 9 in [1]. For this purpose, we consider the following definition:

Definition 1. *Let $\mathcal{F}$ be a family of graphs satisfying the following property: for every graph G in $\mathcal{F}$, $|V(G)| \geq 2$ and, for every $v \in V(G)$, no connected component of $G - v$ is complete. We call* **G** *the class of $\mathcal{F}$-free graphs.*

Some examples of graph classes in **G** are $\{house, hole, gem\}$-free graphs, $\{house, hole, domino\}$-free graphs and $\{house, hole, domino, sun\}$-free graphs. It is worth studying the complexities for the mentioned classes since they all have the parameter clique-width unbounded. For other examples, like distance-hereditary graphs which are $\{house, hole, domino, gem\}$-free graphs, the complexity of both problems is already known since they have the parameter clique-width bounded by a constant.

We can state and prove:

Theorem 2. *Consider the graph class* **G** *in Definition 1. For fixed positive integer k and graph G in* **G**, *$G \otimes K_k \in$* **G**.

Proof. Let k be a fixed positive integer and G be graph in **G**. We will prove that $G \otimes K_k \in$ **G**, i.e. we will prove that $G \otimes K_k$ is $\mathcal{F}$-free. Let G' be a subgraph of $G \otimes K_k$ induced by V' with $|V'| \geq 2$. Then V' is the disjoint union of sets V'_{v_j} with $j \in J$, where $1 \leq |J| \leq |V(G)|$.

When $|J| = 1$, $G' = K_k$ and thus $G' \notin \mathcal{F}$. When $|J| \geq 2$, consider the subgraph G'' of G induced by the vertices $\{v_1, \ldots, v_{|J|}\}$. Since G is $\mathcal{F}$-free, there is a vertex v_r with $r \in J$ and such that $G'' - v_r$ has a complete connected component. From the definition of $G \otimes K_k$, it is not difficult to see that $G' - v'_r$ has a complete connected component, where v'_r is any vertex in V'_{v_r}. Therefore, $G' \notin \mathcal{F}$. Since G' is arbitrary, this proves that $G \otimes K_k$ is $\mathcal{F}$-free, concluding that $G \otimes K_k \in$ **G**. $\square$

As a corollary, knowing from [4] that $L_{\{k\}}(G) = L_k(G \otimes K_k)$ for every graph G and positive integer k, we have:

Corollary 1. *Consider the graph class* **G** *in Definition 1. Then, for fixed positive integer k, $\{k\}$PF is solvable in polynomial time in the class* **G**, *provided that kLP is solvable in polynomial time in* **G**. *Besides, if $\{k\}$PF is NP-complete in* **G**, *then kLP is NP-complete in* **G**.

4   Final Remarks

It remains an open problem to know if there exists a graph class where one of the problems considered in this work is NP-complete and the other can be solved in polynomial time. Corollary 1 helps to keep working on this line of research. Besides, it is an open problem to determine the complexity of kLP for dually chordal graphs, as shown in Table 1, or at least for one of its maximal subclasses constituted by doubly chordal graphs (also shown in Table 1).

Acknowledgements. This work was partially supported by grants PIP CONICET 11220120100277 (2014–2017), 1ING 391 (2012–2016) and MINCyT-MHEST SLO 14/09 (2015–2017).

References

1. Bonomo, F., Durán, G., Marenco, J.: Exploring the complexity boundary between coloring and list-coloring. Ann. Oper. Res. **169**(1), 3–16 (2009)
2. Dobson, M.P., Leoni, V., Nasini, G.: The k-limited packing and k-tuple domination problems in strongly chordal, P4-tidy and split graphs. Electron. Notes Discrete Math. **36**, 559–566 (2010)
3. Dobson, M.P., Leoni, V., Nasini, G.: The multiple domination and limited packing problems in graphs. Inf. Proc. Lett. **111**, 1108–1113 (2011)
4. Leoni, V.A., Hinrichsen, E.G.: $\{k\}$-Packing functions of graphs. In: Fouilhoux, P., Gouveia, L.E.N., Mahjoub, A.R., Paschos, V.T. (eds.) ISCO 2014. LNCS, vol. 8596, pp. 325–335. Springer, Heidelberg (2014)
5. Dobson, M.P., Hinrichsen, E., Leoni, V.: NP-completeness of the $\{k\}$-packing function problem in graphs. Electron. Notes Discrete Math. **50**, 115–120 (2015)
6. Dobson, M.P, Hinrichsen, E., Leoni, V.: On the complexity of the $\{k\}$-packing function problem. Int. Trans. Oper. Res. (2016, in press)
7. Gallant, R., Gunther, G., Hartnell, B., Rall, D.: Limited Packings in graphs. Electron. Notes Discrete Math. **30**, 15–20 (2008)

Exact Solution Methods for the k-Item Quadratic Knapsack Problem

Lucas Létocart[1] and Angelika Wiegele[2(✉)]

[1] Université Paris 13, Sorbonne Paris Cité, LIPN, CNRS, (UMR 7030),
93430 Villetaneuse, France
lucas.letocart@lipn.univ-paris13.fr
[2] Alpen-Adria-Universität Klagenfurt, 9020 Klagenfurt am Wörthersee, Austria
angelika.wiegele@aau.at

Abstract. The purpose of this paper is to solve the 0–1 k-item quadratic knapsack problem $(kQKP)$, a problem of maximizing a quadratic function subject to two linear constraints. We propose an exact method based on semidefinite optimization. The semidefinite relaxation used in our approach includes simple rank one constraints, which can be handled efficiently by interior point methods. Furthermore, we strengthen the relaxation by polyhedral constraints and obtain approximate solutions to this semidefinite problem by applying a bundle method. We review other exact solution methods and compare all these approaches by experimenting with instances of various sizes and densities.

Keywords: Quadratic programming · 0–1 knapsack · k-cluster · Semidefinite programming

1 Introduction

The 0–1 k-item quadratic knapsack problem consists of maximizing a quadratic objective function subject to a linear capacity constraint with an additional equality cardinality constraint:

$$(kQKP) \begin{cases} \max f(x) = \sum_{i=1}^{n} \sum_{j=1}^{n} c_{ij} x_i x_j & \\ \text{s.t.} \ \sum_{j=1}^{n} a_j x_j \leq b & (1) \\ \sum_{j=1}^{n} x_j = k & (2) \\ x_j \in \{0, 1\} \quad j = 1, \ldots, n \end{cases}$$

where n denotes the number of items, and all the data, k (number of items to be filled in the knapsack), a_j (weight of item j), c_{ij} (profit associated with the selection of items i and j) and b (capacity of the knapsack) are nonnegative integers. Without loss of generality, matrix $C = (c_{ij})$ is assumed to be symmetric. Moreover, we assume that $\max_{j=1,\ldots,n} a_j \leq b < \sum_{j=1}^{n} a_j$ in order to avoid either trivial solutions or variable fixing via constraint (1). Let us denote by k_{max} the largest number of items which could be filled in the knapsack, that is the largest number of the smallest a_j whose sum does not exceed b. We can assume that

© Springer International Publishing Switzerland 2016
R. Cerulli et al. (Eds.): ISCO 2016, LNCS 9849, pp. 166–176, 2016.
DOI: 10.1007/978-3-319-45587-7_15

$k \in \{2, \ldots, k_{max}\}$, where k_{max} can be found in $O(n)$ time [2]. Otherwise, either the value of the problem is equal to $\max_{i=1,\ldots,n} c_{ii}$ (for $k = 1$), or the domain of $(kQKP)$ is empty (for $k > k_{max}$).

$(kQKP)$ is an NP-hard problem as it includes two classical NP-hard subproblems, the k-cluster problem [6] by dropping constraint (1), and the quadratic knapsack problem [18] by dropping constraint (2). Even more, the work of Bhaskara et al. [4] indicates that approximating k-cluster within a polynomial factor might be a harder problem than Unique Games. Rader and Woeginger [19] state negative results concerning the approximability of QKP if negative cost coefficients are present.

Applications of $(kQKP)$ cover those found in previous references for k-cluster or classical quadratic knapsack problems (e.g., task assignment problems in a client-server architecture with limited memory), but also multivariate linear regression and portfolio selection. Specific heuristic and exact methods including branch-and-bound and branch-and-cut with surrogate relaxations have been designed for these applications (see, e.g., [3,5,9,17,22]).

The purpose of this paper is twofold.

1. We introduce a new algorithm for solving $(kQKP)$ and
2. we briefly review other state of the art methods and compare the methods by running numerical experiments.

Our new algorithm consists of a branch-and-bound framework using

– a combination of a semidefinite relaxation and polyhedral cutting planes to obtain tight upper bounds and
– fast hybrid heuristics [16] for computing high quality lower bounds.

This paper is structured as follows. In Sect. 2 a semidefinite relaxation is derived, followed by a discussion of solving the semidefinite problems in Sect. 3. The relaxation is used inside a branch-and-bound framework, the various components of this branch-and-bound algorithm are discussed in Sect. 4. Other methods for solving $(kQKP)$ and numerical results are presented in Sect. 5 and Sect. 6 concludes.

Notation. We denote by e the vector of all ones of appropriate size. $\mathrm{diag}(X)$ refers to diagonal of X as a vector and $\mathrm{Diag}(v)$ is the diagonal matrix having diagonal v.

2   A Semidefinite Relaxation of $(kQKP)$

In order to develop a branch-and-bound algorithm for solving $(kQKP)$ to optimality we aim in finding strong upper bounds. Semidefinite optimization proved to provide such strong bounds, see e.g. [1,20,21].

A straightforward way to obtain a semidefinite relaxation is the following. Express all functions involved as quadratic functions, i.e. functions in xx^t, replace the product xx^t by a matrix X and get rid of non-convexities by relaxing $X = xx^t$ to $X \succeq xx^t$.

Hence, we apply the following changes:

- Replace the constraint $e^t x = k$ by the constraint $(e^t x - k)^2 = 0$.
- As for the capacity constraint, define b' to be the sum of the weights of the k smallest items. Clearly, $b' \leq a^t x$ is a valid constraint for $(kQKP)$. Combining this redundant constraint with the capacity constraint we obtain $(b' - a^t x)(b - a^t x) \leq 0$.
- Transform the problem to a ± 1 problem by setting $y = 2x - e$.
- Relax the problem by relaxing $Y = yy^t$ to $Y \succeq yy^t$, i.e., dropping the constraint Y being of rank one.

This procedure yields the following semidefinite problem:

$$
\begin{aligned}
\max \quad & \langle \tilde{C}, Y \rangle \\
\text{s.t.} \quad & \operatorname{diag}(Y) = e \\
& \langle \tilde{E}, Y \rangle = 0 \\
& \langle \tilde{A}, Y \rangle \leq (b - b')^2 \\
& Y \succeq 0
\end{aligned}
\qquad (SDP_1)
$$

with $\tilde{E} = \tilde{e}\tilde{e}^t$, $\tilde{e} = \begin{pmatrix} n - 2k \\ e \end{pmatrix}$, $\tilde{A} = \tilde{a}\tilde{a}^t$, $\tilde{a} = \begin{pmatrix} a^t e - (b + b') \\ a \end{pmatrix}$, and appropriate $\tilde{C}$.

Observation 1. *(SDP_1) has no strictly feasible point and thus Slater's condition does not hold.*

Proof. Note that $\langle \tilde{E}, Y \rangle = \tilde{e}^t Y \tilde{e} = 0$ together with $Y \succeq 0$ implies Y being singular and thus every feasible solution is singular. □

Observe that $\tilde{e} = \begin{pmatrix} n - 2k \\ e \end{pmatrix}$ is an eigenvector to the eigenvalue 0 of every feasible Y. Now consider matrix $V = \begin{pmatrix} \frac{1}{2k-n} e^t \\ I_n \end{pmatrix}$. V spans the orthogonal complement of the span of eigenvector $\tilde{e}$. Set $Y = VXV^t$ to "project out" the 0-eigenvalue and consider the $n \times n$ matrix X instead of the $(n+1) \times (n+1)$ matrix Y. The relationship between X and Y is simply given by

$$
Y = VXV^t = \begin{pmatrix} \frac{1}{(2k-n)^2} e^t X e & \frac{1}{2k-n}(Xe)^t \\ \frac{1}{2k-n} Xe & X \end{pmatrix}.
$$

Looking at the effect of the constraints of (SDP_1) on matrix X, we derive the following conditions.

- From $\mathrm{diag}(Y) = e$ we obtain the constraints

$$e^t X e = (2k - n)^2$$
$$\mathrm{diag}(X) = e$$

- The left-hand side of constraint $\langle \tilde{E}, Y \rangle = 0$ translates into

$$\langle \tilde{E}, Y \rangle = \langle \tilde{E}, VXV^t \rangle = \langle V^t \tilde{E} V, X \rangle = \langle 0, X \rangle = 0$$

and the constraint becomes obsolete.
- Constraint $\langle \tilde{A}, Y \rangle \leq (b - b')^2$ yields the following.

$$\langle \tilde{A}, Y \rangle = \langle \tilde{A}, VXV^t \rangle = \langle V^t \tilde{A} V, X \rangle =$$
$$= \langle (\frac{a^t e - (b + b')}{2k - n} e + a)(\frac{a^t e - (b + b')}{2k - n} e + a)^t, X \rangle$$

Hence,

$$\langle (\frac{a^t e - (b + b')}{2k - n} e + a)(\frac{a^t e - (b + b')}{2k - n} e + a)^t, X \rangle \leq (b - b')^2$$

Defining $\bar{a} = (\frac{a^t e - (b + b')}{2k - n} e + a)$ we finally obtain

$$\begin{aligned}
\max \quad & \langle \bar{C}, X \rangle \\
\text{s.t.} \quad & \mathrm{diag}(X) = e \\
& \langle E, X \rangle = (2k - n)^2 \\
& \langle A, X \rangle \leq (b - b')^2 \\
& X \succeq 0
\end{aligned} \qquad (SDP)$$

where $E = ee^t$, $A = \bar{a}\bar{a}^t$, and appropriate cost matrix $\bar{C}$.

Strengthening the Relaxation. Since we derived a relaxation from a problem in ± 1 variables, we can further tighten the bound by adding the well known *triangle inequalities* to the semidefinite relaxation (SDP). These are for any triple $1 \leq i < j < k \leq n$:

$$\begin{aligned}
x_{ij} + x_{ik} + x_{jk} &\geq -1 \\
-x_{ij} - x_{ik} + x_{jk} &\geq -1 \\
-x_{ij} + x_{ik} - x_{jk} &> -1 \\
x_{ij} - x_{ik} - x_{jk} &\geq -1
\end{aligned} \qquad (1)$$

For several problems formulated in ± 1 variables adding these constraints significantly improves the bound, see e.g. [20]. The set of matrices satisfying all triangle-inequalities is called the *metric polytope* and is denoted by MET. Thus, the strengthend semidefinite relaxation reads

$$\begin{aligned}
\max \quad & \langle \bar{C}, X \rangle \\
\text{s.t.} \quad & \operatorname{diag}(X) = e \\
& \langle E, X \rangle = (2k - n)^2 \\
& \langle A, X \rangle \leq (b - b')^2 \\
& X \in MET \\
& X \succeq 0
\end{aligned} \qquad (SDP_{MET})$$

3 Solving the Semidefinite Relaxations

3.1 Solving the Basic Relaxation (SDP)

The most prominent methods for solving semidefinite optimization problems are interior point methods. The interior point method is an iterative algorithm where in each iteration Newton's method is applied in order to compute new search directions.

Consider the constraints $\mathcal{A}(X) = (\vdots)$ with $\mathcal{A}(X) = \begin{pmatrix} \langle A_1, X \rangle \\ \langle A_2, X \rangle \\ \vdots \\ \langle A_m, X \rangle \end{pmatrix}$. In each iteration we determine a search direction Δy (y are variables in the dual semidefinite problem) by solving the system $M \Delta y = rhs$ where

$$m_{ij} = \operatorname{trace}(Z^{-1} A_j X A_i).$$

Z denotes the (positive definite) matrix variable of the dual semidefinite program.

Forming this system matrix requires $O(mn^3 + m^2 n^2)$ steps and is among the most time-consuming operations inside the interior point algorithm. (The other time-consuming steps are maintaining positive definiteness of the matrices X and Z and linear algebra operations such as forming inverse matrices.)

The primal-dual pair of (SDP) in variables (X, s, y, Z, t) is given as follows.

$$\max \big\{ \langle \bar{C}, X \rangle$$
$$\text{s.t.} \quad \operatorname{diag}(X) = e, \langle E, X \rangle = (2k - n)^2, \langle A, X \rangle + s = (b - b')^2, X \succeq 0, s \geq 0 \big\}$$

$$\min \big\{ e^t y_{1:n} + (n - 2k)^2 y_{n+1} + (b - b')^2 y_{n+2}$$
$$\text{s.t.} \quad \operatorname{Diag}(y_{1:n}) + y_{n+1} E + y_{n+2} A - Z = \bar{C}, y_{n+2} - t = 0, Z \succeq 0, t \geq 0 \big\}$$

Hence, the set of constraints is rather simple and the system matrix M reads

$$\begin{pmatrix} Z^{-1} \circ X & \text{diag}(Z^{-1}EX) & \text{diag}(Z^{-1}AX) \\ \text{diag}(Z^{-1}EX)^t & \langle E, Z^{-1}EX \rangle & \langle E, Z^{-1}AX \rangle \\ \text{diag}(Z^{-1}AX)^t & \langle A, Z^{-1}EX \rangle & \langle A, Z^{-1}AX \rangle + \frac{s}{t} \end{pmatrix}.$$

Even more, all data matrices have rank one which can be exploited when computing the inner products, e.g.,

$$\langle A, Z^{-1}AX \rangle = \text{trace}(\bar{a}\bar{a}^t Z^{-1} \bar{a}\bar{a}^t X) = (\bar{a}^t Z^{-1} \bar{a})(\bar{a}^t X \bar{a})$$

Thus, the computation of the inner products of the matrices simplifies and computing the system matrix can be reduced from $O(mn^3 + m^2 n^2)$ to $O(mn^2 + m^2 n)$. And since $m = n + 2$ in our case, we end up with $O(n^3)$.

Hence, (SDP) can be solved efficiently by interior point methods.

3.2 Solving the Strengthened Relaxation (SDP_{MET})

Problem (SDP_{MET}) has a considerably larger number of constraints than (SDP). Remember that $X \in MET$ is described by $4\binom{n}{3}$ linear inequalities and thus solving (SDP_{MET}) by interior point methods is intractable. An alternative has been proposed in [12]. Therein the concept of *bundle methods* is used, in order to obtain an approximate optimizer on the dual functional and thus getting a valid upper bound on (SDP_{MET}), leading to a valid upper bound on $(kQKP)$.

Bundle methods have been developed to minimize nonsmooth convex functions. To characterize the problem to be solved, an oracle has to be supplied that evaluates the function at a given point and computes an ϵ-subgradient. The set of points, function values, and subgradients is collected in a "bundle", which is used to construct a cutting plane model minorizing the function to be minimized. By doing a sequence of *descent steps* the cutting plane model is refined and one gets closer to the minimizer of the function.

We will apply the bundle method to minimize the dual functional of (SDP_{MET}). Let

$$\mathcal{X} = \{X \succeq 0 \colon \text{diag}(X) = e, \ \langle E, X \rangle = (2k - n)^2, \ \langle A, X \rangle \le (b - b')^2\}$$

i.e., the feasible region of (SDP). We introduce the dual functional

$$\begin{aligned} f(\gamma) &= \max_{X \in \mathcal{X}} \{\langle \bar{C}, X \rangle + \gamma^t(e - T(X))\} \\ &= e^t\gamma + \max_{X \subset \mathcal{X}} \langle \bar{C} - T^t(\gamma), X \rangle \end{aligned} \qquad (2)$$

where $T(X) \le e$ denotes the triangle inequalities (1). Minimizing $f(\gamma)$ over $\gamma \ge 0$ gives a valid upper bound on $(kQKP)$. In fact, any $\tilde{\gamma} \ge 0$ gives a valid upper bound

$$z^* = \min_{\gamma \ge 0} f(\gamma) \le f(\tilde{\gamma}) \text{ for any } \tilde{\gamma} \ge 0.$$

Since we use this bound inside a branch-and-bound framework, this allows us to stop early and prune a node as soon as $f(\tilde{\gamma})$ is smaller than some known lower bound. Furthermore, we do not rely on getting to the optimum. We will stop once we are "close" to optimum and branch, rather than investing time in dropping the bound by a tiny number.

Evaluating function (2) (the most time consuming step in the bundle method) amounts in solving (SDP) (with varying cost matrix), which can be done efficiently as discussed in the previous section. Having the maximizer X^* of (SDP), i.e. the function evaluation, a subgradient is given by $g^* = e - T(X^*)$.

Dynamic Version of the Bundle Method. The number of variables γ in (2) is $4\binom{n}{3}$. This number is substantially larger than the dimension of the problem and we are interested only in those inequalities that are likely to be active at the optimum. Thus, we do not consider *all* triangle inequalities but work with a subset that is updated on a regular basis, say every fifth descent step. The update consists of

1. adding the m inequalities being most violated by the current iterate X and
2. removing constraints with γ close to 0 (an indicator for an inactive constraint).

In this way we are able to efficiently run the bundle algorithm by keeping the size of the variable vector γ reasonably small.

4   Branch and Bound

We develop an exact solution method for solving $(kQKP)$ by designing a branch-and-bound framework using relaxation (SDP_{MET}) discussed above for getting upper bounds.

The remaining tools of our branch-and-bound algorithm are described in this section.

4.1   Heuristics for Obtaining Lower Bounds

We use two heuristics to obtain a global lower bound inside our algorithm: one that is executed at the root node and another one that is called at each other node in the branch-and-bound tree.

As a heuristic method at the root node we chose the primal heuristic denoted by H_{pri} in [16], which is an adaption of a well-known heuristic developed by Billionnet and Calmels [7] for the classical quadratic knapsack problem (QKP). This primal heuristic combines a greedy algorithm with local search.

At each node of the branch-and-bound tree, we apply a variable fixation heuristic inspired from H_{sdp} [16]. This heuristic method uses the solution of the semidefinite relaxation obtained at each node, it fixes variables under some treshold $\epsilon > 0$ to zero and applies the primal heuristic over the reduced problem. It updates the solution by performing a fill-up and exchange procedure over the unreduced problem. This procedure iterates, increasing ϵ at each iteration, until the reduced problem is empty.

Both heuristics, the primal and the variable fixation one, are very fast and take only hundredths of a second for sizes of our interest.

4.2   Branching Rule and Search Strategy

As a branching variable we choose the "most fractional" variable, i.e., $v = \mathrm{argmin}_i |\frac{1}{2} - x_i|$. The vector x is extracted from matrix X given by the semidefinite relaxation.

We traverse the search tree in a *best first search* manner, i.e., we always consider the node in the tree having the smallest upper bound.

4.3   Speed up for Small k

Whenever k, the number of items to be filled in the knapsack, is small, a branch-and-prune algorithm is triggered in order to speed-up the approach. No relaxation is performed at each node of the branch-and-prune tree and a fast depth first search strategy, in priority fixing variables to one, is implemented. We only check the feasibility of the current solution through the cardinality and capacity constraints.

This branch-and-prune approach is very fast, at most a few seconds, for very small k. So we embedded it into our branch-and-bound algorithm and run it at nodes where the remaining number of items to be filled in the current knapsack is very small (less or equal than 5 in practice). To solve the original problem, we can also replace the global branch-and-bound method using this branch-and-prune approach for small initial values of k, in practice we choose $k \leq 10$.

5   Numerical Results

We coded the algorithm in C++. For the function evaluation (i.e., solving (SDP)) we implemented a predictor-corrector variant of an interior point algorithm [14]. We use the ConicBundle Library of Ch. Helmberg [13] as framework for the bundle method to solve (SDP_{MET}).

We compare our method $(B\&C)$ to:

– (Cplex): IBM CPLEX solver, version 12.6.2 [11], with default settings. The original nonconvex 0–1 quadratic problem is given directly to CPLEX which is now able to deal with such a formulation.
– (MIQCR+Cplex): our implementation of the MIQCR method [8]. MIQCR uses a semidefinite relaxation in order to obtain a problem having a convexified objective function; the resulting convex integer problem can then be solved by standard solvers. We use the CSDP solver [10] for solving the semidefinite relaxation to convexify the objective function, and IBM CPLEX 12.6.2 [11] with default settings to solve the reformulated convex problem.

– (BiqCrunch): Also BiqCrunch [15] is an algorithm based on semidefinite and polyhedral relaxations within a branch-and-bound framework. In BiqCrunch a quadratic regularization term is added to the objective function of the semidefinite problem and a quasi-Newton method is used to compute the bounds. We use the BiqCrunch solver enhanced with our primal and variable fixation heuristics described in Sect. 4.1.

All experiments have been performed on an Intel i7-2600 quad core 3.4 GHz with 8 GB of RAM, using only one core. The computational results have been obtained for randomly generated instances from [16] with up to 150 variables. We choose $k \in \{1, \ldots, \lfloor \frac{n}{4} \rfloor\}$, b, a_j and c_{ij} are positive integers. The time limit for each approach is 3 h.

In Table 1 we display the run time of the overall algorithm, the gap at the root node, and the number of nodes produced during the branch-and-bound algorithm for each method. Each line of Table 1 represents average values over 10 instances for each n and δ where n is the number of variables and δ is the density of the profit matrix C. We put the number of instances solved within the time limit into brackets in case not all 10 instances could be solved. Average values are computed only over instances solved within the time limit.

The numerical experiments demonstrate that the methods having semidefinite optimization inside clearly outperform Cplex. In fact, Cplex already fails to solve all instances of size $n = 50$ within the time limit.

Table 1. Numerical results comparing four approaches. (Time limit: 3 h)

n	δ	Cplex Gap root %	Time (s)	#Nodes	MIQCR+Cplex Gap root %	Time (s)	#Nodes	BiqCrunch Gap root %	Time (s)	#Nodes	B&C Gap root %	Time (s)	#Nodes
50	25	102.7	3.7	3426.9	30.5	1.0	621.2	7.4	21.4	79.6	0.9	72.4	11.6
	50	150.6	150.8	77807.9	25.2	1.0	1276.3	4.9	24.9	136.8	1.3	9.1	11.2
	75	230.3	213.1	104419.5	102.0	0.7	656.7	56.1	26.6	98.5	0.6	3.6	9.1
	100	356.5	(8) 53.1	(8) 14228.8	62.7	1.5	3620.0	31.4	23.0	80.6	0.9	73.0	38.1
60	25	60.8	3.0	917.1	127.4	0.9	621.2	123.0	32.2	85.2	0.6	18.3	18.4
	50	93.7	282.4	134246.3	15.1	1.4	1280.3	4.7	39.7	136.8	2.0	110.3	88.1
	75	212.7	(9) 50.9	(9) 8258.3	137.5	3.3	7594.5	131.4	71.3	123.0	1.3	75.8	28.2
	100	284.5	(8) 186.6	(8) 55411.8	61.2	3.2	5808.1	47.8	63.0	147.2	0.3	21.5	18.4
70	25	130.2	23.7	12065.8	37.9	3.4	2884.9	13.8	109.6	147.7	4.5	259.3	42.0
	50	177.1	(6) 213.8	(6) 63859.7	71.7	8.4	11221.8	59.2	141.0	207.4	2.2	128.2	139.7
	75	382.4	(8) 873.0	(8) 105465.6	56.1	16.2	33821.0	17.4	196.2	211.7	3.5	246.8	114.9
	100	252.2	(4) 60.2	(4) 10867.5	59.6	14.6	25809.6	53.0	153.3	243.2	4.0	319.7	338.8
80	25	111.2	226.6	89013.4	33.5	7.8	6115.3	13.0	149.6	195.2	7.5	390.9	86.1
	50	271.6	(8) 872.9	(8) 181325.9	55.0	26.8	36346.3	20.9	373.9	366.2	8.6	544.8	213.6
	75	313.3	(5) 278.7	(5) 14838.5	82.0	47.8	96543.3	70.8	615.1	745.2	2.6	413.4	359.0
	100	473.0	(6) 1469.5	(6) 98024.5	47.0	96.5	216700.0	17.1	717.5	804.6	5.4	1849.4	1219.7
90	25	118.5	(9) 585.9	(9) 693035.0	111.5	23.3	22836.9	107.3	188.6	390.4	3.6	430.6	94.1
	50	248.6	(6) 3708.5	(6) 312105.5	82.2	67.8	99574.6	72.3	532.3	810.0	3.4	729.1	404.9
	75	388.7	(2) 2850.5	(2) 62190.5	37.9	735.1	1348558.3	14.2	1281.2	970.8	8.7	(7) 3234.7	(7) 2233.1
	100	390.0	(3) 146.2	(3) 5047.5	26.6	180.4	282966.1	10.4	1094.7	5644.1	6.5	2740.0	1357.9
100	25	169.4	2308.1	623731.5	74.4	65.4	71449.9	61.6	392.5	617.4	10.9	1583.1	284.0
	50	145.7	(6) 1724.3	(6) 122716.0	17.5	308.0	465749.5	7.6	986.9	882.0	8.0	(9) 3379.6	(9) 1488.6
	75	270.9	(2) 4243.5	(2) 88176.5	58.0	886.8	1322350.0	6.8	980.0	967.8	14.8	(7) 2613.0	(7) 855.6
	100	473.0	(5) 2658.8	(5) 120959.0	98.6	649.7	977246.9	94.0	(9) 723.8	(9) 5166.7	6.0	(7) 318.1	(7) 115.4
110	25	124.0	277.4	36270.2	72.2	327.0	288120.4	64.4	848.8	1003.6	13.5	(8) 2602.9	(8) 810.4
	50	117.5	(3) 661.5	(3) 55327.0	14.3	1188.1	1089556.0	4.7	1010.2	727.8	5.7	(7) 2065.1	(7) 652.3
	75	580.7	(6) 908.8	(6) 35891.8	138.7	(7) 27.4	(7) 37408.8	118.8	(8) 2305.7	(8) 4523.3	10.7	(7) 1062.7	(7) 297.7
	100	332.2	(1) 1911.6	(1) 118852.0	19.6	(8) 758.0	(8) 956886.3	7.1	(8) 1438.1	(8) 1575.0	7.0	(6) 1789.2	(6) 511.7
120	25	55.1	(6) 320.1	(6) 94936.5	95.3	1771.4	1644176.9	111.8	424.3	447.5	5.7	(8) 1872.3	(8) 317.3
	50	288.1	(3) 2995.9	(3) 81429.7	90.3	(7) 1888.9	(7) 1554792.4	82.5	(8) 1073.9	(8) 821	10.2	(6) 1725.3	(6) 427.5
	75	507.6	(5) 305.0	(5) 11101.2	133.8	(6) 484.9	(6) 177043.8	126.7	(9) 3001.6	(9) 7506.3	7.9	(0) 0.8	(0) 0.0
	100	179.7	(3) 41.9	(3) 4166.5	66.2	(6) 61.6	(6) 36075.0	68.7	(9) 1582.6	(9) 969.0	3.6	(6) 683.4	(6) 144.4
130	25	129.1	(6) 2014.5	(5) 383586.2	24.7	1256.6	698089.0	10.6	3341.8	2520.8	7.1	(8) 4194.0	(8) 850.7
	50	411.8	(4) 3246.9	(4) 245787.0	67.3	(7) 493.0	(7) 384516.5	50.5	(8) 1719.6	(8) 2330.7	6.9	(7) 2590.8	(7) 450.8
	75	207.3	(0)	(0)	12.2	(5) 4975.3	(5) 3138617.2	3.8	(9) 2680.0	(9) 1000.3	11.9	(2) 6813.5	(2) 1430.0
	100	383.5	(0)	(0)	21.1	(4) 2250.4	(4) 1170285.3	5.8	(6) 4365.0	(6) 2437.4	14.1	(3) 2012.0	(3) 83.5
140	25	207.9	(5) 15.0	(3) 1180.5	48.8	(8) 1770.8	(8) 654608.8	44.4	2624.2	1693.3	13.3	(4) 2561.8	(4) 298.5
	50	306.2	(1) 2401.8	(1) 106348.0	36.8	(5) 2692.8	(5) 2306370.0	16.4	(7) 4609.5	(7) 1134.1	24.0	(3) 3360.3	(3) 213.3
	75	259.0	(0)	(0)	19.6	(4) 2283.5	(4) 1520163.5	8.0	(5) 4065.2	(5) 1773.4	12.5	(1) 431.0	(1) 0.0
	100	647.8	(2) 64.1	(2) 483.0	49.4	(4) 1042.9	(4) 1238929.3	37.1	(6) 6123.7	(6) 17745.3	13.3	(4) 2561.7	(4) 229.3
150	25	103.7	(4) 1744.3	(4) 202004.0	67.4	(6) 587.1	(6) 552495.8	69.1	(8) 3203.9	(8) 994.8	12.6	(3) 1458.0	(3) 164.3
	50	105.6	(5) 91.8	(5) 5591.7	98.2	(7) 2240.0	(7) 935027.4	101.4	(7) 2761.5	(7) 2349.4	8.8	(5) 2797.7	(5) 3458.7
	75	496.9	(0)	(0)	7.9	(5) 957.2	(5) 300455.4	123.1	(3) 3908.7	(3) 876.3	17.5	(2) 155.0	(2) 0.0
	100	171.1	(1) 1039.3	(1) 24546.0	43.7	(3) 3493.0	(3) 5264462.0	8.3	(4) 4320.1	(4) 2171.0	8.1	(3) 5391.7	(3) 554.3
Avg		258.5	(236) 787.0	(236) 127524.4	59.0	(372) 570.0	(372) 902502.7	45.7	(394) 1215.4	(394) 1487.2	7.4	(328) 1250.5	(328) 433.8

Instances with up to $n = 100$ variables can be solved most efficiently by the MIQCR approach, i.e., finding a convexified problem via semidefinite optimization and then solve the resulting convex problem using Cplex.

For $n > 100$, BiqCrunch performs best in terms of overall run time, but the dominance to MIQCR and our approach is not significant.

Our new approach provides by far the smallest gap at the root node. The high quality of our bound is also reflected in the number of nodes in the branch-and-bound tree. Our method explores a substantial smaller number of nodes than the other approaches.

Our approach is not superior to MIQCR or BiqCrunch in terms of overall computation time, however, the implementation is a prototype and there is room for speeding up the approach by experimenting with different settings in the branch-and-bound framework (such as branching strategies) as well as parameter settings in the bundle algorithm and in the update of the set of triangle inequalities. This is currently under investigation.

6 Conclusion

The 0–1 k-item quadratic knapsack problem is a challenging problem, as it includes two NP-hard problems, namely quadratic knapsack and k-cluster. We review approaches to solve this problem to optimality and introduce a new method, where the bound computation is based on a semidefinite relaxation. The derived basic semidefinite relaxation has only simple constraints, in fact all constraints are of rank one. This can be exploited in interior point methods to efficiently compute the system matrix. We strengthen the relaxation using triangle inequalities and solve the resulting semidefinite problem by a dynamic version of the bundle method.

To have a comparison with state of the art algorithms we implement the convexification algorithm MIQCR [8], use BiqCrunch [15] enhanced with our primal heuristics, and run Cplex. The numerical results prove that CPLEX is clearly outperformed by all the methods based on semidefinite programming. Our new method provides the tightest bound at the root node, while the overall computation time is smallest for MIQCR for $n \leq 100$ and BiqCrunch for larger n. An optimized implementation and a study of the best parameter settings for the various components inside our code is subject of further study.

References

1. Anjos, M.F., Ghaddar, B., Hupp, L., Liers, F., Wiegele, A.: Solving k-way graph partitioning problems to optimality: the impact of semidefinite relaxations and the bundle method. In: Jünger, M., Reinelt, G. (eds.) Facets of combinatorial optimization, pp. 355–386. Springer, Heidelberg (2013)
2. Balas, E., Zemel, E.: An algorithm for large zero-one knapsack problems. Oper. Res. **28**, 1130–1154 (1980)
3. Bertsimas, D., Shioda, R.: Algorithm for cardinality-constrained quadratic optimization. Comput. Optim. Appl. **43**, 1–22 (2009)

4. Bhaskara, A., Charikar, M., Guruswami, V., Vijayaraghavan, A., Zhou, Y.: Polynomial integrality gaps for strong SDP relaxations of densest k-subgraph. In: Proceedings of the Twenty-Third Annual ACM-SIAM Symposium on Discrete Algorithms, pp. 388–405 (2012)
5. Bienstock, D.: Computational study of a family of mixed-integer quadratic programming problems. Math. Programm. **74**, 121–140 (1996)
6. Billionnet, A.: Different formulations for solving the heaviest k-subgraph problem. Inf. Syst. Oper. Res. **43**(3), 171–186 (2005)
7. Billionnet, A., Calmels, F.: Linear programming for the 0–1 quadratic knapsack problem. Eur. J. Oper. Res. **92**, 310–325 (1996)
8. Billionnet, A., Elloumi, S., Lambert, A.: Extending the QCR method to general mixed-integer programs. Math. Programm. **131**(1–2), 381–401 (2012)
9. Bonami, P., Lejeune, M.: An exact solution approach for portfolio optimization problems under stochastic and integer constraints. Oper. Res. **57**, 650–670 (2009)
10. Borchers, B.: CSDP, a C library for semidefinite programming. Optim. Meth. Softw. **11**(1), 613–623 (1999)
11. IBM ILOG CPLEX Callable Library version 12.6.2. http://www-03.ibm.com/software/products/en/ibmilogcpleoptistud/
12. Fischer, I., Gruber, G., Rendl, F., Sotirov, R.: Computational experience with a bundle approach for semidefinite cutting plane relaxations of Max-Cut, equipartition. Math. Programm. Ser. B **105**(2–3), 451–469 (2006)
13. Helmberg, C.: The conicbundle library for convex optimization, August 2015. https://www-user.tu-chemnitz.de/~helmberg/ConicBundle/Manual/index.html
14. Helmberg, C., Rendl, F., Vanderbei, R.J., Wolkowicz, H.: An interior-point method for semidefinite programming. SIAM J. Optim. **6**(2), 342–361 (1996)
15. Krislock, N., Malick, J., Roupin, F.: Improved semidefinite bounding procedure for solving max-cut problems to optimality. Math. Program. Ser. A **143**(1–2), 61–86 (2014)
16. Létocart, L., Plateau, M.-C., Plateau, G.: An efficient hybrid heuristic method for the 0–1 exact k-item quadratic knapsack problem. Pesquisa Operacional **34**(1), 49–72 (2014)
17. Mitra, G., Ellison, F., Scowcroft, A.: Quadratic programming for portfolio planning: insights into algorithmic and computational issues. J. Asset Manage. **8**, 249–258 (2007). Part ii: Processing of Portfolio Planning Models with Discrete Constraints
18. Pisinger, D.: The quadratic knapsack problem: a survey. Discrete Appl. Math. **155**, 623–648 (2007)
19. Rader Jr., D.J., Woeginger, G.J.: The quadratic 0–1 knapsack problem with series-parallel support. Oper. Res. Lett. **30**(3), 159–166 (2002)
20. Rendl, F., Rinaldi, G., Wiegele, A.: Solving max-cut to optimality by intersecting semidefinite, polyhedral relaxations. Math. Program. Ser. A **121**(2), 307–335 (2010)
21. Rendl, F., Sotirov, R.: Bounds for the quadratic assignment problem using the bundle method. Math. Program. Ser. B **109**(2–3), 505–524 (2007)
22. Shawa, D.X., Liub, S., Kopmanb, L.: Lagrangean relaxation procedure for cardinality-constrained portfolio optimization. Optim. Meth. Softw. **23**, 411–420 (2008)

On Vertices and Facets of Combinatorial 2-Level Polytopes

Manuel Aprile$^{(\boxtimes)}$, Alfonso Cevallos, and Yuri Faenza

École Polytechnique Fédérale de Lausanne, Lausanne, Switzerland
{manuel.aprile,alfonso.cevallosmanzano,yuri.faenza}@epfl.ch

Abstract. 2-level polytopes naturally appear in several areas of mathematics, including combinatorial optimization, polyhedral combinatorics, communication complexity, and statistics.

We investigate upper bounds on the product of the number of facets $f_{d-1}(P)$ and the number of vertices $f_0(P)$, where d is the dimension of a 2-level polytope P. This question was first posed in [3], where experimental results showed $f_0(P)f_{d-1}(P) \leq d2^{d+1}$ up to $d = 6$.

We show that this bound holds for all known (to the best of our knowledge) 2-level polytopes coming from combinatorial settings, including stable set polytopes of perfect graphs and all 2-level base polytopes of matroids. For the latter family, we also give a simple description of the facet-defining inequalities. These results are achieved by an investigation of related combinatorial objects, that could be of independent interest.

1 Introduction

Let $P \subseteq \mathbb{R}^d$ be a polytope. We say that P is *2-level* if, for all facets F of P, all the vertices of P that are not vertices of F lie in the same translate of the affine hull of F (Fig. 1). Equivalently, P is 2-level if and only if it has theta-rank 1 [9], or all its pulling triangulations are unimodular [16], or it has a slack matrix with entries that are only 0 or 1 [3]. Those last three definitions appear in papers from the semidefinite programming, statistics, and polyhedral combinatorics communities respectively, showing that 2-level polytopes naturally arise in many areas of mathematics.

Arguably, the most important reasons 2-level polytopes are interesting for researchers in polyhedral combinatorics and theoretical computer science are their connections with the theory of *linear extensions* and the prominent *log-rank conjecture* in communication complexity, since they generalize stable set polytopes of perfect graphs.

Because of all the reasons above, a complete understanding of 2-level polytopes would be desirable. Unfortunately, despite an increasing number of studies [3,9–11], such an understanding has not been obtained yet: we do not have e.g. any decent bound on the number of d-dimensional 2-level polytopes or on their linear extension complexity, nor do we have a structural theory of their slack matrices, of the kind that has been developed for totally unimodular matrices (see e.g. [14]). On the positive side, many properties of 2-level polytopes have

© Springer International Publishing Switzerland 2016
R. Cerulli et al. (Eds.): ISCO 2016, LNCS 9849, pp. 177–188, 2016.
DOI: 10.1007/978-3-319-45587-7_16

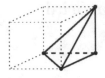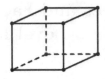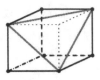

Fig. 1. The first three polytopes (the simplex, the cross-polytope and the cube) are clearly 2-level. The fourth one is not 2-level, due to the highlighted facet (Images excerpted from [9] with permission from the authors).

been shown. For instance, each d-dimensional 2-level polytope is affinely isomorphic to a 0/1 polytope [9], hence it has at most 2^d vertices. Interestingly, one can show that a d-dimensional 2-level polytope has at most 2^d facets [9]. This makes 2-level polytopes quite different from "random" 0/1 polytopes, that have $(d/\log d)^{\Theta(d)}$ facets [2]. In fact, 2-level polytopes seem to be a very restricted subclass of 0/1 polytopes, as experimental results from [3] have shown.

The goal of this paper is to shed some light on the relationship between the number of vertices and the number of facets of a 2-level polytope. Experimental evidence from [3] up to dimension 6 suggests the existence of a trade-off between those two numbers, in a very strong sense: a d-dimensional 2-level polytope can have at most 2^d vertices and facets, but their product seems to be upper bounded by a number much smaller than 2^{2d}. More formally, for a polytope P and $i \in \mathbb{Z}_+$, let $f_i(P)$ be the number of its i-dimensional faces. The following was posed as a question in [3], and we turn it here into a conjecture.

Conjecture 1 (Vertex/Facet Trade-off). *Let P be a d-dimensional 2-level polytope. Then $f_0(P)f_{d-1}(P) \le d2^{d+1}$. Moreover, equality is achieved if and only if P is affinely isomorphic to the cross-polytope or the cube.*

It is immediate to check that the cube and the cross-polytope (its polar) indeed verify $f_0(P)f_{d-1}(P) = d2^{d+1}$. The conjecture above essentially states that those basic polytopes maximize $f_0(P)f_{d-1}(P)$ among all 2-level polytopes of a fixed dimension.

Conjecture 1 has an interesting interpretation as an upper bound on the "size" of slack matrices of 2-level polytopes, since $f_0(P)$ (resp. $f_{d-1}(P)$) is the number of columns (resp. rows) of the (smallest) slack matrix of P. Many fundamental results on linear extensions of polytopes (including the celebrated upper bound on the extension complexity of the stable set polytope of perfect graphs [17]) are based on properties of slack matrices. We believe therefore that answering Conjecture 1 would be an interesting step towards a better understanding of 2-level polytopes.

Our Contribution and Organization of the Paper. We show that Conjecture 1 holds true for all known classes (to the best of our knowledge) of 2-level polytopes coming from combinatorial settings. In most cases, this is deduced from properties of associated combinatorial objects, that are also shown in the current paper and we believe could be of independent interest.

Detailed results and the organization of the paper are as follows. We introduce some common definitions and techniques in Sect. 2: those are enough to show that Conjecture 1 holds for Hanner polytopes. In Sect. 3 we give a simple but surprisingly sharp upper bound on the product of the numbers of stable sets and cliques of a graph. This is used to show that the conjecture holds for stable set polytopes of perfect graphs, order polytopes, and Hansen polytopes. In Sect. 4, we give a non-redundant description of facets of the base polytope of the 2-sum of matroids in terms of the facets of the base polytopes of the original matroids. This is used to obtain a compact description (in the original space) of 2-level base polytopes of matroids and a proof of Conjecture 1 for this class. In Sect. 5, we prove the conjecture for the cycle polytopes of certain binary matroids, which generalizes all cut polytopes that are 2-level. In Sect. 6 we give examples showing that Conjecture 1 does not trivially hold for all "well-behaved" 0/1 polytopes. **NOTE:** Because of space constraints, most proofs and some definitions are deferred to the journal version of the paper.

Related Work. We already mentioned the paper [3] that provides an algorithm based on the enumeration of closed sets to list all 2-level polytopes, as well as papers [9, 11, 16] where equivalent definitions and/or families of 2-level polytopes are given. Among other results, in [9] it is shown that the stable set polytope of a graph G is 2-level if and only if G is perfect. A characterization of all base polytopes of matroids that are 2-level is given in [11], building on the decomposition theorem for matroids that are not 3-connected (see e.g. [13]).

2 Basics

We let $\mathbb{R}_+$ be the set of non-negative real numbers. For a set S and an element e, we denote by $A + e$ and $A - e$ the sets $A \cup \{e\}$ and $A \setminus \{e\}$, respectively. For a point $x \in \mathbb{R}^I$, where I is an index set, and a subset $J \subseteq I$, let $x(J) = \sum_{i \in J} x_i$.

For basic definitions about polytopes and graphs, we refer the reader to [18] and [6], respectively. The d-dimensional *cube* is $[-1, 1]^d$, and the d-dimensional *cross-polytope* is its corresponding polar. Taking the polar of a polytope is a dual operation, that produces a polytope of the same dimension, where the number of vertices and the number of facets are swapped. Thus, a polytope and its polar will simultaneously satisfy or not satisfy Conjecture 1. A 0/1 polytope is the convex hull of a subset of the vertices of $[0, 1]^d$. The following facts will be used many times:

Lemma 2 [9]. *Let P be a 2-level polytope of dimension d. Then*

1. $f_0(P), f_{d-1}(P) \leq 2^d$.
2. *Any face of P is again a 2-level polytope.*

One of the most common operation with polytopes is the *Cartesian product*. Given two polytopes $P_1 \subseteq \mathbb{R}^{d_1}$, $P_2 \subseteq \mathbb{R}^{d_2}$, their Cartesian product is $P_1 \times P_2 = \{(x, y) \in \mathbb{R}^{d_1 + d_2} : x \in P_1, y \in P_2\}$. This operation will be useful to us as it preserves the bound of Conjecture 1.

Lemma 3. *If two 2-level polytopes P_1 and P_2 satisfy Conjecture 1, then so does their Cartesian product.*

2.1 Hanner Polytopes

We start off with an easy example. *Hanner* polytopes can be defined as the smallest family that contains the $[-1, 1]$ segment of dimension 1, and is closed under taking polars and Cartesian products. These polytopes are 2-level and centrally symmetric, and from the previous observations it is straightforward that they verify Conjecture 1.

Theorem 4. *Hanner polytopes satisfy Conjecture 1.*

3 Graph Theoretical 2-Level Polytopes

We present a general result on the number of cliques and stable sets of a graph. Proofs of all theorems from the current section will be based on it.

Theorem 5 (Stable set/clique trade-off). *Let $G = (V, E)$ be a graph on n vertices, $\mathcal{C}$ its family of non-empty cliques, and $\mathcal{S}$ its family of non-empty stable sets. Then $|\mathcal{C}||\mathcal{S}| \leq n(2^n - 1)$. Moreover, equality is achieved if and only if G or its complement is a clique.*

Proof. Consider the function $f : \mathcal{C} \times \mathcal{S} \to 2^V$, where $f(C, S) = C \cup S$. For a set $W \subset V$, we bound the size of its pre-image $f^{-1}(W)$. This will imply a bound for $|\mathcal{C} \times \mathcal{S}| = \sum_{W \subseteq V} |f^{-1}(W)|$. If W is a singleton, the only pair in its pre-image is (W, W). For $|W| \geq 2$, we claim that $|f^{-1}(W)| \leq 2|W|$.

There are at most $|W|$ intersecting pairs (C, S) in $f^{-1}(W)$. This is because the intersection must be a single element, $C \cap S = \{v\}$, and once it is fixed every element adjacent to v must be in C, and every other element must be in S.

There are also at most $|W|$ disjoint pairs in $f^{-1}(W)$, as we prove now. Fix one such disjoint pair (C, S), and notice that both C and S are non-empty proper subsets of W. All other disjoint pairs (C', S') are of the form $C' = C \setminus A \cup B$ and $S' = S \setminus B \cup A$, where $A \subseteq C$, $B \subseteq S$, and $|A|, |B| \leq 1$. Let X (resp. Y) denote the set formed by the vertices of C (resp. S) that are anticomplete to S (resp. complete to C). Clearly, either X or Y is empty. We settle the case $Y = \emptyset$, the other being similar. In this case $\emptyset \neq A \subseteq X$, so $X \neq \emptyset$. If $X = \{v\}$, then $A = \{v\}$ and we have $|S| + 1$ choices for B, with $B = \emptyset$ possible only if $|C| \geq 2$, because we cannot have $C' = \emptyset$. This gives at most $1 + |S| + |C| - 1 \leq |W|$ disjoint pairs (C', S') in $f^{-1}(W)$. Otherwise, $|X| \geq 2$ forces $B = \emptyset$, and the number of such pairs is at most $1 + |X| \leq 1 + |C| \leq |W|$.

We conclude that $|f^{-1}(W)| \leq 2|W|$, or one less if W is a singleton. Thus

$$|\mathcal{C} \times \mathcal{S}| \leq \sum_{k=0}^{n} 2k \binom{n}{k} - n = n2^n - n,$$

where the (known) fact $\sum_{k=0}^{n} 2k\binom{n}{k} = n2^n$ holds since

$$n2^n = \sum_{k=0}^{n}(k+(n-k))\binom{n}{k} = \sum_{k=0}^{n} k\binom{n}{k} + (n-k)\binom{n}{n-k} = 2\sum_{k=0}^{n} k\binom{n}{k}.$$

The bound is clearly tight for $G = K_n$ and $G = \overline{K_n}$. For any other graph, there is a subset W of 3 vertices that induces 1 or 2 edges. In both cases, $|f^{-1}(W)| = 5 < 2|W|$, hence the bound is loose. □

For a graph $G = (V, E)$, its stable set polytope $\mathrm{STAB}(G)$ is the convex hull of the characteristic vectors of all stable sets in G. It is known that $\mathrm{STAB}(G)$ is 2-level if and only if G is a *perfect graph* [9], or equivalently [5] if and only if

$$\mathrm{STAB}(G) = \{x \in \mathbb{R}_+^V : x(C) \leq 1 \text{ for all maximal cliques } C \text{ of } G\}.$$

Theorem 6. *Stable set polytopes of perfect graphs satisfy Conjecture 1.*

Given a $(d-1)$-dimensional polytope P, the *twisted prism* of P is the d-dimensional polytope defined as the convex hull of $\{(x, 1) : x \in P\}$ and $\{(-x, -1) : x \in P\}$. For a perfect graph G with $d-1$ vertices, its *Hansen* polytope $\mathrm{Hans}(G)$ is defined as the twisted prism of $\mathrm{STAB}(G)$. Hansen polytopes are 2-level and centrally symmetric.

Theorem 7. *Hansen polytopes satisfy Conjecture 1.*

Given a poset P on $[d]$, with order relation $<_P$, its *order polytope* $\mathcal{O}(P)$ is:

$$\mathcal{O}(P) = \{x \in [0, 1]^d : x_i \leq x_j \ \forall \ i <_P j\}.$$

A subset $I \subseteq P$ is called an *upset* if $x \in I$ and $x <_P y$ imply $y \in I$. In [15] the following characterization of vertices of an order polytope is given.

Lemma 8. *The vertices of $\mathcal{O}(P)$ are the characteristic vectors of upsets of P. In particular, the number of vertices of $\mathcal{O}(P)$ is the number of upsets of P.*

From this result it is clear that $\mathcal{O}(P)$ is a 2-level polytope. Indeed, if all vertices of a polytope have $0/1$ coordinates and all facet-defining inequalities can be written as $0 \leq c^T x \leq 1$ for integral vectors c, then the polytope is 2-level.

Given a poset P, we say that j *covers* i in P if $i <_P j$ and there is no k in P such that $i <_P k <_P j$. We say that i, j is a *covering pair* if j covers i or i covers j. P can be described by a graph called *Hasse Diagram* $G_P([d], E)$, with $ij \in E$ if and only if i, j is a covering pair. This graphical representation and Theorem 5 are the main ingredients to prove the following.

Theorem 9. *Order polytopes satisfy Conjecture 1.*

4   2-Level Matroid Base Polytopes

We now give a non-redundant description of the base polytopes of the 2-sum $M_1 \oplus_2 M_2$ of matroids in terms of the facets of the base polytopes of M_1 and M_2. We then focus on 2-level matroids. We give an explicit description of the associated base polytopes, and prove that they verify Conjecture 1. For basic definitions and facts about matroids we refer to [13].

4.1   The Base Polytope of the 2-Sum of Matroids

We identify a matroid M by the couple $(E, \mathcal{B})$, where $E = E(M)$ is its ground set, and $\mathcal{B} = \mathcal{B}(M)$ is its base set. Given $M = (E, \mathcal{B})$ and a set $F \subseteq E$, the *restriction* $M|F$ is the matroid with ground set F and independent sets $\mathcal{I}(M|F) = \{I \in \mathcal{I}(M) : I \subseteq F\}$; and the *contraction* M/F is the matroid with ground set $M \setminus F$ and rank function $r_{M/F}(A) = r_M(A \cup F) - r_M(F)$. For an element $e \in E$, the *removal of* e is $M - e = M|(E - e)$. A set $F \subseteq E$ is a *flat* if it is maximal for its rank, i.e. $r(F) < r(F + x)$ for all $x \in E \setminus F$.

Consider matroids $M_1 = (E_1, \mathcal{B}_1)$ and $M_2 = (E_2, \mathcal{B}_2)$, with non-empty base sets. If $E_1 \cap E_2 = \emptyset$, we can define the *direct sum* $M_1 \oplus M_2$ as the matroid with ground set $E_1 \cup E_2$ and base set $\mathcal{B}_1 \times \mathcal{B}_2$. If, instead, $E_1 \cap E_2 = \{p\}$, where p is neither a loop nor a coloop in M_1 or M_2, we let the *2-sum* $M_1 \oplus_2 M_2$ be the matroid with ground set $E_1 \cup E_2 - p$, and base set $\{B_1 \cup B_2 - p : B_i \in \mathcal{B}_i$ for $i = 1, 2$ and $p \in B_1 \triangle B_2\}$. A matroid is *connected* if it cannot be written as the direct sum of two matroids, each with fewer elements.

The *base polytope* $B(M) \subseteq \mathbb{R}^E$ of a matroid $M = (E, \mathcal{B})$ is given by the convex hull of the characteristic vectors of its bases. For a matroid M, the following is known to be a description of $B(M)$.

$$B(M) = \{x \in [0, 1]^E : x(F) \leq r(F) \text{ for } F \subseteq E; \text{ and } x(E) = r(E)\}. \quad (1)$$

When M is connected [7] give the following characterization of the facet-defining inequalities for (1). (We report the statement as it appears in [11])

Theorem 10. *Let $M = (E, \mathcal{B})$ be a connected matroid. For every facet F of $B(M)$ there is a unique $S \subseteq E$, $S \neq \emptyset$, such that $F = B(M) \cap \{x \in \mathbb{R}^E : x(S) = r(S)\}$. Moreover, a non-empty subset S gives rise to a facet of $B(M)$ if and only if one of the these two conditions holds:*

1. *S is a flat such that $M|S$, M/S are connected;*
2. *$S = E - e$ for some $e \in E$ such that $M|S$, M/S are connected.*

The subsets S in 1. are called *flacets*, and they are in 1-to-1 correspondence with the facet-defining inequalities in (1) of the form $x(S) \leq r(S)$, including $x_e \leq 1$ for $e \in E$. For $S = E - e$ satisfying the conditions in 2., we refer to element e as defining a *non-negativity* facet. Indeed it can be easily seen that it defines the same facet as $x_e \geq 0$.

Throughout the rest of the section, we assume that $M_1(E_1, \mathcal{B}_1)$, $M_2(E_2, \mathcal{B}_2)$ are connected matroids, with, $E_1 \cap E_2 = \{p\}$, and we define $M = M_1 \oplus_2 M_2$.

It is well known that under these assumptions M is also connected. By the arguments above, characterizing $B(M)$ essentially boils down to characterizing flacets of $M_1 \oplus_2 M_2$.

Theorem 11. *Let F be a flacet of M. One of the following holds:*

1. $F = E_i \cup F' - p$, *where F' is a flacet of M_j containing p, and $i \neq j \in \{1,2\}$.*
2. F *is a flacet of M_i not containing p for some $i \in \{1,2\}$.*
3. $F = E_i - p$ *for some $i \in \{1,2\}$.*

Conversely, let F_1 be a flacet of M_1, $F_1 \neq \{p\}$. Then

1. *If $p \in F_1$, $F = E_2 \cup F_1 - p$ is a flacet of M.*
2. *If $p \notin F_1$, F_1 is a flacet of M.*
3. *If M_2/p and $M_1 - p$ are connected, then $E_1 - p$ is a flacet of M.*

We remark that a statement similar to the first half of Theorem 11 for an analogous definition of 2-sum and flacets appeared in [4]. However, we were not able to convince ourselves that the proof from [4] is complete, and some of its statements appear to be wrong.

Corollary 12. *The following is a non-redundant description of $B(M)$:*

$$B(M) = \{x \in \mathbb{R}^E :$$

$$
\begin{array}{llll}
x_e & \geq 0 & & e \in E_i - p : M_i - e \text{ connected}, i = 1, 2 \\
x(E_i \cup F - p) & \leq r(E_i \cup F - p) & & F \text{ flacet of } M_j : \{p\} \subsetneq F, i \neq j \in \{1,2\} \\
x(F) & \leq r(F) & & F \text{ flacet of } M_i : p \notin F, i \in \{1,2\} \\
x(E_i - p) & \leq r(E_i - p) & & \text{if } M_i - p, M_j/p \text{ connected}, i \neq j \in \{1,2\} \\
x(E) & = r(E)\}.
\end{array}
$$

$$(2)$$

Corollary 13. *Let us write $f(M) = f_{d-1}(B(M))$, and similarly for M_1, M_2. Then $f(M_1) + f(M_2) - 2 \leq f(M) \leq f(M_1) + f(M_2) + 2$.*

4.2   Linear Description of 2-Level Matroid Base Polytopes

A matroid $M(E, \mathcal{B})$ is *uniform* if $\mathcal{B} = \binom{E}{k}$, where k is the rank of M. We denote the uniform matroid with n elements and rank k by $U_{n,k}$. Notice that, if M_1 and M_2 are uniform matroids with $|E(M_1) \cap E(M_2)| = 1$, then $M_1 \oplus_2 M_2$ is unique up to isomorphism, for any possible common element. Let $\mathcal{M}$ be the class of matroids whose base polytope is 2-level. $\mathcal{M}$ has been characterized in [11]:

Theorem 14. *The base polytope of a matroid M is 2-level if and only if M can be obtained from uniform matroids through a sequence of direct sums and 2-sums.*

The following lemma implies that we can, when looking at matroids in $\mathcal{M}$, decouple the operations of 2-sum and direct sum.

Lemma 15. *Let M be a matroid obtained by applying a sequence of direct sums and 2-sums from the matroids $M_1, \ldots, M_k$. Then $M = M'_1 \times M'_2 \times \ldots \times M'_t$, where each of the M'_i is obtained by repeated 2-sums from some of the matroids $M_1, \ldots, M_k$.*

Since the base polytope of the direct sum of matroids is the Cartesian product of the base polytopes, to obtain a linear description of $B(M)$ for $M \in \mathcal{M}$, we can focus on base polytopes of connected matroids obtained from the 2-sums of uniform matroids. A sequence of 2-sums can be represented via a tree (see Fig. 2): the following is a version of [13, Proposition 8.3.5] tailored to our needs.

Theorem 16. *Let M be obtained by a sequence of 2-sums operations from matroids $M_1, \ldots, M_t$. Then there is a t-vertex tree $T = T(M)$ with edges labelled $e_1, \ldots, e_{t-1}$ and vertices labelled $M_1, \ldots, M_t$, such that*

1. $E(M_1) \cup E(M_2) \cup \cdots \cup E(M_t) = E(M) \cup \{e_1, \ldots, e_{t-1}\};$
2. if the edge e_i joins the vertices M_{j_1} and M_{j_2}, then $E(M_{j_1}) \cap E(M_{j_2}) = \{e_i\};$
3. if no edge joins the vertices M_{j_1} and M_{j_2}, then $E(M_{j_1}) \cap E(M_{j_2}) = \emptyset.$

Moreover, M is the matroid that labels the single vertex of the tree $T/e_1, \ldots, e_{t-1}$ at the conclusion of the following process: contract the edges $e_1, \ldots, e_{t-1}$ of T one by one in order; when e_i is contracted, its ends are identified and the vertex formed by this identification is labeled by the 2-sum of the matroids that previously labeled the ends of e_i.

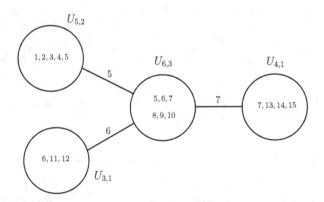

Fig. 2. An example of the tree structure of a matroid M that is a 2-sum of uniform matroids. Note that the elements $5, 6, 7$ will not be present in the ground set of M. From the picture it is easy to see that M is a matroid with 12 elements and rank 4. One basis of M is e.g. $\{1, 8, 9, 14\}$.

Observation 17. *If $M \in \mathcal{M}$ is connected and non-uniform, we can assume without loss of generality that every node in its tree structure given by Theorem 16 is a uniform matroid with at least 3 elements. Each of those uniform matroid has no flacets besides its singletons.*

For a connected matroid $M(E, \mathcal{B}) \in \mathcal{M}$, Theorem 16 reveals a tree structure $T(M)$, where every node represents a uniform matroid, and every edge represents a 2-sum operation. We now give a simple description of the associated base polytope. Let a be an edge of $T(M)$. The removal of a breaks T into 2 connected components C_a^1 and C_a^2. Let E_a^1 (resp. E_a^2) be the set of elements from E that belong to uniform matroids from C_a^1 (resp. C_a^2). All inequalities needed to describe $B(M)$ are the "trivial" inequalities $0 \leq x \leq 1$, plus $x(F) \leq r(F)$, where $F = E_a^1$ or E_a^2 for some edge a of $T(M)$. Thus the number of inequalities is linear in the number of elements.

Theorem 18. *Let $M = (E, \mathcal{B}) \in \mathcal{M}$ be a connected matroid obtained as 2-sums of uniform matroids $U_1 = U_{n_1, k_1}, \ldots, U_t = U_{n_t, k_t}$. Let $T(N, A)$ be the tree structure of M according to Theorem 16. For each $a \in A$, let $C_a^1, C_a^2, E_a^1, E_a^2$ be defined as above. Then*

$$B(M) = \{x \in \mathbb{R}^E : \quad x \geq 0$$
$$x \leq 1$$
$$x(F) \leq r(F) \quad \text{for } F = E_a^i \text{ for some } i \in \{1, 2\} \text{ and } a \in A,$$
$$x(E) = r(E)\}.$$

Moreover, if $F = E_a^i$ for $i \in \{1, 2\}$ and some $a \in A$, then $r(F) = 1 - |C_a^i| + \sum_{j : U_j \in C_a^i} k_j$.

Proof. Let us call a subset $C \subseteq N$ a *valid component* for T if $C = C_a^i$ for some $i \in \{1, 2\}$ and $a \in A$, and denote the set of all valid components of T by $\mathcal{F}$. Each connected subtree of $T(N, A)$ represents a connected matroid obtained as 2-sums of uniform matroids. Thus, we can prove the theorem by induction on t. The statement on the rank is immediate. For $t = 1$, $\mathcal{F}$ is empty and thanks to Observation 17, the remaining inequalities are enough to describe $B(M)$. Now let $t > 1$. Thanks to Theorem 10, to prove the thesis it is enough to show that, if F is a flacet of M with $|F| \geq 2$, then $F \in \mathcal{F}$. First notice that we can write, without loss of generality, $M = M' \oplus_2 U_t$, where U_t corresponds to a leaf v_t of T and M' is obtained as 2-sums of $U_1, \ldots, U_{t-1}$, hence it satisfies the inductive hypothesis. Note that the tree corresponding to M' is then $T - v_t$. Let us denote by v_l the only neighbor of v_t in T. Let $E' + p$, $E(U_t) = E_t + p$ be the ground sets of M', U_t respectively, where $E' = \bigcup_{i=1}^{t-1} E_i$, and $E_i = E \cap E(U_i)$ for $i = 1, \ldots, t$. Clearly $p \in E(U_l)$. Now, since F is a flacet of M, we can apply Theorem 11 to get three possible cases. If F has non-empty intersection with both $E(M')$ and E_t, then we are in case 1 and either $F = E(U_t) \cup F' - p$ or $F = E' \cup F_t - p$, where F', F_t are flacets of M', U_t respectively, containing p. However, the latter case is not possible because of Observation 17, so the only possibility is that $F = E_t \cup F'$. By induction, F' belongs to $\mathcal{F}'$ defined for M' as in the statement of the theorem. Moreover, since F' contains p, its corresponding component C in $T - v_t$ contains v_l and then $C + v_t$ is a valid component for T. Moreover $|F' \cap E_i| \in \{0, |E_i|\}$ for any $i = 1, \ldots, t - 1$, which implies $F \in \mathcal{F}$. Suppose now we are in case 2, i.e., F is strictly contained in one of E', E_t. Then F is a flacet of one of M', U_t, the latter not being possible again due to Observation 17. So F is a flacet of

M' and it does not contain p, hence by induction hypothesis its corresponding component C does not contain v_l. But then C is a valid component of T and again $F \in \mathcal{F}$. Finally, if we are in case 3 then $F = E_t$ or $F = E$, and in both cases $F \in \mathcal{F}$. □

Theorem 19. *2-Level matroid base polytopes satisfy Conjecture 1.*

As the forest matroid of a graph G is in $\mathcal{M}$ if and only if G is series-parallel [11], we deduce the following.

Corollary 20. *Conjecture 1 is true for the spanning tree polytope of series-parallel graphs.*

5   Cut Polytope and Matroid Cycle Polytope

A cycle of a matroid M is a disjoint union of circuits. The *cycle polytope* $C(M)$ is given by the convex hull of the characteristic vectors of its cycles, and it is a generalization of the cut polytope $CUT(G)$ for a graph G [1]. In this section we prove Conjecture 1 for the cycle polytope $C(M)$ of the binary matroids M that have no minor isomorphic to F_7^*, R_{10}, $M_{K_5}^*$ and are 2-level. When those minors are forbidden, a complete linear description of the associated polytope is known (see [1]). This class includes all cut polytopes that are 2-level, and has been characterized in [8]:

Theorem 21. *Let M be a binary matroid with no minor isomorphic to F_7^*, R_{10}, $M_{K_5}^*$. Then $C(M)$ is 2-level if and only if M has no chordless cocircuit of length at least 5.*

Corollary 22. *The polytope $CUT(G)$ is 2-level if and only if G has no minor isomorphic to K_5 and no induced cycle of length at least 5.*

Recall that the cycle space of graph G is the set of its Eulerian subgraphs (subgraphs where all vertices have even degree), and it is known (see for instance [12]) to have a vector space structure over the field $\mathbb{Z}_2$. This statement and one of its proofs generalizes to the cycle space (the set of all cycles) of binary matroids.

Lemma 23. *Let M be a binary matroid with d elements and rank r. Then the cycles of M form a vector space $\mathcal{C}$ over $\mathbb{Z}_2$ with the operation of symmetric difference as sum. Moreover, $\mathcal{C}$ has dimension $d - r$.*

Corollary 24. *Let M be a binary matroid with d elements and rank r. Then M has exactly 2^{d-r} cycles.*

The only missing ingredient is a description of the facets of the cycle polytope for the class of our interest.

Theorem 25 [1]. *Let M be a binary matroid, and let $\overline{C}$ be its family of chordless cocircuits. Then M has no minor isomorphic to F_7^*, R_{10}, $M_{K_5}^*$ if and only if*

$$C(M) = \{x \in [0,1]^E : x(F) - x(C \setminus F) \leq |F| - 1 \text{ for } C \in \overline{C}, F \subseteq C, |F| \text{ odd}\}.$$

Theorem 26. *Let M be a binary matroid with no minor isomorphic to F_7^*, R_{10}, $M_{K_5}^*$ and such that $C(M)$ is 2-level. Then $C(M)$ satisfies Conjecture 1.*

Proof. As remarked in [1,8], the following equations are valid for $C(M)$: (a) $x_e = 0$, for e coloop of M; and (b) $x_e - x_f = 0$, for $\{e, f\}$ cocircuit of M.

The first equation is due to the fact that a coloop cannot be contained in a cycle, and the second to the fact that circuits and cocircuits have even intersection in binary matroids. A consequence of this is that we can delete all coloops and contract e for any cocircuit $\{e, f\}$ without changing the cycle polytope: for simplicity we will just assume that M has no coloops and no cocircuit of length 2. In this case $C(M)$ has full dimension $d = |E|$. Let r be the rank of M. Corollary 24 implies that $C(M)$ has 2^{d-r} vertices. Let now T be the number of cotriangles (i.e., cocircuits of length 3) in M, and S the number of cocircuits of length 4 in M. Thanks to Theorem 25 and to the fact that M has no chordless cocircuit of length at least 5, we have that $C(M)$ has at most $2d + 4T + 8S$ facets. Hence the bound we need to show is:

$$2^{d-r}(2d + 4T + 8S) \leq d2^{d+1}, \quad \text{which is equivalent to} \quad 2T + 4S \leq d(2^r - 1).$$

Since the cocircuits of M are circuits in the binary matroid M^*, whose rank is $d - r$, we can apply Corollary 24 to get $T + S \leq 2^r - 1$, where the -1 comes from the fact that we do not count the empty set. Hence, if $d \geq 4$,

$$2T + 4S \leq 4(T + S) \leq d(2^r - 1).$$

The bound is loose for $d \geq 5$. The cases with $d \leq 4$ can be easily verified, the only tight examples being affinely isomorphic to cubes and cross-polytopes.   □

Corollary 27. *2-level cut polytopes satisfy Conjecture 1.*

6   Conclusions

In this paper, we showed that Conjecture 1 holds true for many important classes of 2-level polytopes. Whether the results and ideas from this paper can be extended to all 2-level polytopes remains open. Another natural question is whether 2-levelness is the "right" assumption for proving $f_{d-1}(P)f_0(P) \leq d2^{d+1}$, and whether this bound is also valid for more general classes of 0/1 polytopes. We provide here two examples showing that spanning tree and forest polytopes – two classes of "well-behaved" 0/1 polytopes – do not verify Conjecture 1.

Example 28 (Forest polytope of $K_{2,d}$). Conjecture 1 implies an upper bound of $d2^{2(d+1)} = O(4 + \varepsilon)^d$ for $f_0(P)f_{d-1}(P)$, with P being the (full-dimensional) forest polytope of $K_{2,d}$ and any $\varepsilon > 0$. Each subgraph of $K_{2,d}$ that takes, for each node v of degree 2, at most one edge incident to v, is a forest. Those graphs are 3^d. Moreover, each induced subgraph of $K_{2,d}$ that takes the nodes of degree d plus at least 2 other nodes is 2-connected, hence it induces a (distinct) facet of P. Those are $2^d - (d+1)$. In total $f_0(P)f_{d-1}(P) = \Omega(6^d)$.

Example 29 (Spanning tree polytope of the skeleton of the 4-dimensional cube). Let G be the skeleton of the 4-dimensional cube, and P the associated spanning tree polytope. Numerical experiments show that $f_0(P)f_{d-1}(P) \geq 1.603 \cdot 10^{11}$, while the upper bound from Conjecture 1 is $\approx 1.331 \cdot 10^{11}$.

References

1. Barahona, F., Grötschel, M.: On the cycle polytope of a binary matroid. J. Comb. Theory, Ser. B **40**(1), 40–62 (1986)
2. Bárány, I., Pór, A.: On 0-1 polytopes with many facets. Adv. Math. **161**(2), 209–228 (2001)
3. Bohn, A., Faenza, Y., Fiorini, S., Fisikopoulos, V., Macchia, M., Pashkovich, K.: Enumeration of 2-level polytopes. In: Bansal, N., Finocchi, I. (eds.) ESA 2015. LNCS, vol. 9294, pp. 191–202. Springer, Heidelberg (2015)
4. Chaourar, B.: On the kth best base of a matroid. Oper. Res. Lett. **36**(2), 239–242 (2008)
5. Chvátal, V.: On certain polytopes associated with graphs. J. Comb. Theory Ser. B **18**, 138–154 (1975)
6. Diestel, R.: Graph Theory. Graduate Texts in Mathematics, vol. 101. Springer, Heidelberg (2005)
7. Feichtner, E.M., Sturmfels, B.: Matroid polytopes, nested sets and bergman fans. Portugaliae Mathematica **62**(4), 437–468 (2005)
8. Gouveia, J., Laurent, M., Parrilo, P.A., Thomas, R.: A new semidefinite programming hierarchy for cycles in binary matroids and cuts in graphs. Math. Program. **133**(1–2), 203–225 (2012)
9. Gouveia, J., Parrilo, P., Thomas, R.: Theta bodies for polynomial ideals. SIAM J. Optim. **20**(4), 2097–2118 (2010)
10. Grande, F., Rué, J.: Many 2-level polytopes from matroids. Discret. Comput. Geom. **54**(4), 954–979 (2015)
11. Grande, F., Sanyal, R.: Theta rank, levelness, and matroid minors (2014). arXiv:1408.1262
12. Gross, J.L., Yellen, J.: Graph Theory and Its Applications. CRC Press, Boca Raton (2005)
13. Oxley, J.G.: Matroid Theory, vol. 3. Oxford University Press, Oxford (2006)
14. Schrijver, A.: Theory of Linear and Integer Programming. Wiley, New York (1998)
15. Stanley, R.: Decompositions of rational convex polytopes. Ann. Discret. Math. **6**, 333–342 (1980)
16. Sullivant, S.: Compressed polytopes and statistical disclosure limitation. Tohoku Math. J. Second Ser. **58**(3), 433–445 (2006)
17. Yannakakis, M.: Expressing combinatorial optimization problems by linear programs. J. Comput. Syst. Sci. **43**, 441–466 (1991)
18. Ziegler, G.: Lectures on Polytopes, vol. 152. Springer, Berlin (1995)

Optimization Problems with Color-Induced Budget Constraints

Corinna Gottschalk[1(✉)], Hendrik Lüthen[2], Britta Peis[1], and Andreas Wierz[1]

[1] RWTH Aachen University, Aachen, Germany
corinna.gottschalk@oms.rwth-aachen.de
[2] TU Darmstadt, Darmstadt, Germany

Abstract. Gabow and Tarjan [9] provided a very elegant and fast algorithm for the following problem: given a matroid defined on a red and blue colored ground set, determine a basis of minimum cost among those with k red elements, or decide that no such basis exists. In this paper, we investigate possible extensions of this result from ordinary matroids to the more general notion of *poset matroids*. Poset matroids (also called *distributive supermatroids*) are defined on the collection of all ideals of an underlying partial order on the ground set. We show that the problem on general poset matroids becomes NP-hard, already if the underlying poset consists of binary trees of height two. On the positive side, we present two polynomial algorithms: one for integer polymatroids, i.e., the case where the poset consists of disjoint chains, and one for the problem to determine a minimum cost ideal of size l with k red elements, i.e., the uniform rank-l poset matroid, on series-parallel posets.

1 Introduction

In a seminal paper [9], Gabow and Tarjan developed a very fast and elegant algorithm to solve the following problem: given an undirected graph $G = (V, E)$ whose edge set is partitioned into red and blue elements $E = R \cup B$, a cost function $c \colon E \to \mathbb{R}_+$, and some integer k, determine a spanning tree $T \subseteq E$ of minimum cost among those with k red edges, or decide that no such tree exists. This algorithm not only works for spanning trees but also for the more general notion of matroid bases [9].

A nonempty family $\mathcal{F} \subseteq 2^E$ defined on a finite ground set E defines a matroid $\mathcal{M} = (E, \mathcal{F})$ if for all $X, Y \subseteq E$ the following two properties are satisfied:

(i) $Y \in \mathcal{F}, X \subset Y$ implies $X \in \mathcal{F}$, and
(ii) $X, Y \in \mathcal{F}, |X| < |Y|$ implies $\exists e \in Y \setminus X$ with $X \cup \{e\} \in \mathcal{F}$.

Sets in $\mathcal{F}$ are called *independent*, inclusionwise maximal sets in $\mathcal{F}$ are called *bases*.

The authors thank the German Research Association (DFG) for funding this work (Research Grants SFB 666 and PE 1434/3-1).

© Springer International Publishing Switzerland 2016
R. Cerulli et al. (Eds.): ISCO 2016, LNCS 9849, pp. 189–200, 2016.
DOI: 10.1007/978-3-319-45587-7_17

Examples of matroids include the collection of cycle-free edge sets of an undirected graph $G = (V, E)$ (the "graphic matroid"), linearly independent column-sets of a given matrix with columns indexed by E (the "linear matroid"), or the subsets of cardinality at most l of a given ground set E (the "uniform rank-l matroid").

Given a matroid $\mathcal{M} = (E, \mathcal{F})$ with basis set $\mathcal{B}$ defined on a red-blue colored ground set $E = R \cup B$, a cost function $c \colon E \to \mathbb{R}_+$, and some integer k, the *color-induced budget-constrained matroid problem* asks for a min-cost basis among those with k red elements. This problem can be formulated as

$$\min\{\sum_{e \in F} c_e \mid F \in \mathcal{B} \text{ and } |F \cap R| = k\}. \tag{1}$$

Variants and Extensions of the Problem. For general weight functions $w \in \mathbb{R}_+^E$, the resulting budget-constrained matroid problem $\min\{c(F) \mid F \in \mathcal{B}, w(F) \le k\}$ immediately becomes NP-hard, already in the special case of graphic matroids (see the NP-hardness proof of length-bounded MST in [1]).

Therefore, when seeking for variants or extensions of the problem, one might consider alternative combinatorial structures $\mathcal{B} \subseteq 2^E$. For example, for the class of perfect matchings and rooted arborescences, color-induced budget constrained optimization problems have been investigated before:

Given a graph with red-blue colored edges and an integer k, it has been shown that the problem to find a maximum matching containing exactly k red edges is solvable in polynomial time by a randomized algorithm due to Mulmuley et al. [13], and deterministically with an additive error of one due to Yuster [17].

Räbiger [15] presented an FPTAS and a pseudopolynomial algorithm for the (k, r)-arborescence problem which asks for a min-cost arborescence in a red-blue colored graph rooted at r containing at most k red edges.

Moreover, a shortest path with k red edges can be found using a multicriteria variant of Dijkstra (see e.g. [12]).

Color-Induced Budget Constrained Poset Matroids. In this paper, we investigate generalizations of the color-induced budget-constrained matroid problem from ordinary matroids to *poset matroids*: Poset matroids, as introduced by Dunstan, Ingleton and Welsh [3], are also known as *distributive supermatroids* and are a special case of *ordered matroids* as introduced by Faigle [5].

For poset matroids, it is assumed that a partial order $P = (E, \preceq)$ on ground set E is given. Recall that a set $I \subseteq E$ is called *ideal* w.r.t. poset P if $e \in I$, $g \preceq e$ implies $g \in I$. We denote by $\mathcal{D}(P)$ the collection of all ideals of P. A nonempty family $\mathcal{F} \subseteq \mathcal{D}(P)$ forms (the independent sets of) a poset matroid if and only if the matroid-defining properties (i) and (ii) hold for two ideals $Y \in \mathcal{F}$, $X \subset Y$ and $X, Y \in \mathcal{F}$, respectively. As before, the inclusion-wise maximal sets in $\mathcal{F}$ are called *bases* of the poset matroid. The set of all bases in $\mathcal{F}$ is denoted by $\mathcal{B}$.

To get an intuition for poset matroids, it is important to observe that the intersection of independent sets of a matroid $(E, \mathcal{F})$ and the ideals of a poset on E (a structure investigated in [7]) does not necessarily yield a poset matroid.

Consider for example the graphic matroid on a multigraph $G = (\{a, b, c\}, \{e_1 = \{a, b\}, e_2 = \{a, b\}, e_3 = \{b, c\}, e_4 = \{b, c\}\})$ with a partial order $e_1 \preceq e_3$, $e_2 \preceq e_4$ on the edges. Then the sets $\{e_1, e_3\}$ and $\{e_2\}$ are independent ideals, but property (ii) is not fulfilled.

Note that there is a one-to-one correspondence between feasible points of an integral polymatroid (we will provide a brief introduction in Sect. 2) and independent sets of poset matroids whose underlying poset consists of the disjoint union of chains. Recall that a set $C \subseteq E$ is a chain in P if any two elements in C are comparable.

Poset matroids admit a generalization of the following two important notions from matroids: (1) the symmetric exchange property [2] and (2) the optimality of the greedy algorithm [4] for $\min\{c(F) \mid F \in \mathcal{B}\}$. The latter requires cost function c to be *consistent*, that is, $c(e) \leq c(g)$ holds whenever $e \preceq g$.

Since Gabow and Tarjan's algorithm [9] heavily relies on these two matroid-characterizing properties, it seems plausible that the algorithm could be adapted to solve the *color-induced budget constrained poset matroid problem*: Given a poset matroid defined on a red-blue colored set $E = R \cup B$ of elements, costs $c \colon E \to \mathbb{R}_+$, and some integer k, find a min-cost basis among those with exactly k red elements or decide that no such basis exists.

Our Contribution. In Sect. 2, we extend Gabow and Tarjan's algorithm to minimize a separable discrete convex function over an integral polymatroid subject to the additional constraint $x(R) := \sum_{i \in R} x_i = k$. This way, we show that the algorithm can be generalized to poset matroids with a consistent cost function c whose underlying poset consists of a disjoint union of chains (Theorem 1).

For poset matroids in general, we prove that even deciding if a feasible solution of the color-induced budget constrained poset matroid problem exists becomes NP-hard. The result holds already on very simple posets, namely, binary trees of height two (Theorem 2).

This motivates the restriction to simple matroid-types, like uniform poset matroids: Note that problem (1) restricted to uniform rank-l poset matroids can equivalently be stated as follows: Given a poset $P = (E, \preceq)$ on a red-blue colored set $E = R \cup B$, a cost function $c \colon E \to \mathbb{R}_+$, and two integers $l, k \in \mathbb{Z}_+$, find a min-cost ideal of cardinality l among those with k red elements. We show that it is still NP-hard to decide if there exists a feasible solution (Theorem 3). We then conclude that the optimization problem is NP-hard even if the cost function is consistent and a trivial feasible solution exists. Our most involved result is a polynomial algorithm, even for general costs, for the special case where the underlying poset is series-parallel (Theorem 4).

2   Color-Induced Budget Constraints on Integer Polymatroids

Recall the polyhedral description of matroids: Given a matroid $\mathcal{M} = (E, \mathcal{F})$, its associated rank function $r \colon 2^E \to \mathbb{Z}_+$ assigns each subset $S \subseteq E$ its *rank*

$r(S) := \max\{|F| \mid F \in \mathcal{F}, F \subseteq S\}$. Matroid rank functions can be characterized by the following properties:

(i) $r(\emptyset) = 0$ ("r is normalized"),
(ii) $r(S) \leq r(S \cup \{e\}) \leq r(S) + 1$ for all $S \subseteq E$ and $e \in E \setminus S$ ("r is unit increasing"),
(iii) $r(S) + r(T) \geq r(S \cap T) + r(S \cup T)$ for all $S, T \subseteq E$ ("r is submodular").

Moreover, the (incidence vectors of) bases of $\mathcal{M}$ are exactly the vertices of the *matroid basis polytope* $\{x \in \mathbb{R}_+^E \mid x(S) \leq r(S) \; \forall S \subseteq E, \; x(E) = r(E)\}$. As a consequence, problem (1) can equivalently be written as

$$\min_{x \in \mathbb{Z}_+^E} \{c(x) \mid x(S) \leq r(S) \; \forall S \subseteq E, \; x(E) = r(E), \; x(R) = k\}, \qquad (2)$$

where $c(x) = \sum_{e \in E} c_e x_e$. If we relax the unit-increase property and only require monotonicity of r in the sense that $S \subseteq T$ implies $r(S) \leq r(T)$ for all $S, T \subseteq E$, the polytope $\bar{P}(r) := \{x \in \mathbb{R}_+^E \mid x(S) \leq r(S) \; \forall S \subseteq E\}$ is called *polymatroid* and the set of integer vectors in $\bar{P}(r)$ is called *integer polymatroid*. A vector x with $x(E) = r(E)$ is a base of an (integer) polymatroid and $r : 2^E \to \mathbb{Z}_+$ is called *polymatroid rank function*. We denote the set of bases of an integer polymatroid by $\mathbb{P}(r)$. For more about polymatroids, we refer to Fujishige's book [8]. In contrast to incidence vectors of matroid bases, elements of $\mathbb{P}(r)$ are not necessarily $\{0, 1\}$-vectors anymore. Instead of separable linear objective functions $c(x) = \sum_{e \in E} c_e x_e$, we now consider more general separable discrete convex functions. Recall that a function $c : \mathbb{Z}_+^E \to \mathbb{R}$ is *(discrete) separable convex* if $c(x) = \sum_{e \in E} c_e(x_e)$ for each $x \in \mathbb{Z}_+^E$, where each function $c_e : \mathbb{Z}_+ \to \mathbb{R}$, $e \in E$ is *discrete convex* in the sense $c_e(x_e + 1) - c_e(x_e) \leq c_e(y_e + 1) - c_e(y_e) \; \forall x_e, y_e \in \mathbb{Z}_+$ with $x_e \leq y_e$.

Any separable discrete convex function can be minimized over $\mathbb{P}(r)$ in a greedy-type manner, e.g., with Faigle's ordered greedy algorithm [4]. While this algorithm is simple and fast, its running time depends on the size of the ordered matroid and is therefore pseudopolynomial. However, [10] also presented a more involved polynomial algorithm. We show in the following how to extend Gabow and Tarjan's algorithm for matroids to minimize a separable discrete convex function over the bases of an integer polymatroid with an additional color-induced budget constraint of type $x(R) = k$.

Theorem 1. *Given an integer polymatroid with rank function $r : 2^E \to \mathbb{Z}_+$, a separable discrete convex cost function $c : \mathbb{Z}^E \to \mathbb{R}$, an integer k, and a coloring $E = R \cup B$, the problem of finding a minimum cost basis x with $x(R) := \sum_{i \in R} x_i = k$ can be solved in polynomial time.*

Proof of Theorem 1. The idea of the algorithm can be sketched as follows: the algorithm starts with a basis vector $\bar{x}$ in $\mathbb{P}(r)$ of minimum cost among those bases x minimizing $x(R)$. This can be done efficiently by minimizing the (separable discrete convex) cost function $\bar{c}$ defined by $\bar{c}_e(x_e) = c_e(x_e) + M$ if $e \in R$ and $\bar{c}_e(x_e) = c_e(x_e)$ otherwise for some sufficiently large constant $M > 0$, over $\mathbb{P}(r)$.

We define for each integer $t \in \{1, \ldots, k\}$ the set $\mathbb{P}_t(r) := \{x \in \mathbb{P}(r) \mid x(R) = t\}$. Hence, $\bar{x}(R)$ is the minimal index t such that $\mathbb{P}_t(r) \neq \emptyset$. In order to prove the theorem, it suffices to compute a cost-minimal basis $\bar{x}_k$ in $\mathbb{P}_k(r)$ or decide that $\mathbb{P}_k(r)$ is empty. If $\bar{x}(R) > k$, clearly $\mathbb{P}_k(r) = \emptyset$. Therefore, we assume $\bar{x}(R) \leq k$ and define $t_{\min} := \bar{x}(R)$.

In analogy to Gabow and Tarjan, we show that, given a cost-minimal basis $\bar{x}_t \in \mathbb{P}_t(r)$, a cost-minimal basis $\bar{x}_{t+1} \in \mathbb{P}_{t+1}(r)$ can in fact be determined in a very simple way if it exists: As we show in Lemma 1 below, it suffices to search for the minimal "swap" which shifts one unit from a blue component to a red component. In general, given a basis $b \in \mathbb{P}(r)$, we call (i, j) a *swap* for b if $i \in R$, $j \in B$, and $b + \chi_i - \chi_j \in \mathbb{P}(r)$. A swap (i, j) for b is *minimal* if the cost difference $c(b + \chi_i - \chi_j) - c(b) = c_i(b_i + 1) - c_i(b) + c_j(b_j - 1) - c_j(b_j)$ is minimized.

Algorithm:

1. Set $\bar{x} \leftarrow \operatorname{argmin}\{\bar{c}(x) \mid x \in \mathbb{P}(r)\}$ and $t = t_{\min}$.
2. While $t < k$ and $P_t(r) \neq \emptyset$ do: determine a minimal swap (i, j) for $\bar{x}_t$ or decide that $P_{t+1}(r) = \emptyset$; then iterate with $\bar{x}_{t+1} = \bar{x}_t + \chi_i - \chi_j$ and $t = t + 1$.
3. If $P_k(r) \neq \emptyset$, return $\bar{x}_k$.

Lemma 1. *Suppose b is a cost-minimal basis in $\mathbb{P}_{t-1}(r)$ and $\mathbb{P}_t(r) \neq \emptyset$. If (i, j) is a minimal swap for b, then $b + \chi_i - \chi_j$ is a cost-minimal basis in $\mathbb{P}_t(r)$.*

Proof. We denote by β_t the collection of cost-minimal bases in $\mathbb{P}_t(r)$. We will show that there always exists a swap (u, v) for b such that $b + \chi_u - \chi_v \in \beta_t$, which implies the lemma. Let $b' \in \beta_t$ be a basis such that $\sum_{i=1}^{|E|} |b_i - b'_i|$ is minimal and let u be an index in R with $b'_u > b_u$. Then the strong basis exchange property for polymatroids (see, e.g., [14], page 101) implies there exists an index v such that $b_v > b'_v$ and both, $\tilde{b} := b + \chi_u - \chi_v$ and $\hat{b} := b' - \chi_u + \chi_v$, are bases in $\mathbb{P}(r)$.

We show that v is blue. For the sake of contradiction, suppose $v \in R$. Then $\tilde{b} \in \mathbb{P}_{t-1}(r)$ and $\hat{b} \in \mathbb{P}_t(r)$. To simplify notation, for a given integer $p \in \mathbb{Z}$ and $i \in E$, we denote by $\Delta_i(p)$ the difference $c_i(p) - c_i(p - 1)$.

Thus, $c(b) \leq c(\tilde{b}) = c(b) + \Delta_u(b_u + 1) - \Delta_v(b_v) \Rightarrow \Delta_u(b_u + 1) \geq \Delta_v(b_v)$ and $c(b') \leq c(\hat{b}) = c(b') - \Delta_u(b'_u) + \Delta_v(b'_v + 1) \Rightarrow \Delta_v(b'_v + 1) \geq \Delta_u(b'_u)$.

Since all functions c_i are discrete convex, we can conclude $\Delta_u(b_u + 1) \geq \Delta_v(b_v) \geq \Delta_v(b'_v + 1) \geq \Delta_u(b'_u) \geq \Delta_u(b_u + 1)$. As a consequence, $c(b') = c(\hat{b})$ and $\hat{b} \in \beta_t$. But $\sum_{i=1}^t |b_i - b'_i| - \sum_{i=1}^t |b_i - \hat{b}_i| = b'_u - b_u + b_v - b'_v - (b'_u - 1 - b_u + b_v - (b'_v + 1)) = 2$, in contradiction to the choice of b'.

Therefore, $v \notin R$ from which $\tilde{b} \in \mathbb{P}_t(r)$ and $\hat{b} \in \mathbb{P}_{t-1}(r)$ follows. The latter implies $c(b) \leq c(\hat{b}) = c(b') - \Delta_u(b'_u) + \Delta_v(b'_v + 1)$ which is equivalent to $c(b') \geq c(b) + \Delta_u(b'_u) - \Delta_v(b'_v + 1) \geq c(b) + \Delta_u(b_u + 1) - \Delta_v(b_v) = c(\tilde{b})$. The last inequality holds by definition of c_i. This shows $\tilde{b} \in \beta_t$. $\qquad \square$

We remark that the Corollaries 3.1, 3.2 and 3.3 from [9] that are used for improving the running time of the algorithm can be generalized directly to our setting using their proofs combined with the techniques used in the proof of Theorem 1. But the subsequent idea from [9] of a *swap sequence* cannot be generalized directly because of the order-dependence in our case.

3   Hardness on Poset Matroids

Gabow and Tarjan's algorithm, as well as its extension described in the previous section, heavily relies on the symmetric exchange property of integer polymatroids, as well as the optimality of the greedy algorithm to determine a cost-minimal basis without color-induced budget constraint.

Barnabei et al. presented a variant of the symmetric exchange property for poset matroids in [2] and Faigle showed in [4] how to obtain a minimum cost basis using a greedy algorithm if the cost function is consistent.

However, as the following theorem states, the color-induced budget constrained poset matroid turns out to be NP-hard, even on very simple posets (see [11]). This is done by reduction from MATROID PARITY. For the proof, we refer to the full version.

Theorem 2. *Let* $\mathcal{P} = (E, \preceq)$ *be a poset on the red-blue colored ground set* $E = R \cup B$ *and* $\mathcal{M} = (E, \mathcal{F})$ *be a poset matroid. Deciding if a basis* T *with* $|T \cap R| = k$ *exists for a given* $k \in \mathbb{N}$ *is NP-hard even if the poset consists of binary trees of height two.*

Since MATROID PARITY is NP-hard in general, but efficiently solvable for several matroid classes, we restrict our considerations to simple poset matroids.

4   Uniform Poset Matroids or the Problem of Finding an Ideal with Color-Induced Constraints

As we have seen in the previous section, deciding if a basis with k red elements exists is NP-hard for poset matroids, even if the poset structure is simple. Therefore, we now consider a simple matroid structure: uniform rank-l poset matroids. That means, for a given poset, any ideal of size at most l is independent. From now on, we will use an equivalent formulation of the problem without referring to poset matroid terminology: Given a poset $\mathcal{P} = (P, \preceq)$ with red-blue colored elements $P = R \cup B$, a weight function $w \colon P \to \mathbb{R}$ and numbers $k \leq l \in \mathbb{N}$. The goal is to find a minimum weight ideal of size l which contains k red elements. We call this problem the *Minimum Colored Ideal Problem* (MCIP).

This problem has been studied without an additional color-induced budget constraint. Faigle and Kern showed that the problem to find a minimum cost ideal of size l is strongly NP-hard for arbitrary nonnegative weights [6]. On the other hand, if the objective function is consistent (and not necessarily nonnegative), this problem can be solved using Faigle's greedy algorithm for ordered matroids [5]. However, the addition of color-induced budget constraints makes even deciding if such an ideal exists NP-hard:

Theorem 3. *Given an instance* $\mathcal{P}, k, l$ *of MCIP. The problem to decide if an ideal of size* l *with* k *red elements exists in* $\mathcal{P}$ *is strongly NP-hard.*

The proof uses a reduction from the clique problem similar to [6]. We give the details in the full version. By reduction from the above existence problem, we obtain:

Corollary 1. *MCIP is strongly NP-hard even if the existence of an ideal of size l with k red elements is guaranteed and the cost function is consistent.*

To verify the corollary, let $\mathcal{P}, k, l$ be an instance of the above existence problem and assign a cost of one to all elements of $\mathcal{P}$. Add l incomparable elements, k of which are red, to the poset. Moreover, we make sure that these new elements are so expensive that they will only be picked if there is no other option.

In the remainder of this section, we are going to show that for a series-parallel poset, MCIP can be solved in polynomial time for arbitrary weights even with a color-induced budget constraint. We remark that the poset used in the proof of Theorem 2 is series-parallel. Thus, for general poset matroids, deciding whether a feasible solution exists is NP-hard even for series-parallel posets.

A poset $\mathcal{P}$ is called *series-parallel* if it can be constructed by the following constructive characterization.

(i) A single element is series-parallel.
(ii) The disjoint union of multiple series-parallel posets is series-parallel *(parallel composition)*.
(iii) For series-parallel posets Q and Q', the poset that arises by adding all comparabilities such that $x \in Q, y \in Q' \Rightarrow x \preceq y$ is series-parallel *(series composition)*.

The definition of series-parallel digraphs is analogous. It is possible to decide in linear time if a given digraph (and thus a given poset) is series-parallel and if so, to obtain the decomposition tree [16]. The decomposition tree is one way to represent the construction of a series-parallel poset using the steps described above. For simplicity, we are going to assume from now on that we are given a fixed *construction sequence*. That is, an order of the above steps which constructs the given poset. Such a sequence can be obtained from the decomposition tree. If we talk about the next composition containing a poset $\mathcal{P}$, we refer to the next step in our fixed construction sequence that contains $\mathcal{P}$. A component in the construction sequence is any poset that can occur during the construction.

Theorem 4. *MCIP can be solved in polynomial time for series-parallel posets.*

In the remainder of this section, we prove this theorem. We start by summarizing the idea: Construct an acyclic digraph with source s, sink t and nodes that correspond to the inclusion or exclusion of certain subsets of P in the ideal to be constructed. Our graph G will be designed in such a way that each s-t-path corresponds to an ideal and each ideal of size at most l containing at most k red elements corresponds to an s-t-path (note that this is not a one-to-one correspondence). The cost $w(p)$ for $p \in P$ will be modeled by costs on ingoing edges. Each vertex will be assigned a number of red elements $r_G(v)$ and total number of elements $l_G(v)$ (we omit the index if it is obvious from the context).

In order to find a minimum cost ideal, we apply a multi-criteria shortest path algorithm in G. We describe the idea, for a more detailed description see e.g. [12]. For each node v and each pair (i,j) with $i \leq k$ and $j \leq l$, the length of a cheapest s-v-path F with $\sum_{v \in V(F)} r(v) = i$ and $\sum_{v \in V(F)} l(v) = j$ is stored: Instead of maintaining a single label that contains the length of a shortest path from s at a node, we store at each node different labels encoding additional side constraints. In our case, there are $k \cdot l$ labels per node. To update a node label, we consider the relevant labels of preceding nodes. For example, for a label (i,j) at v with $r(v) = 2$ and $l(v) = 3$ and an edge (w,v) we consider the label $(i-2, j-3)$ at w. Depending on whether the cost function is positive, we can use Dijkstra's algorithm or Moore-Bellman-Ford for the shortest path computation.

It remains to show how G is constructed and prove the correspondence of s-t-paths and ideals. All paths that we refer to will be directed. As usual we denote by $N_G(S)$ for $S \subset V(G)$ the neighborhood of S in G (not including S).

Let us present the construction formally. For an example of the complete construction see Fig. 3. For a series-parallel poset $\mathcal{P}$, we introduce vertices s, t and x and for each $p \in P$ a vertex p_+. Intuitively, visiting p_+ corresponds to including p in an ideal and x corresponds to $\emptyset$. Then, we introduce additional vertices and edges to construct a graph G depending on the construction sequence of the poset. The graph G is constructed in the same order as the series-parallel poset $\mathcal{P}$, following its construction sequence.

Construction of G: For a single vertex $p \in P$, $G = (\{p_+, x, s, t\}, \{(s,v), (v,t) \mid v \in \{p_+, x\}\})$. We have $r(p_+) = 1$ for $p \in R$, $l(p_+) = 1$ and $r(v) = l(v) = 0$ for all other vertices. Also, $c(s, p_+) = w(p)$ and $c(e) = 0$ for the remaining edges.

Given two series-parallel posets $\mathcal{P}_1, \mathcal{P}_2$ with corresponding graphs G_1 and G_2, we construct graph G from G_1 and G_2:

If the posets $\mathcal{P}_1$ and $\mathcal{P}_2$ are joined by parallel composition (i.e. their elements are incomparable), we set $V(G) := (V(G_1) \setminus \{t\}) \cup (V(G_2) \setminus \{s, x\}) \cup \{v_1, v_2\}$ where the v_i's are two new vertices. Intuitively, visiting vertex v_i tells us that we want to choose some element from the given parallel components but not one in P_i. We set

$$E(G) := E[V(G_1) \setminus \{t\}] \cup E[V(G_2) \setminus \{s\}] \cup \{(s, v_1), (v_2, t)\} \cup \{((N_{G_1}(t) \setminus \{x\}) \cup \{v_1\}) \times ((N_{G_2}(s) \setminus \{x\}) \cup \{v_2\})\}.$$

If the parallel composition joins more than two components, the construction is analogous with a new vertex v_i for each component and only one vertex x in the graph G. For a schematic example see Fig. 1.

For $v \in V(G_i)$ we have $r_G(v) = r_{G_i}(v)$ and $r(v_i) = 0$ since these vertices model the exclusion of P_i (analogously for $l_G(v)$). The edge costs are assigned analogously: $c(e) = 0$ if $e \in \delta^-(v_i)$, otherwise for $e \in \delta_G^-(v)$ let $e' \in \delta_{G_i}^-(v)$ and set $c(e) = c(e')$.

If $\mathcal{P}_1$ and $\mathcal{P}_2$ are joined by series composition (the elements of $\mathcal{P}_2$ are smaller than those of $\mathcal{P}_1$), we set $V(G) := (V(G_1) \setminus \{t\}) \cup (V(G_2) \setminus \{s, x\}) \cup \{u, z\}$.

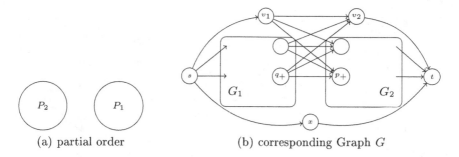

(a) partial order   (b) corresponding Graph G

Fig. 1. Construction for parallel composition

The vertex u will be used as a shortcut. Any path that visits a node in $V(G_1) \setminus \{x, s, t\}$ must visit u without visiting $V(G_2)$. This models the following: If an ideal contains an element of P_1, it also contains all elements of P_2 and thus u corresponds to the inclusion of all of them. Vertex z is used to model the opposite: The ideal should contain some element of P_2 but no element of P_1. Then

$$E(G) := E[V(G_1) \setminus \{t\}] \cup E[V(G_2) \setminus \{s\}] \cup \{(N_{G_1}(t) \setminus \{x\}) \times \{u\}\} \cup \{(s, z), (u, t)\} \cup \{\{z\} \times (N_{G_2}(s) \setminus \{x\})\},$$

$r(u) = |R \cap P_2|$, $l(u) = |P_2|$ and $c(v, u) = \sum_{p \in P_2} w(p)$, $r(z) = l(z) = 0$ and $c(s, z) = 0$, otherwise $r_G(v) = r_{G_i}(v)$ and c is determined as in the parallel composition. The idea is illustrated in Fig. 2.

Therefore, for a parallel composition, the order of P_1 and P_2 can be chosen arbitrarily but for series decomposition, we start with the larger poset (w.r.t. $\preceq$). Moreover, the structure of the graph depends heavily on the chosen construction sequence of the poset.

Since the size of G and the number of labels per node is polynomially bounded, a shortest s-t-path F with $\sum_{v \in V(F)} r(v) = k$ and $\sum_{v \in V(F)} l(v) = l$ can be found in polynomial time if it exists by considering the path corresponding to the label (k, l) at t. Thus, we obtain an optimal solution for MCIP as the following claim shows.

Claim. Any s-t-path F with $\sum_{v \in F} r(v) = i$ and $\sum_{v \in F} l(v) = j$ corresponds to an ideal of size j with cost $c(F)$ and i red elements for $i \leq k$ and $j \leq l$ and vice versa.

Proof of the Claim. We define a function $f : V(G) \to 2^P$ which, for each vertex, maps to the corresponding elements in P. That is, for $p \in P$, $f(p_+) = \{p\}$, for $u \in V(G)$ introduced for a series decomposition where P_i is the lower of the two posets $f(u) = P_i$ and otherwise $f(v) = \emptyset$.

Observe that for any s-t-path F and vertices $u, v \in V(F)$, we have $f(u) \cap f(v) = \emptyset$: Since $\mathcal{P}$ is series-parallel and nodes are always mapped to components in the construction sequence, either $f(u) \cap f(v) = \emptyset$ or w.l.o.g. $f(u) \cap f(v) = f(u)$.

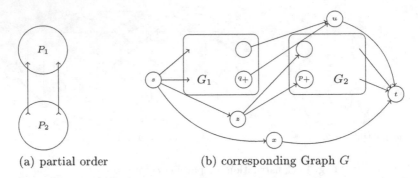

(a) partial order   (b) corresponding Graph G

Fig. 2. Construction for series composition

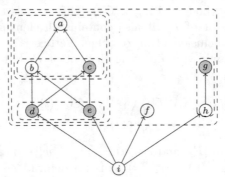

(a) partial order with construction sequence indicated by dashed rectangles and red (light gray) elements

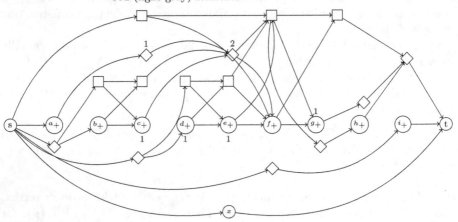

(b) corresponding Graph G with values $r > 0$, square nodes are introduced for parallel and diamond nodes for series compositions

Fig. 3. Example for a given series-parallel partial order

But, for a parallel composition, only elements corresponding to disjoint subsets are connected. For a series composition, a vertex v with $f(v) \neq \emptyset$ is only connected to a vertex v' where $y' \preceq y$ and $y' \neq y$ for all $y \in f(v), y' \in f(v')$. Thus, the statement follows.

To prove the claim, we show that for an s-t-path F in G, the set $I := \cup_{v \in V(F)} f(v)$ is an ideal of cost $c(F)$ with $|I| = \sum_{v \in V(F)} l(v)$ and $\sum_{v \in V(F)} r(v)$ red elements. The claim about the equivalence of costs and r, l-values follows from the above observation.

If $I = \emptyset$, it obviously is an ideal. Therefore, consider $a \in I$ and a direct predecessor $b \preceq a$. We show that $b \in I$. If there exists $v \in V(F)$ such that $\{a, b\} \subseteq f(v)$, this is clearly true, so suppose there is no such v. At some point there must have been a series composition of two components P_a, where a was a smallest, and P_b, where b was a largest element. Let P_{ab} be the resulting component. By the assumption above, F does not visit a vertex v with $P_{ab} \subseteq f(v)$. Thus, $a \in I$ implies that F visits some vertex y in the corresponding graph G_{ab} with $y \notin \{x, z\}$. The construction for series compositions then implies, that F visits $u \in V(G_{ab})$. But $b \in f(u)$ and thus $b \in I$. By transitivity, it follows that I is an ideal.

Conversely, for an ideal $I \subseteq P$, it can be shown by induction over the number of steps in the construction sequence of $\mathcal{P}$ that there is a path corresponding to I.

If the construction sequence is empty, we have $|P| \leq 1$. So an ideal either consists of the element $p \in P$ with corresponding path $\{s, p_+, t\}$ or is empty which corresponds to $\{s, x, t\}$.

Consider a series-parallel poset which is constructed using more than one step. Suppose the last construction step is a parallel composition of $\mathcal{P}_1, \ldots, \mathcal{P}_j$ ordered by the order induced by G. By induction, we know that there is a path F_i in G_i corresponding to $P_i \cap I$ for all $i \leq j$. If $F_i = \{s, x, t\}$ for all i, then $F := \{s, x, t\}$. Otherwise, we concatenate the paths F_i: Start with F_1, end with F_j and for connecting F_i and F_{i+1}, let (y_i, t) be the last edge of F_i, (s, y'_{i+1}) the first edge of F_{i+1}. Then by construction $(y_i, y'_{i+1}) \in E(G)$ and we can use that edge to connect the vertices. There is one exception: If $F_i = \{s, x, t\}$, then we connect to and from the vertex v_i instead.

Now assume that the last construction step is a series composition of $\mathcal{P}_1$ and $\mathcal{P}_2$ such that $\mathcal{P}_2$ contains the smaller elements. Again, if $I = \emptyset$, set $F = \{s, x, t\}$. If $I \cap P_1 = \emptyset$, we know by induction that there exists a path F' corresponding to I in G_2. Let (s, y) be the first edge of F'. Then the path $\{s, z, F'_{[y,t]}\}$ is the desired path in G. Otherwise, by induction there is a path F'' in G_1 corresponding to $I \cap P_1$. Let (y, t) be the last edge of F''. Then $y \neq x$ and thus, $(y, u) \in E(G)$. Therefore, $\{F''_{[s,y]}, u, t\}$ is the desired path since $P_2 \subset I$.   □

5   Concluding Remarks

It seems like color-induced budget constraints generate problems that are on the edge of NP-hardness in the sense that we can find interesting special cases that are in P as well as NP-complete ones. Hence, investigating other poset matroids that have a well-understood structure in this setting would also be interesting. The problem we considered on polymatroids in Sect. 2 where each index is assigned a color is still a special kind of coloring in terms of poset matroids. Since there is a one-to-one correspondence between poset matroids on chains and integer polymatroids, it is natural to consider a more general budget-constraint: Given a poset matroid where the poset $\mathcal{P}$ consists only of chains and a coloring $P = R \cup B$, find a minimum cost basis with k red elements. For non-uniform poset matroids, the complexity of this problem remains open even for consistent weight functions. While Dilworth completion has been successfully used in similar contexts, unfortunately it is not helpful here.

References

1. Aggarwal, V., Aneja, Y.P., Nair, K.: Minimal spanning tree subject to a side constraint. Comput. Oper. Res. **9**(4), 287–296 (1982)
2. Barnabei, M., Nicoletti, G., Pezzoli, L.: The symmetric exchange property for poset matroids. Adv. Math. **102**, 230–239 (1993)
3. Dunstan, F.D.J., Ingleton, A.W., Welsh, D.J.A.: Supermatroids. In: Welsh, D.J.A., Woodall, D.R. (eds.) Combinatorics (Proceedings of the Conference on Combinatorial Mathematics), pp. 72–122. The Institute of Mathematics and its Applications (1972)
4. Faigle, U.: The greedy algorithm for partially ordered sets. Discrete Math. **28**, 153–159 (1979)
5. Faigle, U.: Matroids on ordered sets and the greedy algorithm. Ann. Discrete Math. **19**, 115–128 (1984)
6. Faigle, U., Kern, W.: Computational complexity of some maximum average weight problems with precedence constraints. Oper. Res. **42**(4), 688–693 (1994). http://dx.doi.org/10.1287/opre.42.4.688
7. Fleiner, T., Frank, A., Iwata, S.: A constrained independent set problem for matroids. Technical report TR-2003-01, Egerváry Research Group, Budapest (2003). www.cs.elte.hu/egres
8. Fujishige, S.: Submodular Functions and Optimization. Annals of Discrete Mathematics, vol. 58, 2nd edn. Elsevier, Amsterdam (2005)
9. Gabow, H.N., Tarjan, R.E.: Efficient algorithms for a family of matroid intersection problems. J. Algorithms **5**(1), 80–131 (1984)
10. Groenevelt, H.: Two algorithms for maximizing a separable concave function over a polymatroid feasible region. Eur. J. Oper. Res. **54**, 227–236 (1991)
11. Lüthen, H.: On matroids and shortest path with additional precedence constraints. Master's thesis, Technische Universität Berlin (2012)
12. Martins, E.: On a multicriteria shortest path problem. Eur. J. Oper. Res. **16**, 236–245 (1984)
13. Mulmuley, K., Vazirani, U.V., Vazirani, V.V.: Matching is as easy as matrix inversion. Combinatorica **7**(1), 105–113 (1987)
14. Murota, K.: Discrete Convex Analysis. Soc. Ind. Appl. Math. (SIAM) (2003)
15. Räbiger, D.: Semi-Präemptives transportieren [in German]. Ph.D. thesis, Universität zu Köln (2005)
16. Valdes, J., Tarjan, R.E., Lawler, E.L.: The recognition of series parallel digraphs. In: Proceedings of the Eleventh Annual ACM Symposium on Theory of Computing, STOC 1979, New York, NY, USA, pp. 1–12. ACM (1979)
17. Yuster, R.: Almost exact matchings. Algorithmica **63**, 39–50 (2011)

Strengthening Chvátal-Gomory Cuts
for the Stable Set Problem

Adam N. Letchford[1], Francesca Marzi[2], Fabrizio Rossi[2],
and Stefano Smriglio[2(✉)]

[1] Department of Management Science, Lancaster University, Lancaster, UK
A.N.Letchford@lancaster.ac.uk
[2] Department of Information Engineering, Computer Science and Mathematics,
University of L'Aquila, L'Aquila, Italy
francesca.marzi@graduate.univaq.it,
{fabrizio.rossi,stefano.smriglio}@univaq.it

Abstract. The stable set problem is a well-known $\mathcal{NP}$-hard combinatorial optimization problem. As well as being hard to solve (or even approximate) in theory, it is often hard to solve in practice. The main difficulty is that upper bounds based on linear programming (LP) tend to be weak, whereas upper bounds based on semidefinite programming (SDP) take a long time to compute. We propose a new method to strengthen the LP-based upper bounds. The key idea is to take violated Chvátal-Gomory cuts and then strengthen their right-hand sides. Although the strengthening problem is itself $\mathcal{NP}$-hard, it can be solved reasonably quickly in practice. As a result, the overall procedure proves to be capable of yielding competitive upper bounds in reasonable computing times.

Keywords: Stable Set Problem · Clique inequalities · Chvátal-Gomory cuts · Cutting plane algorithm

1 Introduction

Given an undirected graph $G = (V, E)$, a *stable set* in G is a set of pairwise non-adjacent vertices. The convex hull of the incidence vectors of all stable sets in G is called the *stable set polytope* and denoted by $\mathrm{STAB}(G)$ [19]. The *Stable Set Problem* (SSP) calls for a stable set of maximum cardinality $\alpha(G)$, or, if a weight vector $w \in \mathbb{Q}_+^n$ is given, of maximum weight $\alpha_w(G)$. The SSP is strongly $\mathcal{NP}$-hard even to approximate [22]; and it is naturally stated as the binary program $\max\{\sum_{i \in V} w_i x_i : x_i + x_j \leq 1 \ \forall\{i, j\} \in E, x \subset \{0, 1\}^{|V|}\}$. Optimizing over its continuous relaxation provides very weak upper bounds on $\alpha_w(G)$. Therefore, a great effort has been devoted to improving the basic relaxation $\mathrm{FRAC}(G) = \{x \in [0, 1]^{|V|} : x_i + x_j \leq 1 \ \forall\{i, j\} \in E\}$, by studying valid inequalities for $\mathrm{STAB}(G)$. The first steps are due to Padberg, who introduced the *clique* inequalities [28]. These have the form $\sum_{i \in C} x_i \leq 1$, for any $C \subseteq V$ inducing a *maximal* clique in G, and induce facets of $\mathrm{STAB}(G)$. The polytope defined by all clique and nonnegativity inequalities is denoted by $\mathrm{QSTAB}(G)$ [19].

© Springer International Publishing Switzerland 2016
R. Cerulli et al. (Eds.): ISCO 2016, LNCS 9849, pp. 201–212, 2016.
DOI: 10.1007/978-3-319-45587-7_18

Many other valid inequalities, such as the *odd hole* and *odd antihole*, *web* and *antiweb* inequalities, have been derived. We refer the reader to [2,14,19] for detailed surveys. The Chvátal rank of some of these inequalities with respect to FRAC(G) and QSTAB(G) has been investigated in [23]. On the computational side, after the pioneering experience illustrated in [27] with clique and lifted odd-hole inequalities, more extensive results have been obtained with general *rank inequalities* and some of their (non-rank) lifted versions. These have been generated by *project-and-lift* separation heuristics introduced in [30] and recently improved in [8,9,29]. A study concerned with the exact separation of rank inequalities is described in [10]. Despite the fairly sophisticated techniques explored in these papers, the resulting upper bounds are not yet satisfactory for several graph classes.

Much stronger upper bounds can be obtained by Semidefinite Programming (SDP) relaxations. In the seminal paper [25], Lovász introduced the *theta function*, denoted by $\vartheta(G)$, as the optimal value of a SDP problem (we refer the reader to [19] for a comprehensive introduction). It has been proved that $\vartheta(G)$ dominates the bound obtained by optimizing over QSTAB(G) [19], and it is often much stronger in practice. Some classes of graphs for which this occurs are illustrated in [31], while a computational comparison is documented in [14]. Computational experiments with $\vartheta(G)$, or stronger relaxations obtained by adding valid linear inequalities, are presented in [4,11,15,20,24]. These approaches typically require long computing times. In order to manage this difficulty, ellipsoidal relaxations have been introduced [16], which allow one to obtain useful convex programming relaxations and derive effective cutting planes. In fact, this method allows one to achieve upper bounds close to $\vartheta(G)$ by optimizing over a linear relaxation.

Strong upper bounds have also been obtained by applying the Lovász and Schrijver lift-and-project operators [26] to FRAC(G). The N operator, based on LP, has been tested by Balas *et al.* [3]. The N^+ operator, based on SDP, yields a much stronger relaxation than the N operator, but it is often very hard to solve in practice. Computational experiments are presented in [5]. Finally, the $M(k,k)$ operator has been applied to QSTAB(G) [13,14]: the resulting non-compact linear relaxations turns out to provide upper bounds comparable to those from SDP relaxations at reasonable computational cost.

We propose a new method to strengthen the LP-based upper bounds. The key idea is to take violated Chvátal-Gomory cuts and then strengthen their right-hand sides relative to STAB(G). Although the strengthening problem is itself a SSP, a careful selection of the source cut can make it computationally tractable in practice. We present a cutting-plane algorithm based on the strengthened cuts and show that it is capable of yielding competitive upper bounds in moderate computing times. The algorithm is illustrated in the next section, while the computational experience is described in Sect. 3. Finally, some conclusions are drawn in Sect. 4.

2   Cutting Plane Algorithm

We consider an initial formulation of the SSP based on clique inequalities. In general, G may have exponentially many cliques and the separation problem associated to clique inequalities is strongly $\mathcal{NP}$-hard [27]. Nevertheless, greedy-like separation heuristics perform extremely well: experience shows that a cutting-plane algorithm embedding such a heuristic often achieves upper bounds quite close to those obtained by exactly optimizing over QSTAB(G). We therefore concentrate on the collection $\mathcal{C}$ of cliques generated by such an algorithm (see [13,14] for details) and consider the basic relaxation $\mathcal{Q}(\mathcal{C}) = \{Ax \leq 1, x \geq 0\}$, where $A = A_\mathcal{C}$ is the incidence matrix of the cliques in $\mathcal{C}$ versus the vertices of G. We also let $UB_\mathcal{C} = \{\max w^T x : x \in \mathcal{Q}(\mathcal{C})\}$ be the associated upper bound on $\alpha(G)$.

We experiment with a cutting plane algorithm based on the classical Chvátal-Gomory (CG) cuts [7,17,18]. These have the form

$$\lfloor u^T A \rfloor x \leq \lfloor u^T b \rfloor, \qquad u \in \mathbb{R}_+^m.$$

The cut generation procedure has two main stages:

1. Identify a violated CG cut $\lambda^T x \leq \lambda_0$
2. Strengthen λ_0 relative to STAB(G)

which are described in the following subsections.

2.1   Cut Separation

The choice of $u \in \mathbb{R}_+^m$ is critical for deriving useful inequalities. The classical idea from Gomory [17,18] is based on *basic* solutions. If one considers $\mathcal{Q}(\mathcal{C})$ in standard form (A, I) and a fractional vertex x^* associated with a basis B, any row i of B^{-1} associated with a fractional component of x^* determines a vector of multipliers u such that the resulting CG inequality cuts off x^*. In what follows, we refer to these inequalities as CG cuts from the tableau.

A different approach to obtaining useful CG cuts has been proposed in [12], where the separation problem associated to CG cuts (for a general MIP) is formulated as a MIP. Let x^* be the current fractional point and denote by $J(x^*) = \{j \in \{1, \ldots, n\} : 0 < x_j^* < 1\}$ the associated fractional support. Let also $\lambda^T x \leq \lambda_0$ be the CG cut to be generated, where $\lambda = \lfloor u^T A \rfloor$ and $\lambda_0 = \lfloor u^T b \rfloor$ for some multiplier $u \in \mathbb{R}_+$. The MIP-CG separation model has the form

$$\max \left(\sum_{j \in J(x^*)} \lambda_j x_j^* - \lambda_0 \right) - \sum_{i=1}^m \gamma_i u_i$$

$$\text{s.t.}$$

$$f_j = u^T A_j - \lambda_j, \quad j \in J(x^*)$$

$$f_0 = u^T b - \lambda_0$$

$$0 \leq f_j \leq 1 - \delta, \qquad j \in J(x^*) \cup \{0\}$$

$$0 \leq u_i \leq 1 - \delta, \qquad i = 1, \ldots, m$$

$$\lambda_j \text{ integer} \qquad j \in J(x^*) \cup \{0\}.$$

Some parts of this model are redundant and have been introduced in [12] for technical reasons. In detail, the explicit slack variables $f_j = u^T A_j - \lfloor u^T A_j \rfloor$, along with the parameter δ (fixed to 0.01), improve numerical tractability; and the constraints $u_i \leq 1 - \delta$ reduce the chance of generating dominated cuts. Our experience confirmed that these arrangements are indeed helpful. Another key feature of MIP-CG deals with the objective function: besides cut violation $\sum_{j \in J(x^*)} \lambda_j x_j^* - \lambda_0$, it includes a penalty on multipliers $\sum_{i=1}^m \gamma_i u_i$, with $\gamma_i = 10^{-4}$, which helps to make the cut sparser and stronger. According to [12], the penalty term γ_i has to be applied only to tight constraints, that is, those for which $s_i^* = 1 - a_i^T x^* = 0$. Sparsity is, in general, an important feature for a cutting plane to be numerically well-behaved. In our development it is also crucial to make the strengthening stage tractable, as discussed later.

Due to all of these modifications, MIP-CG is not an exact separation oracle any more. However, as pointed out in [12], exceptions are likely to occur only in pathological cases. We solve MIP-CG by the commercial MIP solver IBM Cplex 12.6.3 (default settings). The computation is stopped using two parameters cutviolation and septlim. In detail, the solver halts if: (i) a feasible solution of value greater than or equal to cutviolation has been found; (ii) the elapsed cpu time reaches septlim. In both cases, we store the whole catalog of violated inequalities corresponding to feasible solutions contained in Cplex pool at termination.

2.2 Cut Strengthening

Andersen and Pochet [1] present a general method to strengthen the left-hand side (lhs) coefficients and right-hand side (rhs) of inequalities relative to the mixed-integer hull. Their theoretical development suggests that the rhs should be strengthened before the lhs coefficients are strengthened. In our context, strengthening the rhs of a cut $\lambda^T x \leq \lambda_0$ relative to the integer hull amounts to solving the problem $\lambda_0^* = \max\{\lambda^T x : x \in \text{STAB}(G)\}$ and replacing λ_0 by λ_0^*. This can be translated into the integer program $\{\max \lambda^T x : x \in \mathcal{P}, x \in \{0,1\}\}$, referred to as MIP-RHS = MIP-RHS($\mathcal{P}$), where $\mathcal{P}$ is any linear formulation of the problem. Notice that the number of nonzero entries of λ determines the actual size of the subproblem MIP-RHS to be solved. In other words, the sparser the cut, the easier the strengthening problem. Again, MIP-RHS is solved by Cplex to which a time limit rhstlim is imposed. When it is reached, λ_0^* is set equal to the best upper bound returned by the solver at the time limit.

The overall cutting-plane algorithm is summarized in Algorithm 1. The parameter niter determines the maximum number of iterations (i.e., the maximum number of cuts generated); maxpercnz establishes the maximum percentage cut density allowed for CG cuts from the tableau; and minimprove stops the algorithm when tailing-off is reached: it establishes the minimum improvement of the objective value between consecutive iterations required to proceed.

In the separation stage of Algorithm 1 a first CG cut, namely, the one with the smallest *support* (number nnz(λ) of nonzero coefficients), is obtained from the tableau. Ties are broken by selecting the cut that forms the minimum angle with the objective function. Then, if this cut turns out to be too dense, the exact MIP-CG separation is invoked with the aim of detecting a sparser one, with the exception of the first iteration, when the cut from the tableau is always preferred. Notice that MIP-CG always generates rank-1 CG cuts of relaxation $Q(C)$, as the generated cutting planes are never added to it. This turned out to be useful to keep safe the sparsity of the cuts. The rationale for this policy is that, although MIP-CG may be time consuming, it is typically largely counterbalanced by the saving yielded by a sparser source cut when solving MIP-RHS.

Algorithm 1. Cutting plane algorithm

Input: Formulation $Q(C)$
Output: An updated formulation P, the upper bound CG-S;
Parameters: `niter,cutviolation,septlim,`
                `rhstlim,maxpercnz,minimprove`

$P \leftarrow Q(C)$
Optimize over P, get x^*
for ($i := 1$ to `niter` **and** x^* is fractional) **do**
    Evaluate **all violated** Gomory cuts from the current tableau
    Select the sparsest cut (λ, λ_0)
    if (nnz (λ) > `maxpercnz` $* |V|$) **and** $i > 1$ **then**
        Solve MIP-CG($Q(C)$, `cutviolation,septlim`)
        Select the **sparsest** violated cut (β, β_0)
        if nnz (λ) > nnz (β) **then**
            $\lambda := \beta, \lambda_0 := \beta_0$
        end if
    end if
    $\lambda_0^* \leftarrow$ Solve MIP-RHS(P, `rhstlim`)
    $P \leftarrow P \cup \{\lambda^T x \leq \lambda_0^*\}$
    $\bar{x} \leftarrow x^*$
    Optimize over P, get x^*
    if $(w^T \bar{x} - w^T x^*)$ < `minimprove` **then**
        return CG-S = $w^T x^*$
    end if
end for
return CG-S = $w^T x^*$

---

3 Computational Achievements

The upper bound CG-S computed by Algorithm 1 is now compared to those achieved by other methods in the literature. The test-bed consists of the DIMACS Second Challenge (Johnson and Trick [21]) benchmark collection, available at the web site [6], representing the standard benchmark for evaluating MSS and max-clique algorithms. We consider the complemented version of these graphs, as they were originally created for the max-clique problem. We include all the graphs with $n \leq 400$ except the "easy" ones, i.e., those for which that upper bound $UB_{\mathcal{C}}$ is close to the integer optimum $\alpha(G)$. The latter include the whole family of johnson graphs and most of the c-fat, hamming and san graphs. All instances are unweighted instances, as these tend to be the most difficult in practice. The computations are run on a machine with processor AMD Opteron 6376 (64 cores) clocked at 1.4 GHz with 64 GB RAM. The LP-MIP solver is IBM CPLEX 12.6.3 (using 32 threads): settings are default for MIP-CG, while mipemphasis is set to *moving best bound* for CG-RHS. The parameter settings for Algorithm 1 are as follows: niter $= 30$, cutviolation $= 0.2$, septlim $\in \{5, 50\}$, rhstlim $\in \{100, 150\}$, maxpercnz $\in \{0.9, 0.7\}$, minimprove $= 0.01$. Pairs $\{x, y\}$ of values indicate that the parameter assumes the value x for $|V| \leq 200$ and y otherwise. Before going through the evaluation of CG-S we analyze the strength of the first Chvátal closure of $\mathcal{Q}(\mathcal{C})$.

3.1   On the Strength of the First Chvátal Closure of QSTAB(G)

Let us denote by $\mathcal{Q}^1(\mathcal{C})$ the first Chvátal closure of $\mathcal{Q}(\mathcal{C})$ and by $UB_{\mathcal{Q}^1(\mathcal{C})} = \{\max \mathbb{1}^T x : x \in \mathcal{Q}^1(\mathcal{C})\}$. A close approximation to $UB_{\mathcal{Q}^1(\mathcal{C})}$ (unless pathological cases) is obtained by a cutting plane algorithm which uses MIP-CG as separation oracle. Table 1 compares $UB_{\mathcal{Q}(\mathcal{C})}$ and $UB_{\mathcal{Q}^1(\mathcal{C})}$ and shows the percentage gap closed by the first Chvátal closure.

In 16 out of 26 cases the gap closed is less than 2 %, in 21 cases less than 6 % and only in two cases greater than 10 %. Overall it turns out to that $\mathcal{Q}^1(\mathcal{C})$ is almost as tight as $\mathcal{Q}(\mathcal{C})$. This also gives a strong pointer about the strength of the Chvátal closure of QSTAB(G), which includes well known inequalities, such as odd-hole, odd-antihole and antiweb inequalities [23]. These results provide a benchmark to demonstrate the remarkable effect of cut strengthening, as documented below.

3.2   Evaluation of CG-S

Table 2 compares the upper bound CG-S returned by Algorithm 1 to the following upper bounds: $UB_{\mathcal{C}}$; $\vartheta(G)$; BCS, obtained by separation algorithms for rank inequalities and local cuts [29]; CMDK, obtained by separation algorithms for rank and non-rank inequalities [8]; MKK, computed by the Lovász and Schrijver $M(k, k)$ lifting operator applied to QSTAB(G) [14]; BLP derived from MKK through projection [13]; GR, obtained by strengthening the Lovász theta relaxation with odd circuit and triangle inequalities [20]; DR, obtained by tailored SDP algorithms to compute ϑ^+ [11]; BV, computed by optimizing over the Lovász and Schrijver lifting operator M_+ applied to FRAC(G) [5]; L, computed by strengthening the ϑ bound with non valid inequalities [24]; ELL achieved by outer approximation of ellipsoidal relaxations [16]. An asterisk in the DR or BV columns means that result was not reported.

The results clearly show the high quality of CG-S. In 10 out of 26 cases CG-S is the (unique) best upper bound while this holds 9 times for $\vartheta(G)$ and MKK.

Table 1. Upper bounds from the first Chvátal closure

Graph	$\alpha(G)$	$UB_{\mathcal{Q}(\mathcal{C})}$	$UB_{\mathcal{Q}^1(\mathcal{C})}$	$\frac{UB_{\mathcal{Q}(\mathcal{C})}-UB_{\mathcal{Q}^1(\mathcal{C})}}{UB_{\mathcal{Q}(\mathcal{C})}-\alpha(G)}\%$
brock200_1	21	38.02	37.83	1.12
brock200_2	12	21.21	21.12	0.98
brock200_3	15	27.3	27.22	0.65
brock200_4	17	30.66	30.54	0.88
brock400_1	27	63.96	63.92	0.11
brock400_2	29	64.39	64.34	0.14
brock400_3	31	64.18	64.13	0.15
brock400_4	33	64.21	64.16	0.16
C125.9	34	43.05	42.59	5.08
C250.9	44	71.39	70.99	1.46
c-fat200-5	58	66.67	65.76	10.5
DSJC125.1	34	43.16	42.64	5.68
DSJC125.5	10	15.39	15.25	2.6
mann_a9	16	18	17	50
mann_a27	126	135	134.15	9.44
hamming6-4	4	5.33	5.23	7.52
keller4	11	14.82	14.76	1.57
p_hat300-1	8	15.26	15.24	0.28
p_hat300-2	25	33.59	33.54	0.58
p_hat300-3	36	54.33	54.18	0.82
san200_0.7-2	18	20.36	20.28	3.39
san200_0.9-3	42	45.13	44	36.1
sanr200_0.7	18	33.34	33.17	1.11
sanr200_0.9	42	59.82	59.39	2.41
sanr400_0.5	13	41.29	41.26	0.11
sanr400_0.7	21	57.02	56.96	0.17

In 11 out of 24 cases it outperforms $\vartheta(G)$ (to the best of our knowledge $\vartheta(G)$ has never been computed for the two sanr400 graphs) and in all the remaining cases it is quite close to $\vartheta(G)$. Looking at the other bounds, CG-S is the best bound ever computed for all instances in the hard classes brock, p_hat and sanr. It is also evident that BCS and CMDK, obtained by inequalities with a combinatorial structure, tend to be weaker in general. However, CMDK performed pretty well in some specific instances.

Table 3 reports the times in seconds required to compute several among the tightest upper bounds. The computing times of MKK, BLP, ELL are reported as in the original papers. These refer to different computers: all of them have

Table 2. Comparison among upper bounds

Graph	α(G)	UB_C	ϑ(G)	BCS	CMDK	MKK	BLP	GR	DR	BV	L	ELL	CG-S
brock200_1	21	38.20	27.5	*	34.85	30.25	33.59	*	*	27.98	*	28.53	24.56
brock200_2	12	21.53	14.22	20.99	16.29	16.09	18.27	*	*	17.08	*	15.16	13.79
brock200_3	15	27.73	18.82	*	23.26	21.16	23.55	*	*	20.79	*	19.6	16.97
brock200_4	17	30.84	21.29	29.93	26.81	23.8	26.77	*	*	22.8	*	22.35	20.34
brock400_1	27	68.47	39.70	*	*	*	*	*	*	*	*	41.83	45.08
brock400_2	29	68.28	39.56	63.84	63.27	*	*	*	*	*	*	42.82	44.90
brock400_3	31	68.42	39.48	*	*	*	*	*	*	*	*	41.05	45.71
brock400_4	33	64.21	39.60	63.89	63.17	*	*	*	*	*	*	41.26	43.77
C.125.9	34	43.06	37.89	41.26	38.84	36.53	37.81	*	*	*	37.06	38.48	38.72
C.250.9	44	71.50	56.24	69.76	65.35	59.96	63.95	*	*	*	*	58.09	56.45
c-fat200-5	58	66.67	60.34	58.89	58	58	58	*	*	58.17	*	*	62.27
DSJC125.1	34	43.15	38.39	*	39.41	36.99	38.22	*	*	*	*	39.16	38.66
DSJC125.5	10	15.60	11.47	*	11.97	11.41	13.21	*	11.4	*	11.46	12.24	11.69
mann_a9	16	18.50	17.47	*	18	16.85	17.11	17.47	*	17.17	17.29	*	16.89
mann_a27	126	135.00	132.76	*	135	131.39	132.44	132.76	*	*	*	*	132.05
hamming6-4	4	5.33	5.33	*	4	4	4.64	*	4	4.54	*	*	4.47
keller4	11	14.82	14.01	14.83	14.2	13.17	14.29	*	*	15.41	13.15	14.55	13.60
p.hat300_1	8	15.68	10.1	*	12.43	11.4	13.45	*	*	18.66	*	11.13	11.39
p.hat300_2	25	34.01	27	33.81	33.5	30	30.73	*	*	30.1	*	28.5	28.60
p.hat300_3	36	54.74	41.16	54.12	51.82	47.32	49.79	*	*	43.32	*	43.02	41.08
san200_0.7-2	18	21.14	18	18.5	18.54	18	18	*	*	20.01	*	*	18.70
san200_0.9-3	44	45.13	44	44	44	44	*	44	*	44.4	*	*	44
sanr200_0.7	18	33.48	23.8	*	30.21	26.12	29.45	*	*	24.97	*	24.75	22.58
sanr200_0.9	42	60.04	49.3	*	54.18	50.73	54.52	49.27	*	49.31	*	50.6	47.74
sanr400_0.5	13	41.30	*	*	*	*	*	*	*	*	*	*	24.08
sanr400_0.7	21	57.02	*	*	*	*	*	*	*	*	*	*	38.61

Table 3. Computational times (sec.) and details

Graph	MKK	BLP	ELL	CG-S	MIP-CG time	MIP-RHS time	#cuts	#cuts tableau	Average frac	$|\mathcal{C}|$	$UB_\mathcal{C}$ time
brock200_1	17,670	373	9.88	87.95	2.61	82.71	16	14	116.75	2,127	1.10
brock200_2	26,501	190	5.8	93.03	14.92	72.74	12	1	101	3,902	0.65
brock200_3	22,386	338	14.37	115.8	0	111.82	16	16	97.81	2,867	3.84
brock200_4	25,362	196	20.12	92.88	7.11	83.03	15	9	109.8	2,567	2.21
brock400_1	*	*	11.43	661.99	110.93	510.07	18	1	269.28	5,676	15.93
brock400_2	*	*	5.88	838.44	110.29	681.67	21	1	268.24	5,713	17.26
brock400_3	*	*	5.44	923.55	172.79	683.1	26	1	278.69	5,673	16.89
brock400_4	*	*	5.83	1,091.41	143.9	896.7	16	1	285.5	7,325	12.74
C.125.9	227	391	0.6	9.48	4.28	5.1	15	8	111.33	489	0.06
C.250.9	9,397	8,908	5.02	323.94	11.71	309.63	20	9	201.85	1,724	0.40
c-fat200-5	265	45	*	16.17	0	6.85	16	16	191.12	7561	1.34
DSJC125.1	274	297	0.44	6.48	1.76	4.62	14	11	106.93	464	0.06
DSJC125.5	377	27	2.27	19.55	3.55	15.21	11	5	81.09	1,522	0.17
mann_a9	0.41	0.26	*	3.95	0.14	3.83	18	4	40.44	48	0.01
mann_a27	393	120	*	1.63	0	1.55	20	20	370.05	468	0.07
hamming6-4	4	5	*	48.5	43.89	4.53	16	1	30.06	149	0.02
keller4	15,324	9,586	0.64	30.07	7.51	22.12	13	1	86.85	868	0.54
p_hat300_1	4,910	767	4.23	66.26	5.61	59.54	12	1	78.25	1,124	1.78
p_hat300_2	24,337	2,207	3.35	95.92	13.76	80.69	11	5	127.09	2,016	1.03
p_hat300_3	46,408	2,419	25.94	255.36	18.68	228.91	12	7	192.17	4,074	9.41
san200_0.7-2	300	151	*	47.93	20.2	26.36	14	1	138.43	1,537	1.16
san200_0.9-3	143	*	*	0.24	0	0.23	1	1	152	1,143	0.14
sanr200_0.7	9,971	762	6.02	64.95	5.29	57.46	14	8	114.43	2,280	1.64
sanr200_0.9	8,483	949	1.42	56.68	5.15	50.87	15	8	156.6	1,150	0.25
sanr400_0.5	*	*	*	1,637.08	137.21	1,454.1	23	1	207.65	4,886	9.80
sanr400_0.7	*	*	*	1,551.21	285.11	1,152.83	25	1	267.84	8,540	25.36

CPU-s with higher clock frequency but a smaller number of cores. Although the comparison is not rigorously documented, these values can be considered reliable enough for a general judgment. In the column CG-S the total time required by Algorithm 1 is reported, while the successive two columns contain the overall time spent for solving MIP-CG and MIP-RHS respectively. The remaining columns report: the number of cuts generated; the number of cuts generated from the tableau; the average size of the fractional support and, finally, the number of clique inequalities in the initial formulation $Q(\mathcal{C})$ and the time required to construct it and compute $UB_\mathcal{C}$.

Table 3 shows that the strong bounds are achieved by a few cuts. Indeed, the very first cuts turn out to close a significant portion of the integrality gap. Notice also that the number of cuts generated by solving MIP-CG is large, which highlights that the adjustments of MIP-CG towards sparsification play a role. In 12 cases only one cut from the tableau is selected. It is indeed the first cut: at the first iteration even the cuts from MIP-CG are very dense and are not worth generating.

The average time for a single cut separation is quite reasonable and, as one can expect, most of this time is spent in solving MIP-RHS. These facts suggest that strengthened CG cuts can be cost-effective when embedded in a branch-and-cut framework.

Looking at computing times, the proposed method is outperformed only by ELL [16], while it is competitive with the other LP-based methods. Even if a direct comparison cannot be done, methods GR, DR, BV, L, based on sophisticated SDP approaches tend to be slower.

Another important fact is that the size of the average fractional support is often around $0.5|V| - 0.7|V|$. In our experience this nice effect is rarely observable with other cutting planes which typically keep $|J(x^*)| \simeq |V|$. This of course impacts on the efficiency of the method, which turns out to practical for graphs with 400 vertices.

It is worthwhile to remark that these experiments were carried out with general-purpose parameter settings, and that results on specific graphs may improve significantly with dedicated tuning.

The overall picture of this experience is that strengthened CG cuts can be quite effective even for a very structured combinatorial optimization problem such as the SSP. In fact, they seem to be competitive with inequalities that are derived from polyhedral studies. Notice also that the latter tend to be sparser than general CG cuts, a feature that usually guarantees a better numerical behaviour. Nevertheless, the above results, along with our previous experience with other general cutting planes, show that some denser cuts are required in order to achieve very strong bounds. This is a key issue for the development of IP algorithms for the SSP and deserve further investigation.

4   Conclusions

We showed that strengthening the right-hand-side of rank-1 CG cuts from a clique relaxation relative to the stable set polytope is extremely effective. In particular, the upper bounds obtained are competitive to those from sophisticated SDP approaches. This is so even though our implementation of the strengthening procedure is rather rudimentary and has significant room for improvement. In fact, one major research direction deals with speeding up the strengthening stage, either by using a combinatorial solver for the weighted SSP instances, or by using upper bounds on the weighted stability number that are faster to compute. Overall, our feeling is that the method can be improved so as to tackle larger graphs. A natural development will also be testing these cutting planes in a branch-and-cut framework. Finally, a theoretical study of strengthened rank-1 CG cuts would be interesting. It can be shown, for example, that the odd hole, odd antihole, web and antiweb inequalities, along with certain lifted versions of them, are cuts of this type.

References

1. Andersen, K., Pochet, Y.: Coefficient strengthening: a tool for reformulating mixed-integer programs. Math. Program. **122**, 121–154 (2010)
2. Borndörfer, R.: Aspects of Set Packing, Partitioning and Covering. Doctoral thesis, Technical University of Berlin (1998)
3. Balas, E., Ceria, S., Cornuéjols, G., Pataki, G.: Polyhedral methods for the maximum clique problem. In: Johnson, D.S., Trick, M.A. (eds.) Cliques, Coloring and Satisfiability. DIMACS Series in Discrete Mathematics and Theoretical Computer Science, vol. 26, pp. 11–28 (1996)
4. Bomze, I.M., Frommlet, F., Locatelli, M.: Copositivity cuts for improving SDP bounds on the clique number. Math. Program. **124**, 13–32 (2010)
5. Burer, S., Vandenbussche, D.: Solving lift-and-project relaxations of binary integer programs. SIAM J. Optim. **16**, 726–750 (2006)
6. DIMACS repository. ftp://dimacs.rutgers.edu/pub/challenge/graph/benchmarks/clique
7. Chvátal, V.: Edmonds polytopes and a hierarchy of combinatorial problems. Discr. Math. **4**, 305–337 (1973)
8. Corrêa, R.C., Delle Donne, D., Koch, I., Marenco, J.: General cut-generating procedures for the stable set polytope (2015). arXiv:1512.08757v1
9. Corrêa, R.C., Delle Donne, D., Koch, I., Marenco, J.: A strengthened general cut-generating procedure for the stable set polytope. Elec. Notes Discr. Math. **50**, 261–266 (2015)
10. Coniglio, S., Gualandi, S.: On the exact separation of rank inequalities for the maximum stable set problem. Optimization (2014). http://www.optimization-online.org/DB_HTML/2014/08/4514.html
11. Dukanovic, I., Rendl, F.: Semidefinite programming relaxations for graph coloring and maximal clique problems. Math. Program. **109**, 345–365 (2007)
12. Fischetti, M., Lodi, A.: Optimizing over the first Chvátal closure. Math. Program. **110**, 3–20 (2007)
13. Giandomenico, M., Rossi, F., Smriglio, S.: Strong lift-and-project cutting planes for the stable set problem. Math. Program. **141**, 165–192 (2013)
14. Giandomenico, M., Letchford, A., Rossi, F., Smriglio, S.: An application of the Lovász-Schrijver $M(K, K)$ operator to the stable set problem. Math. Program. **120**, 381–401 (2009)
15. Giandomenico, M., Letchford, A.N., Rossi, F., Smriglio, S.: Approximating the Lovász theta function with the subgradient method. Elec. Notes Discr. Math. **41**, 157–164 (2013)
16. Giandomenico, M., Letchford, A.N., Rossi, F., Smriglio, S.: Ellipsoidal relaxations of the stable set problem: theory and algorithms. SIAM J. Optim. **25**(3), 1944–1963 (2015)
17. Gomory, R.E.: Outline of an algorithm for integer solutions to linear programs. Bull. Amer. Math. Soc. **64**, 275–278 (1958)
18. Gomory, R.E.: An algorithm for integer solutions to linear programs. In: Graves, R.L., Wolfe, P. (eds.) Recent Advances in Mathematical Programming, pp. 269–302. McGraw-Hill, New York (1963)
19. Grötschel, M., Lovász, L., Schrijver, A.J.: Geometric Algorithms in Combinatorial Optimization. Wiley, New York (1988)
20. Gruber, G., Rendl, F.: Computational experience with stable set relaxations. SIAM J. Optim. **13**, 1014–1028 (2003)

21. Johnson, D.S., Trick, M.A. (eds.): Cliques, Coloring, Satisfiability: Observation of Strains: The 2nd DIMACS Implementation Challenge. American Mathematical Society, Providence (2011)

22. Håstad, J.: Clique is hard to approximate within $n^{1-\epsilon}$. Acta Math. **182**, 105–142 (1999)

23. Holm, E., Torres, L.M., Wagler, A.K.: On the Chvátal rank of linear relaxations of the stable set polytope. Int. Trans. Oper. Res. **17**, 827–849 (2010)

24. Locatelli, M.: Improving upper bounds for the clique number by non-valid inequalities. Math. Prog. **150**, 511–525 (2015)

25. Lovász, L.: On the Shannon capacity of a graph. IEEE Trans. Inform. Theor. **25**, 1–7 (1979)

26. Lovász, L., Schrijver, A.J.: Cones of matrices and set-functions and 0–1 optimization. SIAM J. Optim. **1**, 166–190 (1991)

27. Nemhauser, G.L., Sigismondi, G.: A strong cutting plane/branch-and-bound algorithm for node packing. J. Oper. Res. Soc. **43**, 443–457 (1992)

28. Padberg, M.W.: On the facial structure of set packing polyhedra. Math. Program. **5**, 199–215 (1973)

29. Rebennack, S., Oswald, M., Theis, D.O., Seitz, H., Reinelt, G., Pardalos, P.M.: A branch and cut solver for the maximum stable set problem. J. Comb. Opt. **21**, 434–457 (2011)

30. Rossi, F., Smriglio, S.: A branch-and-cut algorithm for the maximum cardinality stable set problem. Oper. Res. Lett. **28**, 63–74 (2001)

31. Juhász, F.: The asymptotic behaviour of lovász' θ function for random graphs. Combinatorica. **2**(2), 153–155 (1982)

Scheduling Personnel Retraining: Column Generation Heuristics

Oliver G. Czibula[✉], Hanyu Gu, and Yakov Zinder

School of Mathematical and Physical Sciences,
University of Technology Sydney, 15 Broadway, Ultimo, NSW 2007, Australia
oliver.g.czibula@students.uts.edu.au

Abstract. Many organisations need periodic retraining of staff. Due to certain requirements on the composition of study groups, the planning of training sessions is an NP-hard problem. The paper presents linear and nonlinear mathematical programming formulations of this problem together with three column generation based heuristic optimisation procedures. The procedures are compared by means of computational experiments that use data originating from a large Australian electricity distributor with several thousand employees.

1 Introduction

This research is motivated by the problem of planning periodic training at Australia's largest electricity distributor, Ausgrid. Australian law mandates that all workers in such organisations must undertake regular safety and technical training. Ausgrid provides training to thousands of its employees and contractors, and also to third parties.

Keeping in mind the primal goal of the training provider, it is reasonable to refer to the people undergoing training as workers, although such terms as students and trainees, justified by the broad variety of participants, are often used in practice, and in the context of this paper have the same meaning. Many of these workers have different learning styles, different levels of education, and different requirements to the learning outcome. Therefore, it is desired to form separate classes for each category of trainees. However, due to the cost of training and the scarcity of resources, it is often not possible to run segregated classes.

The considered situation can be modelled by introducing for each pair of trainees a cost (penalty) for assigning this pair to the same training session (class). Such a penalty may not only be associated with different categories of trainees but can also reflect a variety of other factors, e.g. work requirements to the staff availability which restrict who can undertake training simultaneously.

Most of the training at Ausgrid has a limited period in which it is valid. Workers are only permitted to work in roles for which they have up-to-date training. Therefore the company incurs a certain cost each time an employee is not permitted to work due to the expiration of required training. Furthermore, training sessions (classes) have different suitability for a trainee not only because

© Springer International Publishing Switzerland 2016
R. Cerulli et al. (Eds.): ISCO 2016, LNCS 9849, pp. 213–224, 2016.
DOI: 10.1007/978-3-319-45587-7_19

they are held at different dates but also because they differ by location, teaching mode, etc. Instead of specifying for each class all these attributes, the models below involve a cost (penalty) for allocating a trainee to a particular class.

Several factors such as the cost of training, nature of the presented material, teaching methods, premises, etc. impose lower and upper bounds on the number of trainees for each training session. Another considered restriction is the lower and upper bounds on the number of different types of trainees in the same class. All these bounds vary from training session to training session. The sessions that do not satisfy these restrictions are cancelled.

The considered problem is a problem of minimising the objective function that is a weighted sum of the total cost of assigning pairs of trainees to the same class and the total cost of assigning each individual trainee to the respective class. This objective function is to be minimised subject to the above mentioned restrictions on the composition of the training sessions. The considered problem is NP-hard in the strong sense.

The remainder of the paper is organised as follows. Section 2 briefly reviews the related literature. Section 3 presents quadratic programming and integer linear programming formulations. Section 4 describes three optimisation procedures that are based on column generation. Section 5 gives the methodology and results of testing these optimisation procedures. The concluding remarks can be found in Sect. 6.

2   Related Literature

Without the restrictions on the number of student types in training sessions, the considered class formation problem can be viewed as a graph partitioning problem (GPP) where the nodes of a given undirected graph are to be partitioned into several given clusters. Some clusters can occur empty as a result of a partition.

If a cluster is not empty, then the number of assigned nodes must be between the upper and lower limits specified for this cluster. The nodes of the graph represent trainees whereas the clusters represent available training sessions. In this formulation, the cost associated with an edge is the penalty for assigning the pair of trainees, corresponding to the end nodes of the edge, to the same class.

Although graph partitioning problems were considered in the literature, to our knowledge, the above mentioned graph partitioning problem is new. For instance, the two most relevant publications, [3,8], do not consider a penalty for assigning a node to a cluster. As far as the solution methods are concerned, the first of these two papers uses valid inequalities, whereas the second adopted a column generation approach.

The Edge-Partition Problem (EPP), which is related to the GPP, is the problem of covering the edges of a graph with subgraphs that contain at most k edges [6]. The NP-hardness of the EPP is shown in [6], and also a linear time approximation algorithm with performance guarantee is proposed. [10] discuss a combination of integer and constraint programming method for the stochastic EPP, and present a two-stage cutting plane algorithm.

The considered problem can also be viewed as an extension of the Quadratic Multiple Knapsack Problem (QMKP) which is a combination of the multiple knapsack and quadratic knapsack problems. The QMKP received little attention in the literature until recently. The suggested methods include metaheuristics [2, 5,7,9], Lagrangian relaxation [1], and greedy algorithms [9].

3   Problem Formulation

Let $\mathcal{N} = \{1, \cdots, N\}$, $\mathcal{M} = \{1, \cdots, M\}$, and $\mathcal{K} = \{1, \cdots, K\}$ be the set of available classes, the set of students to be assigned, and the set of student types respectively. Denote the cost of assigning student $j \in \mathcal{M}$ to class $i \in \mathcal{N}$ by $c_{i,j}$, and the cost of pairing student types $k \in \mathcal{K}$ and $l \in \mathcal{K}$ together in the same class by $b_{k,l}$. Each student has exactly one type, and the set of students who are of type k is represented by T_k, $k \in \mathcal{K}$. Each student must be assigned to exactly one class, but not all classes must be run. Each class $i \in \mathcal{N}$ that is run must contain at least a_i and at most b_i students, and at least p_i and at most q_i student types. Students or student types cannot be assigned to classes that are not run. The binary variable $X_{i,j}$ is defined to be 1 if student j is assigned to class i, or 0 otherwise; the binary variable $Y_{i,k}$ is defined to be 1 if student type k is assigned to class i, or 0 otherwise; The binary variable Z_i is defined to be 1 if class i is run, or 0 otherwise. The following Quadratic Program describes the problem:

$$\text{(QP) Minimise:} \quad \alpha \sum_{i=1}^{N} \sum_{k=1}^{K} \sum_{l=1}^{K} b_{k,l} Y_{i,k} Y_{i,l} + \beta \sum_{i=1}^{N} \sum_{j=1}^{M} c_{i,j} X_{i,j} \tag{1}$$

$$\text{Subject To:} \quad \sum_{i=1}^{N} X_{i,j} = 1 \quad j = 1, \dots, M \tag{2}$$

$$a_i Z_i \le \sum_{j=1}^{M} X_{i,j} \le b_i Z_i \quad i = 1, \dots, N \tag{3}$$

$$p_i Z_i \le \sum_{k=1}^{K} Y_{i,k} \le q_i Z_i \quad i = 1, \dots, N \tag{4}$$

$$X_{i,j} \le Y_{i,k} \quad i = 1, \dots, N; k = 1, \dots, K; j \in T_k \tag{5}$$

$$Y_{i,k} \le \sum_{j \in T_k} X_{i,j} \quad i = 1, \dots, N; k = 1, \dots, K \tag{6}$$

$$X_{i,j} \in \{0,1\} \quad i = 1, \dots, N; j = 1, \dots, M \tag{7}$$

$$Y_{i,k} \in \{0,1\} \quad i = 1, \dots, N; k = 1, \dots, K \tag{8}$$

$$Z_i \in \{0,1\} \quad i = 1, \dots, N \tag{9}$$

The quadratic term in (1) represents the cost of pairing student types together, and the linear term represents the cost of assigning students to classes, weighted by coefficients α and β, respectively.

The constraints (2) express the requirement that each student must be assigned to exactly one class. The constraints (3) and (4) express the requirement that each running class must have between a_i and b_i students, and between p_i and q_i student types, respectively, if the class is run, or zero otherwise. The constraints (5) express the requirement that a student may only be assigned to a class if that student's type has also been assigned to that class. The constraints (6) ensure that a class can only be assigned a type if at least one student of that type is in the class.

It is possible to linearise the quadratic term in (1) by introducing $\hat{Y}_{i,k,l} = Y_{i,k}Y_{i,l}$ together with constraints:

$$\hat{Y}_{i,k,l} \leq Y_{i,k} \quad i = 1,\ldots,N; k = 1,\ldots,K; l = 1,\ldots,K \quad (10)$$

$$\hat{Y}_{i,k,l} \leq Y_{i,l} \quad i = 1,\ldots,N; k = 1,\ldots,K; l = 1,\ldots,K \quad (11)$$

$$\hat{Y}_{i,k,l} \geq Y_{i,k} + Y_{i,l} - 1 \quad i = 1,\ldots,N; k = 1,\ldots,K; l = 1,\ldots,K \quad (12)$$

to give the linearised model:

$$\text{(LQP) Minimise: } \alpha \sum_{i=1}^{N} \sum_{k=1}^{K} \sum_{l=1}^{K} b_{k,l}\hat{Y}_{i,k,l} + \beta \sum_{i=1}^{N} \sum_{j=1}^{M} c_{i,j}X_{i,j} \quad (13)$$

$$\text{Subject To: } (2) - (12)$$

$$\hat{Y}_{i,k,l} \in \{0,1\} \quad i = 1,\ldots,N; k = 1,\ldots,K; l = 1,\ldots,K. \quad (14)$$

An augmented model (AQP) can be constructed by relaxing constraints (2) by introducing variables $S_j \in \mathbb{Z}^+$ and $T_j \in \mathbb{Z}^+$, where $\mathbb{Z}^+$ is the set of nonnegative integers, which together represent the deviation in the number of classes to which student j is assigned. The sum of these variables is then heavily penalised in the objective function:

$$\text{(AQP) Minimise: } \alpha \sum_{i=1}^{N} \sum_{k=1}^{K} \sum_{l=1}^{K} b_{k,l}Y_{i,k}Y_{i,l} + \beta \sum_{i=1}^{N} \sum_{j=1}^{M} c_{i,j}X_{i,j} + \gamma(\sum_{j=1}^{M} S_j + T_j)$$

$$(15)$$

$$\text{Subject To: } \sum_{i=1}^{N} X_{i,j} + S_j - T_j = 1 \quad j = 1,\ldots,M \quad (16)$$

$$(3) - (9)$$

$$S_j \in \mathbb{Z}^+ \quad j = 1,\ldots,M \quad (17)$$

$$T_j \in \mathbb{Z}^+ \quad j = 1,\ldots,M \quad (18)$$

where γ is very large, typically many orders of magnitude greater than α and β. The (LQP) model can be modified in the same way to form the (ALQP) augmented linearised model.

The advantage of the augmented model is that feasible solutions are significantly easier to find as violations of (2) allow (3) and (4) to be satisfied more easily. In the event that a solution cannot be found in which $\max_j S_j = \max_j T_j = 0$, the organisation can then decide what steps to take next, for example to create additional classes, to modify class size constraints, etc.

The time required for a general purpose MIP/QP solver to find an optimal solution to each of the models discussed so far grows rapidly. Even very small test cases with just a few dozen trainees can take many hours to solve with a powerful computer and a commercial optimisation solver. As real world problem instances are significantly larger than this, we propose to use a heuristic approach.

4 Column Generation Approaches

Define $\mathcal{P}_i$ to be the set of all feasible sets of trainees that can be assigned to class i. We define $p \in \mathcal{P}_i$ to be a *pattern*.

We can then define the set-covering formulation:

$$\text{(M) Minimise:} \quad \sum_{i=0}^{N} \sum_{p \in \mathcal{P}_i} c_p^i X_p^i \tag{19}$$

$$\text{Subject To:} \quad \sum_{p \subset \mathcal{P}_i} X_p^i \leq 1 \quad i = 1, \dots, N \tag{20}$$

$$\sum_{i=1}^{N} \sum_{p \in \mathcal{P}_i} s_{p,j}^i X_p^i = 1 \quad j = 1, \dots, M \tag{21}$$

$$X_p^i \in \{0,1\} \quad i = 1, \dots, N; p \in \mathcal{P}_i \tag{22}$$

where the binary variable X_p^i is 1 if pattern p is selected for class i or zero otherwise; c_p^i is the cost of selecting pattern p for class i, and $s_{p,j}^i$ is 1 if student j exists in pattern p or 0 otherwise.

The constraints (20) express the requirement that each class can have at most one pattern selected, and the constraints (21) express the requirement that each trainee must be assigned to exactly one class.

For problems of practical size, there are far too many patterns in the master problem (M) to consider. For all such problems, it is useful to consider the reduced master problem (RM) that is identical to (M), but contains only a subset of patterns $P_i \subseteq \mathcal{P}_i$.

The (RM) objective function and constraints equivalent to (21) can easily be modified to incorporate the S_j and T_j variables, as with (AQP), to give the augmented reduced master problem (ARM).

According to the column generation approach [4], new columns (patterns) are iteratively added to P_i for each class i. First, the linear relaxation of the reduced master problem is solved for an initial set of patterns. If we solve the augmented variant, the initial set of patterns can be empty. At each iteration, new patterns can be generated by solving with an IP solver, for each class i, the subproblem:

$$(\text{SP}_i)\,\text{Minimise:}\quad \alpha \sum_{k=1}^{K} \sum_{l=1}^{K} b_{k,l} Y_{i,k} Y_{i,l} + \beta \sum_{j=1}^{M} c_{i,j} X_{i,j} - \pi_{1,i} - \sum_{j=1}^{M} \pi_{2,j} X_{i,j} \quad (23)$$

$$\text{Subject To:}\quad a_i \le \sum_{j=1}^{M} X_{i,j} \le b_i \quad\quad (24)$$

$$p_i \le \sum_{k=1}^{K} Y_{i,k} \le q_i \quad\quad (25)$$

$$X_j \le Y_{i,k} \quad k = 1,\ldots,K; j \in T_k \quad\quad (26)$$

$$Y_{i,k} \le \sum_{j \in T_k} X_k \quad k = 1,\ldots,K \quad\quad (27)$$

$$X_{i,j} \in \{0,1\} \quad j = 1,\ldots,M \quad\quad (28)$$

$$Y_{i,k} \in \{0,1\} \quad k = 1,\ldots,K \quad\quad (29)$$

where $\pi_{1,i}$ is the dual variable corresponding to constraint (20) for class i, and $\pi_{2,j}$ is the dual variable corresponding to constraint (21) for student j from the linear relaxation of (ARM). The objective of (SP$_i$) is therefore to find the pattern with the most negative reduced cost to add to P_i.

At each iteration, the linear relaxation of (RM) or (ARM) is solved subject to the set of available patterns P_i, and (SP$_i$) is solved for each class i to find new patterns with the most negative reduced cost. If (SP$_i$) cannot produce a pattern with negative reduced cost, then no new patterns are available for class i in this iteration. When no new patterns are available for any class, the column generation procedure has reached its conclusion and can be terminated.

The above-mentioned column generation procedure will most likely not yield a feasible integer solution to the original problem as it is solved for the linear relaxation of (RM) or (ARM). We now present three solution approaches, incorporating the column generation procedure, to produce integer solutions.

4.1   Reduced Master Heuristic

The most simple solution approach is to solve (RM) or (ARM) with their original, unrelaxed integer variables, subject to the patterns generated according to the column generation approach. This approach is not guaranteed to yield an optimal solution since the set of generated patterns may not contain the necessary patterns required for an optimal solution. Moreover, the (RM) may not have a feasible integer solution if the generated patterns cannot satisfy (21), however the (ARM) problem will always yield a feasible integer solution, regardless of the available patterns.

Solving (RM) or (ARM) with many patterns can still be computationally challenging. In cases where the number of patterns is very large, solving (RM) or (ARM) may not be possible in acceptable time, and alternative approaches may need to be considered.

4.2   Fix Columns

Another solution approach based on column generation is to make assumptions about which patterns will run in the final solution using information from the linear relaxation of the master problem. Those patterns whose corresponding variables in the linear relaxation of the master problem have the greatest value are assumed to run, i.e. the corresponding X_p^i variable is set to 1 in (RM) or (ARM). We refer to these as accepted patterns. If the corresponding variable's value in the solution to the linear relaxation was already 1, then setting this variable equal to 1 will not affect the objective value since no change has been made. If, however, the value of the variable in the solution was less than 1, then setting it equal to 1 may cause the objective value to deteriorate. The process of finding these accepted patterns is applied iteratively, with more and more patterns assumed in the final solution. Whenever the objective value deteriorates by more than a factor of τ since the previous iteration, typically around 1 %, the column generation procedure will be run again to generate new patterns subject to the set of accepted patterns. Since there are only N classes and at most one pattern can be run per class, at most N patterns can be fixed.

Pattern-Fixing Heuristic

Step 1. Generate the initial set of patterns P by the column generation procedure described above. Initialise the empty set of accepted patterns $\hat{P}$. Initialise the iteration counter c to 1.

Step 2. Solve the linear relaxation to (ARM) subject to the patterns P, with the additional assumption that the X_p^i variables corresponding to each pattern in $\hat{P}$ must have value 1. The optimal LP solution is denoted by σ_c^*.

Step 3. Find a pattern $\tilde{p}$ from $P \setminus \hat{P}$ whose corresponding X_p^i variable in σ_c^* has greatest value. Add $\tilde{p}$ to $\hat{P}$, and remove all patterns from P that are mutually exclusive with $\tilde{p}$.

Step 4. If the objective value of σ_c^* is worse than σ_{c-1}^* by more than a factor of τ then generate additional patterns by running the column generation procedure described above, with the assumption in (ARM) that the X_p^i variables corresponding to each pattern in $\hat{P}$ must have value 1. Add these new patterns to P.

Step 5. If some students remain unallocated in $\hat{P}$, increment c and return to Step 2.

The only parameter in the pattern fixing heuristic is τ, which determines how much objective value deterioration is required to trigger the subroutine to generate additional patterns. In practice, however, it is useful to impose a time limit and/or iteration limit on both the subroutine to generate more patterns and also the algorithm as a whole. When the time or iteration limit is reached in the pattern generating subroutine, the subroutine terminates with the patterns generated so far. When the time or iteration limit is reached on the whole algorithm between Step 2 and Step 4, the current (ARM) is solved as an IP to produced a solution.

Aside from the time and iteration limit, the pattern generating subroutine normally terminates when no more patterns can be generated, i.e. when there are no more patterns with reduced cost less than zero. In practice, this subroutine can sometimes continue generating a very large number of new patterns for a long time, with reduced costs very close to zero. While having more patterns allows more possible solutions in (ARM), having too many patterns results in a problem with many variables, which can be computationally difficult to solve. When the number of patterns grows very large, typically above ten thousand, even the linear relaxation can take a minute or more to solve. For this reason, the pattern generating subroutine should also terminate when no patterns can be generated with reduced cost with absolute value less than some small ϵ.

4.3   Student Clustering

Another solution approach based on the column generation procedure is to make assumptions about which pairs of students will be assigned to classes together using information from the linear relaxation of the master problem. Let $p_{j_1,j_2} = \sum_{p \in \tilde{P}_{(j_1,j_2)}} X_p^{i*}$, for set of patterns $\tilde{P}_{(j_1,j_2)} \subseteq P$ containing student pair (j_1,j_2) and solution values X_p^{i*}, which can be interpreted as an indication that students j_1 and j_2 should appear in the same pattern in the final solution. The pairs with highest p_{j_1,j_2} values will be enforced in the solution process. We refer to these as accepted pairs of students. The process of finding accepted pairs of students is applied iteratively, with more and more pairs of students assumed in the final solution. As with the pattern-fixing heuristic from Sect. 4.2, when the objective value deteriorates by more than a factor of τ since the previous iteration, new patterns are generated subject to the pairs of students that have been fixed thus far.

Student-Clustering Heuristic

Step 1. Generate the initial set of patterns P by the column generation procedure described above. Initialise the empty set of accepted student pairs $\hat{S}$. Initialise the iteration counter c to 1.

Step 2. Solve the linear relaxation to (ARM) subject to the patterns P. The optimal LP solution is denoted by σ_c^*.

Step 3. Find a pair of students (j_1, j_2) not already in $\hat{S}$ where p_{j_1,j_2} has greatest value. Add this pair of students to $\hat{S}$ and delete all patterns from P that contain j_1 but not j_2, or contain j_2 but not j_1.

Step 4. If the objective value of σ_c^* is worse than σ_{c-1}^* by more than a factor of τ then generate additional patterns by the column generation procedure described above, with the assumption in each (SP$_i$) that $X_{j_1} = X_{j_2}$ for all pairs $(j_1, j_2) \in \hat{S}$, and increment c.

Step 5. If some students remain unpaired in $\hat{S}$, increment c and return to Step 2. Otherwise, solve (AQP) subject to the pairs specified in $\hat{S}$.

As with the pattern-fixing heuristic described in Sect. 4.2, the only parameter is τ, but it is useful to impose time and iteration limits to the main algorithm and the pattern generating subroutine. It is again also useful to terminate the pattern generating subroutine when no new patterns can be generated whose reduced cost has absolute value greater than some small ϵ.

5 Computational Experiments

In order to test the proposed solution approaches, a set of 27 test cases were randomly generated with similar characteristics to real-world cases from Ausgrid.

The three column generation-based solution approaches described in Sects. 4.1, 4.2, and 4.3 were computationally tested and compared, and the results are presented in this section. All test cases and corresponding solution files are available for download at https://goo.gl/S4b305.

We used an Intel i7-4790K quad core CPU with 16 GB RAM, running Microsoft Windows 10 64-bit. Code was written in C# 4.0, and we used IBM ILOG CPLEX 12.5.0.0 64-bit using the ILOG Concert API to solve the mathematical programming models. In all cases, except where otherwise specified, we solved the augmented model variants, i.e. those models with the S and T variables so that feasible solutions could be found more easily. For the objective functions, we used $\alpha = \beta = 1$, and $\gamma = 10^6$. For the heuristics, we used $\tau = 1\%$ and $\epsilon = 10^{-3}$. The time limit we chose for each heuristic was given by $(|\mathcal{K}| + |\mathcal{M}| + |\mathcal{K}|) \times 18\,\text{s}$, meaning the smallest test case would be allowed $(10 + 40 + 100) \times 18 = 2700\,\text{s}$, and the largest test case would be allowed $(40 + 120 + 400) \times 18 = 10080\,\text{s}$.

Table 1. The parameters of the 27 test cases.

Ca	Ty	Cl	St	Va	Cs	Ca	Ty	Cl	St	Va	Cs	Ca	Ty	Cl	St	Va	Cs
01	10	40	100	6440	10060	10	20	40	100	12640	27860	19	40	40	100	37040	99460
02	10	40	200	10640	14160	11	20	40	200	16840	31960	20	40	40	200	41240	103560
03	10	40	400	19040	22360	12	20	40	400	25240	40160	21	40	40	400	49640	111760
04	10	80	100	12680	20020	13	20	80	100	25080	55620	22	40	80	100	73880	198820
05	10	80	200	20880	28120	14	20	80	200	33280	63720	23	40	80	200	82080	206920
06	10	80	400	37280	44320	15	20	80	400	49680	79920	24	40	80	400	98480	223120
07	10	160	100	25160	39940	16	20	160	100	49960	111140	25	40	160	100	147560	397540
08	10	160	200	41360	56040	17	20	160	200	66160	127240	26	40	160	200	163760	413640
09	10	160	400	73760	88240	18	20	160	400	98560	159440	27	40	160	400	196160	445840

Table 1 outlines the 27 test cases used for the computational experimentation. For each test case (Ca), we report the total number of student types (Ty), the total number of classes (Cl), the total number of students (St), the total number of variables (Va) and constraints (Cs) for the corresponding (ALQP) model.

Table 2 presents the remainder of the results of the computational experimentation. For each test case (Ca), we report the objective value (denoted by Z) of the best solution without the contribution of the augmenting variables S and T, the total number of violations $\sum_{j=1}^{M} S_j + T_j$ (denoted by V), and the total time, in seconds, (denoted by T) required to obtain the solution. The solution approaches compared are solving the (ARM) model with integer variables as described in Sect. 4.1 (Z_{MP}), (V_{MP}), and (T_{MP}); by the Pattern-fixing heuristic described in Sect. 4.2 (Z_{PF}), (V_{PF}), and (T_{PF}); the Student-clustering heuristic described in Sect. 4.3 (Z_{SC}), (V_{SC}), and (T_{SC}); and straightforward solution of the (AQP) model (Z_{AQP}) and (V_{AQP}). When solving the (AQP) model, default

Table 2. A comparison of the solution approaches for each of the test cases.

Ca	Z_{MP}	V_{MP}	T_{MP}	Z_{PF}	V_{PF}	T_{PF}	Z_{SC}	V_{SC}	T_{SC}	Z_{AQP}	V_{AQP}
01	30.21	3	2700	49.93	0	164	34.83	0	1632	34.83	0
02	139.6	23	4500	218.89	4	1009	152.16	0	4502	180.40	0
03	560.62	94	8101	1232.87	45	5407	552.6	0	8101	572.43	0
04	12.33	0	0	15.48	0	55	12.33	0	127	12.33	0
05	49.58	7	5220	55.3	0	930	46.17	0	5227	73.56	0
06	323.36	86	8821	191.58	0	8826	184.67	0	8833	218.53	0
07	2.94	0	0	3.53	0	16	2.94	0	12	2.98	0
08	12.38	0	2	14.48	0	199	12.38	0	1576	20.87	0
09	71.5	30	10261	55.46	0	10030	55.29	0	10266	105.39	0
10	44.89	1	2880	46.86	0	338	41.68	0	2890	62.16	0
11	183.31	25	4680	310	4	2250	199.74	0	4690	268.01	0
12	724.55	107	8281	1841.08	22	5673	704.81	0	8283	704.81	0
13	14.71	0	0	16.22	0	148	14.71	0	2179	17.15	0
14	57.12	7	5400	66.16	0	4660	69.59	0	5410	140.68	0
15	357.13	83	9001	273.4	0	9003	339.8	0	9000	481.30	0
16	3.81	0	0	4.09	0	25	3.81	0	38	3.81	0
17	12.91	0	3	17.96	0	1085	15.72	0	6865	20.69	0
18	87.99	28	10441	101.34	0	6305	104.93	0	10441	170.37	235
19	56.89	1	3240	68.77	0	925	61.4	0	3244	96.62	4
20	248.01	28	5040	258.88	0	3691	302.85	0	5040	468.74	84
21	921.31	112	8641	1910.95	58	5571	1191.44	0	8640	533.58	284
22	13.28	0	0	22.57	0	93	19.17	0	3967	79.49	1
23	94.76	7	5760	84.31	0	2580	126.68	0	5761	294.86	31
24	599.69	103	9361	501.42	0	9361	454.14	0	9366	705.62	241
25	3.35	0	0	4.68	0	44	3.35	0	182	4.33	0
26	14.81	0	3	21.54	0	1016	21.32	0	7202	93.13	79
27	143.13	29	10801	144.29	0	10804	161.85	0	10802	0.00	400
Avg	177.19	28.67	4560.63	278.96	4.93	3341.04	181.12	0	5343.56	201.65	50.33

CPLEX settings were used, except for a time limit equal to the largest of the three heuristic times for that test case (i.e. $\max\{T_{MP}, T_{PF}, T_{SC}\}$). The bottom row shows the average values for the column above.

It is clear from the results in Table 2 that the approach of solving the integer reduced master problem, subject to generated patterns, does not produce consistently good solutions. While this method did outperform the other heuristics in a few test cases, there were many instances where the number of constraint violations was much higher than for the other methods. This is not to say that the set of generated patterns did not allow for an integer solution with no violations, but that this approach could not find such a solution within the allowed time.

For most of the tested cases, the pattern-fixing approach outperformed the integer reduced master problem approach in terms of time and solution quality. In most cases the pattern-fixing approach produced solutions with no constraint violations, and the approach often terminated well before the time limit was reached.

The student-clustering approach performed significantly better than the other two approaches. While it required more time than the pattern-fixing approach, the student-clustering approach was able to consistently produce solutions with no constraint violations. In the 22 cases where both the pattern-fixing and student-clustering approaches produced solutions without constraint violations, the latter approach produced superior solutions for all but six test cases.

The approach of solving the (AQP) model directly, subject to the time limit, produced results that were generally poorer than those provided by the heuristics, especially for the larger test cases. For the smaller test cases, good solutions were produced with no constraint violations. For 9 out of the 10 largest test cases, solving the (AQP) model, subject to the time limit, produced solutions with many constraint violations.

6   Conclusion

The paper presents three column generation based heuristics for the optimisation problem of assigning employees to classes for the purpose of technical and safety training and retraining under restrictions on the composition of these classes. The considered problem is common to many large organisations.

The proposed heuristics were tested by computational experiments on a number of randomly generated test cases, based on data supplied by Ausgrid. The basic approach of solving the integer reduced master problem subject to generated columns did not perform well, most often producing solutions with many constraint violations. The approach of preferentially fixing patterns performed much better, but could not always produce solutions without constraint violations. The approach of preferentially clustering students performed the best, on average, producing good quality solutions free of constraint violations in acceptable time. The straight forward approach of solving the augmented quadratic programming model did not perform well on the larger test cases.

References

1. Caprara, A., Pisinger, D., Toth, P.: Exact solution of the quadratic knapsack problem. INFORMS J. Comput. **11**(2), 125–137 (1999)
2. Chen, Y., Hao, J.K.: Iterated responsive threshold search for the quadratic multiple knapsack problem. Ann. Oper. Res. **226**, 101–131 (2015)
3. Chopra, S., Rao, M.R.: The partition problem. Math. Program. **59**(1–3), 87–115 (1993)
4. Desrosiers, J., Lübbecke, M.E.: A Primer in Column Generation. Springer, New York (2005)
5. García-Martínez, C., Rodriguez, F., Lozano, M.: Tabu-enhanced iterated greedy algorithm: a case study in the quadratic multiple knapsack problem. Eur. J. Oper. Res. **232**, 454–463 (2014)
6. Goldschmidt, O., Hochbaum, D.S., Levin, A., Olinick, E.V.: The sonet edge-partition problem. Networks **41**(1), 13–23 (2003)
7. Hiley, A., Julstrom, B.A.: The quadratic multiple knapsack problem and three heuristic approaches to it. In: Proceedings of the 8th Annual Conference on Genetic and Evolutionary Computation, pp. 547–552. ACM (2006)
8. Johnson, E.L., Mehrotra, A., Nemhauser, G.L.: Min-cut clustering. Math. Programm. **62**(1–3), 133–151 (1993)
9. Julstrom, B.A.: Greedy, genetic, and greedy genetic algorithms for the quadratic knapsack problem. In: Proceedings of the 7th Annual Conference on Genetic and Evolutionary Computation, pp. 607–614. ACM (2005)
10. Taşkın, Z.C., Smith, J.C., Ahmed, S., Schaefer, A.J.: Cutting plane algorithms for solving a stochastic edge-partition problem. Discrete Optim. **6**(4), 420–435 (2009)

Diagonally Dominant Programming
in Distance Geometry

Gustavo Dias and Leo Liberti[✉]

CNRS LIX, École Polytechnique, 91128 Palaiseau, France
{dias,liberti}@lix.polytechnique.fr

Abstract. Distance geometry is a branch of geometry which puts the concept of distance at its core. The fundamental problem of distance geometry asks to find a realization of a finite, but partially specified, metric space in a Euclidean space of given dimension. An associated problem asks the same question in a Euclidean space of *any* dimension. Both problems have many applications to science and engineering, and many methods have been proposed to solve them. Unless some structure is known about the structure of the instance, it is notoriously difficult to solve these problems computationally, and most methods will either not scale up to useful sizes, or will be unlikely to identify good solutions. We propose a new heuristic algorithm based on a semidefinite programming formulation, a diagonally-dominant inner approximation of Ahmadi and Hall's, a randomized-type rank reduction method of Barvinok's, and a call to a local nonlinear programming solver.

1 Introduction

The main problem studied in this paper is the

DISTANCE GEOMETRY PROBLEM (DGP). Given an integer $K \geq 1$ and a simple, edge-weighted, undirected graph $G = (V, E, d)$, where $d : E \rightarrow \mathbb{R}_+$, verify the existence of a *realization function* $x : V \rightarrow \mathbb{R}^K$, i.e. a function such that:

$$\forall \{i, j\} \in E \quad \|x_i - x_j\| = d_{ij}. \tag{1}$$

A recent survey on the DGP with the Euclidean norm is given in [15]. The DGP is **NP**-hard, by reduction from PARTITION using 2-norms [21]. Three well-known applications are to clock synchronization ($K = 1$), sensor network localization ($K = 2$), and protein conformation ($K = 3$). If distances are Euclidean, the problem is called Euclidean DGP (EDGP) — but, given the preponderance of Euclidean distances in the DGP literature w.r.t. other distances, if the norm is not specified, it is safe to assume the 2-norm is used.

G. Dias—Financially supported by a CNPq PhD thesis award.

L. Liberti—Partly supported by the ANR "Bip:Bip" project under contract ANR-10-BINF-0003.

© Springer International Publishing Switzerland 2016

R. Cerulli et al. (Eds.): ISCO 2016, LNCS 9849, pp. 225–236, 2016.
DOI: 10.1007/978-3-319-45587-7_20

A related problem, the DISTANCE MATRIX COMPLETION PROBLEM (DMCP), asks whether a partially defined matrix can be completed to a distance matrix. The difference is that while K is part of the input in the DGP, it is part of the output in the DMCP, in that a realization to a Euclidean space of *any* dimension satisfying (1) provides a certificate. When the completion is required to be to a Euclidean distance matrix (EDM), i.e. where distances are given by 2-norms, this problem is called Euclidean DMCP (EDMCP). It is remarkable that, albeit the difference between EDGP and EDMCP is seemingly minor, it is not known whether the EDMCP is in **P** or **NP**-hard (whereas the EDGP is known to be **NP**-hard). The EDMCP is currently thought to be "between the two classes".

In this paper we propose a new heuristic algorithm designed to be accurate yet solve instances of sufficiently large sizes. Our motivation in proposing new heuristics based on Mathematical Programming (MP) formulations is that they can be easily adapted to uncertainty on the distances d_{ij} expressed as intervals, i.e. they can also solve the problem

$$\forall \{i,j\} \in E \quad d_{ij}^L \leq \|x_i - x_j\| \leq d_{ij}^U. \tag{2}$$

This is in contrast to some very fast combinatorial-type algorithms such as the Branch-and-Prune (BP) [13,14], which are natively limited to solving Eq. (1). In fact, this paper is in support of a study which is auxiliary to the development of the BP algorithm, namely to endow the BP with an ability to treat at least some fraction of the distances being given as intervals, which appears to be the case in practice for protein conformation problems from distances.

Our heuristic has three main ingredients: Diagonally Dominant Programming (DDP), very recently proposed by Ahmadi et al. [1,18]; a randomized rank-reduction method of Barvinok's [4]; and a call to a general-purpose local Nonlinear Programming (NLP) solver.

DDP is a technique for obtaining a sequence of inner approximating Linear Programs (LP) or Second-Order Cone Programs (SOCP) to semidefinite programming (SDP) formulations. In this paper we only consider the LP variant, since LP solution technology is more advanced than SDP or SOCP. DDP has been proposed in very general terms; its adaptation to (dual) SDP formulations for the DGP yields a valid LP relaxation for the DGP. In this paper, since we are proposing a heuristic method, and need feasible solutions, we apply DDP to a primal SDP formulation of the DGP.

Note that SDP solutions are square symmetric matrices of any rank, whereas a feasible solution of the DGP must have the given rank K. Although there are many rank reduction techniques, most of them do not have guaranteed properties, even in probability, of preserving the feasibility of the solution. In fact, in the case of the DGP, it is exactly this rank constraint which makes the problem hard, so we can hardly hope in an efficient rank reduction technique that works infallibly. We found the next best thing to be a probabilistic rank reduction technique proposed by Barvinok: although it does not exactly preserve feasibility, it gives a probabilistic guarantee that it will place the reduced rank solution fairly close to all of the manifolds $\mathcal{X}_{ij}$ of realizations x satisfying $\|x_i - x_j\| = d_{ij}$ (for each $\{i, j\} \in E$). At this point, we attempt to achieve feasibility via a single call to a local NLP solver.

Our computational results are preliminary and are simply designed as validation, as this is work-in-progress. We compare the heuristic sketched above to the same, with DDP replaced by SDP. With this limited set-up, we found that the DDP approach exhibits its large-scale potential as the instance sizes increase.

The rest of this paper is organized as follows. We give some technical notation and background in Sect. 1.1. We propose some existing and new SDP formulations of the DGP and EDMCP in Sect. 2. We explain DDP and give a new DDP formulation for the DGP and EDMCP in Sect. 3. We discuss our new DGP heuristic in Sect. 4. We present our preliminary results in Sect. 5. We sketch our roadmap ahead in Sect. 6.

1.1   Relevant Background

The EDGP calls for a solution to the set of nonlinear equations

$$\forall \{i, j\} \in E \quad \|x_i - x_j\|_2 = d_{ij}, \tag{3}$$

where $x_i \in \mathbb{R}^K$ for all $i \le n = |V|$. Usually, the squared version of Eq. (3) is employed, for two reasons: first, since the vast majority of algorithmic implementations employ floating point representations, there is a risk that $\sum_k (x_{ik} - x_{jk})^2 = 0$ might be represented by a tiny negative floating point scalar, resulting in a computational error when extracting the square root. Secondly, as pointed out in [6], the squared EDM $D^2 = (d_{ij}^2)$ has rank at most $K + 2$, a fact which can potentially be exploited. Obviously, solving the squared system yields exactly the same set of solutions as the original system.

Most methods for solving Eq. (3) do not address the original system explicitly, but rather a penalty function:

$$\sum_{\{i,j\} \in E} (\|x_i - x_j\|_2^2 - d_{ij}^2)^2, \tag{4}$$

which has global optimum x^* with value zero if and only if x^* satisfies Eq. (3). This formulation is convenient since most local NLP solvers find it easier to improve the cost of a feasible non-optimal solution, rather than achieving feasibility from an infeasible point. This is relevant since such solvers are often employed to solve EDGP instances. Equation (4) can be easily adjusted to deal with imprecise distances represented by intervals (see e.g. [17]).

There are several Semidefinite Programming (SDP) relaxations of the EDGP [2, 19, 22], mostly based on linearizing the constraint

$$\forall \{i, j\} \in E \quad \|x_i\|_2^2 + \|x_j\|_2^2 - 2x_i \cdot x_j = d_{ij}^2$$

into

$$\forall \{i, j\} \in E \quad X_{ii} + X_{jj} - 2X_{ij} = d_{ij}^2 \tag{5}$$

and then relaxing the rank constraint $X = xx^\top$ to $X \succeq xx^\top$, which, via the Schur complement, can be written as the semidefinite constraint

$$Y = \begin{pmatrix} I_K & x^\top \\ x & X \end{pmatrix} \succeq 0. \tag{6}$$

Such formulations mostly come from the application to sensor networks, for which $K = 2$. In his EE3920 course 2003, Y. Ye proposes the objective function:

$$\min \operatorname{tr}(Y), \tag{7}$$

motivated by a probabilistic interpretation of the solution of the SDP. Purely based on (unpublished) empirical observations, we found what is possibly a better objective function (at least for some protein conformation instances), discussed in Sect. 2 below.

Several methods aim to decompose large graphs into rigid components [5,9, 11], since many rigid graphs can be realized efficiently [7,16]. Each rigid subgraph realization is then "stiched up" consistently by either global optimization [9] or SDP [5,11].

One notable limitation of SDP for practical purposes is that current technology still does not allow us to scale up to large-scale instance sizes. More or less, folk-lore says that interior point methods (IPM) for SDP are supposed to work well up to sizes of "around" 1000 variables, i.e. a matrix variable of around 33×33, which is hardly "large-scale". As remarked, a technique which can address this limitation is the very recent DDP [1,18]. Since all diagonally dominant (DD) matrices are positive semidefinite (PSD), any DDP obtained from an SDP by replacing the PSD constraint with a DD one is an inner approximation of the original SDP. The interesting feature of DDP is that it can be reformulated to an LP, which current technology can solve with up to millions of variables.

Once a solution $\bar{X}$ of an SDP relaxation has been found, the problem of finding another solution of the correct rank, which satisifies $X = xx^\top$ is called *rank reduction*. Possibly the most famous rank reduction algorithm is the Goemans-Williamson algorithm for MAX CUT [8]. Other ideas, connected with the concentration of measure phenomenon, have been proposed in [4] in order to find a solution x which is reasonably close, on average and with high probability, from the manifolds $\mathcal{X}_{ij}$ described in Eq. (3).

Although being "reasonably close to a manifold" is certainly no guarantee that a local NLP solver will move the reasonably close point to the manifold itself, there is a good hope of this being the case.

2 SDP Formulations for DG

We represent a realization x in matrix form by an $n \times K$ matrix where $n = |V|$, and where each of the n rows is a vector $x_i \in \mathbb{R}^K$ which gives the position of vertex $i \in V$. We discussed a well known SDP for the EDGP in Sect. 1.1, which we recall here without the objective function, for later reference.

$$\left. \begin{array}{c} \forall \{i,j\} \in E \; X_{ii} + X_{jj} - 2X_{ij} = d_{ij}^2 \\ Y = \begin{pmatrix} I_K & x^\top \\ x & X \end{pmatrix} \succeq 0. \end{array} \right\} \tag{8}$$

2.1 A Better Objective for Protein Conformation

The empirical evidence collected by Ye about Eq. (8) with $\min \mathrm{tr}(Y)$ as objective concerns the application of EDGP to the localization of sensor networks. Our own (unpublished and preliminary) computations on protein conformation instances with the above objective were not particularly encouraging. We found relatively better results with a different objective function:

$$\left. \begin{array}{c} \min \displaystyle\sum_{\{i,j\} \in E} (X_{ii} + X_{jj} - 2X_{ij}) \\ \forall \{i,j\} \in E \qquad X_{ii} + X_{jj} - 2X_{ij} \geq d_{ij}^2 \\ Y = \begin{pmatrix} I_K & x^\top \\ x & X \end{pmatrix} \succeq 0. \end{array} \right\} \tag{9}$$

Note that Eq. (9) can be trivially derived as the natural SDP relaxation of the nonconvex NLP:

$$\left. \begin{array}{c} \min \displaystyle\sum_{\{i,j\} \in E} \|x_i - x_j\|_2^2 \\ \forall \{i,j\} \in E \qquad \|x_i - x_j\|_2^2 \geq d_{ij}^2, \end{array} \right\} \tag{10}$$

which is an exact reformulation of Eq. (4) since, if Eq. (3) has a solution x^*, at x^* all of the inequality constraints of Eq. (10) are tight, and therefore the objective cannot be further decreased. Conversely, if there was an x' with lower objective function value, at least one of the constraints would be violated.

For the EDMCP, where the rank is of no importance, we only require that X should be the Gram matrix of a realization x (of any rank). Since the Gram matrices are exactly the PSD matrices, Eq. (9) can be simplified to:

$$\left. \begin{array}{c} \min \displaystyle\sum_{\{i,j\} \in E} (X_{ii} + X_{jj} - 2X_{ij}) \\ \forall \{i,j\} \in E \qquad X_{ii} + X_{jj} - 2X_{ij} \geq d_{ij}^2 \\ X \succeq 0. \end{array} \right\} \tag{11}$$

Note that objective functions for the EDGP are often (though not always [3]) a matter of preference and empirical experience on sets of instances, which makes sense since the EDGP is a pure feasibility problem (other possible objectives include adding slack variables which are then minimized). From here onwards, therefore, we shall simply discuss pure feasibility formulations expressed with equality constraints, each of which can be turned into an optimization problem at need, with equality constraints possibly changed into inequalities, and/or by additional slacks and surplus variables to be minimized.

3   Diagonally Dominant Programming

One serious drawback of SDP is that current solving technology is limited to instances of fairly low sizes. Ahmadi and Hall recently remarked [1] that diagonal dominance provides a useful tool for inner approximating the PSD cone. A matrix (Y_{ij}) is DD if

$$\forall i \leq n \quad Y_{ii} \geq \sum_{j \neq i} |Y_{ij}|. \tag{12}$$

It follows from Gershgorin's theorem that all DD matrices are PSD (the converse does not hold, hence the inner approximation). This means that

$$\left. \begin{array}{c} \forall \{i,j\} \in E \ X_{ii} + X_{jj} - 2X_{ij} = d_{ij}^2 \\ Y = \begin{pmatrix} I_K & x^\top \\ x & X \end{pmatrix} \text{ is DD} \end{array} \right\} \tag{13}$$

is a DDP formulation with a feasible region which is an inner approximation of that of Eq. (8).

The crucial observation is that Eq. (12) is easy to linearize exactly, as follows:

$$\forall i \leq n \quad \sum_{j \neq i} T_{ij} \leq Y_{ii}$$

$$\forall i,j \leq n \quad -T_{ij} \leq Y_{ij} \leq T_{ij}.$$

We exploit this idea to derive a new DDP formulation related to the EDGP, which is in fact an LP for the EDGP.

$$\left. \begin{array}{c} \forall \{i,j\} \in E \ X_{ii} + X_{jj} - 2X_{ij} = d_{ij}^2 \\ \begin{pmatrix} I_K & x^\top \\ x & X \end{pmatrix} = Y \\ \forall i \leq n+K \qquad \sum_{\substack{j \leq n+K \\ j \neq i}} T_{ij} \leq Y_{ii} \\ -T \leq Y \leq T. \end{array} \right\} \tag{14}$$

Note that, previous to Eq. (14), the only existing LP formulation for the EDGP was the relaxation of Eq. (4) in which every monomial $m(x)$ of the quartic polynomial in the objective is linearized to a variable μ subject to linear convex and concave relaxations of the nonconvex constraint $\mu = m(x)$. It is known [12] that, for large enough variable bounds, this relaxation is much weaker than the obvious lower bound 0. We hope that the new formulation Eq. (14) will improve the situation.

3.1   DDP from the Dual

Since Eq. (14) is an inner approximation of Eq. (8), there might conceivably be cases where the feasible region of Eq. (14) is empty while the feasible region of Eq. (8) is non-empty (quite independently of whether the original EDGP instance has a solution or not). For such cases, Ahmadi and Hall recall that the dual of any SDP is another SDP (moreover, strong duality holds). So it suffices to derive a DDP from the dual of the SDP relaxation Eq. (8) in order to obtain a new, valid LP relaxation of the EDGP.

3.2   Iterative Improvement of the DDP Formulation

Ahmadi and Hall also provide an iterative method to improve the DDP inner approximation for general SDPs, which we adapt here to Eq. (14). For any symmetric $n \times n$ matrix U, we have $U^\top U \succeq 0$ since any Gram matrix is PSD. By the same reason, $U^\top X U \succeq 0$ for any $X \succeq 0$. This implies that

$$\mathcal{D}(U) = \{U^\top A U \mid A \text{ is DD}\} \tag{15}$$

is a subset of the PSD cone. We can therefore replace the constraint "Y is DD" by $Y \in \mathcal{D}(U)$ in Eq. (13). Note that this means the LP formulation is now parametrized on U, which offers the opportunity to choose U so as to improve the approximation. More precisely, we define a sequence of DDP formulations:

$$\left.\begin{array}{c} \forall \{i,j\} \in E \; X_{ii} + X_{jj} - 2X_{ij} = d_{ij}^2 \\ Y = \begin{pmatrix} I_K & x^\top \\ x & X \end{pmatrix} \in \mathcal{D}(U^h), \end{array}\right\} \tag{16}$$

for each $h \in \mathbb{N}$, with

$$U^0 = I$$
$$U^h = \mathsf{factor}(\bar{Y}^{h-1}),$$

where $\mathsf{factor}(\cdot)$ indicates a factor of the argument matrix (Ahmadi and Hall suggest using Choleski factors for efficiency), and $\bar{Y}^h$ is the solution of Eq. (16) for a given h.

The iterative method ensures that, for each h, the feasible region of Eq. (16) contains the feasible region for $h - 1$. This is easily seen to be the case since, if U^h is a factor of $\bar{Y}^{h-1}$, we trivially have $(U^h)^\top I U^h = (U^h)^\top U^h = \bar{Y}^{h-1}$, and since I is trivially DD, $\bar{Y}^{h-1} \in \mathcal{D}(U^h)$. Moreover, $\bar{Y}^{h-1}$ is feasible in Eq. (16), which proves the claim.

The transformation of the constraint $Y \in \mathcal{D}(U)$ into a set of linear constraints is also straightforward. $Y \in \mathcal{D}(U)$ is equivalent to "$Y = U^\top Z U$ and Z is DD", i.e.

$$\forall i \leq n + K \quad \sum_{\substack{j \leq n+K \\ j \neq i}} T_{ij} \leq Z_{ii}$$
$$-T \leq Z \leq T$$
$$U^\top Z U = Y,$$

as observed above.

4   A New Heuristic for the DGP

In this section we use some of the techniques discussed above in order to derive a new heuristic algorithm which will hopefully be able to solve large-scale EDGP instances.

1. Solve a DDP approximation to an SDP relaxation of the DGP (see previous sections) to yield $\bar{X}$. If $\mathsf{rank}(\bar{X}) \leq K$, factor $\bar{X} = \bar{x}\bar{x}^\top$ and return $\bar{x}$.
2. We now have $\bar{X}$ with $\mathsf{rank}(\bar{X}) > K$. We run Barvinok's randomized rank reduction algorithm [4]:
   (a) sample $y \in \mathbb{R}^{nK}$ from a multivariate normal distribution $\mathcal{N}^{nK}(0,1)$
   (b) let $T = \mathsf{factor}(\bar{X})$
   (c) let $x' = Ty$ (optionally repeat a given number of times from Step 2a and choose best x').
3. Call any local NLP solver with x' as a starting point, and hope to return a rank K solution $x^* \in \mathbb{R}^{nK}$ which is feasible in Eq. (3).

Barvinok proves that there is concentration of measure for the randomized rank reduction in Steps 2a–c, so that, if κ is the least number such that $m = |E| \leq n^\kappa$, there is n_0 large enough such that, if $n \geq n_0$, we have:

$$\mathsf{Prob}\left(\forall \{i,j\} \in E \quad \mathsf{dist}(x', \mathcal{X}_{ij}) \leq c(\kappa) \sqrt{\|\bar{X}\|_2 \ln n} \right) \geq p, \qquad (17)$$

where $\mathsf{dist}(x, \mathcal{X}_{ij})$ is the Euclidean distance from x to the manifold $\mathcal{X}_{ij}$, $\|X\|_2$ is the largest eigenvalue of $\bar{X}$, $c(\kappa)$ is a constant depending only on κ, and p is given in [4] as $p = 0.9$.

Note that we can actually solve an SDP relaxation of the DGP, in Step 1, rather than a DDP approximation thereof. This variant of the heuristic will be used to obtain a computational comparison in Sect. 5.

We remark that we are actually mis-using Barvinok's rank reduction algorithm, which was originally developed only for $K = 1$. Concentration of measure phenomena, however, are based on average behaviour being what one would expect; the most important part of the work is always to prove that large distortions from the mean are controllably improbable. We therefore believe we are justified in our mis-appropriation, at the risk of the probability being somewhat lower than advertised; but since we have no good estimations for c, this vagueness is not overly detrimental. Essentially, most concentration of measure results are often used qualitatively in algorithmic design, as a statement that, for large enough sizes, the expected behaviour is going to happen with ever higher probability.

On the other hand, mis-using a theoretical result is not to be taken lightly, even if justified by common sense. This is why we also obtained some additional computational experiments (not reported here) with SDPs and DDPs derived from writing realizations as vectors in $\mathbb{R}^{nK}$ rather than $n \times K$ matrices, i.e. precisely the setting of Barvinok's theorem. We found that these results yielded similar outcomes to our heuristic, but in much slower times, due to the much larger size $O(nK \times nK)$ of the involved matrices.

5   Preliminary Computational Assessment

We implemented the proposed heuristic in Python 2.7 and tested it on a Darwin Kernel 15.3 (MacOSX "El Capitan") running on an Intel i7 dual-core (virtual quad-core) CPU at 3.1 GHz with 16 GB RAM.

These are very preliminary experiments, and should be taken as a token of validation of our ideas, not as sound empirical evidence that our idea is computationally the best for the task. As concerns the task, we aim at finding solutions for Eq. (1). Although our stated motivation is to be able to solve Eq. (2), we would like our heuristic to be able to handle both equalities and inequalities, and, for this work, all we had time for was the former.

We tested two variants of our heuristic for comparison: the original one, with Step 1 solving a DDP, and the variant where Step 1 solves an SDP.

Our heuristic is configured as follows.

1. We solved DDP formulations with CPLEX 12.6 [10] (default configuration), which automatically exploits all the cores.
2. We implemented Barvinok's rank approximation heuristic in Python, which only runs on a single core. Steps 2a–c are repeated five times.
3. After testing a few local NLP solvers, we decided to use the L-BFGS implementation given in the Python module `scipy.optimize` in its default configuration, as it gives a good trade-off between speed and solution quality.

We decided *not* to use the iterative DDP approximation method (Sect. 3.2) because of an implementation issue of the Python modelling API we used [20].

For each instance and solution method we record the (scaled) largest distance error (LDE) of the solution x, defined as

$$\mathsf{lde}(x) = \max_{\{i,j\} \in E} (|\, \|x_i - x_j\|_2 - d_{ij}\,|/d_{ij}),$$

the (scaled) mean distance error (MDE)

$$\mathsf{mde}(x) = \frac{1}{|E|} \sum_{\{i,j\} \in E} (|\, \|x_i - x_j\|_2 - d_{ij}\,|/d_{ij}),$$

and the CPU time. All CPU times have been computed in Python using the `time` module. They indicate the CPU time used by the Python process as well as its spawned sub-processes (including CPLEX and the local NLP solver) to reach termination

We tested some randomly generated instances as well as some protein instances taken from the Protein Data Bank (PDB). In the latter, only edges smaller than 5 Å were kept, which is realistic w.r.t. Nuclear Magnetic Resonance (NMR) experiments.

Our first test (see Table 1) aims at solving DGPs for $K = 2$ on three groups of instances.

- Small toy instances, infeasible for $K = 2$.
- A set of instances named euclid-n_p, generated randomly as follows:
  1. place n points in a square, uniformly at random;
  2. generate the cycle $1, \ldots, n$ to ensure biconnectedness;
  3. for each other vertex pair i, j, decide whether $\{i, j\} \in E$ with probability p;
  4. record the Euclidean distance d_{ij} between pairs of points in E;
  obviously, all such instances are feasible.
- Two protein instances 1b03 and 1crn, obviously infeasible for $K = 2$.

Table 1. Tests for $K = 2$.

| Instance | | | LDE | | MDE | | CPU | |
Name	$|V|$	$|E|$	SDP	DDP	SDP	DDP	SDP	DDP
test1	4	6	**0.06**	0.58	**0.03**	0.21	0.07	**0.06**
test2	4	6	0.44	0.44	0.09	0.09	0.10	**0.08**
test3	4	6	**0.06**	0.55	**0.02**	0.18	0.11	**0.05**
random-8_0.5	8	19	**0.79**	1.00	0.17	**0.15**	**0.16**	0.19
cl3	10	23	2.97	2.97	0.53	0.53	0.30	**0.28**
dmdgp-3_10	10	24	0.81	**0.56**	**0.13**	0.15	**0.27**	0.33
dmdgp-3_20	20	54	0.80	0.80	0.15	**0.13**	1.24	**1.17**
testrandom	100	1008	0.98	**0.97**	0.21	0.21	**35.75**	67.96
euclid-10_0.5	10	26	0*	0*	0*	0*	**0.32**	0.41
euclid-20_0.5	20	111	2.32	**0***	0.14	**0***	**1.54**	1.70
euclid-30_0.5	30	240	0*	0*	0*	0*	**3.20**	3.98
euclid-40_0.5	40	429	0*	0*	0*	0*	**6.21**	9.57
euclid-50_0.2	50	290	**0***	12.56	**0***	0.25	**8.10**	10.29
euclid-50_0.3	50	412	0*	0*	0*	0*	**8.23**	12.77
euclid-50_0.4	50	535	0*	0*	0*	0*	**11.96**	12.28
euclid-50_0.5	50	642	0*	0*	0*	0*	**11.51**	16.43
euclid-60_0.2	60	407	**6.38**	14.78	**0.08**	0.22	**11.42**	14.72
euclid-60_0.5	60	938	0*	0*	0*	0*	**17.31**	20.75
euclid-60_0.6	60	1119	0*	0*	0*	0*	**20.07**	21.61
euclid-70_0.5	70	1212	0*	0*	0*	0*	**29.05**	33.43
euclid-80_0.5	80	1639	0*	0*	0*	0*	38.17	**44.69**
euclid-90_0.5	90	1959	0*	0*	0*	0*	**62.03**	72.70
1b03	89	456	0.98	**0.85**	0.13	0.13	63.18	**53.53**
1crn	138	846	1.20	**1.12**	0.14	0.14	481.13	**214.50**

0* indicates values of $O(10^{-5})$ or less.

Our heuristic, when based on SDP, outperforms the DDP-based one on smaller instances, but, time-wise, the technological edge of solving LPs is visible in the larger instances.

Our second test (Table 2) is more realistic, and finds realizations of (feasible) protein instances in $K = 3$. The tests validate our expectations: SDP provides a tighter bound than DDP, and hence the SDP-based heuristic yields better quality solutions, but at the expense of CPU time.

Table 2. Tests on proteins for $K = 3$.

Instance			LDE		MDE		CPU	
Name	$|V|$	$|E|$	SDP	DDP	SDP	DDP	SDP	DDP
C0700.odd.G	36	308	0^*	0.42	0^*	0.01	59.84	**37.79**
C0700.odd.H	36	308	0^*	0.42	0^*	0.02	35.07	**32.89**
C0150alter.1	37	335	0^*	0.58	0^*	0.06	61.80	**14.69**
C0080create.1	60	681	0^*	0^*	0^*	0^*	172.55	**121.15**
C0080create.2	60	681	0^*	0.55	0^*	0.04	149.40	**83.92**
1b03	89	456	**0.28**	0.48	**0.02**	0.04	255.75	**103.22**
1crn	138	846	**0.73**	0.88	0.03	0.03	874.17	**469.99**
1guu-1	150	959	**0.83**	0.89	**0.02**	0.04	2767.71	**978.39**

0^* indicates values of $O(10^{-5})$ or less.

6 Conclusion

We propose a new heuristic algorithm for the DGP, based on diagonally-dominant programming, a randomized rank reduction algorithm, and a local NLP solver. Although our computational test set-up is preliminmnary, we believe our results are promising, and give an indication that the computational bottle-neck of SDP can be overcome by diagonal dominance and LP.

References

1. Ahmadi, A., Hall, G.: Sum of squares basis purouit with linear and second order cone programming. Technical report 1510.01597v1, arXiv (2015)
2. Alfakih, A., Khandani, A., Wolkowicz, H.: Solving Euclidean distance matrix completion problems via semidefinite programming. Comput. Optim. Appl. **12**, 13–30 (1999)
3. Barvinok, A.: Problems of distance geometry and convex properties of quadratic maps. Discrete Comput. Geom. **13**, 189–202 (1995)
4. Barvinok, A.: Measure concentration in optimization. Math. Program. **79**, 33–53 (1997)

5. Cucuringu, M., Singer, A., Cowburn, D.: Eigenvector synchronization, graph rigidity and the molecule problem. Inf. Infer. J. IMA **1**, 21–67 (2012)
6. Dokmanić, I., Parhizkar, R., Ranieri, J., Vetterli, M.: Euclidean distance matrices: essential theory, algorithms and applications. IEEE Sig. Process. Mag. **32**, 12–30 (2015)
7. Eren, T., Goldenberg, D., Whiteley, W., Yang, Y., Morse, A., Anderson, B., Belhumeur, P.: Rigidity, computation, and randomization in network localization. In: IEEE Infocom Proceedings, pp. 2673–2684 (2004)
8. Goemans, M., Williamson, D.: Improved approximation algorithms for maximum cut and satisfiability problems using semidefinite programming. J. ACM **42**(6), 1115–1145 (1995)
9. Hendrickson, B.: The molecule problem: exploiting structure in global optimization. SIAM J. Optim. **5**, 835–857 (1995)
10. IBM: ILOG CPLEX 12.6 User's Manual. IBM (2014)
11. Krislock, N., Wolkowicz, H.: Explicit sensor network localization using semidefinite representations and facial reductions. SIAM J. Optim. **20**, 2679–2708 (2010)
12. Lavor, C., Liberti, L., Maculan, N.: Computational experience with the molecular distance geometry problem. In: Pintér, J. (ed.) Global Optimization: Scientific and Engineering Case Studies, pp. 213–225. Springer, Berlin (2006)
13. Lavor, C., Liberti, L., Maculan, N., Mucherino, A.: The discretizable molecular distance geometry problem. Comput. Optim. Appl. **52**, 115–146 (2012)
14. Liberti, L., Lavor, C., Maculan, N.: A branch-and-prune algorithm for the molecular distance geometry problem. Int. Trans. Oper. Res. **15**, 1–17 (2008)
15. Liberti, L., Lavor, C., Maculan, N., Mucherino, A.: Euclidean distance geometry and applications. SIAM Rev. **56**(1), 3–69 (2014)
16. Liberti, L., Lavor, C., Mucherino, A.: The discretizable molecular distance geometry problem seems easier on proteins. In: Mucherino, A., Lavor, C., Liberti, L., Maculan, N. (eds.) Distance Geometry: Theory, Methods, and Applications. Springer, New York (2013)
17. Liberti, L., Lavor, C., Mucherino, A., Maculan, N.: Molecular distance geometry methods: from continuous to discrete. Int. Trans. Oper. Res. **18**, 33–51 (2010)
18. Majumdar, A., Ahmadi, A., Tedrake, R.: Control and verification of high-dimensional systems with DSOS and SDSOS programming. In: Conference on Decision and Control, vol. 53, pp. 394–401. IEEE, Los Angeles (2014)
19. Man-Cho So, A., Ye, Y.: Theory of semidefinite programming for sensor network localization. Math. Program. B **109**, 367–384 (2007)
20. Sagnol, G.: PICOS: a python interface for conic optimization solvers. Zuse Institut Berlin (2016). http://picos.zib.de
21. Saxe, J.: Embeddability of weighted graphs in k-space is strongly NP-hard. In: Proceedings of 17th Allerton Conference in Communications, Control and Computing, pp. 480–489 (1979)
22. Yajima, Y.: Positive semidefinite relaxations for distance geometry problems. Jpn. J. Indus. Appl. Math. **19**, 87–112 (2002)

A Decomposition Approach for Single Allocation Hub Location Problems with Multiple Capacity Levels

Borzou Rostami[(✉)], Christopher Strothmann, and Christoph Buchheim

Fakultät für Mathematik, TU Dortmund, Dortmund, Germany
brostami@mathematik.tu-dortmund.de

Abstract. In this paper we consider an extended version of the classical capacitated single allocation hub location problem in which the size of the hubs must be chosen from a finite and discrete set of allowable capacities. We develop a Lagrangian relaxation approach that exploits the problem structure and decomposes the problem into a set of smaller subproblems that can be solved efficiently. Upper bounds are derived by Lagrangian heuristics followed by a local search method. Moreover, we propose some reduction tests that allow us to decrease the size of the problem. Our computational experiments on some challenging benchmark instances from literature show the advantage of the decomposition approach over commercial solvers.

Keywords: Hub location · Capacity decisions · Lagrangian relaxation

1 Introduction

Given a complete graph $G = (N, A)$, where N represents the origins, destinations and possible hub locations, and A is the edge set. Hub location problems consider the location of hubs and the allocation of origin-destination nodes to hub nodes in order to route the flow w_{ij} from each origin $i \in N$ to each destination $j \in N$. Hub nodes are used to sort, consolidate, and redistribute flows and their main purpose is to realize economies of scale: while the construction and operation of hubs and the resulting detours lead to extra costs, the bundling of flows decreases costs. The economies of scale are usually modelled as being proportional to the transport volume, defined by multiplication with a discount factor $\alpha \in [0, 1]$.

Depending on the way in which non-hub nodes may be assigned to hub nodes, hub location problems can be classified as either multiple allocation [7] or single allocation [6, 13, 15, 16] hub location problems. In multiple allocation problems, the flow of the same non-hub node can be routed through different hubs, while in single allocation problems, each non-hub node is assigned to exactly one hub. In addition, each of these problems can be classified as capacitated or uncapacitated

The first author has been supported by the German Research Foundation (DFG) under grant BU 2313/2.

© Springer International Publishing Switzerland 2016
R. Cerulli et al. (Eds.): ISCO 2016, LNCS 9849, pp. 237–248, 2016.
DOI: 10.1007/978-3-319-45587-7_21

depending on various types of capacity restrictions. In particular, there can be limitations on the total flow routed on a hub-hub link [12] or on the volume of flow into the hub nodes [8]. For recent overviews on hub location problems we refer the reader to [1,3]. Hub location problems have important applications including, among others, telecommunication systems [11], airline services [10], postal delivery services [6], and public transportation [14].

Due to the importance of capacity restrictions in real-world hub location problems, many papers can be found in the literature that address this type of problems in both multiple and single allocation cases [2,8]. In what follows, we concentrate on different variants of the *capacitated single allocation hub location problem* (CSAHLP). The classical mixed integer linear programming (MILP) formulation for the CSAHLP was proposed by Campbell [2]. It allows a limit on the incoming flow at the hubs coming from both non-hub and hub nodes and defines set-up costs for establishing each of the hubs. Motivated by a postal delivery application, Ernst and Krishnamoorthy [8] studied a variant of the CSAHLP with capacity constraints on the incoming flow at the hubs coming only from non-hub nodes. They proposed an MILP formulation, two heuristics for obtaining upper bounds, and a branch and bound method. Labbé et al. [12] study a CSAHLP where for each hub there is a limit on the total flow traversing it. They studied some polyhedral properties of the problem and propose a branch-and-cut method. Costa et al. [9] proposed a bi-criteria approach to deal with the CSAHLP where the second objective function either minimizes the time that hubs take for processing the flow or minimizes the maximum service time at the hubs. Contreras et al. [4] present a branch-and-price approach for the CSAHLP where lower bounds are obtained using Lagrangian relaxation.

As an extension of the above models, in this paper, we consider a CSAHLP where the choice of capacity levels is explicitly included in the model. This problem was introduced by Correia et al. [5] and is called the *capacitated single allocation hub location problem with multiple capacity levels* (CSAHLPM). In CSAHLPM the capacity restrictions are applied only on incoming flow from origins and each capacity level available incurs a specific set-up cost. All aforementioned CSAHLPs consider a discrete set of potential hub locations, with each hub location having an exogenously defined maximum capacity, while in CSAHLPM individual capacity levels can be installed for each hub location. Accordingly, not only have the hub nodes to be chosen but also the capacity level at which each of them will operate. In [5], the authors propose some MILP formulations for the problem, compare them in terms of the linear programming relaxation, and use state-of-the art optimization software to solve the problem.

In this paper we consider a general form of the CSAHLPM where distances between possible hub locations are not necessarily Euclidean distances in the plane. Starting from a natural quadratic binary program, we first provide a reformulation of the problem which shifts the quadratic term from the objective function to a set of constraints. This allows us to deal with an even more general form of the problem where transportation costs do not need to be linear anymore [17,19]. We develop a Lagrangian relaxation scheme of a path-based

MILP formulation by relaxing the assignment constraints and also constraints that link the assignment variables with the path variables. The Lagrangian function exploits the problem structure and decomposes the problem into a set of smaller subproblems that can be solved efficiently. Some of the latter can be reduced to continuous knapsack problems that can be solved quickly. Since the proposed Lagrangian relaxation does not have the so-called integrality property, the obtained bound will be stronger than the one given by the continuous relaxation of the MILP. To calculate feasible solutions we propose a two-phase heuristic where a greedy algorithm is used to construct an initial solution in the first phase and a local search scheme tries to improve the initial solution in the second phase. Finally, we present some reduction tests that allow us to decrease the size of the problem without affecting the set of optimal solutions and, accordingly, to obtain tighter bounds with less computational effort.

2 Problem Formulations

Let a directed graph $G = (N, A)$ be given, where the set N contains the nodes, representing the origins, destinations and possible hub locations, and A is the edge set. Let w_{ij} be the amount of flow to be transported from node i to node j. We denote by $O_i = \sum_{j \in N} w_{ij}$ and $D_i = \sum_{j \in N} w_{ji}$ the total outgoing flow from node i and the total incoming flow to node i, respectively. For each $k \in N$, we consider $Q_k = \{1, 2, \ldots, s_k\}$ as a set of different capacity levels available for a potential hub to be installed at node k. For each $k \in N$ and each $\ell \in Q_k$, let $f_{k\ell}$ and $\Gamma_{k\ell}$ represent the fixed set-up cost and the capacity of hub k associated with capacity level ℓ. The capacity of a hub represents an upper bound on the total incoming flow that can be processed in the hub. Thus, it refers only to the sum of the flow generated at the nodes that are assigned to the hub and not taking into account the inter-hub flow. The cost per unit of flow for each path $i - k - m - j$ from an origin node i to a destination node j which passes hubs k and m respectively, is $\chi d_{ik} + \alpha d_{km} + \delta d_{mj}$, where χ, α, and δ are the nonnegative collection, transfer and distribution costs respectively, and d_{ij} represents the distance between nodes i and j. Note that we do not require that the distances satisfy the triangle inequality. The CSAHLPM now consists in selecting a subset of nodes as hubs with specific capacity levels and assigning the remaining nodes to these hubs such that each non-hub node is assigned to exactly one hub node without exceeding its chosen capacity, with the minimum overall cost.

2.1 Quadratic Binary Formulation

In order to model the problem as an integer quadratic program, we define binary variables x_{ik} indicating whether a source/sink i is allocated to a hub k. In particular, the variables x_{kk} are used to indicate whether k becomes a hub. Moreover, for each $k \in N$ and $\ell \in Q_k$, we define a binary variable $t_{k\ell}$ indicating whether node k receives a hub with capacity level ℓ. For ease of presentation, we set $c_{ik} := d_{ik}(\chi O_i + \delta D_i)$. The CSAHLPM can then be formulated as follows:

$$\text{P:} \quad \min \quad \sum_i \sum_k c_{ik}x_{ik} + \sum_i \sum_j \sum_k \sum_m \alpha d_{km}w_{ij}\, x_{ik}x_{jm} + \sum_k \sum_{\ell \in Q_k} f_{k\ell}t_{k\ell}$$

$$\text{s.t.} \quad \sum_k x_{ik} = 1 \quad (i \in N) \tag{1}$$

$$x_{ik} \leq x_{kk} \quad (i,k \in N) \tag{2}$$

$$\sum_i O_i x_{ik} \leq \sum_{\ell \in Q_k} t_{k\ell}\Gamma_{k\ell} \quad (k \in N) \tag{3}$$

$$\sum_{\ell \in Q_k} t_{k\ell} = x_{kk} \quad (k \in N) \tag{4}$$

$$\sum_k x_{kk} \geq p \tag{5}$$

$$x_{ik} \in \{0,1\} \quad (i,k \in N) \tag{6}$$

$$t_{\ell k} \in \{0,1\} \quad (k \in N, \ell \in Q_k) \tag{7}$$

where the objective function measures the total transport costs consisting of the collection and distribution costs of nonhub-hub and hub-nonhub connections, the hub-hub transfer costs, as well as the set-up costs of the hubs. Constraints (1) force every node to be allocated to precisely one hub node. Constraints (2) state that i can only be allocated to k if k is chosen as a hub. Constraints (3) are capacity constraints and ensure that the overall incoming flow of nodes assigned to a hub does not exceed its capacity. Constraints (4) assure that if a hub is installed at a node then exactly one capacity level is chosen.

Finally, Constraint (5) sets a lower bound p on the number of chosen hubs. We use this constraint only to strengthen the lower bounds given by this model, choosing a value for p that is a lower bound on the number of hubs already due to the remaining constraints, so that Constraint (5) is redundant. For details about the computation of p we refer the reader to [5].

Now, let us consider a reformulation of the above model in which the quadratic term from the objective function is shifted to the set of constraints. To this end, we define a new continuous variable z_{km} that models the traffic on the hub-hub connection $(k,m) \in A$. This allows us to rewrite the above model as:

$$\text{P1:} \quad \min \quad \sum_i \sum_k c_{ik}\, x_{ik} + \sum_k \sum_m \alpha\, d_{km}\, z_{km} + \sum_k \sum_{\ell \in Q_k} f_{k\ell}t_{k\ell}$$

$$\text{s.t.} \quad \sum_i \sum_j w_{ij}\, x_{ik}x_{jm} \leq z_{km} \quad (k,m \in N) \tag{8}$$

$$(1)-(7)$$

where the traffic variable z_{km} for $(k,m) \in A$ is determined by Constraints (8). Note that since all data are non-negative, there exists an optimal solution for problem P1 where all constraints (8) are tight. Therefore, the problems P and P1 are equivalent.

2.2   Linearization

All formulations proposed in [5] for the CSAHLPM are, in fact, based on the classical path based formulation of [2,18] and the flow based formulation of [8].

Moreover, in [5] the authors assume that the distances between nodes satisfy the triangle inequality. However, if the distances considered are, for example, road distances which do not necessarily satisfy the triangle inequality, then the flow based formulations cannot be applied.

To linearize Problem P1, we follow the path based formulation of [18] for uncapacitated hub location problems and define a set of binary variables y_{ikjm} for $i, k, j, m \in N$ to indicate whether the flow from node i to node j travels via hubs located at nodes k and m or not. The resulting formulation is as follows:

$$\text{ILP1:} \quad \min \quad \sum_i \sum_k c_{ik} x_{ik} + \sum_k \sum_m \alpha\, d_{km}\, z_{km} + \sum_k \sum_{\ell \in Q_k} f_{k\ell} t_{k\ell}$$

$$\text{s.t.} \quad \sum_i \sum_j w_{ij}\, y_{ikjm} \leq z_{km} \quad (k, m \in N) \tag{9}$$

$$\sum_k y_{ikjm} = x_{jm} \quad (i, j, m \in N) \tag{10}$$

$$\sum_m y_{ikjm} = x_{ik} \quad (i, j, k \in N) \tag{11}$$

$$0 \leq y_{ikjm} \leq 1 \quad (i, k, j, m \in N) \tag{12}$$

$$(1) - (7).$$

This problem is equivalent to Problem P. However, if the integrality restrictions on variables x or t are relaxed, it is no longer equivalent to P, but only provides a lower bound on its objective function value.

3   Solution Method

3.1   Lagrangian Relaxation

Due to the large number of variables and constraints, solving the linear relaxation of ILP1 requires considerable running time as the size of the instances increases. To overcome this problem, we develop a Lagrangian relaxation approach based on relaxing Constraints (1), (10) and (11) of the ILP1 formulation. Using Lagrangian multipliers π, λ, and μ, respectively, we obtain the following Lagrangian function:

$$L(\pi, \lambda, \mu): \quad \min \quad \sum_i \pi_i + \sum_i \sum_k \bar{c}_{ik} x_{ik} + \sum_k \sum_{\ell \in Q_k} f_{k\ell} t_{k\ell}$$
$$+ \sum_k \sum_m \alpha\, d_{km}\, z_{km} + \sum_i \sum_j \sum_k \sum_m (\lambda_{ijm} + \mu_{jik}) y_{ikjm}$$

$$\text{s.t.} \quad (2) - (7), (9), (12)$$

where

$$\bar{c}_{ik} := c_{ik} - \pi_i - \sum_j (\lambda_{jik} + \mu_{jik}) \quad (i, k \in N).$$

The best lower bound is then obtained by solving the Lagrangian dual problem given as $\max_{\pi, \lambda, \mu} L(\pi, \lambda, \mu)$.

Considering the independence between the two groups of variables (x, t) and (y, z) in $L(\pi, \lambda, \mu)$, we can first decompose the latter into two subproblems $L_{xt}(\pi, \lambda, \mu)$ and $L_{yz}(\lambda, \mu)$:

$$L_{xt}(\pi, \lambda, \mu): \quad \min \quad \sum_i \sum_k \bar{c}_{ik} x_{ik} + \sum_k \sum_{\ell \in Q_k} f_{k\ell} t_{k\ell}$$
$$\text{s.t.} \quad (2)-(6) \tag{13}$$
$$L_{yz}(\lambda, \mu): \quad \min \quad \sum_k \sum_m d_{km} z_{km} + \sum_i \sum_j \sum_k \sum_m (\lambda_{ijm} + \mu_{jik}) y_{ikjm}$$
$$\text{s.t.} \quad (9), (12). \tag{14}$$

Solving $L_{xt}(\pi, \lambda, \mu)$: To solve Subproblem (13), let us suppose that node k receives a hub with capacity level ℓ, that is, $t_{k\ell} = 1$. Then the remaining nodes that will be assigned to hub k can be found by solving the following binary knapsack problem:

$$\xi_{k\ell} = \min \quad \sum_{i \neq k} \bar{c}_{ik} x_{ik}$$
$$\text{s.t.} \quad \sum_{i \neq k} O_i x_{ik} \leq \Gamma_{k\ell} - O_k$$
$$x_{ik} \in \{0, 1\} \quad (i \in N, \ i \neq k).$$

Now suppose that a hub is located at $k \in N$, such that $x_{kk} = 1$. Then the following problem just selects the best capacity level of the hub from Q_k:

$$\varepsilon_k = \min \quad \sum_{\ell \in Q_k} (f_{k\ell} + \xi_{k\ell}) t_{k\ell}$$
$$\text{s.t.} \quad \sum_{\ell \in Q_k} t_{k\ell} = 1$$
$$t_{k\ell} \in \{0, 1\} \quad (\ell \in Q_k).$$

Finally, solving the following problem gives the optimal value of $L_{xt}(\pi, \lambda, \mu)$:

$$\min \quad \sum_k (\varepsilon_k + \bar{c}_{kk}) x_{kk}$$
$$\text{s.t.} \quad \sum_k x_{kk} \geq p$$
$$x_{kk} \in \{0, 1\} \quad (k \in N).$$

This problem can be solved easily by sorting and choosing at least the p hubs with smallest objective coefficient $\varepsilon_k + \bar{c}_{kk}$.

Solving $L_{yz}(\lambda, \mu)$: Subproblem (14) can be further decomposed into n^2 subproblems, one for each pair (k, m), as follows:

$$\max \quad -\alpha d_{km} z_{km} - \sum_i \sum_j (\lambda_{ijm} + \mu_{jik}) y_{ikjm}$$
$$\text{s.t.} \quad \sum_i \sum_j w_{ij} y_{ikjm} \leq z_{km}$$
$$0 \leq y_{ikjm} \leq 1 \quad (i, j \in N)$$

where we may assume $\lambda_{ijm} + \mu_{jik} < 0$ for all $i, j \in N$. This problem is a special knapsack problem where the capacity of the knapsack is part of the decision making process with per unit cost αd_{km}. An optimal solution for this problem can be found by adding only those items to the knapsack whose profit per unit of weight exceeds the cost of one unit of capacity. More precisely, for each item $i, j \in N$, if

$$-(\lambda_{ijm} + \mu_{jik}) y_{ikjm} / w_{ij} > \alpha d_{km}$$

we add this item to the knapsack and set $y_{ikjm} = 1$, otherwise we set $y_{ikjm} = 0$.

3.2   Primal Heuristics

To obtain a valid upper bound for our given formulation, we aim at creating a feasible solution in a two step process, starting from the current solution of the Lagrangian relaxation procedure. In the first step, we try to derive a feasible solution and, if we found one, we reassign hubs with a local search algorithm to improve the solution.

Let (x, t, z) be the solution of the Lagrangian relaxation in the current iteration. As this solution does not need to be feasible, we first try to find valid variable assignment. Firstly, we will only consider the current solution if the opened capacity is large enough to meet the demand. The procedure to heuristically generate a feasible solution $(\bar{x}, \bar{t}, \bar{z})$ is as follows: whenever $x_{kk} = 1$ in the Lagrangian relaxation, node k will be a hub in the heuristic solution, i.e., $\bar{x}_{kk} = 1$, and we set $\bar{t}_{k\ell} = t_{k\ell}$ for all $\ell \in Q_k$. If a node i is assigned to more than one hub, we only assign it to the hub with the least cost. The remaining unassigned nodes will now be sorted according to their weights O_i and we consider the free capacity γ_k in each hub given by

$$\gamma_k := \sum_{\ell \in Q_k} \Gamma_{k\ell} \bar{t}_{k\ell} - \sum_{i=1}^{n} O_i \bar{x}_{ik} .$$

Now we iteratively assign each remaining node to a hub, starting with the one with the highest O_i. The node with the highest weight will be assigned to the hub with the lowest cost and so on. If we cannot assign every node to a hub in this way, we stop our heuristic.

In the second step we try to improve the upper bound found in the first step by means of a local search algorithm. For that, we only consider reassigning nodes if the calculated upper bound from the first step is not considerably worse than the best known upper bound. The local search phase then tries to find possible shift or swap moves that improve the current solution. While shifting tries to assign nodes to different hubs with enough free capacity, swapping exchanges two assigned hubs. Note that when we reassign a node we only need to calculate the estimated cost Δ of the new feasible solution in terms of the old upper bound, corrected by the influence of the new assignment. For example in the shifting step, when node i is reassigned from hub k to hub m, the value of Δ is computed as:

$$\Delta = \bar{z}_{km} + \bar{z}_{mk} + c_{im} - c_{ik}$$
$$+ \sum_{j \neq m} \alpha(d_{jm}\bar{z}_{jm} + d_{mj}\bar{z}_{mj}) + \sum_{j \neq k} \alpha(d_{jk}\bar{z}_{jk} + d_{kj}\bar{z}_{kj}).$$

If $\Delta < 0$, then the so found upper bound is smaller than the original one and we update the solution accordingly.

3.3   Reduction Test

The size of ILP1 may be reduced by eliminating the hub variables which do not appear in any optimal solution of a given instance. For this, we use information obtained from the Lagrangian function at a given iteration for any given

Lagrangian multipliers π, λ and μ. Our reduction test is based on testing if a hub k with a specific level ℓ is going to be excluded in the optimal solution of a given instance. The main idea of fixing a variable to zero is to check if including this variable in the solution will lead to a lower bound greater than the best upper bound found so far. If so, the variable cannot belong to an optimal solution. We consider variables $t_{k\ell}$, $k \in N, \ell \in Q_k$, that are not already fixed and are equal to zero in the current iteration. We thus impose an additional constraint $t_{k\ell} = 1$ to the current Lagrangian function. Let $LB^{k\ell}$ represent the new value of the Lagrangian function. If $LB^{k\ell} \geq UB$, then we fix $t_{k\ell} = 0$ and add it to the list of fixed variables.

The main question arising here is how to compute $LB^{k\ell}$ without resolving the Lagrangian function. To answer this question, we distinguish between the following two situations: (i) If $x_{kk} = 1$, then there exists a level $\ell' \in Q_k$ such that $t_{k\ell'} = 1$. Therefore, to open hub k with different capacity level ℓ, we need to exclude hub k with capacity level ℓ'. Hence we have:

$$LB^{k\ell} = LB - (f_{k\ell'} + \xi_{k\ell'}) + (f_{k\ell} + \xi_{k\ell}).$$

(ii) If $x_{kk} = 0$, we consider two different scenarios. If the number of open hubs agrees with the lower bound p and all values $\varepsilon_m + \bar{c}_{mm}$ of open hubs are non-positive, or the number of open hubs is strictly greater than p, we do not need to close any hubs in order to open hub k, which means:

$$LB^{k\ell} = LB + (\bar{c}_{kk} + f_{k\ell} + \xi_{k\ell}) \tag{15}$$

Otherwise, if for some open hubs the values $\varepsilon_m + \bar{c}_{mm}$ are positive, we close the most expensive one and subtract $\varepsilon_{m^*} + \bar{c}_{m^*m^*}$ from $LB^{k\ell}$, i.e.,

$$LB^{k\ell} = LB + (\bar{c}_{kk} + f_{k\ell} + \xi_{k\ell}) - (\varepsilon_{m^*} + \bar{c}_{m^*m^*})$$

where $m^* = \text{argmax}\{\varepsilon_m + \bar{c}_{mm} \mid x_{mm} = 1\}$ is the most expensive open hub. Note that if for every k and all $\ell \in Q_k$ we have $t_{k\ell} = 0$, we can fix all x_{kk} and x_{ik} to zero as well.

In our computational experiments we performed this reduction test in any iteration of the subgradient method. When the set of potential hubs or the set of capacity levels for a given hub was reduced, we updated Subproblems (13) and (14) accordingly.

4   Extension to Link Capacities

The reformulation of Problem P as Problem P1 allows to extend the CSAHLPM in order to include link capacities. In particular, consider an application in telecommunication networks containing set-up cost s_{ij} for installing needed capacities on the connection $(i,j) \in A$; see [19]. The capacity is provided by the installation of an integer number of links of a fixed capacity q. The resulting problem is as follows:

MILP1:   min $\sum_i \sum_k c_{ik} x_{ik} + \sum_k \sum_m s_{km} z_{km} + \sum_{k,\ell \in Q_k} f_{k\ell} t_{k\ell}$

s.t. $\sum_i \sum_j w_{ij} y_{ikjm} \leq q z_{km}$ $(k, m \in N)$ (16)

$(1) - (7), (9) - (12)$

$z_{ij} \geq 0, \text{integer}$ $(i, j) \in A$

where z_{ij} indicates the number of installed links on $(i, j) \in A$. Constraints (16) relate the number of installed links on each hub-hub connections to the total flow passing the link and state that the flow on each hub-hub connection cannot exceed the capacity of the link. Moreover, since the total incoming and outgoing flow for each node i is given, the cost factors c_{ik} can be precomputed. Note that if we divide all w_{ij} by q and set $s_{km} = \alpha d_{km}$, and if we relax the integrality of the z variables, Problem MILP1 agrees with Problem ILP1.

Such stepwise cost functions make the problem much harder to solve. They have been considered also in [17] for the uncapacitated single allocation hub location problem arising in transportations where the stepwise function results from the integrality of the number of vehicles on hub-hub-connections.

Note that our decomposition approach is still valid for problem MILP1 with a simple modification in Subproblem (14) to include the integrality of z variables. An efficient solution method can be found in [17].

5   Computational Experiments

In this section we present our computational experiments on the lower bound and upper bound computations for the CSAHLPM. We apply our decomposition approach to both problems ILP1 and MILP1, and compare the results with those obtained from the linear relaxation of these models solved by Cplex 12.6. To compute the optimal (or near-optimal) Lagrangian multipliers, we use a subgradient optimization method with a maximum number of 2000 iterations. We implemented the algorithms in C++ and performed all experiments on an Intel Xeon processor running at 2.5 GHz.

For our numerical tests, we considered the Australian Post data set (AP) from the OR library[1], which is used frequently for different hub location problems. We followed the pattern proposed in [5] to generate hub levels: for the highest hub level, the capacity and the set-up cost are equal to the values that are included in the AP data set. Additional levels are produced recursively, starting from the second highest level, according to the formulae

$$\Gamma_{k\ell} = 0.7 \cdot \Gamma_{k\,\ell+1} \quad \text{and} \quad f_{k\ell} = \rho \cdot 0.7 \cdot f_{k\,\ell+1} ,$$

where $\rho = 1.1$ or 1.2 is a factor to model economies of scale. We consider instances with either three or five levels, with the highest capacity level equal to the

---

[1] http://people.brunel.ac.uk/~mastjjb/jeb/orlib/phubinfo.html.

Table 1. Computational results for the original CSAHLPM.

Instance			Lagrangian relaxation				Cplex	
Type	Level	Ub	Gap(%)	Time(s)	ch(%)	chl(%)	Gap(%)	Time(s)
$\rho = 1.1$								
25LL	3	216557.6	0.1	18.0	4.0	66.7	0.1	15.9
	5	202773.6	8.3	14.6	0.0	5.6	7.3	6.0
25LT	3	278062.7	1.0	16.8	8.0	56.0	1.1	11.5
	5	260850.8	2.2	17.1	4.0	42.4	3.5	11.7
50LL	3	217183.0	1.1	405.2	6.0	59.3	1.0	993.2
	5	200016.4	1.5	408.5	0.0	45.6	2.1	815.9
50LT	3	290376.8	1.9	390.4	18.0	54.0	2.2	1837.3
	5	272592.7	1.1	384.4	24.0	58.0	1.9	1651.4
$\rho = 1.2$								
25LL	3	227274.4	0.1	16.4	64.0	86.7	4.8	12.1
	5	211538.1	0.3	17.3	0.0	69.6	11.2	4.9
25LT	3	289881.1	0.6	16.7	16.0	58.7	0.9	11.7
	5	282810.3	0.8	16.9	8.0	56.8	2.3	14.6
50LL	3	229339.1	1.4	407.2	4.0	41.3	6.3	1414.6
	5	219429.2	1.0	406.1	2.0	46.8	10.7	1205.2
50LT	3	303998.6	1.5	393.7	12.0	43.3	1.3	1715.3
	5	291289.2	0.2	372.0	56.0	85.2	1.5	1522.3

loose (L) capacity for the potential hub in the corresponding instance in the AP data set. For the highest set-up cost level we use both the tight (T) and the loose (L) set-up costs from the AP data set. For the data used in MILP1 we followed [16]: for all $i, j \in N$, we set $s_{ij} = \alpha \, q \, d_{ij}$, where $q = \sum_{i,j} w_{ij}/p^2$.

Tables 1 and 2 present the results for problems ILP1 and MILP1, respectively. Each of these tables is divided into two parts where we separately report results for $\rho = 1.1$ or 1.2. In both tables, the first columns indicate the problem type ($|N|Lx$) with $|N| \in \{25, 50\}$ and $x \in \{L, T\}$. The next two columns give the number of levels (level) and the best upper bound (Ub) obtained with our primal heuristic. The next columns present the results of our Lagrangian relaxation and the results of Cplex. For each algorithm, gap(%), and time(s) represent the relative gap in percent and the total required time (in seconds), respectively. The formula we used to compute the relative gaps is $100 \times (Ub - Lb)/Ub$, where Lb stands for the value of the lower bound. The columns under the headings ch(%) and chl(%) of the Lagrangian relaxation present, respectively, the percent of closed hubs and closed hub levels by our reduction tests.

As we can observe from Table 1, the Lagrangian relaxation almost always outperforms Cplex in terms of both the bound tightness and running times. More precisely, the duality gaps for instances with $\rho = 1.1$ and $\rho = 1.2$ are,

Table 2. Computational results for the CSAHLPM with link capacities.

Instance			Lagrangian relaxation				Cplex	
Type	Level	Ub	Gap(%)	Time(s)	ch(%)	chl(%)	Gap(%)	Time(s)
$\rho = 1.1$								
25LL	3	292025.9	3.4	62.2	0.0	14.7	25.9	11.4
	5	292025.9	3.5	70.1	0.0	13.6	35.6	4.7
25LT	3	353996.0	4.6	54.5	4.0	13.3	22.3	11.1
	5	362634.8	6.6	56.7	4.0	5.6	30.6	11.1
50LL	3	288545.2	3.4	1144.1	2.0	14.7	25.5	772.1
	5	284577.4	2.1	1167.4	2.0	18.4	31.2	751.6
50LT	3	339367.0	7.2	957.0	8.0	20.7	16.3	1152.4
	5	350838.8	9.8	883.3	4.0	10.0	23.8	1616.1
$\rho = 1.2$								
25LL	3	292025.9	3.5	55.2	4.0	17.3	22.3	11.8
	5	292025.9	3.4	60.8	0.0	12.8	28.2	7.7
25LT	3	360757.7	6.0	58.5	4.0	6.7	20.3	11.4
	5	360757.7	5.9	60.8	4.0	7.2	23.4	14.4
50LL	3	292622.1	5.4	1169.5	2.0	4.7	23.1	1123.3
	5	289010.6	3.4	1193.7	2.0	13.2	26.1	804.2
50LT	3	358772.3	11.7	894.6	8.0	10.7	16.4	1462.5
	5	359860.4	12.1	1010.3	6.0	8.0	20.3	1427.7

on average, 2.1 % and 0.7 % for the Lagrangian relaxation, and 2.4 % and 4.9 % for Cplex. The results reported in columns 6 and 7 show the effectiveness of the proposed reduction tests in closing hub levels: on average, 49 % and 61 % of levels have been closed for instances with $\rho = 1.1$ and $\rho = 1.2$, respectively. This has a significant positive effect on the required computational times of our Lagrangian relaxation.

Table 2 reports the results for the CSAHLPM with link capacities (MILP1). As we can observe, in principle, the problem is much more difficult than ILP1. However, the Lagrangian relaxation outperforms Cplex significantly: the average duality gap is 5.7 % for the Lagrangian relaxation, compared with an average duality gap of 24.5 % for Cplex.

References

1. Alumur, S.A., Kara, B.Y.: Network hub location problems: the state of the art. Eur. J. Oper. Res. **190**(1), 1–21 (2008)
2. Campbell, J.F.: Integer programming formulations of discrete hub location problems. Eur. J. Oper. Res. **72**(2), 387–405 (1994)
3. Campbell, J.F., O'Kelly, M.E.: Twenty-five years of hub location research. Transp. Sci. **46**(2), 153–169 (2012)

4. Contreras, I., Díaz, J.A., Fernández, E.: Branch and price for large-scale capacitated hub location problems with single assignment. INFORMS J. Comput. **23**(1), 41–55 (2011)
5. Correia, I., Nickel, S., Saldanha-da Gama, F.: Single-assignment hub location problems with multiple capacity levels. Transp. Res. Part B Method. **44**(8), 1047–1066 (2010)
6. Ernst, A.T., Krishnamoorthy, M.: Efficient algorithms for the uncapacitated single allocation p-hub median problem. Location Sci. **4**(3), 139–154 (1996)
7. Ernst, A.T., Krishnamoorthy, M.: Exact and heuristic algorithms for the uncapacitated multiple allocation p-hub median problem. Eur. J. Oper. Res. **104**(1), 100–112 (1998)
8. Ernst, A.T., Krishnamoorthy, M.: Solution algorithms for the capacitated single allocation hub location problem. Ann. Oper. Res. **86**, 141–159 (1999)
9. da Graça Costa, M., Captivo, M.E., Clímaco, J.: Capacitated single allocation hub location problem-a bi-criteria approach. Comput. Oper. Res. **35**(11), 3671–3695 (2008)
10. Jaillet, P., Song, G., Yu, G.: Airline network design and hub location problems. Location Sci. **4**(3), 195–212 (1996)
11. Klincewicz, J.G.: Hub location in backbone/tributary network design: a review. Location Sci. **6**(1), 307–335 (1998)
12. Labbé, M., Yaman, H., Gourdin, E.: A branch and cut algorithm for hub location problems with single assignment. Math. Program. **102**(2), 371–405 (2005)
13. Meier, J.F., Clausen, U., Rostami, B., Buchheim, C.: A compact linearisation of Euclidean single allocation hub location problems. Electronic Notes in Discrete Mathematics Accepted for Publication (2015)
14. Nickel, S., Schöbel, A., Sonneborn, T.: Hub location problems in urban traffic networks. In: Pursula, M., Niittymäki, J. (eds.) Mathematical Methods on Optimization in Transportation Systems, pp. 95–107. Springer, New York (2001)
15. O'Kelly, M.E.: A quadratic integer program for the location of interacting hub facilities. Eur. J. Oper. Res. **32**(3), 393–404 (1987)
16. Rostami, B., Buchheim, C., Meier, J.F., Clausen, U.: Lower bounding procedures for the single allocation hub location problem. Electronic Notes in Discrete Mathematics Accepted for Publication (2015)
17. Rostami, B., Meier, J.F., Buchheim, C., Clausen, U.: The uncapacitated single allocation p-hub median problem with stepwise cost function. Technical report, Optimization Online (2015). http://www.optimization-online.org/DB_HTML/2015/07/5044.html
18. Skorin-Kapov, D., Skorin-Kapov, J., O'Kelly, M.: Tight linear programming relaxations of uncapacitated p-hub median problems. Eur. J. Oper. Res. **94**(3), 582–593 (1996)
19. Yaman, H., Carello, G.: Solving the hub location problem with modular link capacities. Comput. Oper. Res. **32**(12), 3227–3245 (2005)

An Algorithm for Finding a Representation of a Subtree Distance

Kazutoshi Ando[1(✉)] and Koki Sato[2]

[1] Faculty of Engineering, Shizuoka University,
Johoku 3-5-1, Hamamatsu 432-8561, Japan
ando.kazutoshi@shizuoka.ac.jp
[2] DAITO GIKEN, INC., Chuo-ku Kyobashi 3-1-1, Tokyo 104-0031, Japan

Abstract. Generalizing the concept of tree metric, Hirai (2006) introduced the concept of subtree distance. A mapping $d : X \times X \to \mathbb{R}_+$ is called a subtree distance if there exist a weighted tree T and a family $\{T_x \mid x \in X\}$ of subtrees of T indexed by the elements in X such that $d(x, y) = d_T(T_x, T_y)$, where $d_T(T_x, T_y)$ is the distance between T_x and T_y in T. Hirai (2006) gave a characterization of subtree distances which corresponds to Buneman's four-point condition (1974) for the tree metrics. Using this characterization, we can decide whether or not a given matrix is a subtree distance in $O(n^4)$ time. However, the existence of a polynomial time algorithm for finding a tree and subtrees representing a subtree distance has been an open question. In this paper, we show an $O(n^3)$ time algorithm that finds a tree and subtrees representing a given subtree distance.

Keywords: Tree metrics · Phylogeny · Realization algorithm

1 Introduction

Let X be a finite set. A mapping $d : X \times X \to \mathbb{R}_+$ is said to be a *dissimilarity mapping* on X if for each $x, y \in X$ we have $d(x, x) = 0$ and $d(x, y) = d(y, x)$. A dissimilarity mapping d on X is called a *tree metric* if there exist a tree $T = (V, E)$ with $X \subseteq V$ and a length function $l : E \to \mathbb{R}_+$ such that for each $x, y \in X$ we have $d(x, y) = d_T(x, y)$, where $d_T(x, y)$ is the length of the unique path in T connecting x and y.

Buneman [2] gave a characterization of tree metrics as follows.

Proposition 1 (Buneman [2]). *A dissimilarity mapping d on X is a tree metric if and only if for each $x, y, z, w \in X$ we have*

$$d(x, y) + d(z, w) \le \max\{d(x, z) + d(y, w), d(x, w) + d(y, z)\}. \tag{1}$$

K. Ando—This work was supported by JSPS KAKENHI Grant Number 15K00033.

© Springer International Publishing Switzerland 2016
R. Cerulli et al. (Eds.): ISCO 2016, LNCS 9849, pp. 249–259, 2016.
DOI: 10.1007/978-3-319-45587-7_22

Inequality (1) is called the *four-point condition*. By Proposition 1, the problem of deciding whether or not a given dissimilarity mapping is a tree metric can be solved in $O(n^4)$ time, where $n = |X|$. If $d : X \times X \to \mathbb{R}_+$ is a tree metric, a tree $T = (V, E)$ with $X \subseteq V$ and a length function $l : E \to \mathbb{R}_+$ such that $d(x, y) = d_T(x, y)$ is called a *representation* of d. Given a tree metric, we can find its representation by Neighbor-Joining [6] in $O(n^3)$ time or by the algorithm [3] in $O(n^2)$ time.

A connected subgraph of a tree is called a *subtree*. Generalizing the concept of tree metric, Hirai [4] introduced the concept of subtree distance. A dissimilarity mapping d on X is called a *subtree distance* if there exist a tree $T = (V, E)$ with length function $l : E \to \mathbb{R}_+$ and a family $\{T_x \mid x \in X\}$ of subtrees of T indexed by the elements in X such that $d(x, y) = d_T(T_x, T_y)$, where

$$d_T(T_x, T_y) = \min\{d_T(v_x, v_y) | v_x \in V(T_x), v_y \in V(T_y)\}.$$

In this case, we call T and $\{T_x | x \in X\}$ *represent* d. For example, a representation of a subtree distance d in (2), where $X = \{a, b, c, e, f, g, h, i\}$, is shown in Fig. 1.

$$
\begin{array}{c}
\begin{array}{cccccccc}
a & b & c & e & f & g & h & i
\end{array} \\
\begin{array}{c}
a \\ b \\ c \\ e \\ f \\ g \\ h \\ i
\end{array}
\left(
\begin{array}{cccccccc}
0 & 3 & 3 & 1 & 6 & 4 & 7 & 8 \\
3 & 0 & 4 & 2 & 7 & 5 & 8 & 9 \\
3 & 4 & 0 & 1 & 7 & 5 & 8 & 9 \\
1 & 2 & 1 & 0 & 4 & 2 & 5 & 6 \\
6 & 7 & 7 & 4 & 0 & 1 & 5 & 6 \\
4 & 5 & 5 & 2 & 1 & 0 & 1 & 1 \\
7 & 8 & 8 & 5 & 5 & 1 & 0 & 3 \\
8 & 9 & 9 & 6 & 6 & 1 & 3 & 0
\end{array}
\right)
\end{array}
\tag{2}
$$

We note two applications of subtree distances. It was proved by Ando [1] that the computation of the Shapley value of minimum cost spanning tree games is #P-hard in general. However, in the same paper [1], it was shown that if the cost function defining a minimum cost spanning tree game is a subtree distance, then the Shapley value of the game can be computed in polynomial time. Also, in [5], Hirai considered a node-capacitated multiflow problem where a multiflow is weighted by a subtree distance. He showed the duality between such multiflow problems and tree-shaped facility location problems on trees establishing a combinatorial min-max theorem for the multiflow problems.

Hirai [4] gave the following characterization of subtree distances.

Theorem 1 (Hirai [4]). *A dissimilarity mapping d on X is a subtree distance if and only if for each $x, y, z, w \in X$ we have*

$$d(x, y) + d(z, w) \le \max \left\{ \begin{array}{l} d(x, z) + d(y, w), d(x, w) + d(y, z), \\ d(x, y), d(z, w), \\ \frac{d(x,y)+d(y,z)+d(z,x)}{2}, \frac{d(x,y)+d(y,w)+d(w,x)}{2}, \\ \frac{d(x,z)+d(z,w)+d(w,x)}{2}, \frac{d(y,z)+d(z,w)+d(w,y)}{2} \end{array} \right\}. \tag{3}$$

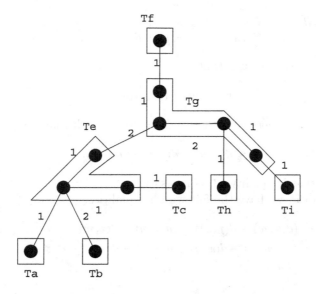

Fig. 1. A weighted tree T and a family $\{T_x | x \in X\}$ of subtrees of T.

If d satisfies the triangular inequality, then the condition (3) is reduced to the four-point condition.

Using Theorem 1, we can decide in $O(n^4)$ time whether or not a given dissimilarity mapping is a subtree distance. However, no polynomial time algorithm for finding a representation of a subtree distance was known so far. In this paper, we present an $O(n^3)$ time algorithm for finding a representation of a subtree distance. By using this algorithm, one can decide in $O(n^3)$ time whether or not a given dissimilarity mapping is a subtree distance.

The rest of the paper is organized as follows. In Sect. 2, we describe an algorithm for finding a representation of a subtree distance. In Sect. 3, we show the validity and the time complexity of the algorithm. In Sect. 4, we give conclusions of this paper.

2   An Algorithm for Finding a Representation of a Subtree Distance

In this section, we present an $O(n^3)$ time algorithm for finding a representation of a subtree distance. We first show the construction of the underlying weighted tree. Then, we show the construction of the subtrees.

2.1   Construction of an Underlying Tree

We begin with the following fundamental result.

Lemma 1. *If $d : X \times X \to \mathbb{R}_+$ is a subtree distance, then there exists an $x \in X$ such that for all $y, z \in X$ we have $d(x, y) + d(x, z) \geq d(y, z)$.*

Proof. Choose $x, w \in X$ such that

$$d(x, w) = \max\{d(u, v) | u, v \in X\}. \tag{4}$$

On the contrary, suppose that there exist $y, z \in X$ such that

$$0 \leq d(x, y) + d(x, z) < d(y, z). \tag{5}$$

We will show that inequality (3) is not satisfied.

By (4) and (5), we have the following inequalities.

$$
\begin{aligned}
&(d(x, w) + d(y, z)) - (d(x, y) + d(z, w)) \\
&> d(x, w) + d(x, y) + d(x, z) - d(x, y) - d(z, w) \\
&\geq d(x, z) \\
&\geq 0,
\end{aligned} \tag{6}
$$

$$
\begin{aligned}
&(d(x, w) + d(y, z)) - (d(x, z) + d(y, w)) \\
&> d(x, w) + d(x, y) + d(x, z) - d(x, z) - d(y, w) \\
&\geq d(x, y) \\
&\geq 0.
\end{aligned} \tag{7}
$$

Also, by (4) and (5), we have

$$(d(x, w) + d(y, z)) - d(x, w) = d(y, z) > 0, \tag{8}$$
$$(d(x, w) + d(y, z)) - d(y, z) = d(x, w) > 0, \tag{9}$$

respectively. By (4) and (5), we have

$$
\begin{aligned}
&(d(x, w) + d(y, z)) - \frac{d(x, w) + d(w, y) + d(y, x)}{2} \\
&= \frac{d(x, w) - d(w, y)}{2} + d(y, z) - \frac{d(y, x)}{2} \\
&\geq d(y, z) - \frac{d(y, x)}{2} \\
&> d(x, y) + d(x, z) - \frac{d(x, y)}{2} \\
&= \frac{d(x, y)}{2} + d(x, z) \\
&\geq 0,
\end{aligned} \tag{10}
$$

$$(d(x,w) + d(y,z)) - \frac{d(x,w) + d(w,z) + d(z,x)}{2}$$

$$= \frac{d(x,w) - d(w,z)}{2} + d(y,z) - \frac{d(z,x)}{2}$$

$$\geq d(y,z) - \frac{d(z,x)}{2}$$

$$> d(x,y) + d(x,z) - \frac{d(x,z)}{2}$$

$$= d(x,y) + \frac{d(x,z)}{2}$$

$$\geq 0. \tag{11}$$

By (5), we have

$$(d(x,w) + d(y,z)) - \frac{d(x,y) + d(y,z) + d(z,x)}{2}$$

$$= d(x,w) + \frac{d(y,z) - d(x,y) - d(z,x)}{2}$$

$$> d(x,w)$$

$$> 0. \tag{12}$$

Finally, by (4) and (5), we obtain

$$(d(x,w) + d(y,z)) - \frac{d(w,y) + d(y,z) + d(z,w)}{2}$$

$$= d(x,w) + \frac{d(y,z)}{2} - \frac{d(w,y) + d(z,w)}{2}$$

$$\geq \frac{d(y,z)}{2}$$

$$> 0. \tag{13}$$

It follows from (6)–(13) that inequality (3) is not satisfied, and hence, by Theorem 1 that d is not a subtree distance, which is a contradiction.

For a subtree distance $d\colon X \times X \to \mathbb{R}$, define V_0 by

$$V_0 = \{x \in X \mid \forall y, z \in X, d(x,y) + d(x,z) > d(y,z)\}. \tag{14}$$

We have $V_0 \neq \emptyset$ by Lemma 1.

For a dissimilarity mapping d on X and a subset $Y \subseteq X$, the *restriction* $d|Y$ of d to Y is a dissimilarity mapping on Y defined by $(d|Y)(x,y) = d(x,y)$ $(x,y \in Y)$. Since $d|V_0$ satisfies the triangular inequality, we have the following.

Proposition 2. *If* $d\colon X \times X \to \mathbb{R}_+$ *is a subtree distance, then* $d|V_0$ *is a tree metric, where* V_0 *is defined by* (14).

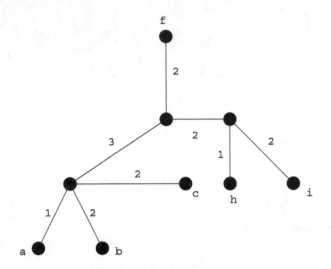

Fig. 2. Weighted tree T representing $d|V_0$.

Since $d|V_0$ is a tree metric by Proposition 2, there exists a weighted tree which represents $d|V_0$. For example, for the subtree distance d given by (2), we see that $V_0 = \{a, b, c, f, h, i\}$ and that a tree T in Fig. 2 represents $d|V_0$.

2.2   Construction of Subtrees

Let $d\colon X \times X \to \mathbb{R}_+$ be a subtree distance. For $v \in X - V_0$ define binary relation $\overset{v}{\sim}$ on V_0 by

$$x \overset{v}{\sim} x' \Leftrightarrow d(x, x') \leq d(x, v) + d(x', v) \quad (x, x' \in V_0). \tag{15}$$

Lemma 2. *Let* $d\colon X \times X \to \mathbb{R}_+$ *be a subtree distance and* $v \in X - V_0$. *Then, the binary relation* $\overset{v}{\sim}$ *defined by* (15) *is an equivalence relation.*

Proof. Let d and v be as stated in the present lemma. It is clear that $\overset{v}{\sim}$ satisfies the reflexivity and the symmetry. We will show that $\overset{v}{\sim}$ satisfies the transitivity.

For $x, x', x'' \in V_0$ suppose that $x \overset{v}{\sim} x'$ and $x' \overset{v}{\sim} x''$. Then, by definition of the binary relation, we have

$$d(x, x') \leq d(x, v) + d(x', v), \tag{16}$$

$$d(x', x'') \leq d(x', v) + d(x'', v). \tag{17}$$

First, we consider the case of $d(x', v) = 0$. By (16) and (17), we have

$$d(x, x') + d(x', x'') \leq d(x, v) + d(x'', v). \tag{18}$$

Since $x' \in V_0$, we have

$$d(x, x'') \leq d(x, x') + d(x', x''). \tag{19}$$

It follows from (18) and (19) that

$$d(x, x'') \leq d(x, v) + d(x'', v),$$

and hence, $x \overset{v}{\sim} x''$.

Next, suppose that $d(x', v) > 0$. On the contrary, suppose that

$$d(x, x'') > d(x, v) + d(x'', v). \tag{20}$$

We will show that inequality (3) does not hold for $x = x, y = x'', z = x'$ and $w = v$.

By (16) and (20), we have

$$
\begin{aligned}
&(d(x, x'') + d(x', v)) - (d(x, x') + d(x'', v)) \\
&\geq d(x, x'') + d(x', v) - (d(x, v) + d(x', v) + d(x'', v)) \\
&= d(x, x'') - (d(x, v) + d(x'', v)) \\
&> 0.
\end{aligned}
\tag{21}
$$

By (17) and (20), we have

$$
\begin{aligned}
&(d(x, x'') + d(x', v)) - (d(x, v) + d(x', x'')) \\
&\geq d(x, x'') + d(x', v) - (d(x, v) + d(x', v) + d(x'', v)) \\
&= d(x, x'') - (d(x, v) + d(x'', v)) \\
&> 0.
\end{aligned}
\tag{22}
$$

By the assumption $d(x', v) > 0$ and (20), we have the followings.

$$(d(x, x'') + d(x', v)) - d(x, x'') = d(x', v) > 0, \tag{23}$$
$$(d(x, x'') + d(x', v)) - d(x', v) = d(x, x'') > 0. \tag{24}$$

By (16), (17) and (20), we have

$$
\begin{aligned}
&(d(x, x'') + d(x', v)) - \frac{d(x, x'') + d(x', x'') + d(x, x')}{2} \\
&\geq d(x, x'') + d(x', v) - \frac{d(x, x'') + d(x, v) + 2d(x', v) + d(x'', v)}{2} \\
&= \frac{d(x, x'') - (d(x, v) + d(x'', v))}{2} \\
&> 0.
\end{aligned}
\tag{25}
$$

By (20), we have

$$(d(x,x'') + d(x',v)) - \frac{d(x,x'') + d(x'',v) + d(x,v)}{2}$$
$$= d(x',v) + \frac{d(x,x'') - d(x'',v) - d(x,v)}{2}$$
$$> d(x',v)$$
$$\geq 0, \tag{26}$$

and by (16) and (20), we have

$$(d(x,x'') + d(x',v)) - \frac{d(x,x') + d(x',v) + d(x,v)}{2}$$
$$\geq d(x,x'') + d(x',v) - (d(x,v) + d(x',v))$$
$$= d(x,x'') - d(x,v)$$
$$> 0. \tag{27}$$

Finally, by (17) and (20), we have

$$(d(x,x'') + d(x',v)) - \frac{d(x',x'') + d(x',v) + d(x'',v)}{2}$$
$$\geq d(x,x'') + d(x',v) - (d(x',v) + d(x'',v))$$
$$= d(x,x'') - d(x'',v)$$
$$> 0. \tag{28}$$

It follows from (21)–(28) that the inequality (3) does not hold for $(x,y,z,w) = (x, x'', x', v)$, and hence, by Theorem 1, d is not a subtree distance. This is a contradiction. Therefore, we must have $x \overset{v}{\sim} x''$.

For a subtree distance $d \colon X \times X \to \mathbb{R}_+$ and $v \in X - V_0$, let us denote by $\mathcal{C}_v$ the equivalence classes induced by the equivalence relation $\overset{v}{\sim}$. We see that the number of the equivalence classes is at least two for each $v \in X - V_0$.

Proposition 3. *Let d be a subtree distance and $\overset{v}{\sim}$ be defined by (15). Then, for each $v \in X - V_0$, we have $|\mathcal{C}_v| \geq 2$.*

Let T be a tree representing the tree metric $d|V_0$. We can assume without loss of generality that V_0 is the leaves of T and that lengths of all edges T are positive except for those incident to a leaf by adding an edge of length zero. For a vertex set W of T, the minimum subtree of T whose vertex set contains W is called the *subtree of T spanned by W* and is denoted by T_W.

For $C \in \mathcal{C}_v$, let z_C be a vertex of T_C which is adjacent to a vertex outside T_C. The existence of such a vertex z_C is guaranteed by Proposition 3. Note that if $|C| > 1$ then, the degree of z_C in T_C is at least two.

To define the subtree T_v of T for each $v \in X - V_0$, we use vertices z_C $(C \in \mathcal{C}_v)$ of T. The following lemma describes properties of such vertices z_C.

Lemma 3. *Let $v \in X - V_0$ and $C \in \mathcal{C}_v$. We have the followings.*

(a) *There exists a unique vertex z_C in T_C which is adjacent to a vertex outside T_C.*

(b) *For each $i \in C$, we have $d(i, v) \geq d_T(i, z_C)$.*

(c) *If there exist two edges outside T_C which are incident to z_C, then we have $d_T(i, z_C) = d(i, v)$ $(i \in C)$.*

(d) *If an edge outside T_C which is incident to z_C is unique, then, letting $\{z_C, z'_C\}$ be such an edge, we have $d_T(i, z_C) \leq d(i, v) \leq d_T(i, z'_C)$ $(i \in C)$.*

Let $v \in X - V_0$. For each $C \in \mathcal{C}_v$, if there exist edges outside T_C incident to z_C we let $w_C = z_C$. Otherwise, divide the unique edge $\{z_C, z'_C\}$ outside T_C incident to z_C into two edges $\{z_C, w_C\}, \{w_C, z'_C\}$ and define the lengths for them so that $d(i, v) = d_T(i, w_C)$ and $l(z_C, z'_C) = l(z_C, w_C) + l(w_C, z'_C)$, where $i \in C$ is arbitrary. Then, define T_v as the subtree of T spanned by $\{w_C | C \in \mathcal{C}_v\}$. Also, for each $y \in V_0$ we define $T_y = (\{y\}, \emptyset)$.

For subtree distance (2) we saw that $V_0 = \{a, b, c, f, h, i\}$ and the tree metric $d|V_0$ is represented by the tree T in Fig. 2. For $e \in X - V_0$, V_0 is partitioned into the equivalence classes $\mathcal{C}_e = \{C, D, E\}$, where $C = \{a, b\}$, $D = \{c\}$ and $E = \{f, h, i\}$. The vertices w_C, w_D and w_E and the subtree T_e spanned by them are depicted in Fig. 3.

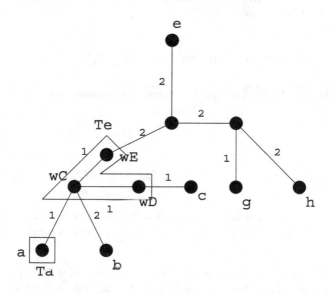

Fig. 3. Vertices w_C, w_D and w_E for $C = \{a, b\}$, $D = \{c\}$ and $E = \{f, h, i\}$ and the subtree T_e spanned by them.

3   The Validity and the Complexity of the Algorithm

We summarize our algorithm for finding a representation of a subtree distance in Algorithm 1. As the following lemma shows, the distances between those subtrees defined by the algorithm constitute the representation of the given subtree distance.

---

1 find V_0 defined by (14);
2 construct a tree representing $d|V_0$;
3 **foreach** $y \in V_0$ **do**
4 $\quad$ let $T_y = (\{y\}, \emptyset)$;
5 **end**
6 **foreach** $v \in X - V_0$ **do**
7 $\quad$ find the partition $\mathcal{C}_v$;
8 $\quad$ find z_C for each $C \in \mathcal{C}_v$;
9 $\quad$ **foreach** $C \in \mathcal{C}_v$ **do**
10 $\quad\quad$ **if** *there exist at least two edges outside T_C incident to z_C* **then**
11 $\quad\quad\quad$ let $w_C = z_C$;
12 $\quad\quad$ **else**
13 $\quad\quad\quad$ divide the unique edge $\{z_C, z'_C\}$ outside T_C incident to z_C into two edges $\{z_C, w_C\}, \{w_C, z'_C\}$ and define the lengths for them so that $d(i, v) = d_T(i, w_C)$ and $l(z_C, z'_C) = l(z_C, w_C) + l(w_C, z'_C)$, where $i \in C$ is arbitrary;
14 $\quad\quad$ **end**
15 $\quad$ **end**
16 $\quad$ let T_v be the subtree of T spanned by $\{w_C | C \in \mathcal{C}_v\}$;
17 **end**

Algorithm 1. Find a representation of a subtree distance.

Lemma 4

(a) *For each $y \in V_0$ and $v \in X - V_0$, we have $d_T(T_v, T_y) = d(v, y)$.*
(b) *For each $v_1, v_2 \in X - V_0$, we have $d_T(T_{v_1}, T_{v_2}) = d(v_1, v_2)$.*

Theorem 2. *Given an arbitrary subtree distance $d : X \times X \to \mathbb{R}_+$, Algorithm 1 finds a weighted tree T and a family of subtrees $\{T_x | x \in X\}$ which represent d in $O(n^3)$ time.*

Proof. The validity of the algorithm is clear from Lemma 4 and the fact that T represents $d|V_0$. We show the running time of the algorithm is $O(n^3)$.

It is clear that V_0 can be find in $O(n^3)$ time. A tree T representing $d|V_0$ can be found in $O(n^3)$ time by Neighbor-Joining [6]. For each $v \in X - V_0$, $\mathcal{C}_v$ can be found in $O(n^2)$ time. It is easy to see that for a $v \in X - V_0$ all z_C $(C \in \mathcal{C}_v)$ can be found in $O(n^2)$ time. For each $C \in \mathcal{C}_v$, w_C can be found in $O(1)$ time. Also, by $|X - V_0| \le n$, for each $v \in X - V_0$, the subtree T_v of T can be computed in $O(n^2)$ time. Therefore, the overall computational time of Algorithm 1 is $O(n^3)$.

Corollary 1. *There exists an algorithm which decides whether or not a given dissimilarity mapping d on X is a subtree distance in $O(n^3)$ time.*

4   Conclusion

The concept of subtree distance is a generalization of that of tree metric and had been introduced by Hirai [4]. So far, no efficient algorithm for finding representation of a subtree distance has been known. In this paper, we showed an $O(n^3)$ time algorithm that finds a representation of a given subtree distance on X, where $n = |X|$. It follows that we have an $O(n^3)$ algorithm for deciding whether a given dissimilarity mapping is a subtree distance or not. A future research direction is to modify the algorithm to have a better time complexity, namely $O(n^2)$.

Acknowledgments. The authors are grateful to Professor Hiroshi Hirai for useful suggestions for the design of the algorithm presented in this paper. Thanks are also due to the anonymous referees for their useful comments which improved the presentation of our results.

References

1. Ando, K.: Computation of the Shapley value of minimum cost spanning tree games: #P-hardness and polynomial cases. Jpn. J. Ind. Appl. Math. **29**, 385–400 (2012)
2. Buneman, P.: A note on metric properties of trees. J. Comb. Theory Ser. B **17**, 48–50 (1974)
3. Culberson, J.C., Rudnicki, P.: A fast algorithm for constructing trees from distance matrices. Inf. Process. Lett. **30**, 215–220 (1989)
4. Hirai, H.: Characterization of the distance between subtrees of a tree by the associated tight span. Ann. Comb. **10**, 111–128 (2006)
5. Hirai, H.: Half-integrality of node-capacitated multiflows and tree-shaped facility locations on trees. Math. Program. Ser. A **137**, 503–530 (2013)
6. Saitou, N., Nei, M.: The neighbor-joining method: a new method for reconstructing phylogenetic trees. Mol. Biol. Evol. **4**, 405–425 (1987)

A Set Covering Approach for the Double Traveling Salesman Problem with Multiple Stacks

Michele Barbato$^{(\boxtimes)}$, Roland Grappe, Mathieu Lacroix, and Roberto Wolfler Calvo

Université Paris 13, Sorbonne Paris Cité, LIPN, CNRS (UMR 7030), 93430 Villetaneuse, France
{barbato,grappe,lacroix,wolfler}@lipn.univ-paris13.fr

Abstract. In the double TSP with multiple stacks, a vehicle with several stacks performs a Hamiltonian circuit to pick up some items and stores them in its stacks. It then delivers every item by performing another Hamiltonian circuit while satisfying the last-in-first-out policy of its stacks. The consistency requirement ensuring that the pickup and delivery circuits can be performed by the vehicle is the major difficulty of the problem. This requirement corresponds, from a polyhedral standpoint, to a set covering polytope. When the vehicle has two stacks this polytope is obtained from the description of a vertex cover polytope. We use these results to develop a branch-and-cut algorithm with inequalities derived from the inequalities of the vertex cover polytope.

Keywords: Double traveling salesman problem with multiple stacks · Polytope · Set cover · Vertex cover · Odd hole

The *traveling salesman problem* (*TSP*) is the problem of finding a Hamiltonian circuit of minimum cost in a complete weighted digraph. The TSP is a well-known NP-hard problem. Nevertheless, one of the greatest advances in combinatorial optimization has been the design of algorithms that made possible to practically solve TSP instances of considerable size [1].

In this paper, we study a generalization of the TSP, namely the *double TSP with multiple stacks* (*DTSPMS*). In this problem, n items have to be picked up in one city, stored in a vehicle having s identical stacks of finite capacity, and delivered to n customers in another city. We assume that the pickup and the delivery cities are far from each other, thus the pickup phase has to be completed before the delivery phase starts. The pickup (resp. delivery) phase consists in a Hamiltonian circuit performed by the vehicle which starts from a depot and visits the n pickup (resp. delivery) locations exactly once before coming back to the depot. Each time a new item is picked up, it is stored on the top of an available stack of the vehicle and no rearrangement of the stacks is allowed. During the delivery circuit the stacks are unloaded by following a last-in-first-out policy: only the items currently on the top of their stack can be delivered. The goal is

© Springer International Publishing Switzerland 2016
R. Cerulli et al. (Eds.): ISCO 2016, LNCS 9849, pp. 260–272, 2016.
DOI: 10.1007/978-3-319-45587-7_23

to find a pickup and a delivery circuit that are *s*-consistent and that minimize the total traveled distance — a pickup and a delivery Hamiltonian circuits are *s-consistent* if a vehicle with *s* stacks can perform both while satisfying the last-in-first-out policy and the capacity of the stacks.

The DTSPMS has been recently introduced in [22] and has received since considerable attention. Several heuristics [6,7,9,22], combinatorial exact methods [15,16] and branch-and-cut algorithms [17,21] have been proposed for its resolution. When the vehicle has two stacks, the best algorithms [2,5] solve to optimality instances with up to 16 items, but mostly fail from 18 items. The main conclusion that can be drawn is that the DTSPMS is extremely hard to solve in practice. We emphasize that, as noted in [2], the finiteness of the capacity is not the major computational difficulty.

An explanation of the fact that exact approaches fail to solve the DTSPMS efficiently is the following. The combinatorial structure behind the consistency of the two circuits has not been deeply addressed. In contrast, the routing part associated with TSP circuits is well understood.

In this paper, we enhance the approach of [2] to overcome this difficulty. More precisely, by focusing on the consistency requirements, we reveal a strong polyhedral connection between the formulation of [2] and set cover problems. This allows us to derive new valid inequalities for the DTSPMS which are embedded into a competitive branch-and-cut algorithm.

The approach of [2] mainly considers the variant of the problem where the stacks have an infinite capacity. The authors develop theoretical results which are implemented in a branch-and-cut framework. A second version of their algorithm is developed with additional features to handle stacks of finite capacity. As we focus on the consistency requirements, we restrict our attention to the problem with stacks of infinite capacity. Indeed, the features of [2] to handle the finite capacities can also be added to our framework. We refer to [2] for more details on these additional features. For a sake of clarity, DTSPMS will now refer to the variant where the stacks have an infinite capacity.

This paper is organized as follows. In Sect. 1 we recall the formulation for the DTSPMS introduced in [2] and the known results about the routing part associated with this formulation. In Sect. 2 we study the set covering polytope that arises from the consistency requirements. In Sect. 3 we consider the case of the DTSPMS with two stacks. We show that in this case the set covering polytope associated with the consistency requirements corresponds to a vertex cover polytope. By using this observation, we derive valid inequalities for the DTSPMS with two stacks. Finally, we test these inequalities in a branch-and-cut algorithm.

1   Formulation of the DTSPMS

In this section, we first describe the DTSPMS in terms of graphs and then present the integer linear formulation for the DTSPMS introduced in [2].

An instance of the DTSPMS with *n* items is given by a complete digraph, two cost vectors defined on its arcs and a positive integer. The complete digraph

$D = (V, A)$ with $V = \{0, \ldots, n\}$ and $A = \{(i, j) : i \neq j \in V\}$ models both cities. The depot is vertex 0. Item i has to be picked up from vertex i of the first city, and delivered to vertex i of the second city. The vectors $c^1 \in \mathbb{R}^{|A|}$ and $c^2 \in \mathbb{R}^{|A|}$ represent the distances between the locations of the pickup and delivery cities, respectively. The positive integer s is the number of stacks of the vehicle. Hence, the DTSPMS consists in finding a pair of s-consistent Hamiltonian circuits C_1 and C_2 whose cost $c^1(C_1) + c^2(C_2)$ is minimum.

Each Hamiltonian circuit of D induces a linear order on $V \setminus \{0\}$ corresponding to the order in which the vertices of $V \setminus \{0\}$ are visited starting from 0. Since the pickup and delivery circuits are Hamiltonian circuits of D, the following proposition characterizes the s-consistency thanks to the linear orders they induce:

Proposition 1 ([4,6,25]). *A pickup circuit and a delivery circuit are s-consistent if and only if no $s + 1$ vertices of $V \setminus \{0\}$ appear in the same order in the linear orders induced by the two circuits.*

Our starting point is the formulation of the DTSPMS of [2], which we now explain. First, the Hamiltonian circuits of D are represented with *arc variables* $x \in \mathbb{R}^{|A|}$ which model the arcs of the Hamiltonian circuits, and *precedence variables* $y \in \mathbb{R}^{n(n-1)}$ which model the associated linear orders. They are described by the following constraints [24]:

$$\sum_{j \in V \setminus \{i\}} x_{ij} = 1 \quad \text{for all } i \in V, \tag{1}$$

$$\sum_{i \in V \setminus \{j\}} x_{ij} = 1 \quad \text{for all } j \in V, \tag{2}$$

$$y_{ij} + y_{ji} = 1 \quad \text{for all distinct } i, j \in V \setminus \{0\}, \tag{3}$$

$$y_{ij} + y_{jk} + y_{ki} \geq 1 \quad \text{for all distinct } i, j, k \in V \setminus \{0\}, \tag{4}$$

$$x_{ij} \leq y_{ij} \quad \text{for all distinct } i, j \in V \setminus \{0\}, \tag{5}$$

$$y_{ij} \in \{0, 1\} \quad \text{for all distinct } i, j \in V \setminus \{0\}, \tag{6}$$

$$x_{ij} \in \{0, 1\} \quad \text{for all distinct } i, j \in V. \tag{7}$$

By the integrality constraints (6) and (7), constraints (1) and (2) ensure that each vertex has exactly one leaving and one entering arc. Inequalities (3) and (4) are the antisymmetry and transitivity constraints respectively, and each binary vector y satisfying them represents a linear order on $V \setminus \{0\}$ [13]. Finally, constraints (5) imply that if the arc (i, j) is in the Hamiltonian circuit then i precedes j in the associated linear order.

Therefore, the DTSPMS can be formulated as follows. Let $(x^1, y^1, x^2, y^2) \in \mathbb{R}^{|A|} \times \mathbb{R}^{n(n-1)} \times \mathbb{R}^{|A|} \times \mathbb{R}^{n(n-1)}$. The variables (x^1, y^1) will correspond to the arc and precedence variables associated with the pickup circuit whereas (x^2, y^2) will refer to the arc and precedence variables associated with the delivery circuit.

The solutions to the DTSPMS are described by the following constraints:[1]

$$(x^t, y^t) \text{ satisfies } (1) - (7) \text{ for } t = 1, 2, \tag{8}$$

$$\sum_{i=1}^{s} (y^1_{v_i v_{i+1}} + y^2_{v_i v_{i+1}}) \geq 1 \text{ for all distinct } v_1, \ldots, v_{s+1} \in V \setminus \{0\}. \tag{9}$$

Inequalities (8) ensure that (x^t, y^t) corresponds to a Hamiltonian circuit for $t = 1, 2$. Inequalities (9) imply that the two Hamiltonian circuits are s-consistent. Indeed, if a constraint (9) is not satisfied, then the vertices $v_{s+1}, \ldots, v_1$ associated with this constraint appear in this order in both the pickup and delivery circuits — a contradiction to Proposition 1. Proposition 1 being an equivalence, the correctness of the above formulation follows.

In the rest of the paper, we will denote by $DTSPMS_{n,s}$ the convex hull of the solutions to (8)–(9). Moreover, $ATSP_n$ will denote the convex hull of the solutions to (1)–(7).

The above formulation makes apparent that the DTSPMS may be separated into two parts: a routing part associated with (8) and a consistency part associated with (9). Every valid inequality for $ATSP_n$ can be used to strengthen the linear relaxation of the DTSPMS. Actually, every facet of $ATSP_n$ gives two facets of $DTSPMS_{n,s}$, as expressed in the following theorem.

Theorem 2 ([2]). *For $n \geq 5$ and $s \geq 2$, if $ax + by \geq c$ defines a facet of $ATSP_n$, then $ax^t + by^t \geq c$ defines a facet of $DTSPMS_{n,s}$, for $t = 1, 2$.*

Theorem 2 characterizes a super-polynomial number of facets of $DTSPMS_{n,s}$ since $ATSP_n$ has a super-polynomial number of facets [10]. Unfortunately, none of these facets relies on the consistency part of the problem. This part has actually not been well studied, and the next section will address this matter.

2 A Set Covering Approach for the s-consistency

As stated in the previous section, there is a one-to-one correspondence between Hamiltonian circuits of D and linear orders on $V \setminus \{0\}$. Thus, the projection onto the precedence variables y^1, y^2 of the solutions to the DTSPMS corresponds to the couples of linear orders on $V \setminus \{0\}$ satisfying (9). When focusing on the consistency part of the problem, we will consider only the consistency constraints (9). In this case, we are interested in the following polytope:

$$SC_{n,s} - conv\{(y^1, y^2) \in \{0, 1\}^{n(n-1)} \times \{0, 1\}^{n(n-1)} : (9) \text{ are satisfied}\}.$$

Clearly, we have $proj_{(y^1, y^2)}(DTSPMS_{n,s}) \subseteq SC_{n,s}$. Moreover, $SC_{n,s}$ is a *set covering polytope*, that is a polytope of the form $conv\{x \in \{0, 1\}^d : Ax \geq 1\}$, with A being a 0,1-matrix. Set covering polytopes have been intensively studied — see for instance [3].

---

[1] In the rest of the paper, the DTSPMS will refer to either the problem and the integer linear formulation depending on the context.

In constraints (9), the coefficients associated with y_{ij}^1 and y_{ij}^2 are the same for all $i \neq j \in V \setminus \{0\}$, and hence $SC_{n,s}$ has a specific form. Indeed, it turns out that all facets of $SC_{n,s}$ can be obtained by studying the following polytope, hereafter called *restricted set covering polytope*:

$$RSC_{n,s} = conv\{y \in \{0,1\}^{n(n-1)}:$$

$$\sum_{i=1}^{s} y_{v_i v_{i+1}} \geq 1 \text{ for all distinct } v_1, \ldots, v_{s+1} \in V \setminus \{0\}\}.$$

Moreover, as shown in the following lemma, the vertices of $RSC_{n,s}$ are connected to the ones of $SC_{n,s}$.

These results are not surprising, yet we did not find them in the literature, thus we provide our own proof. In our proofs we often implicitly use the fact that a binary point of a binary polytope is one of its vertices.

Lemma 3. *For* $y = (y^1, y^2) \in \mathbb{R}^{n(n-1)} \times \mathbb{R}^{n(n-1)}$, *define* $f(y) \in \mathbb{R}^{n(n-1)}$ *by* $f(y)_{ij} = \max\{y_{ij}^1, y_{ij}^2\}$, *for all distinct* $i, j \in \{1, \ldots, n\}$. *Then,*

$$RSC_{n,s} = conv\{f(y) \in \mathbb{R}^{n(n-1)}: y \text{ is a vertex of } SC_{n,s}\}.$$

Proof. Let $P = conv\{f(y) \in \mathbb{R}^{n(n-1)}: y \text{ is a vertex of } SC_{n,s}\}$.

To show $P \subseteq RSC_{n,s}$, let $\bar{v}$ be a vertex of P. By construction, $\bar{v} = f(\bar{y})$ for some vertex $\bar{y}$ of $SC_{n,s}$. Since $\bar{y}$ is binary, so is $\bar{v}$. In addition, $\bar{v}_{j_i j_{i+1}} = 0$ if and only if $\bar{y}_{j_i j_{i+1}}^1 = \bar{y}_{j_i j_{i+1}}^2 = 0$. The vector $\bar{v}$ being binary, $\sum_{i=1}^{s} \bar{v}_{j_i j_{i+1}} < 1$ if and only if $\bar{v}_{j_i j_{i+1}} = 0$ for all $i = 1, \ldots, s$. But this can happen only if $\sum_{i=1}^{s} (\bar{y}_{j_i j_{i+1}}^1 + \bar{y}_{j_i j_{i+1}}^2) = 0$, which is impossible by $\bar{y} \in SC_{n,s}$ and (9). Hence, $\bar{v} \in RSC_{n,s}$. As this holds for every vertex $\bar{v}$ of P, convexity implies $P \subseteq RSC_{n,s}$.

We prove now that $RSC_{n,s} \subseteq P$. Given a vertex $\bar{v}$ of $RSC_{n,s}$, we define $\bar{y} = (\bar{y}^1, \bar{y}^2) \in \{0,1\}^{n(n-1)} \times \{0,1\}^{n(n-1)}$ as follows.

$$\bar{y}_{j_i j_{i+1}}^1 = \bar{y}_{j_i j_{i+1}}^2 = 1 \text{ if } v_{j_i j_{i+1}} = 1,$$

$$\bar{y}_{j_i j_{i+1}}^1 = \bar{y}_{j_i j_{i+1}}^2 = 0 \text{ otherwise.}$$

For distinct $j_1, \ldots, j_{s+1}$, we have $\sum_{i=1}^{s} (\bar{y}_{j_i j_{i+1}}^1 + \bar{y}_{j_i j_{i+1}}^2) = 0$ if and only if $\bar{v}_{j_1 j_2} = \bar{v}_{j_2 j_3} = \cdots = \bar{v}_{j_s j_{s+1}} = 0$. The latter is impossible since $\bar{v} \in RSC_{n,s}$. Hence, since $\bar{y}$ is binary, it satisfies (9). Thus $\bar{y}$ is a vertex of $SC_{n,s}$. By construction, we have $\bar{v} = f(\bar{y})$, therefore $\bar{v}$ is a vertex of P. This holds for every vertex $\bar{v}$ of $RSC_{n,s}$, hence $RSC_{n,s} \subseteq P$ by convexity. □

The next proposition shows how the linear description of $SC_{n,s}$ can be deduced from the one of $RSC_{n,s}$. Inequalities that consist in 0,1 bounds on the variables are called *trivial*.

Proposition 4. *Every non-trivial facet-defining inequality of* $SC_{n,s}$ *is of the form* $ay^1 + ay^2 \geq b$, *where* $ay \geq b$ *is a non-trivial facet-defining inequality of* $RSC_{n,s}$.

Proof. Well-known results about set covering polytopes — see *e.g.*, [19] — immediately imply the following:

(i) $SC_{n,s}$ is full dimensional.
(ii) Inequalities $y_{ij}^t \leq 1$ define facets of $SC_{n,s}$ for all distinct $1 \leq i, j \leq n$ and $t = 1, 2$.
(iii) If $a^1 y^1 + a^2 y^2 \geq b$ is non-trivial and defines a facet of $SC_{n,s}$, then $b > 0$ and $a_{ij}^t \geq 0$ for all distinct $1 \leq i, j \leq n$ and $t = 1, 2$.

We first show that all facets of $RSC_{n,s}$ define facets of $SC_{n,s}$.

Claim. If $ay \geq b$ is a non-trivial facet-defining inequality of $RSC_{n,s}$, then $ay^1 + ay^2 \geq b$ is a facet-defining inequality of $SC_{n,s}$.

Proof. We first prove that $ay^1 + ay^2 \geq b$ is valid for $SC_{n,s}$. Let $\gamma = (\gamma^1, \gamma^2)$ be a vertex of $SC_{n,s}$ and suppose that $a\gamma^1 + a\gamma^2 < b$. By Lemma 3, $f(\gamma)$ is a vertex of $RSC_{n,s}$. From $\gamma \geq 0$, we get $f(\gamma)_{ij} \leq \gamma_{ij}^1 + \gamma_{ij}^2$ for all distinct $1 \leq i, j \leq n$. Since, by (iii), $a_{ij} \geq 0$, we get $af(\gamma) \leq a\gamma^1 + a\gamma^2 < b$, contradicting the validity of $ay \geq b$ for $RSC_{n,s}$.

We now prove that $ay^1 + ay^2 \geq b$ defines a facet of $SC_{n,s}$. Let F' denote the facet of $RSC_{n,s}$ defined by $ay \geq b$ and $\{\xi^1, \ldots, \xi^{n(n-1)}\}$ be an affine base of F'. Since $b > 0$ these vectors are linearly independent. Thus the $2n(n-1)$ vectors $\{(\xi^\ell, 0), (0, \xi^\ell)\}_{\ell=1,\ldots,n(n-1)}$ are linearly independent points of $SC_{n,s}$, satisfying $ay^1 + ay^2 \geq b$ with equality. ∎

We now show that non-trivial facet-defining inequalities of $SC_{n,s}$ have a symmetric structure:

Claim. Let $a^1 y^1 + a^2 y^2 \geq b$ be a non-trivial facet-defining inequality of $SC_{n,s}$. Then $a^1 = a^2$.

Proof. Let us fix $i, j \in \{1, \ldots, n\}$ with $i \neq j$ and let us write for convenience the vectors $\gamma \in \mathbb{R}^{2n(n-1)}$ as $(\bar{\gamma}, \gamma_{ij}^1, \gamma_{ij}^2)$. By contradiction, we suppose that $a_{ij}^1 > a_{ij}^2$. By (iii), we get $a_{ij}^1 > 0$. If $(\bar{\gamma}, 1, 1)$ is a vertex of $SC_{n,s}$, then so are $(\bar{\gamma}, 1, 0)$ and $(\bar{\gamma}, 0, 1)$, since, in each of constraints (9), the coefficients of y_{ij}^1 and y_{ij}^2 are the same.

Let $F = SC_{n,s} \cap \{a^1 y^1 + a^2 y^2 = b\}$ be the facet defined by the given inequality and B a base of F. It is not restrictive to assume B is composed of vertices of $SC_{n,s}$. Then, no element of B has the form $(\bar{\gamma}, 1, 1)$, as otherwise, by $\bar{a}\bar{\gamma} + a_{ij}^1 + a_{ij}^2 = b$ and $a_{ij}^1 > 0$, we would get that $(\bar{\gamma}, 0, 1)$ violates the given inequality. Given that F arises from a non-trivial facet-defining inequality of $SC_{n,s}$, there exists $(\bar{\gamma}, 1, 0) \in B$ as otherwise, $F \subseteq SC_{n,s} \cap \{y_{ij}^1 = 0\}$. This implies that $(\bar{\gamma}, 0, 1)$ violates the facet-defining inequality. We deduce that $a_{ij}^1 \leq a_{ij}^2$. Symmetrically, $a_{ij}^2 \leq a_{ij}^1$ and the desired equality follows. ∎

We finally prove that all the facets of $RSC_{n,s}$ can be obtained from those of $SC_{n,s}$.

Claim. If $ay^1 + ay^2 \geq b$ is a non-trivial facet-defining inequality of $SC_{n,s}$, then $ay \geq b$ is a non-trivial facet-defining inequality of $RSC_{n,s}$.

Proof. The point $(\gamma, \mathbf{0})$ is a vertex of $SC_{n,s}$ whenever γ is a vertex of $RSC_{n,s}$. Thus the validity of $ay \geq b$ for $RSC_{n,s}$ follows from the validity of $ay^1 + ay^2 \geq b$ for $SC_{n,s}$.

Now, let us suppose, by contradiction, that $ay \geq b$ does not define a facet of $RSC_{n,s}$. Then there exists an integer $f \geq 2$ such that $a = \sum_{i=1}^{f} \lambda_i a^i$ and $b = \sum_{i=1}^{f} \lambda_i b^i$, where $\lambda_i > 0$ and $a^i y \geq b^i$ is a facet of $RSC_{n,s}$ for every $0 \leq i \leq f$. Thus, the inequalities $a^i y^1 + a^i y^2 \geq b^i$ are valid for $SC_{n,s}$. However, $(a, a) = \sum_{i=1}^{f} \lambda_i (a^i, a^i)$, contradicting the fact that $ay^1 + ay^2 \geq b$ defines a facet of $SC_{n,s}$.   ■

□

Proposition 4 asserts that the linear description of $RSC_{n,s}$ immediately gives the description of $SC_{n,s}$. Since $proj_{(y^1,y^2)}(DTSPMS_{n,s}) \subseteq SC_{n,s}$, the s-consistency of two Hamiltonian circuits can be modeled by using inequalities which are valid for $RSC_{n,s}$. Our goal is to use such inequalities to better capture the s-consistency in a branch-and-cut algorithm to solve the DTSPMS.

3   Focus on Two Stacks

In this section we first observe that, in the special case of the DTSPMS with two stacks, the restricted set covering polytope is a vertex cover polytope. This result allows us to derive valid inequalities for the DTSPMS. These inequalities are then embedded in a branch-and-cut algorithm, described at the end of the section together with the corresponding experimental results.

3.1   A Vertex Cover Approach

As explained in the previous section, the linear relaxation of our formulation can be strengthened by studying facet-defining inequalities of $RSC_{n,s}$. When considering only two stacks, the polytope $RSC_{n,2}$ is:

$$conv\{y \in \{0,1\}^{n(n-1)} : y_{ij} + y_{jk} \geq 1 \text{ for all distinct } i, j, k \in V \setminus \{0\}\}.$$

As it turns out, $RSC_{n,2}$ can be expressed as a vertex cover polytope. Let $G_n = (U, E)$ be the graph whose vertices are u_{ij} for all distinct $i, j \in V \setminus \{0\}$ and the edges are $\{u_{ij}, u_{jk}\}$ for all distinct $i, j, k \in V \setminus \{0\}$. A *vertex cover* of a graph is a set S of vertices such that each edge contains a vertex of S. The *vertex cover polytope* of a graph is the convex hull of the incidence vectors of its vertex covers.

Please note that $RSC_{n,2}$ and the vertex cover polytope of G_n have the same variables. Moreover, each non-trivial inequality of $RSC_{n,2}$ contains two variables which correspond to the extremities of an edge of G_n. Therefore $RSC_{n,2}$ is nothing but the vertex cover polytope of G_n.

The vertex cover polytope has been intensively studied. Many families of valid inequalities are known. We will more specifically use the so-called odd hole inequalities to derive new valid inequalities for the DTSPMS with two stacks.

Odd Hole Inequalities. An *odd hole* of a graph $G = (W, F)$ is a vertex subset $H = \{v_1, \ldots, v_{2k+1}\}$ such that $\{v_i, v_j\} \in F$ if and only if $|i-j| = 1$ or $|i-j| = 2k$ for all distinct $i, j \in \{1, \ldots, n\}$. The following inequalities are valid for the vertex cover polytope of G [20]:

$$y(H) \geq \frac{|H| + 1}{2} \text{ for all odd holes } H \text{ of } G. \tag{10}$$

Corollary 5. *Inequalities*

$$y^1(H) + y^2(H) \geq \frac{|H| + 1}{2} \text{ for all odd holes } H \text{ of } G_n, \tag{11}$$

are valid for $DTSPMS_{n,2}$.

There is a one-to-one correspondence between the vertices of G_n and the arcs of D. However, if every odd circuit of D provides an odd hole of G_n, the converse is not true. Thus, inequalities (11) generalize the odd circuit inequalities introduced in [2].

3.2   A Branch-and-Cut Algorithm

This section presents a branch-and-cut algorithm for the DTSPMS with two stacks. The reader interested in an exhaustive description of branch-and-cut methods can refer to *e.g.,* [18].

Initialization. The linear program we start with for computing the lower bounds is the one given by inequalities (1)–(3) and (5) and the trivial inequalities. Since the available instances are symmetrical, we add the constraint $y^1_{12} = 1$ to our starting formulation. This trick halves the number of solutions to our problem without affecting the correctness of the algorithm. In addition, we provide our algorithm with the upper bound given by the heuristic algorithm of [7].

Separation. To strengthen the routing part we consider the so-called GDDL inequalities [12] and the 2-simple cut inequalities [11]. The separation phase is as follows. The families of inequalities are separated in this order:

- 2-consistency constraints (9),
- GDDL inequalities,
- 2-simple cut inequalities,
- transitivity constraints (4),[2]
- odd hole inequalities (11).

Constraints (4) and (9) are separated by enumeration. For the 2-simple cut inequalities we use the exact separation algorithms given in [11]. We also use for separating the GDDL inequalities the algorithm of [11] which we restrict to the

---

[2] We use the lifted version $y_{ij} + y_{jk} + y_{ki} - x_{ji} \geq 1$, for all distinct $i, j, k \in V \setminus \{0\}$.

most promising cases to speed it up. Finally we apply the heuristic separation algorithm given in [23] for the odd hole inequalities to the point $\bar{y} = \bar{y}^1 + \bar{y}^2$, where $(\bar{y}^1, \bar{y}^2)$ are the precedence variables of the current solution. The separation of each family is performed when separating the previous ones yielded no violated constraint. Moreover we mention that, since inequalities (4) and (9) are problem-defining, we always separate them on integer current points.

3.3 Experimental Results

The branch-and-cut algorithm described above is a first and preliminary implementation of the vertex cover approach for the DTSPMS with two stacks. The algorithm has been coded in C++ using CPLEX 12.5 [8]. The graph-based routines have been coded with the COIN-OR library LEMON [14]. The algorithm is tested over the benchmark instances introduced in [22], with a CPU time limit of 3 h. Tests are run in a Linux environment, using a 3.4 GHz Intel Core i7 processor, in sequential mode (1 thread).

Since we test our algorithm only for two stacks of unlimited capacity, we present a comparison with the approach given in [2]. However, please recall the conclusion of [2] stating that the capacity of the stacks has little impact on the performance of the algorithm, in terms of CPU time and enumerated nodes of the search tree.[3] Both versions of the algorithm of [2] with finite and infinite capacity for the stacks outperform all other exact approaches.

Table 1 presents the results obtained by the branch-and-cut algorithm described in this paper and those obtained in [2]. Each row of the table corresponds to a tested instance. The first two columns contain the information relative to each instance: its name given in [22] and the number of items it involves. For both algorithms the remainder of the table consists of five columns. Columns UB and LB respectively contain the value of the best integer feasible solution obtained for that instance, and the best lower bound obtained by the algorithm within the 3 h. Columns CPU and Nodes respectively report the time spent (in seconds) and the number of nodes of the branch-and-cut tree. Finally, column Gap reports the gap for each instance, calculated as $100 \cdot (UB-LB)/UB$.

The algorithm proposed in this paper solves all the instances up to 16 items to optimality. Moreover, it solves nine out of the 20 instances with 18 items. For the instances not solved to optimality, the average gap is 1.94 % for 18 items.

Compared with [2], our current algorithm exhibits a better performance. More precisely, it needs respectively 5.9 %, 33.7 % and 7.3 % less time to solve the instances with 14, 16 and 18 items. Moreover, it solves within the time limit one instance more with 18 items with respect to the algorithm of [2]. Finally, we mention that the algorithm presented in this paper solves at optimality two instances with 20 items, within the time limit.

---

[3] Note that the optimal values can differ when passing from the finite capacity case to the infinite capacity case.

Table 1. Computational results of our algorithm and comparison with the results of [2].

Instance	Items	Our B&C					B&C of [2]				
		UB	LB	CPU	Nodes	Gap	UB	LB	CPU	Nodes	Gap
R00	14	766	766.00	147.99	1717	0.00	766	766.00	118,39	1544	0.00
R01	14	761	761.00	22.80	239	0.00	761	761.00	27.97	346	0.00
R02	14	690	690.00	68.65	833	0.00	690	690.00	129.33	1648	0.00
R03	14	791	791.00	28.77	336	0.00	791	791.00	52.13	593	0.00
R04	14	756	756.00	606.73	8305	0.00	756	756.00	509.33	6918	0.00
R05	14	773	773.00	87.35	958	0.00	773	773.00	127.46	1589	0.00
R06	14	811	811.00	16.10	167	0.00	811	811.00	28.71	304	0.00
R07	14	693	693.00	24.13	239	0.00	693	693.00	28.21	319	0.00
R08	14	824	824.00	288.28	3749	0.00	824	824.00	259.09	3573	0.00
R09	14	733	733.00	9.15	67	0.00	733	733.00	5.93	58	0.00
R10	14	733	733.00	95.29	1267	0.00	733	733.00	99.86	1330	0.00
R11	14	719	719.00	362.73	4359	0.00	719	719.00	238.89	2975	0.00
R12	14	803	803.00	86.78	1088	0.00	803	803.00	59.10	722	0.00
R13	14	743	743.00	28.04	319	0.00	743	743.00	36.56	508	0.00
R14	14	747	747.00	193.64	2207	0.00	747	747.00	353.82	4847	0.00
R15	14	765	765.00	29.90	308	0.00	765	765.00	32.47	484	0.00
R16	14	685	685.00	37.69	411	0.00	685	685.00	31.57	376	0.00
R17	14	818	818.00	142.82	1591	0.00	818	818.00	246.35	2992	0.00
R18	14	774	774.00	68.06	920	0.00	774	774.00	94.40	1325	0.00
R19	14	833	833.00	211.86	2472	0.00	833	833.00	237.57	3002	0.00
Average				*127.84*	*1577.60*	*0.00*			*135.86*	*1772.65*	*0.00*
R00	16	795	795.00	1346.11	10356	0.00	795	795.00	1498.13	12002	0.00
R01	16	794	794.00	104.99	686	0.00	794	794.00	169.58	1467	0.00
R02	16	752	752.00	5239.07	40516	0.00	752	752.00	6688.66	51700	0.00
R03	16	855	855.00	2431.51	18037	0.00	855	855.00	1879.71	13641	0.00
R04	16	792	792.00	3350.76	26204	0.00	792	792.00	6616.13	52883	0.00
R05	16	820	820.00	1203.36	9616	0.00	820	820.00	4248.95	32078	0.00
R06	16	900	900.00	813.29	5930	0.00	900	900.00	988.01	8057	0.00
R07	16	756	756.00	87.90	624	0.00	756	756.00	130.26	958	0.00
R08	16	907	907.00	1057.53	9036	0.00	907	907.00	1526.68	12634	0.00
R09	16	796	796.00	67.47	535	0.00	796	796.00	99.46	789	0.00
R10	16	755	755.00	357.42	2791	0.00	755	755.00	664.12	5300	0.00
R11	16	759	759.00	1095.26	8151	0.00	759	759.00	909.18	7377	0.00
R12	16	825	825.00	348.77	2661	0.00	825	825.00	653.00	5264	0.00
R13	16	824	824.00	427.94	3051	0.00	824	824.00	719.47	5878	0.00
R14	16	823	823.00	2764.04	20967	0.00	823	823.00	5892.60	41223	0.00
R15	16	807	807.00	934.73	6731	0.00	807	807.00	568.39	4549	0.00
R16	16	781	781.00	462.52	3850	0.00	781	781.00	2347.62	18234	0.00
R17	16	852	852.00	1584.47	12029	0.00	852	852.00	2136.11	16101	0.00
R18	16	846	846.00	1674.27	13835	0.00	846	846.00	1289.01	10532	0.00
R19	16	882	882.00	1566.98	11750	0.00	882	882.00	1509.37	12501	0.00
Average				*1345.92*	*10367.80*	*0.00*			*2030.75*	*15658.40*	*0.00*
R00	18	839	839.00	3485.49	17926	0.00	839	839.00	5128.95	28232	0.00
R01	18	825	825.00	1101.54	5129	0.00	825	825.00	1574.57	7119	0.00
R02	18	793	759.81	10800.00	47666	4.19	793	750.06	10800.00	46046	5.42
R03	18	896	864.13	10800.00	44448	3.56	896	848.67	10800.00	43700	5.28
R04	18	832	781.29	10800.00	41852	6.09	832	781.50	10800.00	44790	6.07

Table 1. *Continued.*

Instance	Items	Our B&C					B&C of [2]				
		UB	LB	CPU	Nodes	Gap	UB	LB	CPU	Nodes	Gap
R05	18	873	858.42	10800.00	55248	1.67	873	847.60	10800.00	50545	2.91
R06	18	930	930.00	6454.46	33943	0.00	930	930.00	9257.50	44850	0.00
R07	18	805	805.00	1686.81	9072	0.00	805	805.00	1488.97	7918	0.00
R08	18	962	922.29	10800.00	47664	4.13	962	907.68	10800.00	43758	5.65
R09	18	815	815.00	254.36	1354	0.00	815	815.00	448.44	2510	0.00
R10	18	856	820.18	10800.00	47890	4.18	856	825.04	10800.00	44155	3.62
R11	18	813	795.99	10800.00	55568	2.09	823	788.97	10800.00	51234	4.13
R12	18	871	871.00	2650.59	12942	0.00	871	871.00	10800.00	4291.89	21560	0.00
R13	18	845	845.00	3415.79	17689	0.00	845	845.00	3455.85	19047	0.00
R14	18	862	830.62	10800.00	47245	3.64	873	813.67	10800.00	40037	6.80
R15	18	869	840.90	10800.00	48243	3.23	869	834.64	10800.00	47370	3.95
R16	18	811	811.00	3195.68	16843	0.00	811	811.00	5499.46	28197	0.00
R17	18	900	862.50	10800.00	43859	4.17	900	840.50	10800.00	38099	6.61
R18	18	883	867.22	10800.00	50824	1.79	883	867.33	10800.00	47342	1.77
R19	18	909	909.00	7982.98	37904	0.00	909	893.13	10800.00	51974	1.75
Average				*7451.39*	*34165.45*	*1.94*			*8037.28*	*35424.15*	*2.70*

4   Concluding Remarks

In this paper we have considered the DTSPMS. We have focused on the *s*-consistency requirements ensuring that both the pickup and delivery circuits can be performed by a vehicle with *s* stacks satisfying the last-in-first-out policy conditions. We have considered the polytope defined by the consistency constraints and the trivial inequalities. It is a relaxation of the convex hull of the solutions to the DTSPMS but it catches most of the difficulty of the problem and every valid inequality for this polytope can be used to reinforce the DTSPMS. This polytope is a set covering polytope and we have shown that when we have only two stacks, this latter can be reduced to a vertex cover polytope.

We used these results to develop a branch-and-cut algorithm to solve the DTSPMS with two stacks of infinite capacity. This algorithm uses the inequalities derived from the odd hole inequalities which are valid for the vertex cover polytope. This branch-and-cut algorithm is competitive with respect to the existing algorithms for the DTSPMS. We believe that strengthening the formulation using inequalities derived from the vertex cover approach will provide an efficient algorithm to solve instances of a larger size.

Apart from these algorithmic questions, one can wonder whether the relaxation we have considered is far from the convex hull of the solutions to the DTSPMS. A way to answer this question is to determine which facets of the set covering polytope define facets of the convex hull. This is another direction of our future work.

References

1. Applegate, D.L., Bixby, R.E., Chvátal, V., Cook, W.J.: The Traveling Salesman Problem: A Computational Study. Princeton University Press, Princeton (2007)
2. Barbato, M., Grappe, R., Lacroix, M., Wolfler Calvo, R.: Polyhedral results and a branch-and-cut algorithm for the double traveling salesman problem with multiple stacks. Discrete Optim. **21**, 25–41 (2016)
3. Borndörfer, R.: Aspects of set packing, partitioning, and covering. Doctoral dissertation, Ph. D. thesis, Technischen Universität Berlin, Berlin, Germany (1998)
4. Borne, S., Grappe, R., Lacroix, M.: The uncapacitated asymmetric traveling salesman problem with multiple stacks. In: Mahjoub, A.R., Markakis, V., Milis, I., Paschos, V.T. (eds.) ISCO 2012. LNCS, vol. 7422, pp. 105–116. Springer, Heidelberg (2012)
5. Carrabs, F., Cerulli, R., Speranza, M.G.: A branch-and-bound algorithm for the double travelling salesman with two stacks. Networks **61**, 58–75 (2013)
6. Casazza, M., Ceselli, A., Nunkesser, M.: Efficient algorithms for the double travelling problem with multiple stacks. Comput. Oper. Res. **39**, 1044–1053 (2012)
7. Côté, J.-F., Gendreau, M., Potvin, J.-Y.: Large neighborhood search for the pickup and delivery traveling salesman problem with multiple stacks. Networks **60**, 19–30 (2012)
8. IBM ILOG® CPLEX Optimization Studio. http://www-03.ibm.com/software/products/en/ibmilogcpleoptistud (2012)
9. Felipe, A., Ortuno, M.T., Tirado, G.: The double traveling salesman problem with multiple stacks: a variable neighborhood search approach. Comput. Oper. Res. **36**, 2983–2993 (2009)
10. Fiorini, S., Massar, S., Pokutta, S., Tiwary, H.R., De Wolf, R.: Exponential lower bounds for polytopes in combinatorial optimization. J. ACM **62**, 17:1–17:23 (2015)
11. Gouveia, L., Pesneau, P.: On extended formulations for the precedence constrained asymmetric traveling salesman problem. Networks **48**, 77–89 (2006)
12. Gouveia, L., Pires, J.M.: The asymmetric travelling salesman problem and a reformulation of the Miller-Tucker-Zemlin constraints. Eur. J. Oper. Res. **112**, 134–146 (1999)
13. Grötschel, M., Jünger, M., Reinelt, G.: Facets of the linear ordering polytope. Math. Programm. **33**, 43–60 (1985)
14. LEMON-library for efficient modeling and optimization in networks (2010). http://lemon.cs.elte.hu/
15. Lusby, R.M., Larsen, J., Ehrgott, M., Ryan, D.: An exact method for the double TSP with multiple stacks. Int. Trans. Oper. Res. **17**, 637–652 (2010)
16. Lusby, R.M., Larsen, J.: Improved exact method for the double TSP with multiple stacks. Networks **58**, 290–300 (2011)
17. Martínez, A.A.M., Cordeau, J.-F., Dell'Amico, M., Iori, M.: A branch-and-cut algorithm for the double traveling salesman problem with multiple stacks. INFORMS J. Comput. **25**, 41–55 (2011)
18. Mitchell, J.E.: Branch-and-cut algorithms for combinatorial optimization problems. In: Pardalos, P.M., Resende, M.G.C. (eds.) Handbook of Applied Optimization. Oxford University Press, New York (2002)
19. Nobili, P., Sassano, A.: Facets and lifting procedures for the set covering polytope. Math. Programm. **45**, 111–137 (1989)
20. Padberg, M.W.: On the facial structure of set packing polyhedra. Math. Programm. **5**, 199–215 (1973)

21. Petersen, H.L., Archetti, C., Speranza, M.G.: Exact solutions to the double travelling salesman problem with multiple stacks. Networks **56**, 229–243 (2010)
22. Petersen, H.L., Madsen, O.B.G.: The double travelling salesman problem with multiple stacks - formulation and heuristic solution approaches. Eur. J. Oper. Res. **198**, 139–147 (2009)
23. Rebennack, S., Oswald, M., Theis, D.O., Seitz, H., Reinelt, G., Pardalos, P.M.: A Branch and Cut solver for the maximum stable set problem. J. Comb. Optim. **21**, 434–457 (2011)
24. Sarin, S.C., Sherali, H.D., Bhootra, A.: New tighter polynomial length formulations for the asymmetric traveling salesman problem with and without precedence constraints. Oper. Res. Lett. **33**, 62–70 (2005)
25. Toulouse, S., Wolfler Calvo, R.: On the complexity of the multiple stack TSP, kSTSP. In: Chen, J., Cooper, S.B. (eds.) TAMC 2009. LNCS, vol. 5532, pp. 360–369. Springer, Heidelberg (2009)

Shared Multicast Trees in Ad Hoc Wireless Networks

Marika Ivanova[(✉)]

University of Bergen, Bergen, Norway
Marika.Ivanova@uib.no
http://www.uib.no/en/persons/Marika.Ivanova

Abstract. This paper addresses a problem of shared multicast trees (SMT), which extends a recently studied problem of shared broadcast trees (SBT). In SBT, a common optimal tree for a given set of nodes allowing broadcasting from any node to the rest of the group is searched. In SMT, also nodes that neither initiate any transmission, nor act as destinations are considered. Their purpose is exclusively to relay messages between nodes. The optimization criterion is to minimize the energy consumption. The present work introduces this generalization and devises solution methods. We model the problem as an integer linear program (ILP), in order to compute the exact solution. However, the size of instances solvable by ILP is significantly limited. Therefore, we also focus on inexact methods allowing us to process larger instances. We design a fast greedy method and compare its performance with adaptations of algorithms solving related problems. Numerical experiments reveal that the presented greedy method produces trees of lower energy than alternative approaches, and the solutions are close to the optimum.

Keywords: Ad hoc wireless network · Steiner tree · Multicast communication · ILP model · Heuristic algorithm

1 Introduction

The purpose of a multicast communication in a wireless ad-hoc network is to route information from a source to a set of destinations. Given a set of devices and distances between them, the task is to assign power to each device (node), so that the demands of the network are met and the energy consumption is as low as possible, assuming their locations are fixed. Power efficiency is an important aspect in constructing ad-hoc wireless networks since the devices are typically heavily energy-constrained. Individual devices work as transceivers, which means that they are able to both transmit and receive a signal. Moreover, the power level of a device can be dynamically adjusted during a multicast session.

Unlike wired networks, nodes in ad-hoc wireless networks use omnidirectional antennas, and hence a message reaches all nodes within the communication range of the sender. This range is determined by the power assigned to the sender, which is the maximum rather than the sum of the powers necessary to reach

© Springer International Publishing Switzerland 2016
R. Cerulli et al. (Eds.): ISCO 2016, LNCS 9849, pp. 273–284, 2016.
DOI: 10.1007/978-3-319-45587-7_24

all intended receivers. This feature is often referred to as the wireless multicast advantage [18].

The problems of finding power-minimizing trees in wired networks are generalizations of the minimum Steiner tree problem (e.g. [14]). Many wireless network design tasks are NP-hard [7,12]. The following problems are relevant to our work.

Minimum Energy Broadcast (MEB) is the problem of constructing an optimal arborescence for broadcasting from a given source to all remaining nodes. In order to be able to perform a broadcast session from different sources, a separate tree has to be stored for each source. A generalized multicast version assumes that the message is intended for a predefined subset of vertices. Remaining vertices can be used as intermediate nodes forwarding the message to other nodes, and possibly reduce the total cost. Such nodes are referred to as *Steiner nodes*.

Range Assignment Problem (RAP) concerns the problem of assigning transmission powers of minimum sum to the wireless nodes while ensuring network connectivity [1,6]. Unlike MEB, the resulting links formed by the energy assignment are undirected. A generalization of the problem considers the strong connectivity within a nonempty subset of nodes.

Shared Broadcast Tree (SBT). A crucial drawback of MEB is the necessity of storing one tree for each source. The basic idea of SBT [16,19] is to construct a common tree that is source independent and hence simplifies routing, as the relaying node does not need to know the original source in order to adjust its corresponding power level. Instead, the power level depends merely on the immediate neighbour from which the message was received. This idea is based on the observation that a signal that is being forwarded by a node does not have to reach the neighbour from which it originally came. So, if a signal comes from the most distant neighbour, the relaying power must correspond to the second most distant neighbour. When, on the other hand, a message comes from one of the closer neighbours, it has to be forwarded with the power necessary to reach the most distant one. With this conception, we get two power levels, and their selection involves only a single binary decision making.

This work introduces the shared multicast tree (SMT) problem, a generalization of SBT. Analogously to the multicast versions of MEB and RAP, in SMT we assume that there are two types of nodes, called destinations and non-destinations, respectively. Destinations can initiate a transmission, and must receive every transmission initiated by other destinations. Non-destinations can relay a message, but do not initiate any transmission. Neither do they have to receive any transmission. Passing messages via non-destinations is thus optional, and is chosen only if it saves energy, which is the main motivation for SMT. The goal is to find a common source-independent tree that connects the destinations while minimizing the power.

The decentralized nature of wireless ad-hoc networks implies its suitability for applications, where it is not possible to rely on central nodes, or where network infrastructure does not exist. This is typical for various short-term events like conferences or fixtures. Simple maintenance makes them useful in applications such as emergency situations, disaster relief, military command and control, and most recently, in the mobile commerce sector.

2   Related Work

MEB was introduced in [18], where the authors considered three heuristic algorithms of which most cited is Broadcast Incremental Power (BIP), a greedy $\mathcal{O}(N^2 \log N)$ approximation algorithm. To our best knowledge, the most recent results for lower and upper bounds on the approximation ratio are 4.6 [3] and 6 [2], respectively. Much is written about refinements of fast sub-optimal methods, for instance [10,11,13,17]. In [6], the authors study RAP and compare cases when the resulting graph is required to be strongly and weakly connected. Several heuristic approaches are proposed (e.g. in [4,5]). Many works are also dedicated to mathematical programming techniques. Various ILP models for both MET and RAP are presented in [8,9,12,15]. A special case where the transmission ranges of the stations ensure the communication between any pair of stations in at most h hops is investigated in [7].

The first work concerning SBT is [16], where the idea of a single source-independent tree embedding N broadcast trees for different sources is introduced. The authors show that using the same broadcast tree does not result in widely varying total powers for different sources. Another contribution of [16] is a polynomial-time approximation algorithm to construct a single broadcast tree, including an analysis of its performance. In [19], the authors present an ILP formulation and apply a dual decomposition method. This approach enables solving larger instances than an explicit formulation can solve, and with less than 3 % performance gap to global optimality.

3   Notation and Assumptions

Let $H = (V_H, E_H)$ be an undirected graph and $u \in V_H$. The degree of u in H is denoted by $\deg_H(u)$. The input and output degree of $v \in V_K$ in a directed graph $K = (V_K, A_K)$ is denoted by $\deg_K^-(v)$ and $\deg_K^+(v)$, respectively. Let $H' = (V_{H'}, E_{H'})$ be a subgraph of H. Then, $H \setminus H'$ denotes the graph induced by the node set $V_H \setminus V_{H'}$.

A wireless network is modelled as a complete graph $G = (V, E)$, where V corresponds to the network nodes, and the edges E correspond to potential direct links between them, i.e. $\forall i, j \in V, i \neq j : \{i, j\} \in E$. The set $A = \{(i, j) : i, j \in V, i \neq j\}$ contains all arcs derived from E. Next, $D \subseteq V$ is a nonempty set of destinations with $N = |V|$ and $M = |D|$.

Let $\mathbf{z} \in \{0, 1\}^E$ be a vector with components corresponding to edges in E. The undirected graph induced by $\mathbf{z}$ is defined as $G_{\mathbf{z}} = (V, E_{\mathbf{z}})$, where $\{i, j\} \in E_{\mathbf{z}} \Leftrightarrow z_{ij} = 1$. The directed graph induced by $\mathbf{x} \in \{0, 1\}^A$ is defined analogously.

For $i, j \in V$, the power requirement for transmission from i to j is denoted by p_{ij}, and depends on the distance d_{ij} between i and j and environmental properties. More precisely, $p_{ij} = \kappa d_{ij}^{\alpha}$, where α is an environment-dependent parameter (typically valued between 2 and 4) and κ is a constant. In this work, the power requirements p_{ij} are referred to as the *arc costs*. It follows from $d_{ij} = d_{ji}$ that the power requirements are symmetric.

If $\{i, j\}$ is an edge in a tree $T = (V_T, E_T)$, where $V_T \subseteq V$, $E_T \subseteq E$, we use $T_{i/j}$ to denote the subtree of T consisting of all vertices k such that the path from k to j visits i, as introduced in [19]. Additionally, we define a function $\text{nod}(T_{i/j})$ that returns the number of destinations in $T_{i/j}$.

Neighbours of i in T are denoted i_1^T, i_2^T, i_3^T, ... in non-increasing order of distance from i. If there is no risk of confusion, we simply omit the superscript T. The highest and second highest power levels of i are defined by its neighbours i_1 and i_2, respectively. If i is a leaf, we set $p_{ii_2} = 0$. The contribution $c_T(i)$ of i in T to the total cost depends on i's power levels:

$$c_T(i) = \text{nod}(T_{i_1/i})p_{ii_2} + \text{nod}(T \setminus T_{i_1/i})p_{ii_1}. \tag{1}$$

The total cost $P(T)$ of the tree T is then determined as

$$P(T) = \sum_{i \in V} c_T(i). \tag{2}$$

Problem 1. (SMT): Find a tree T in G minimizing $P(T)$ such that T spans D.

The most costly two incident edges of each node contribute to the objective function value. This reflects the nature of SBT/SMT, when the power level of a node is determined by the most costly link along which a message has to be forwarded. The power requirement of the link is multiplied by the number of senders whose transmissions are relayed through this link, which captures how often the link is used.

4   Discrete Optimization Model

We present an integer programming model for SMT, extending the model in [19] by non-destinations. In this setting we consider a set of destinations $D \subseteq V$ where a broadcast session takes place. Variables are defined as follows:

$$z_{ij} = \begin{cases} 1 & \text{if edge } \{i, j\} \in E \text{ is in } T, \\ 0 & \text{otherwise,} \end{cases}$$

$$x_{ij}^s = \begin{cases} 1 & \text{if arc } (i, j) \in A \text{ is used to transmit messages from } s \in D, \\ 0 & \text{otherwise,} \end{cases}$$

$$y_{ij}^s = \begin{cases} 1 & \text{if node } i \in V \text{ uses power } p_{ij} \text{ to transmit messages from } s \in D, \\ 0 & \text{otherwise.} \end{cases}$$

Let $\mathbf{x}^s \in \{0, 1\}^A$ denote the vector consisting of variables x_{ij}^s for all $(i, j) \in A$. The ILP model is presented below. The x-variables induce $|D|$ directed trees, that are encapsulated into a single spanning tree induced by the z-variables. The power levels are determined by the y-variables.

$$\min \sum_{(i,j)\in A} \sum_{s\in D} p_{ij} y_{ij}^s \tag{3a}$$

s.t.

$$\sum_{\{i,j\}\in E} z_{ij} \le N - 1 \tag{3b}$$

$$\sum_{i\in V\setminus\{j\}} x_{ij}^s = 1 \qquad\qquad j, s \in D, j \ne s, \tag{3c}$$

$$x_{jk}^s \le \sum_{i\in V\setminus\{j\}} x_{ij}^s \le 1 \qquad j \in V \setminus D, s \in D, k \in V \setminus \{j\}, \tag{3d}$$

$$\sum_{i\in V\setminus\{j\}} x_{ij}^s \le \sum_{k\in V\setminus\{j\}} x_{jk}^s \qquad s \in D, j \in V \setminus D, \tag{3e}$$

$$x_{ij}^s + x_{ji}^s = z_{ij} \qquad\qquad \{i,j\} \in E, s \in D, \tag{3f}$$

$$x_{ij}^j = 0 \qquad\qquad j \in D, i \in V \setminus \{j\}, \tag{3g}$$

$$x_{ij}^s \le \sum_{k\in V: p_{ik}\ge p_{ij}} y_{ik}^s \qquad s \in D, (i,j) \in A, \tag{3h}$$

$$\mathbf{z} \in \{0,1\}^E, \qquad \mathbf{x}, \mathbf{y} \in \{0,1\}^{A\times D}. \tag{3i}$$

By constraint (3b), we express the upper bound on the number of edges in the Steiner tree. There is also a lower bound $M - 1$ on the size of the spanning tree, but addition of this constraint would neither reduce the space of feasible solutions nor increase the strength of the model. If the tree does not contain any Steiner nodes, its size is the lower bound, while if all nodes are used (either as Steiner nodes, or $D = V$), its size equals the upper bound. By (3c), we ensure that a message from source s reaches a destination j from exactly one neighbour $i \in V$. Next, constraint (3d) covers the case when $j \in V \setminus D$: If a non-destination j forwards a message from s towards k, the message must come from exactly one neighbour i. Note that assuming there is no outgoing arc from a non-destination j, constraint (3d) does not prevent j from being a leaf in $G_{\mathbf{x}^s}$. We make such undesired solutions impossible by adding constraint (3e), which reduces the set of feasible solutions. However, (3e) is not necessary, because a solution, where a non-destination that does not relay any message is assigned a non-zero power, would be filtered out by the minimization procedure. The expression (3f) says that for any edge $\{i, j\}$ in $G_{\mathbf{z}}$ a message from s is transferred via either arc (i, j) or arc (j, i). The next constraint (3g) expresses that a transmission initiated by $s \in D$ cannot reach s again, which implies non-existence of a directed cycle containing s. Finally, by (3h), we define a relation between x-variables and y-variables. When arc (i, j) is used for transmission of a message from $s \in D$, vertex i relaying the message must be assigned power at least p_{ij}. The remainder of this section justifies that the model is a correct formulation of SMT. Proofs of all claims can be found in Appendix A.

Lemma 1. *Let* $(\mathbf{x}, \mathbf{z})$ *satisfy (3b)–(3i). A replacement of all directed arcs in* $G_{\mathbf{x}^s}$ *by undirected ones yields graph* $G_{\mathbf{z}}$, *for all* $s \in D$.

Lemma 2. *Let* $(\mathbf{x}, \mathbf{z})$ *satisfy (3b)–(3i) and* $s \in D$. *All arcs in a path* $(s = u_1, u_2, \ldots, u_k)$ *in digraph* $G_{\mathbf{x}^s}$ *are directed from* s *towards* u_k.

Proposition 1. *Let* $(\mathbf{x}, \mathbf{z})$ *satisfy constraints (3b)–(3i). If* Q *is a connected component in* $G_{\mathbf{z}}$ *such that* $V_Q \cap D \neq \emptyset$, *then,* Q *does not contain any cycle.*

Proposition 2. *If* $(\mathbf{x}, \mathbf{z})$ *satisfies constraints (3b)–(3i), then there exists a path in* $G_{\mathbf{z}}$ *between any two destinations* $s, t \in D$.

The optimal solution to (3a)–(3i) is a graph $G_{\mathbf{z}}$ with one connected component containing all destinations. Non-destinations outside of this component are isolated vertices, as any potential links between them would be eliminated by the optimization.

Proposition 3. *If* $(\mathbf{x}, \mathbf{y}, \mathbf{z})$ *is an optimal solution to (3a)–(3i), then one of the connected components of* $G_{\mathbf{z}}$ *is an optimal tree in Problem 1, and* $\sum_{(i,j) \in A} \sum_{s \in D} p_{ij} y_{ij}^s = P(G_{\mathbf{z}})$.

5   Inexact Methods

Solving the ILP model presented in the previous section provides the optimal power assignment, but the computation in large instances takes prohibitively long time. Hence, we now focus on algorithms with better trade-off between optimality and runtime. Any tree T in G spanning D is a feasible solution to *SMT*. We study the following methods:

1. Construction by MST (minimum-weight spanning tree), where all vertices are considered as destinations, and a MST is constructed over the set V.
2. Construction by BIP, where we regard the set of vertices as an instance of MEB, and apply the BIP algorithm over V.
3. Greedy Anticipating SMT algorithm described in the following Sect. 5.1.

The global impact of a local change suggests that the first two algorithms are rather myopic for our purposes. Unlike MST and MEB, the nature of SBT/SMT implies that an addition of an edge does not cause only a local change of power levels. Every time a new edge is appended, all nodes already included in the tree increase their contributions to the resulting cost, because the addition of an edge also increases the size of corresponding subtrees of every interior node. Therefore, the new cost cannot be calculated in constant time.

In general, the algorithms work in two phases. The first phase, construction, creates a spanning tree according to a certain strategy. Further improvements can be achieved in the second phase (refinement) which can be applied regardless of what construction method is used.

5.1   Greedy SMT Approach

We present in Algorithm 1 GASMT, a greedy algorithm aimed to construct a Steiner tree $T = (V_T, E_T)$, with low SMT cost. The algorithm starts with T containing only a pre-defined root, and iteratively expands T by an edge until all destinations are present in V_T. The selection of the new edge is based on an anticipation of the entire resulting tree.

Algorithm 1. Greedy Anticipating SMT (GASMT)

Input: Complete graph $G = (V, E)$, root $r \in V$, destinations $D \subseteq V$
Output: Steiner tree $T = (V_T, E_T)$, $D \subseteq V_T \subseteq V$, $E_T \subseteq E$
1: **procedure** BUILDTREE(G)
2:     $T \leftarrow (\{r\}, \emptyset)$
3:     **while** $D \not\subseteq V_T$ **do**
4:         $bestCost \leftarrow \infty$
5:         $Cand \leftarrow \{\{i, j\} : i \in V_T, j = \arg\min\{d_{ik} : k \in V \setminus V_T\}\}$
6:         **for each** $\{i, j\} \in Cand$ **do**
7:             $T' \leftarrow$ ANTICIPATETREE(T, i, j)
8:             **if** $P(T') < bestCost$ **then**
9:                 $bestCost \leftarrow P(T')$
10:                 $\{i^*, j^*\} \leftarrow \{i, j\}$
11:         $T \leftarrow (V_T \cup \{j^*\}, E_T \cup \{\{i^*, j^*\}\})$
        return T
12:
13: **procedure** ANTICIPATETREE(T, i, j)
14:     $T' \leftarrow (V_T \cup \{j\}, E_T \cup \{\{i, j\}\})$
15:     $Disconnected \leftarrow V \setminus V_{T'}$
16:     **for each** $v \in Disconnected \cap D$ **do**
17:         $u \leftarrow \arg\min\{d_{kv} : k \in V_{T'}\}$
18:         $T' \leftarrow (V_{T'} \cup \{v\}, E_{T'} \cup \{\{u, v\}\})$
        return T'

Before a new edge is appended, we determine a set $Cand$ of potential edges that can be selected: for every $u \in V_T$, we remember a potential edge linking u to the closest $v \in V \setminus V_T$ (line 5 in Algorithm 1). For each candidate edge $\{i, j\}$, we build an anticipated tree spanning D. The edge $\{i, j\}$ is temporarily appended to T, which produces tree T'. Subsequently, all destinations that are not yet included in T' are connected one by one to the growing anticipated T' using the shortest possible edges. The candidate link resulting in the cheapest anticipated tree is then selected and added permanently to T. The purpose of the anticipation procedure is to predict the sizes of individual subtrees in the final tree. This allows a more realistic estimation of the resulting objective value, in contrast to the construction by MST and BIP. Non-destinations are disregarded in the anticipation procedure, because they do not alter the subtree sizes.

5.2   Refinement

Any construction algorithm can be followed by an additional phase refining the existing tree T. In particular, this phase handles non-destinations, and replaces expensive transmissions by cheaper ones.

Although the use of non-destinations may reduce the cost, the construction phase does not guarantee that all non-destinations do. Thus, cost reductions can be achieved by removing non-destinations that actually deteriorate the tree. How a non-destination v is processed depends on its degree:

- $\deg(v) = 1$: Non-destination leaves can immediately be deleted recursively.
- $\deg(v) = 2$: Let $(v_1, \ldots, v_m)$ be a maximal path in T such that $\deg(v_1) = \cdots = \deg(v_m) = 2$ and $v_1, \ldots, v_m \in V \setminus D$, and consider the two connected components T_1 and T_2 arising when the path is deleted from T. If there exists an edge $e \in E$ connecting T_1 and T_2 such that $P(T') < P(T)$, where $T' = (V_{T_1} \cup V_{T_2}, E_{T_1} \cup E_{T_2} \cup \{e\})$, the path is replaced by the best choice of e. If no such edge exists, T.
- $\deg(v) \geq 3$: Let $E(v)$ be the set of edges incident to v in T and let $T' = (V_T \setminus \{v\}, (E_T \setminus E(v)) \cup E_{\mathrm{MST}})$, where E_{MST} is the set of edges of a MST constructed over the set of v's neighbours in T. If $P(T') < P(T)$, the current tree is updated to T'.

The cost of the tree can be further improved by eliminating unnecessary transmissions by means of so called "sweep" operations [18]. After removal of an edge e, the vertices are partitioned into a cut. We then select and include the edge across the cut leading to the cheapest tree, possibly e itself. This procedure can be done for all edges, or only for selected ones - for example it makes sense to test only edges longer than a certain threshold.

6   Experimental Evaluation

We have implemented the ILP model as well as the inexact algorithms and compared numerically their performance in terms objective value and runtime.

The input parameters of the procedure generating individual instances are the number of all vertices and the number of destinations. It generates instances with the intended number of destinations and non-destinations with random coordinates uniformly distributed on a square. Finally, the power requirements are determined using $p_{ij} = \kappa d_{ij}^{\alpha}$ with $\kappa = 1$ and $\alpha = 2$. All experiments were run on an Intel Core 2 Quad CPU at 2.83 GHz and 7 GB RAM.

6.1   ILP Model

The integer programming formulation was implemented in AMPL and submitted to solver CPLEX [20] which computed optimal solutions as well as LP relaxations of the generated instances. The running time of determining the optimal solution for instances containing more than 22 vertices becomes excessively long, and so

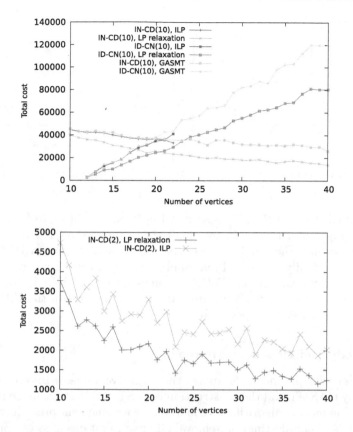

Fig. 1. Dependence of the total cost on the number of all vertices

we computed only the corresponding LP relaxations, which gives lower bounds on the objective function value.

Two instance settings were tested. In the first setting, we kept the number of non-destinations $|V \setminus D|$ constant, while increasing the number of destinations $|D|$. The abbreviation ID-CN is used to refer to this series of experiments, followed by $|V \setminus D|$ whenever it needs to be specified. In the second setting, $|D|$ was constant while $|V \setminus D|$ was increasing (IN-CD).

The first series of experiments concerns the change of the objectve function value with growing instance size. It is obvious from the graphs in Fig. 1, that in ID-CN, increasing the number of vertices also increases the total SMT cost. On the other hand, in IN-CD, the total cost decreases. This decline gradually mitigates, and it can be assumed that the average cost converges to a constant value. By way of contrast, the first graph in Fig. 1 also contains the costs obtained by the inexact algorithm. The difference between the optimum and the result of GASMT is almost negligible.

The second series of experiments shows how fast the CPU time grows with increasing number of nodes. It is apparent that the time used by the solver grows

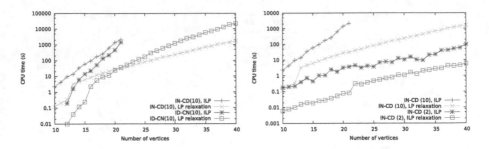

Fig. 2. Dependence of the solution time on the total number of vertices

faster in ID-CN than in IN-CD, as seen in Fig. 2. Nevertheless, in both settings, the time grows exponentially. In the smallest instances, the CPU time is longer for IN-CD, because the number of destinations is higher than in ID-CN. This difference is gradually reduced. From approximately 20 vertices on, the solving time of the LP-relaxations in ID-CN becomes longer. In IN-CD(10) and ID-CN(10), every added destination causes an average increase in the ILP solution time by 89 % and 208 %, respectively.

6.2 Greedy and Heuristic Approach

The next set of experiments compares the objective value of inexact solutions produced by GASMT and the construction by MST and BIP discussed in Sect. 5. Each run of an inexact algorithm was followed by a refinement procedure, namely two iterations of non-destination removal and two iterations of sweep operations for every edge. The graphs in Fig. 3 show the results of two experimental settings, IN-CD(10) (left) and ID-CN(10) (right). Each column in the graphs corresponds to the average SMT cost calculated for 100 instances with fixed $|V|/|D|$ ratio.

It can be seen that the cost of the solutions produced by the construction by MST and BIP are similar, but GASMT is always perceptibly better. Nevertheless, worse time complexity of GASMT becomes apparent while processing

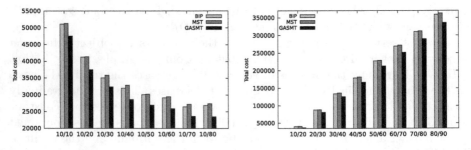

Fig. 3. Comparison of the greedy methods. The fractions on the x-axis determine the corresponding $|D|/|V|$ values.

instances of around 100 nodes, when the time spent on one instance is approximately units of minutes. On the contrary, the other two methods return the solution almost immediately.

7   Conclusion and Future Work

We have introduced a multicast version of SBT, a natural generalization of the problem. We have proposed a discrete optimization model together with the proof of its correctness. Due to the limited size of instances solvable in a practical time, heuristic and greedy approaches are also developed.

Moreover, we have conducted several numerical experiments using the CPLEX solver. It turned out that the presented ILP model can be used for solving instances with up to around 20 vertices. An increasing number of destinations causes a much faster growth of the solution time than an increasing number of non-destinations. In addition, these experiments give us insight into the cost reduction as a function of increasing number of non-destinations.

Further experiments involved the inexact methods. The GASMT algorithm presented in this work gives better results than the construction by MST and BIP applied on SMT. Moreover, the solutions provided by GASMT are close to the optimal ones determined by solving the ILP model.

A subject of continued research is a detailed theoretical study of the inexact methods. There are several interesting questions regarding this topic, like whether any inexact method is an approximation algorithm for SBT/SMT, or whether any method performs consistently better than others. There is also a substantial room for further improvements of the ILP model so that it can be applicable for larger instances. In particular, methods like strong valid inequalities, lazy constraints and user cuts could serve for this purpose.

References

1. Althaus, E., Calinescu, G., Mandoiu, I.I., Prasad, S., Tchervenski, N., Zelikovsky, A.: Power efficient range assignment in ad-hoc wireless networks. Wirel. Commun. Networking **3**, 1889–1894 (2003)
2. Ambühl, C.: An optimal bound for the MST algorithm to compute energy efficient broadcast trees in wireless networks. In: Caires, L., Italiano, G.F., Monteiro, L., Palamidessi, C., Yung, M. (eds.) ICALP 2005. LNCS, vol. 3580, pp. 1139–1150. Springer, Heidelberg (2005)
3. Bauer, J., Haugland, D., Yuan, D.: New results on the time complexity and approximation ratio of the broadcast incremental power algorithm. Inf. Process. Lett. **109**(12), 615–619 (2009)
4. Bein, D., Zheng, S.Q.: Energy efficient all-to-all broadcast in all-wireless networks. Inf. Sci. **180**(10), 1781–1792 (2010)
5. Bhukya, W.N., Singh, A.: An effective heuristic for construction of all-to-all minimum power broadcast trees in wireless networks. In: Advances in Computing, Communications and Informatics, pp. 74–79 (2014)

6. Blough, D.M., Leoncini, M., Resta, G., Santi, P.: On the symmetric range assignment problem in wireless ad hoc networks. In: Proceedings of the IFIP 17th World Computer Congress - TC1 Stream / 2nd IFIP International Conference on Theoretical Computer Science: Foundations of Information Technology in the Era of Networking and Mobile Computing, pp. 71–82 (2002)

7. Clementi, E.F., Penna, P., Silvestri, R.: On the power assignment problem in radio networks. Mob. Netw. Appl. Discrete Algorithms Methods Mob. Comput. Commun. **9**(2), 125–140 (2004)

8. Das, A.K., Marks, R.J., El-sharkawi, M.: Minimum power broadcast trees for wireless networks: integer programming formulations. In: The 22nd Annual Joint Conference of the IEEE Computer and Communications Societies, pp. 245–248 (2003)

9. Das, A.K., Marks II, R.J., El-Sharkawi, M., Arabshahi, P., Gray, A.: Optimization methods for minimum power bidirectional topology construction in wireless networks with sectored antennas. In: Global Telecommunications Conference, vol. 6, pp. 3962–3968 (2004)

10. Das, A.K., Marks, R.J., El-sharkawi, M., Arabshahi, P., Gray, A.: r-Shrink: a heuristic for improving minimum power broadcast trees in wireless networks. In: Proceedings of the IEEE GLOBECOM 2003, vol. 1, pp. 523–527 (2003)

11. Das, A.K., Marks, R.J., El-sharkawi, M., Arabshahi, P., Gray, A.: e-Merge: A heuristic for improving minimum power broadcast trees in wireless networks, Technical report. University of Washington, Department of Electrical Engineering (2003)

12. Haugland, D., Yuan, D.: Wireless Network Design: Optimization Models and Solution Procedures. International Series in Operations Research & Management Science, vol. 158. Springer, New York (2011). pp. 219–246

13. Liang, W.: Constructing minimum-energy broadcast trees in wireless ad hoc networks. In: Proceedings of the 3rd ACM International Symposium on Mobile Ad Hoc Networking & Computing, pp. 112–122 (2002)

14. Hwang, F.K., Richards, D.S., Winter, P.: The Steiner Tree Problem. Annals of Discrete Mathematics. Elsevier Science, Netherlands (1992)

15. Montemanni, R., Gambardella, L.M.: Exact algorithms for the minimum power symmetric connectivity problem in wireless networks. Comput. Oper. Res. **32**(11), 2891–2904 (2005)

16. Papadimitriou, I., Georgiadis, L.: Minimum-energy broadcasting in multi-hop wireless networks using a single broadcast tree. Mob. Networks Appl. **11**(3), 361–375 (2006)

17. Wan, P., Clinescu, G., Yi, C.: Minimum-power multicast routing in static ad hocwireless networks. IEEE/ACM Trans. Networking (TON) **12**(3), 507–514 (2004)

18. Wieselthier, J.E., Nguyen, G.D., Ephremides, A.: On the construction of energy-efficient broadcast and multicast trees in wireless networks. In: Proceedings of the Nineteenth Annual Joint Conference of the IEEE Computer and Communications Societies, vol. 2, pp. 585–594 (2000)

19. Yuan, D., Haugland, D.: Dual decomposition for computational optimization of minimum-power shared broadcast tree in wireless networks. IEEE Trans. Mob. Comput. **12**(11), 2008–2019 (2012)

20. IBM ILOG CPLEX V12.1 User's Manual for CPLEX (2009)

Two-Level Polytopes with a Prescribed Facet

Samuel Fiorini, Vissarion Fisikopoulos, and Marco Macchia$^{(\boxtimes)}$

Département de Mathématique, Université libre de Bruxelles, ULB CP 216,
Boulevard du Triomphe, 1050 Brussels, Belgium
{sfiorini,vfisikop,mmacchia}@ulb.ac.be

Abstract. A (convex) polytope is said to be *2-level* if for every facet-defining direction of hyperplanes, its vertices can be covered with two hyperplanes of that direction. These polytopes are motivated by questions, e.g., in combinatorial optimization and communication complexity. We study 2-level polytopes with one prescribed facet. Based on new general findings about the structure of 2-level polytopes, we give a complete characterization of the 2-level polytopes with some facet isomorphic to a sequentially Hanner polytope, and improve the enumeration algorithm of Bohn *et al.* (ESA 2015). We obtain, for the first time, the complete list of d-dimensional 2-level polytopes up to affine equivalence for dimension $d = 7$. As it turns out, geometric constructions that we call suspensions play a prominent role in both our theoretical and experimental results. This yields exciting new research questions on 2-level polytopes, which we state in the paper.

1 Introduction

We start by giving a formal definition of 2-level polytopes and reasons why we find that they are interesting objects.

Definition 1 (2-level polytope). *A polytope P is said to be 2-level if each hyperplane Π defining a facet of P has a parallel Π' such that every vertex of P is on either Π or Π'.*

Motivation. First of all, many famous polytopes are 2-level. To name a few, stable set polytopes of perfect graphs [3], *twisted* prisms over those — also known as Hansen polytopes [15] — Birkhoff polytopes, and order polytopes [21] are all 2-level polytopes. Of particular interest in this paper are a family of polytopes interpolating between the cube and the cross-polytope.

Definition 2 (sequentially Hanner polytope). *We call a polytope $H \subseteq \mathbb{R}^d$ a* sequentially Hanner polytope *if either $H = H' \times [-1,1]$ or $H = \mathrm{conv}(H' \times \{0\} \cup \{-e_d, e_d\})$, where H' is a sequentially Hanner polytope in $\mathbb{R}^{d-1}$ in case $d > 1$, or $H = [-1,1]$ in case $d = 1$. We call H' the* base *of H.*

We acknowledge support from ERC grant *FOREFRONT* (grant agreement no. 615640) funded by the European Research Council under the EU's 7th Framework Programme (FP7/2007-2013).

© Springer International Publishing Switzerland 2016
R. Cerulli et al. (Eds.): ISCO 2016, LNCS 9849, pp. 285–296, 2016.
DOI: 10.1007/978-3-319-45587-7_25

The name comes from the fact that these polytopes are Hanner polytopes [14] but not all Hanner polytopes are sequentially Hanner. Hanner polytopes are related to some famous conjectures such as Kalai's 3^d conjecture [16] and the Mahler conjecture [19].

More motivation for the study of 2-level polytopes comes from combinatorial optimization, since 2-level polytopes have minimum positive semidefinite extension complexity [9]. Moreover, a finite point set has theta rank 1 if and only if it is the vertex set of a 2-level polytope. This result was proved in [7], and answered a question of Lovász [17]. We already mentioned the stable set polytopes of perfect graphs as prominent examples of 2-level polytopes. To our knowledge, the fact that these polytopes have small positive semidefinite extended formulations is the only known reason why one can efficiently find a maximum stable set in a perfect graph [13]. Moreover, 2-level polytopes are also related to nice classes of matrices such as totally unimodular or balanced matrices that are central in integer programming, see e.g. [20].

Finally, 2-level polytopes are also of interest in communication complexity since they provide interesting instances to test the *log-rank conjecture* [18], one of the fundamental open problems in that area. Indeed, every d-dimensional 2-level polytope has a *slack matrix* that is a 0/1-matrix of rank $d + 1$. If the log-rank conjecture were true, the communication problem associated to any such matrix should admit a deterministic protocol of complexity polylog(d), which is open. Returning to combinatorial optimization, the log-rank conjecture for slack matrices of 2-level polytopes is (morally) equivalent to the statement that every 2-level polytope has an extended formulation with only $2^{\text{polylog}(d)}$ inequalities.

Goal. Despite the motivation described above, we are far from understanding the structure of general 2-level polytopes. This paper offers results in this direction. The recurring theme here is how much local information about the geometry of a given 2-level polytope determines its global structure. For instance, it is fairly easy to prove that if a 2-level polytope has a simple vertex, then it is necessarily isomorphic[1] to the stable set polytope of a perfect graph — this observation generalizes a result of [7]. Here, we study 2-level polytopes with a prescribed facet.

Prescribing facets of 2-level polytopes is a natural way to enumerate 2-level polytopes. Indeed, since every facet of a 2-level polytope is also 2-level, in order to enumerate all d-dimensional 2-level polytopes one could go through the list of all $(d-1)$-dimensional 2-level polytopes P_0 and enumerate all 2-level polytopes P having P_0 as a facet. The enumeration algorithm [2] builds on this strategy. It gave the complete list of 2-level polytopes up to dimension $d = 6$. However, the method in [2] was by far not able to compute the list of 2-level polytopes in $d = 7$.

---

[1] While in general two polytopes can be combinatorially equivalent without being affinely equivalent, for 2-level polytopes these two notions coincide [2]. We simply say that two 2-level polytopes are *isomorphic* whenever they are combinatorially (or affinely) equivalent.

Contribution and Outline. The contribution of this paper is twofold.

First, we revisit the enumeration algorithm from [2] and propose a new and significantly more efficient variant based on a more geometric interpretation of the algorithm. We implemented the new algorithm and computed for the first time the complete list of 2-level polytopes up to isomorphism for $d = 7$. The algorithm and the results are exposed in Sect. 2.

Second, we characterize 2-level polytopes with a cube or cross-polytope facet and more generally with a sequentially Hanner facet, see Sect. 3. We give an informal statement (Theorem 1) of the result there and illustrate the proof strategy with the special case of the cube and cross-polytope. The full statement and proof can be found in the appendix.

Our main tool to obtain these results is a certain polyhedral subdivision that one can define given any prescribed facet, see below in the present section. In addition to this, we make use of the lattice decomposition property, a general property of 2-level polytopes that is tightly related to the integer decomposition property.

Finally, in Sect. 4, we discuss suspensions of 2-level polytopes. A *suspension* of a polytope $P_0 \subseteq \{x \in \mathbb{R}^d \mid x_1 = 0\}$ is any polytope P obtained as the convex hull of P_0 and P_1, where $P_1 \subseteq \{x \in \mathbb{R}^d \mid x_1 = 1\}$ is the translate of some non-empty face of P_0. The prism and the pyramid over a polytope P_0 are special cases of suspensions. As an outcome of our results, we found that suspensions seem to play an important role in understanding the structure of general 2-level polytopes. We conclude Sect. 4 by stating promising new research questions on 2-level polytopes that are inspired by this.

Now, we describe our approach in more detail and then give further pointers to related work.

Approach. Given any 2-level $(d - 1)$-polytope P_0 that we wish to prescribe as a facet of a 2-level d-polytope P, we define a new polytope that we call the "master polytope" and a polyhedral subdivision of this master polytope.

Definition 3 (Master polytope, polyhedral subdivision). *Let P_0 be a $(d-1)$-dimensional 2-level polytope embedded in $\{x \in \mathbb{R}^d \mid x_1 = 0\} \simeq \mathbb{R}^{d-1}$. Since P_0 is 2-level, each facet-defining hyperplane Π^- has a parallel hyperplane Π^+ such that Π^- and Π^+ together contain all the vertices of P_0. Let v^- and v^+ be vertices of P_0 on Π^- and Π^+ respectively. Consider the three hyperplanes $\Pi^- - v^+$, $\Pi^- - v^- = \Pi^+ - v^+$ and $\Pi^+ - v^-$. Let $Q(P_0) \subseteq \{x \in \mathbb{R}^d \mid x_1 = 0\}$ denote the polytope bounded by the "outer" hyperplanes $\Pi^- - v^+$ and $\Pi^+ - v^-$ obtained for each facet-defining direction. We call $Q = Q(P_0)$ the master polytope of P_0. The "middle" hyperplanes $\Pi^- - v^- = \Pi^+ - v^+$ define a polyhedral subdivision of the master polytope Q, which we denote by $\mathcal{S}(P_0)$. See Fig. 1 for an illustration for $d = 3, 4$.*

The improved enumeration algorithm that we propose in Sect. 2 is based on three new ideas, two of which are related to the polyhedral subdivision $\mathcal{S}(P_0)$: (i) we enumerate the possible vertex sets of the top face P_1 in the whole polyhedral subdivision $\mathcal{S}(P_0)$, instead of branching prematurely and miss the opportunity

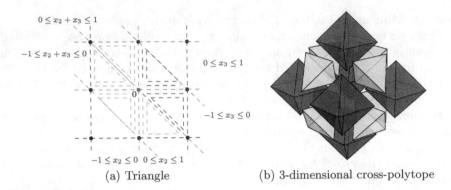

(a) Triangle   (b) 3-dimensional cross-polytope

Fig. 1. (a) The polyhedral subdivision $\mathcal{S}(P_0)$ in case P_0 is a triangle. Coloured sets correspond to the 2-level polytopes P computed by the enumeration algorithm. Red sets yield suspensions. (b) polyhedral subdivision $\mathcal{S}(P_0)$ in case P_0 is the 3-cross-polytope; the blue cells yield prisms, all the faces yield suspensions except from yellow cells that yield quasi-suspensions (figure from `Wikipedia`). (Color figure online)

to discard redundant computations; (ii) we exploit the fact that many possible vertex sets for P_1 are related to each other by translations within $\mathcal{S}(P_0)$, which further decreases the number of cases to consider; (iii) we use an ordering on the set of prescribed 2-level facets that allows the algorithm to compute significantly less convex hulls.

In order to prove our main theoretical result stated in Sect. 3, we embed the given sequentially Hanner facet $P_0 = H$ in $\{x \in \mathbb{R}^d \mid x_1 = 0\}$, and consider the polyhedral subdivision $\mathcal{S}(H)$. Up to isomorphism, every 2-level polytope P that has H as a facet is the convex hull of $P_0 = H$ and some 2-level (possibly low-dimensional) polytope P_1 in $\{x \in \mathbb{R}^d \mid x_1 = 1\}$. We prove that P_1 is always a translate of some face of $\mathcal{S}(H)$, a fact that we repeatedly use in our analysis. This uses the fact that for sequentially Hanner polytopes, facet directions exactly correspond to 2-level directions. Although the structure of $\mathcal{S}(H)$ for a sequentially Hanner facet H seems quite wild, we are able to characterize which points of $\{x \in \mathbb{R}^d \mid x_1 = 1\}$ can possibly appear as vertices of P, assuming that e_1 is a vertex of P. There, we use the lattice decomposition property.

Then, we analyse the vertex set of P_1 through an associated *bidirected graph* which determines the projection of P_1 to a subset of the coordinates, namely, those that correspond to prism operations in the sequentially Hanner facet H. We prove that this bidirected graph can always be assumed to be a star (possibly with some parallel edges). In order to reconstruct P_1 from its bidirected graph, we show that every bidirected edge of our bidirected graph has corresponding face in P_1, which is an axis-parallel cube. Next, we characterize the choices of cubes that lead to a 2-level polytope and conclude that P is a generalization of a suspension, which we call *quasi-suspension*. Finally, we complete the characterization by proving that quasi-suspensions of sequentially Hanner polytopes are always 2-level.

Related Work. The enumeration of all combinatorial types of point configurations and polytopes is a fundamental problem in discrete and computational geometry. Latest results in [4] report complete enumeration of polytopes for dimension $d = 3, 4$ with up to 8 vertices and $d = 5, 6, 7$ with up to 9 vertices. For 0/1-polytopes this is done completely for $d \leqslant 5$ and $d = 6$ with up to 12 vertices [1].

Regarding 2-level polytopes, recent related results include an excluded minor characterization of 2-level matroid base polytopes [12], a $O(c^d)$ lower bound on the number of 2-level matroid d-polytopes [11], and a complete classification of polytopes with minimum positive semidefinite rank, which generalize 2-level polytopes, in dimension 4 [8].

2   Enumeration of 2-Level Polytopes

Preliminaries. We start by sketching the main ideas of the algorithm of [2] along with definitions and useful tools.

Definition 4 (Slack matrix). *The* slack matrix *of a polytope $P \subseteq \mathbb{R}^d$ with m facets $F_1, \ldots, F_m$ and n vertices $v_1, \ldots, v_n$ is the $m \times n$ nonnegative matrix $S = S(P)$ such that S_{ij} is the slack of the vertex v_j with respect to the facet F_i, that is, $S_{ij} = g_i(v_j)$ where $g_i : \mathbb{R}^d \to \mathbb{R}$ is any affine form such that $g_i(x) \geqslant 0$ is valid for P and $F_i = \{x \in P \mid g_i(x) = 0\}$. The slack matrix of a polytope is defined up to scaling its rows by positive reals.*

A polytope is 2-level if and only if each row of its slack matrix takes exactly two values (namely, 0 and some positive number that can may vary from row to row). When dealing with 2-level polytopes, we will always assume the slack matrices to be 0/1, which may be always achieved by scaling the rows of the matrix with positive scalars.

Definition 5 (Simplicial core). *A* simplicial core *for a d-polytope P is a $(2d + 2)$-tuple $(F_1, \ldots, F_{d+1}; v_1, \ldots, v_{d+1})$ of facets and vertices of P such that each facet F_i does not contain vertex v_i but contains vertices $v_{i+1}, \ldots, v_{d+1}$.*

Every d-polytope P admits a simplicial core and this fact can be proved by a simple induction on the dimension, see, e.g., [9, Proposition 3.2]. We use simplicial cores to define two types of embeddings that are full-dimensional. Let P be a 2-level d-polytope with m facets and n vertices, and let $\Gamma := (F_1, \ldots, F_{d+1}; v_1, \ldots, v_{d+1})$ be a simplicial core for P. From now on, we assume that the rows and columns of the slack matrix $S(P)$ are ordered compatibly with the simplicial core, so that the ith row of $S(P)$ corresponds to facet F_i for $1 \leqslant i \leqslant d + 1$ and the j-th column of $S(P)$ corresponds to vertex v_j for $1 \leqslant j \leqslant d + 1$.

Definition 6 ($\mathcal{V}$- and $\mathcal{H}$-embedding). *The $\mathcal{H}$-embedding with respect to Γ is defined by mapping each v_j to the unit vector e_j of $\mathbb{R}^d$ for $1 \leqslant j \leqslant d$, and v_{d+1} to the origin. In the $\mathcal{H}$-embedding of P, facet F_i for $1 \leqslant i \leqslant m$ is defined* by the inequality $\sum_{j \in [d], S_{ij}=1} x_j \geqslant 0$ if $v_{d+1} \in F_i$ and by $\sum_{j \in [d], S_{ij}=0} x_j \leqslant 1$ if $v_{d+1} \notin F_i$. In the $\mathcal{V}$-embedding of P with respect to Γ, vertex v_j is the point of $\mathbb{R}^d$ whose ith coordinate is S_{ij}, for $1 \leqslant j \leqslant n$ and $1 \leqslant i \leqslant d$.

Equivalently, the $\mathcal{V}$-embedding can be defined via the transformation $x \mapsto Mx$, where the *embedding matrix* $M = M(\Gamma)$ is the top left $d \times d$ submatrix of $S(P)$ and $x \in \mathbb{R}^d$ is a point in the $\mathcal{H}$-embedding.

The next result is fundamental for the enumeration algorithm.

Proposition 7 ([2]). *In the $\mathcal{H}$-embedding P of a 2-level d-polytope with respect to any simplicial core Γ, the vertex set of P equals $P \cap M^{-1} \cdot \{0,1\}^d \subseteq \mathbb{Z}^d$, where $M = M(\Gamma)$ is the embedding matrix of Γ.*

The algorithm computes a complete set L_d of non-isomorphic 2-level d-polytopes, from a similar set L_{d-1} of 2-level $(d-1)$-polytopes. For a given polytope $P_0 \in L_{d-1}$, define $L(P_0)$ to be the set of all 2-level polytopes that have P_0 as a facet. Since every facet of a 2-level polytope is 2-level, the union of these sets $L(P_0)$ over all polytopes $P_0 \in L_{d-1}$ is our desired set L_d. The main loop of the algorithm is as follows (Algorithm 1, lines 2–18): given some $P_0 \in L_{d-1}$, embed it in the hyperplane $\{x \in \mathbb{R}^d \mid x_1 = 0\} \simeq \mathbb{R}^{d-1}$. Then compute a collection $\mathcal{A} \subseteq \{x \in \mathbb{R}^d \mid x_1 = 1\}$ of point sets, such that for each 2-level polytope $P \in L(P_0)$, there exists $A \in \mathcal{A}$ with $P \simeq \text{conv}(P_0 \cup \{e_1\} \cup A)$. For each $A \in \mathcal{A}$, compute $P = \text{conv}(P_0 \cup \{e_1\} \cup A)$ and, in case it is 2-level and not isomorphic to any polytope in the current set L_d, add P to L_d (Algorithm 1, lines 11–18).

The efficiency of this approach depends greatly on how the collection $\mathcal{A}$ is chosen. In [2], $\mathcal{A}$ is constructed using a proxy for 2-level polytopes in terms of closed sets with respect to a closure operator.

Definition 8 (Closure operator). *An operator* cl $: 2^{\mathcal{X}} \to 2^{\mathcal{X}}$ *over a ground set $\mathcal{X}$ is a closure operator if it is: (i) idempotent (i.e., $cl(cl(A)) = cl(A)$), (ii) $(\preccurlyeq, \sqsubseteq)$-monotone (i.e., $A \preccurlyeq B \Rightarrow cl(A) \sqsubseteq cl(B)$) and (iii) $\preccurlyeq$-expansive (i.e., $A \preccurlyeq cl(A)$), where $A, B \subseteq \mathcal{X}$, $A \preccurlyeq B \Leftrightarrow A = B$ or $\max((A \cup B) \setminus (A \cap B)) \in B$, $A \sqsubseteq B \Leftrightarrow A \subseteq B$ and $\max(B \setminus A) \leqslant \min A$, and $\subseteq$ is the usual containment.*

The reader can verify that $A \sqsubseteq B \Rightarrow A \subseteq B \Rightarrow A \preccurlyeq B$. A set $A \subseteq \mathcal{X}$ is said to be *closed* with respect to cl if $cl(A) = A$.

In [2] the closure operator $cl_G \circ cl_{(\mathcal{X},\mathcal{F})}$ is used, where $\mathcal{F} \subseteq \mathbb{R}^d$ is a finite set of points disjoint from $\mathcal{X}$ and G is an "incompatibility graph". Then the algorithm of [5] is used, which is a polynomial delay algorithm for enumerating all the closed sets of a given closure operator.

New Enumeration Algorithm. We propose a new variant of the algorithm described above, Algorithm 1, inspired by a more geometric understanding of the enumeration method, relying on polyhedral subdivisions. There are three main improvements. They are described below.

In the first improvement we change the way the algorithm constructs the ground set $\mathcal{X}$ whose subsets are candidate point sets for the collection $\mathcal{A}$. In [2] M_d is computed — using (1) — for each possible bit-vector $b \in \{0,1\}^{d-2}$ and

---

Algorithm 1. Enumeration algorithm

---

1   Set $L_d := \varnothing$;

2   **foreach** $P_0 \in L_{d-1}$ *with simplicial core* $\Gamma_0 := (F'_2, \ldots, F'_{d+1}; v_2, \ldots, v_{d+1})$ **do**

3   Construct the $\mathcal{H}$-embedding of P_0 in $\{0\} \times \mathbb{R}^{d-1} \simeq \mathbb{R}^{d-1}$ w.r.t. Γ_0;

4   Let $M_{d-1} := M(\Gamma_0)$ and $\mathcal{X} := \emptyset$;

5   **foreach** *bit vector* $b \in \{0,1\}^{d-2}$ **do**   /* Improv. 1 */

6   Complete M_{d-1} to a $d \times d$ matrix in the following way:

$$M_d := \begin{pmatrix} 1 & 0 \cdots 0 \\ 0 & \\ b_1 & \\ \vdots & M_{d-1} \\ b_{d-2} & \end{pmatrix} \tag{1}$$

7       $\mathcal{X} := \mathcal{X} \cup M_d^{-1} \cdot (\{1\} \times \{0,1\}^{d-1}) \smallsetminus \{e_1\}$;

8   Let $\mathcal{F} := \mathrm{vert}(P_0) \cup \{e_1\}$;

9   Let G be the incompatibility graph on $\mathcal{X}$ w.r.t. P_0 and M_d;

10   Using the Ganter-Reuter algorithm [5], compute the list $\mathcal{A}$ of closed sets of the closure operator $cl_{(G,\mathcal{X},\mathcal{F})}$ (see Equation (2)) ;   /* Improv. 2 */

11   **foreach** $A \in \mathcal{A}$ **do**

12   **if** P_0 *has as many vertices as every adjacent facet in* $\mathrm{conv}(A \cup \mathcal{F})$ **then**   /* Improv. 3 */

13   Let $P := \mathrm{conv}(A \cup \mathcal{F})$;

14   **if** P *is 2-level and not isomorphic to any polytope in* L_d **then**

15   Let $F_1 := P_0$ and $v_1 := e_1$;

16   **for** $i = 2, \ldots, d+1$ **do**

17   Let F_i be the facet of P distinct from F_1 s.t. $F_i \supseteq F'_i$;

18   Add P to L_d with $\Gamma := (F_1, \ldots, F_{d+1}; v_1, \ldots, v_{d+1})$;

$\mathcal{X} := M_d^{-1} \cdot (\{1\} \times \{0,1\}^{d-1}) \smallsetminus \{e_1\}$. Here we construct a *larger ground set* $\mathcal{X}$ as the union of all the old $\mathcal{X}$ sets. See Algorithm 1, lines 5–7.

To illustrate the difference of approaches in $d = 2$ note that in Fig. 1 the old approach would construct two ground sets of four points each (corresponding to the two small squares to the right), while the new approach constructs a single ground set of six points. What we gain is that Algorithm 1 avoids enumerating many times a set A in the intersection of cells of $\mathcal{S}(P_0)$ (blue sets in Fig. 1). In this section, to simplify things, we translate the master polytope $Q(P_0)$ and its subdivision $\mathcal{S}(P_0)$ in the $\{x \in \mathbb{R}^d \mid x_1 = 1\}$ hyperplane, so that the origin is translated to e_1.

The second improvement is to exploit symmetries in a more sophisticated way. The symmetries we have in mind are translations "within" $\mathcal{S}(P_0)$. Note that the closure operator used in [2] satisfies stronger properties than those in Definition 8, in particular, it is idempotent, $(\subseteq, \subseteq)$-monotone, and $\subseteq$-expansive.

Letting $A \ddagger a := ((A \cup \{e_1\}) + e_1 - a) \smallsetminus \{e_1\}$ for $a \in A$, we define a new closure operator as follows:

$$cl_{(G,\mathcal{X},\mathcal{F})}(A) := \max_{\preccurlyeq}\{cl_G \circ cl_{(\mathcal{X},\mathcal{F})}(A \ddagger a) \mid a \in A, \ A \ddagger a \subseteq \mathcal{X}\}, \qquad (2)$$

where $A \subseteq \mathcal{X}$. This new operator returns a single representative from an *equivalence class* of point sets A in the polyhedral subdivision $\mathcal{S}(P_0)$ (green sets in Fig. 1). The idea is that since all sets of points A in the equivalence class yield the same polytope, we only need a single representative, which we define as the maximum within the equivalence class with respect to $\preccurlyeq$. The new closure operator is used in line 10 of Algorithm 1.

The third improvement consists in considering a *partial order* on the set L_{d-1} of $(d-1)$-dimensional 2-level polytopes, which is based on the number of vertices. The idea is that if the d-dimensional 2-level polytope P contains a facet having strictly more vertices than the current prescribed facet P_0, then it has been already enumerated before. We choose an ordering of L_{d-1} that is consistent with this.

Actually, the algorithm does not check that no facet of $P = \text{conv}(A \cup \mathcal{F})$ has strictly more vertices than P_0, because we want to avoid unnecessary convex hull computations as much as possible. Instead, we check that this condition holds only for the facets of P that are adjacent to P_0, which is possible without computing any new convex hull, since we know already all facets of P_0. This improvement is implemented in line 12 of Algorithm 1.

Finally, in Theorem 9 we prove the correctness of Algorithm 1.

Theorem 9. *Algorithm 1 outputs the list of all combinatorial types of 2-level d-polytopes, each with a simplicial core.*

Implementation and Experiments. We implement the skeleton of Algorithm 1 in `Perl`. For demanding computations, such as isomorphism tests, convex hull computations, and linear algebra operations we use `polymake` [6], a standard library for polyhedral computation. The implementation is based upon and improves the implementation presented in [2]. The improvements described above in current section yield a significant speed-up in the algorithm, which is ×12 for $d = 6$. There are 447362 convex hulls (i.e. 96 % of total convex hulls) avoided in $d = 6$ yielding a 0.065 ratio of the number of computed 2-level polytopes over the total number convex hulls computed. More interestingly, we enumerate all 2-level

Table 1. Experimental results of enumeration algorithms (time is sequential).

Method	d	2-level	closed sets	not 2-level	time(sec)	$\frac{\text{2-level}}{\text{closed sets}}$
Algorithm from [2]	6	1150	$4.1 \cdot 10^6$	$3.5 \cdot 10^6$	$6.9 \cdot 10^5$	$3.0 \cdot 10^{-4}$
Algorithm 1	6	1150	$4.6 \cdot 10^5$	$1.1 \cdot 10^4$	$5.5 \cdot 10^4$	$2.5 \cdot 10^{-3}$
	7	27291	$1.9 \cdot 10^8$	$1.1 \cdot 10^6$	$2.1 \cdot 10^7$	$1.4 \cdot 10^{-4}$

polytopes in one dimension higher than in [2], namely $d = 7$. See Table 1 for more details on experimental results[2].

On the technical part, for $d = 7$ we create 1150 jobs one for each 6-dimensional 2-level polytope and submit them to a computer cluster[3]. The vast majority of the jobs, namely 1132 finish in less than a day. The remaining 18 finished in a range from a week to a month. The use of high performance computing is crucial for this computation since the corresponding sequential time for this experiment is more than 5 years! The most time demanding job is the one corresponding to the simplex, which however corresponds to the known case of simplicial 2-level polytopes. Simplicial 2-level polytopes have been characterized in [10]. By applying the result in $d = 7$ there exist exactly two simplicial 2-level 7-polytopes: the simplex and the cross-polytope.

3 Prescribing a Sequentially Hanner Facet

We start by the following property, which plays an important role in our analysis.

Definition 10 (Strong separation property). *Let P be a d-dimensional polytope. We say that P satisfies the* strong separation property *if for every ordered pair K_1, K_2 of non-empty disjoint faces of P, there exists a facet F of P such that $F \supseteq K_1$ and $F \cap K_2 = \varnothing$.*

In general, it is not true that all suspensions of a given 2-level polytope Q are 2-level. However, this is true when Q has the strong separation property.

Proposition 11. *Let $Q \subseteq \mathbb{R}^{d-1}$ be a full-dimensional 2-level polytope that satisfies the strong separation property. Let G be one of its non-empty faces, and let $P \subseteq \mathbb{R}^d$ denote the suspension of Q with respect to G. Then every facet of P either is parallel to $\{0\} \times Q$ or intersects $\{0\} \times Q$ in a facet. In particular, P is a 2-level polytope.*

All sequentially Hanner polytopes have the strong separation property.

Prescribing a Cubical Facet. Consider a 2-level d-polytope P one of whose facets is the $(d-1)$-cube $P_0 = \{0\} \times [-1,1]^{d-1} \subseteq \{x \in \mathbb{R}^d \mid x_1 = 0\}$. Let $P_1 \subseteq \{x \in \mathbb{R}^d \mid x_1 = 1\}$ denote the face of P opposite to P_0. Without loss of generality, assume that e_1 is a vertex of P_1. The master polytope $Q(P_0)$ is the cube $2P_0$. This larger cube is subdivided by the coordinate hyperplanes in the polyhedral subdivision $\mathcal{S}(P_0)$. The cells of $\mathcal{S}(P_0)$ are 2^{d-1} translated copies of P_0. It is easy to prove that P_1 is the translate of some face of $\mathcal{S}(P_0)$, and thus the translate of some face of P_0. In other words, P is a suspension. Combining this with Proposition 11, we obtain:

---

[2] The computed polytopes in `polymake` format and more experimental results are available online http://homepages.ulb.ac.be/~vfisikop/data/2-level.html.
[3] Hydra balanced cluster: http://cc.ulb.ac.be/hpc/hydra.php.

Proposition 12. *A d-dimensional 2-level polytope P has a facet isomorphic to a $(d-1)$-cube if and only if it is isomorphic to some suspension of a $(d-1)$-cube.*

Prescribing a Cross-Polytope Facet. This time, consider a 2-level d-polytope P one of whose facets is the $(d-1)$-cross-polytope $P_0 := \text{conv}\{\pm e_2, \ldots, \pm e_d\} \subseteq \{x \in \mathbb{R}^d \mid x_1 = 0\}$ and define $P_1 \subseteq \{x \in \mathbb{R}^d \mid x_1 = 1\}$ as before, so that $P = \text{conv}(P_0 \cup P_1)$. Again, assume that e_1 is a vertex of P_1. Using the lattice decomposition property, we can prove that the vertices of P_1 are all of the form $e_1 + w_1 + w_2$ where w_1 and w_2 are vertices of the base P_0.

Using the properties of the embedding of P, we construct a bidirected graph $G = G(P)$ that encodes the vertices of the top face P_1. The node set of G is $V := \{2, \ldots, d\}$, and for every vertex $x = e_1 \pm e_i \pm e_j$, $i, j \in V$, of P_1 that is distinct from e_1 we create an edge in G with endpoints i and j, each endowed with a sign that coincides with the signs of the corresponding coordinate of x.

The rest of the analysis is done by establishing properties of the bidirected graph G. The most important is the fact that G has no two disjoint edges. Next, we establish a form of sign-consistency for G: every two edges of G have the same sign at exactly one of their one or two common endpoints.

These two properties put extreme restrictions on the bidirected graph G. One possible case arises when all the edges of G have a common endpoint, which has the same sign in all the edges. This forces P to be a suspension. Moreover, the presence of a pair of parallel edges or loop automatically leads to the case of a prism. In the remaining case G is a triangle without pair of parallel edges or loop. This leads to the sporadic case of the hypersimplex $\Delta(4,2)$. We obtain the following result.

Proposition 13. *A d-dimensional 2-level polytope P has a facet isomorphic to a $(d-1)$-cross-polytope if and only if it is isomorphic to some suspension of a $(d-1)$-cross-polytope or to the hypersimplex $\Delta(4,2)$.*

Prescribing a Sequentially Hanner Facet. The following result generalizes Propositions 12 and 13.

Theorem 1 (Informal). *The 2-level polytopes with a facet isomorphic to a sequentially Hanner polytope essentially coincide with the suspensions of sequentially Hanner polytopes.*

4   Discussion

In this last section, we discuss suspensions of 2-level polytopes (called just suspensions below). As is supported by the theoretical and experimental results of this paper, suspensions seem to play an important role towards a broader understanding of the structure of general 2-level polytopes.

Since there are suspensions that are not 2-level, it is natural to ask what is the class of 2-level polytopes whose suspensions are always 2-level. Proposition 11 sheds some light in this direction by providing a sufficient condition for any

Table 2. Number of 2-level suspensions $s(d)$, 2-level polytopes $\ell(d)$, ratio of number of 2-level suspensions to 2-level polytopes, and maximum number of faces of 2-level polytopes $f(d)$ for low dimensions d.

d	3	4	5	6	7
$\ell(d)$	5	19	106	1150	27291
$s(d)$	4	15	88	956	23279
$\frac{s(d)}{l(d)}$	.8	.789	.830	.831	.853
$f(d)$	28	82	304	940	3496

suspension of a 2-level polytope to be 2-level. It remains open to find a necessary and sufficient condition.

Another related question is the following: what fraction of 2-level d-polytopes are suspensions of $(d-1)$-polytopes? Table 2 gives the ratio for small dimension d. Excluding dimension 3, we observe that this fraction increases with the dimension. Using notation from Table 2, is true that $\ell(d) = O(s(d))$?

If one could prove that this is true, this would have strong consequences on $\ell(d)$. Let $c > 1$ be any constant such that $\ell(d) \leqslant c \cdot s(d)$. Since 2-level d-polytopes have at most 2^d vertices, each of them being affinely equivalent to 0/1-polytope, we have $f(d) \leqslant c^{d^2}$ provided we choose c large enough. Now assume that $\ell(d-1) < c^{(d-1)^3}$ (this would be our induction hypothesis). Then we would have: $\ell(d) \leqslant c \cdot s(d) \leqslant c \cdot \ell(d-1) \cdot f(d-1) \leqslant c \cdot c^{(d-1)^3} \cdot c^{(d-1)^2} \leqslant c^{d^3}$. In fact, a singly exponential upper bound on $f(d)$ would imply $\ell(d) = 2^{O(d^2)}$. This would not contradict any known lower bound, since all known constructions of 2-level polytopes are ultimately based on graphs and do not imply more than a $2^{\Omega(d^2)}$ lower bound on $\ell(d)$. For instance, stable sets of bipartite graphs show $\ell(d) \geqslant 2^{\frac{d^2}{4} - o(1)}$. Can one show at least $\ell(d) \leqslant 2^{\text{poly}(d)}$? Independently of this, is it true that $f(d) = 2^{O(d)}$?

References

1. Aichholzer, O.: Extremal properties of 0/1-polytopes of dimension 5. In: Ziegler, G., Kalai, G. (eds.) Polytopes - Combinatorics and Computation, pp. 111–130. Birkhäuser, Basel (2000)
2. Bohn, A., Faenza, Y., Fiorini, S., Fisikopoulos, V., Macchia, M., Pashkovich, K.: Enumeration of 2-level polytopes. In: Bansal, N., Finocchi, I. (eds.) ESA 2015. LNCS, vol. 9294, pp. 191–202. Springer, Heidelberg (2015). doi:10.1007/978-3-662-48350-3_17
3. Chvátal, V.: On certain polytopes associated with graphs. J. Comb. Theory Ser. B **18**, 138–154 (1975)
4. Fukuda, K., Miyata, H., Moriyama, S.: Complete enumeration of small realizable oriented matroids. Discrete Comput. Geom. **49**(2), 359–381 (2013)
5. Ganter, B., Reuter, K.: Finding all closed sets: a general approach. Order **8**(3), 283–290 (1991)

6. Gawrilow, E., Joswig, M.: Polymake: an approach to modular software design in computational geometry. In: International Symposium on Computational Geometry (SOCG), pp. 222–231. ACM Press (2001)

7. Gouveia, J., Parrilo, P., Thomas, R.: Theta bodies for polynomial ideals. SIAM J. Optim. **20**(4), 2097–2118 (2010)

8. Gouveia, J., Pashkovich, K., Robinson, R.Z., Thomas, R.R.: Four dimensional polytopes of minimum positive semidefinite rank (2015)

9. Gouveia, J., Robinson, R., Thomas, R.: Polytopes of minimum positive semidefinite rank. Discrete Comput. Geom. **50**(3), 679–699 (2013)

10. Grande, F.: On k-level matroids: geometry and combinatorics. Ph.D. thesis, Free University of Berlin (2015)

11. Grande, F., Rué, J.: Many 2-level polytopes from matroids. Discrete Comput. Geom. **54**(4), 954–979 (2015)

12. Grande, F., Sanyal, R.: Theta rank, levelness, matroid minors. arXiv preprint (2014). arXiv:1408.1262

13. Grötschel, M., Lovász, L., Schrijver, A.: Geometric Algorithms and Combinatorial Optimization. Algorithms and Combinatorics, vol. 2. Springer, Heidelberg (1988)

14. Hanner, O.: Intersections of translates of convex bodies. Mathematica Scandinavica **4**, 65–87 (1956)

15. Hansen, A.: On a certain class of polytopes associated with independence systems. Mathematica Scandinavica **41**, 225–241 (1977)

16. Kalai, G.: The number of faces of centrally-symmetric polytopes. Graphs Comb. **5**(1), 389–391 (1989)

17. Lovász, L.: Semidefinite programs and combinatorial optimization. In: Reed, B.A., Sales, C.L. (eds.) Recent Advances in Algorithms and Combinatorics. CMS Books in Mathematics, pp. 137–194. Springer, Heidelberg (2003)

18. Lovász, L., Saks, M.: Lattices, mobius functions and communications complexity. In: Proceedings of the 29th Annual Symposium on Foundations of Computer Science, SFCS 1988, pp. 81–90. IEEE Computer Society, Washington (1988)

19. Mahler, K.: Ein übertragungsprinzip für konvexe körper. Časopis Pěst. Mat. Fys. **68**, 93–102 (1939)

20. Schrijver, A.: Theory of Linear and Integer Programming. Wiley, New York (1986)

21. Stanley, R.: Two poset polytopes. Discrete Comput. Geom. **1**(1), 9–23 (1986)

Optimum Solution of the Closest String Problem via Rank Distance

Claudio Arbib[1], Giovanni Felici[2(✉)], Mara Servilio[2], and Paolo Ventura[2]

[1] Dipartimento di Scienze/Ingegneria dell'Informazione e Matematica,
Università degli Studi dell'Aquila, via Vetoio, Coppito, 67010 L'Aquila, Italy
claudio.arbib@univaq.it

[2] Istituto di Analisi dei Sistemi e Informatica, Consiglio Nazionale delle Ricerche,
via dei Taurini 19, 00185 Roma, Italy
{giovanni.felici,mara.servilio,paolo.ventura}@iasi.cnr.it

Abstract. The Closest String Problem (CSP) calls for finding an n-string that minimizes its maximum distance from m given n-strings. Integer linear programming (ILP) proved to be able to solve large CSPs under the *Hamming distance*, whereas for the *Levenshtein distance*, preferred in computational biology, no ILP formulation has so far be investigated. Recent research has however demonstrated that another metric, *rank distance*, can provide interesting results with genomic sequences. Moreover, CSP under rank distance can easily be modeled via ILP: optimal solutions can then be certified, or information on approximation obtained via dual gap. In this work we test this ILP formulation on random and biological data. Our experiments, conducted on strings with up to 600 nucleotides, show that the approach outperforms literature heuristics. We also enforce the formulation by cover inequalities. Interestingly, due to the special structure of the rank distance between two strings, cover separation can be done in polynomial time.

1 Introduction

Let A be an alphabet with p symbols. The CLOSEST STRING — or CENTER STRING — PROBLEM (CSP) calls for finding a string $\mathbf{x} \in A^n$ that best approximates a given set S of strings $\mathbf{s}^1, \ldots, \mathbf{s}^m \in A^n$. Approximation is measured with a distance function $d(.,.)$. A center (optimal solution of the CSP) is an $\mathbf{x}^*$ that, among all strings $\mathbf{x} \in A^n$, minimizes the maximum distance $d(\mathbf{x}, \mathbf{s}^i)$ from any $\mathbf{s}^i \in S$.

When d is the *Hamming distance*, d returns the number of different components in the two strings. A different metric, the *rank distance* recently proposed by Dinu and Ionescu [6] and Dinu and Popa [7], seems to provide more interesting information in DNA sequence comparisons, with respect to the Hamming distance. Similarly to *Levenshtein distance* (a special type of edit distance, [10]) and unlike Hamming, the rank distance is in fact able to take into account, via specific penalties, symbol insertions or deletions. In the sequel, we will call *rank-central* a string $\mathbf{x}^*$ that minimizes the maximum rank-distance from the strings of S.

© Springer International Publishing Switzerland 2016
R. Cerulli et al. (Eds.): ISCO 2016, LNCS 9849, pp. 297–307, 2016.
DOI: 10.1007/978-3-319-45587-7_26

To compute the rank distance between two strings, one has first to enucle-
ate from each string the substrings formed by identical symbols (here generally referred to α-*substrings* when formed by symbols equal to α). For instance,
$\mathbf{s} = abbcbba$ contains the a-substring a_1a_2, the b-substring $b_1b_2b_3b_4$ and the c-
substring c_1, whereas $\mathbf{s}' = aaccbab$ contains $a_1a_2a_3$, b_1b_2 and c_1c_2. The subscript
of each character indicates the rank that the symbol has in the relevant sub-
string. If two α-substrings of $\mathbf{s}$ and $\mathbf{s}'$ have different lengths l, l', then we say
they have $|l - l'|$ *out-of-ranks*.

The total rank distance $d(\mathbf{s}, \mathbf{s}')$ sums up the distances between the positions
of identical symbols, one in $\mathbf{s}$ and the other in $\mathbf{s}'$, that have the same rank, plus
a penalty for each out-of-rank in $\mathbf{s}$ and $\mathbf{s}'$. In [6], such a penalty is assumed equal
to the position of the out-of-rank in the string. Precisely, let

- R denote the set of index pairs hk such that $s_h = s'_k$ and the symbol has the
  same rank in the respective substrings.
- O, O' denote the positions of the out-of-ranks in $\mathbf{s}$, $\mathbf{s}'$.

Then

$$d(\mathbf{s}, \mathbf{s}') = \sum_{hk \in R} |h - k| + \sum_{k \in O \cup O'} k \ .$$

In our example, we first rewrite $\mathbf{s} = a_1b_1b_2c_1b_3b_4a_2$, $\mathbf{s}' = a_1a_2c_1c_2b_1a_3b_2$;
then we compare the positions of α_k in both strings for $\alpha = a, b, c$, and add
penalties for out-of-ranks (b_3, b_4 in $\mathbf{s}$ and c_2, a_3 in $\mathbf{s}'$). As a result we get (Fig. 1)

$$d(\mathbf{s}, \mathbf{s}') = \underbrace{(|1 - 1| + |7 - 2| + 6)}_{a} + \underbrace{(|2 - 5| + |3 - 7| + 11)}_{b} + \underbrace{(|4 - 3| + 4)}_{c} = 34$$

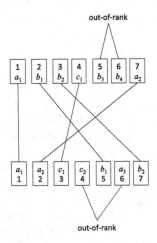

Fig. 1. Computation of rank distance.

Integer linear programming (ILP) formulations have been proposed to solve the CSP with the Hamming metric, see [4], and ILP is a key factor of success for state-of-the-art heuristics [2,5]. Aim of the present research is to extend the ILP approach to the rank metric. In fact another advantage of rank distance is that it is not difficult to formulate the problem of finding a rank-central string as an integer linear program.

The paper is organized as follows. In Sect. 2 we formulate the CSP via rank distance in terms of bipartite multi-weight matching. A computational experience based on both random and biological data is reported and commented in Sect. 3.

2 Rank Distance Optimization: A Matching Model

In this section we formulate as an integer linear program the problem of finding a string $\mathbf{x}^* \in A^n$ that is rank-central with respect to a given set $S = \{\mathbf{s}^1, \ldots, \mathbf{s}^m\}$ of strings in A^n. The formulation basically follows [7].

Solution encoding

Let $l^i(\alpha)$ denote the length of the α-substring of $\mathbf{s}^i$, $\alpha \in A$ and $i = 1, \ldots, m$, and define $l(\alpha) = \max_{i=1,\ldots,m}\{l^i(\alpha)\}$. The model encodes a solution $\mathbf{x}$ by 0-1 assignment variables y_{kr}^α associated with the arcs of a bipartite graph $G = (U \cup V, E)$, where (see Fig. 2):

- U contains n nodes, one per component of $\mathbf{x}$.
- V contains $l = \sum_{\alpha \in A} l(\alpha) \geq n$ nodes.

Decision variables

For any $\alpha \in A$, $r = 1, \ldots, l(\alpha)$ and $k = r, \ldots, n$, we set $y_{kr}^\alpha = 1$ if $x_k = \alpha$ and, specifically, the k-th component of $\mathbf{x}$ is matched with the r-th character of the α-string $\alpha_1 \ldots \alpha_{l(\alpha)}$. Otherwise, we set $y_{kr}^\alpha = 0$. Notice that variables are undefined for $k < r$, that is: a symbol of rank r in any α-substring cannot be matched to any component of $\mathbf{x}$ before the r-th. Each triple (α, r, k) indexing a y-variable identifies an arc of G (see Fig. 2).

A solution $\mathbf{x}$ may not contain all the characters of a substring, that is, the r-th occurrence of a symbol α may remain unmatched (out-of-rank). We express this by a further 0-1 variable $y_{n+1,r}^\alpha$, defined for all $\alpha \subset A$ and $r = 1, \ldots, l(\alpha)$.

Encoding constraints

Symbol α_r is assigned to either one or no position in $\mathbf{x}$:

$$\sum_{k=r}^{n+1} y_{kr}^\alpha = 1 \qquad \alpha \in A \text{ and } r = 1, \ldots, l(\alpha) \tag{1}$$

On the other hand, every component of $\mathbf{x}$ needs to be defined, thus:

$$\sum_{\alpha \in A} \sum_{r=1}^{\min\{k,l(\alpha)\}} y_{kr}^{\alpha} = 1 \qquad\qquad k = 1,\ldots,n \qquad\qquad (2)$$

Finally, by definition, the y_{kr}^{α} fulfill upper/lower bounds and integrality clauses

$$y_{kr}^{\alpha} \geq 0 \qquad\qquad (3)$$
$$-y_{kr}^{\alpha} \geq -1 \qquad\qquad (4)$$
$$y_{kr}^{\alpha} \quad \text{integer} \qquad \alpha \in A; r = 1,\ldots,l(\alpha); k = r,\ldots,n+1 \qquad (5)$$

Distance constraints

To write the rank distance $d(\mathbf{x},\mathbf{s}^i)$ from a target string $\mathbf{s}^i$, we introduce $k_r^{\alpha,i}$ as the (known) position that the r-th character of the α-substring has in $\mathbf{s}^i$, $r = 1,\ldots,l^i(\alpha)$. So $k_1^{a,i}$ is the position of the first a of $\mathbf{s}^i$ etcetera. For example, referring to $\mathbf{s}^i = \mathbf{s} = acbabca$, $k_1^a = 1$, $k_3^a = 7, k_1^b = 3$ and so on. The distance is the sum of penalties and misplacement costs $c_{kr}^{\alpha,i}$. Penalties are of two types: $p_k^{\alpha,i}$, accounting for an α out-of-rank in $\mathbf{x}$; $q_r^{\alpha,i}$ for an α out-of-rank in the target string. Misplacement costs measure instead the distance between the positions of matched pairs, when α is not out-of-rank in $\mathbf{x}$:

$$c_{kr}^{\alpha,i} = \begin{cases} |k - k_r^{\alpha,i}| & 1 \leq r \leq l^i(\alpha) \\ p_k^{\alpha,i} & l^i(\alpha) < r \leq l(\alpha) \\ q_r^{\alpha,i} & k = n+1, 1 \leq r \leq l^i(\alpha) \end{cases}$$

for any i (in [6], it is assumed $p_k^{\alpha,i} = |0 - k| = k, q_r^{\alpha,i} = |0 - k_r^{\alpha,i}| = k_r^{\alpha,i}$). Note that misplacement costs and penalties $p_r^{\alpha,i}$ are associated with variables $y_{kr}^{\alpha,i}$, $k \leq n$ (therefore with the arcs of G), whereas penalties $q_r^{\alpha,i}$ correspond to the $y_{n+1,r}^{\alpha,i}$.

To limit the rank distance between $\mathbf{x}$ and $\mathbf{s}^i$ we can then write

$$d(\mathbf{x},\mathbf{s}^i) = \sum_{\alpha \in A} \sum_{r=1}^{l(\alpha)} \sum_{k=r}^{n} c_{kr}^{\alpha,i} y_{kr}^{\alpha} + \sum_{\alpha \in A} \sum_{r=1}^{l^i(\alpha)} c_{n+1,r}^{\alpha,i} y_{n+1,r}^{\alpha} \leq d \qquad (6)$$

where d is a convenient upper bound.

Additional constraints

To respect the arrangement in each α-substring, no two arcs of G can cross each other. This constraint, strengthened by lifting as in [1], can be enforced in this way:

$$\sum_{s=r}^{k} y_{sr}^{\alpha} - \sum_{s=r-1}^{k-1} y_{s,r-1}^{\alpha} \leq 0 \qquad\qquad (7)$$

for $\alpha \in A, r = 2,\ldots,l(\alpha)$ and $k = r,\ldots,n$.

Moreover, constraints on variables $y^\alpha_{n+1,r}$ should be used to prevent the r-th character of an α-substring from being matched when the previous is not:

$$y^\alpha_{n+1,r-1} \le y^\alpha_{n+1,r} \qquad \alpha \in A \text{ and } r = 1, \ldots, l(\alpha)$$

Note, however, that the above constraints are implied by other inequalities. In fact, by (7) and (3)

$$\sum_{s=r}^{n} y^\alpha_{sr} \le \sum_{s=r-1}^{n-1} y^\alpha_{s,r-1} \le \sum_{s=r-1}^{n} y^\alpha_{s,r-1}$$

from which, using (1), we get the previous inequalities.

Additionally, based on the definition of the rank distance, we also note that inequalities (7) are not necessary, since it is easy to verify that

Proposition 1. *With the rank distance defined as in* [6], $c^{\alpha,i}_{hr} + c^{\alpha,i}_{ks} \le c^{\alpha,i}_{hs} + c^{\alpha,i}_{kr}$ *whenever* $r < s$ *and* $h < k \le n$.

Example

Let us explain model (1)–(6) via an example. Let $\mathbf{s}^1 = aabab$, $\mathbf{s}^2 = accba$. In this case

$$l^1(a) = 3 \qquad l^1(b) = 2 \qquad l^1(c) = 0$$
$$l^2(a) = 2 \qquad l^2(b) = 1 \qquad l^2(c) = 2$$

thus $l(a) = 3$ and $l(b) = l(c) = 2$.

Figure 2 shows graph G. The feasible matching $M = \{1b_1, 2a_1, 3b_2, 4c_1, 5c_2\}$ (thick arcs) encodes $\mathbf{x} = babcc$. Let us illustrate the computation of the rank distance between $\mathbf{x}$ and $\mathbf{s}^2 = accba$ via matching M, adopting for penalties the convention proposed in [6]. The arc weights in M give the first term of $d(\mathbf{x}, \mathbf{s}^2)$:

$$c^{b,2}_{11} = |1 - k^b_1| = |1 - 4| = 3 \quad c^{a,2}_{21} = |2 - k^a_1| = |2 - 1| = 1$$
$$c^{b,2}_{32} = p^b_3 = |3 - 0| = 3 \quad c^{c,2}_{41} = |4 - k^c_1| = |4 - 2| = 2$$
$$c^{c,2}_{52} = |5 - k^c_2| = |5 - 3| = 2$$

These weights measure the symbol misplacements, plus an out-of-rank penalty for the second $b \in \mathbf{x}$.

The second term of $d(\mathbf{x}, \mathbf{s}^2)$ sums, up to $l^2(\alpha)$, the out-of-rank penalties in $\mathbf{s}^2$: these correspond to nodes of $\mathbf{s}^2$ that are uncovered by M. There is just one of these nodes: a_2 (in fact, $a_3 \not\subset \mathbf{s}^2$), its weight is $q^{u,2}_2 = |0 - k^a_2| = 5$. Therefore, $d(\mathbf{x}, \mathbf{s}^2) = 16$.

Generally speaking, string $\mathbf{s}^2$ introduces the distance constraint

$$\begin{aligned}
y^a_{21} + 2y^a_{31} + 3y^a_{41} + 4y^a_{51} + 3y^a_{22} + 2y^a_{32} + y^a_{42} + 3y^a_{33} + 4y^a_{43} + 5y^a_{53} \\
+ \; 3y^b_{11} + 2y^b_{21} + y^b_{31} + y^b_{51} + 2y^b_{22} + 3y^b_{32} + 4y^b_{42} + 5y^b_{52} \\
+ \; y^c_{11} + y^c_{31} + 2y^c_{41} + 3y^c_{51} + y^c_{22} + y^c_{42} + 2y^c_{52} \\
+ \; y^a_{61} + 5y^a_{62} + 4y^b_{61} + 2y^c_{61} + 3y^c_{62} \; \le \; d
\end{aligned}$$

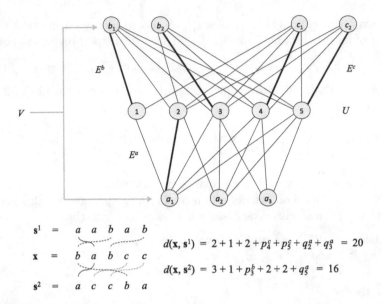

$$\mathbf{s}^1 \;=\; a \quad a \quad b \quad a \quad b$$

$$\mathbf{x} \;=\; b \quad a \quad b \quad c \quad c$$

$$\mathbf{s}^2 \;=\; a \quad c \quad c \quad b \quad a$$

$$d(\mathbf{x}, \mathbf{s}^1) = 2 + 1 + 2 + p_4^c + p_5^c + q_2^a + q_3^a = 20$$

$$d(\mathbf{x}, \mathbf{s}^2) = 3 + 1 + p_3^b + 2 + 2 + q_5^a = 16$$

Fig. 2. Graph G associated with $\mathbf{s}^1 = aabab, \mathbf{s}^2 = accba$: $l(a) = 3, l(b) = l(c) = 2$; the matching encodes $\mathbf{x} = babcc$ (as implied by thick arcs in G). Distances from the positions of homologous characters of $\mathbf{x}, \mathbf{s}^1$ and $\mathbf{x}, \mathbf{s}^2$ are shown down left.

No-cross constraints can be written

$$y_{22}^a + y_{31}^a \leq 1 \qquad y_{32}^a + y_{41}^a \leq 1 \qquad y_{42}^a + y_{31}^a \leq 1 \qquad y_{52}^a + y_{41}^a \leq 1$$

$$y_{33}^a + y_{42}^a + y_{51}^a \leq 1 \qquad\qquad y_{43}^a + y_{52}^a \leq 1$$

$$y_{22}^b + y_{31}^b \leq 1 \qquad\qquad y_{32}^b + y_{41}^b \leq 1 \qquad\qquad y_{42}^b + y_{51}^b \leq 1$$

$$y_{22}^c + y_{31}^c \leq 1 \qquad\qquad y_{32}^c + y_{41}^c \leq 1 \qquad\qquad y_{42}^c + y_{51}^c \leq 1$$

or in the lifted form (7). But as observed (Proposition 1), these inequalities are useless for the rank distance defined in [6].

Valid inequalities

As shown by Kaparis and Letchford [9], multiple knapsack constraints (6) can be strengthened by *global cover inequalities* [3,8]. Let T^i contain all the triples $t = (\alpha, r, k)$ indexing the y-variables occurring in the i-th inequality (6). A cover is a set $C \subseteq T^i$ such that

$$\sum_{t \in C} c_t > d$$

and is said to be minimal if $C - \{t\}$ is not a cover for any $t \in C$. A *local cover inequality* is a constraint of the form

$$\sum_{t \in C} y_t \leq |C| - 1 \qquad\qquad (8)$$

where C is a minimal cover.

Let $\tilde{\mathbf{y}}$ be a fractional solution of (1)–(6). Separating $\tilde{\mathbf{y}}$ by a cover inequality (8) means finding $C \subseteq T^i$ such that (8) is violated by $\tilde{\mathbf{y}}$. The violation is

$$viol(\tilde{\mathbf{y}}) \;=\; \sum_{t \in C} \tilde{y}_t - |C| + 1 \;=\; 1 - \sum_{t \in C}(1 - \tilde{y}_t)$$

Defining $v_t = 1 - \tilde{y}_t \geq 0$, we reduce the problem of finding a maximally violated inequality (8) to the 0-1 knapsack problem

$$\max \quad \sum_{t \in T^i} v_t z_t \tag{9}$$

$$\sum_{t \in T^i} c_t z_t \leq \sum_{t \in T^i} c_t - d - 1 = b^i$$

$$z_t \in \{0, 1\} \qquad\qquad t \in T^i$$

a solution $\mathbf{z}$ of which is the incidence vector of the triples in T^i that are *not* part of the cover. The non-negativity of v_t, c_t guarantees that an optimal solution identifies a minimal cover.

Proposition 2. *A cover inequality that is maximally violated by a given fractional solution to (1)–(6) can be found in time $O(n^5)$.*

Proof. Problem (9) is solved by the following recursion:

$$v(t, \beta) = \max\{v(t - 1, \beta), v(t - 1, \beta - c_t) + v_t\}$$
$$v(t, 0) = 0 \tag{10}$$

for $t \in T^i$ and $\beta = 1, \ldots, b^i$. The theorem immediately derives from observing that $|T^i| \leq l(n + 1) \leq 4n(n + 1)$ and the weights c_t fulfill $c_t \leq n$ for any $t \in T^i$. □

Indeed, the minimal covers are subject to assignment conditions (1)–(2). However, such conditions are always fulfilled by a violated inequality: in fact, should more than one variable occur both in (8) and in one of (1)–(2), no more than $|C| - 1$ variables out of those indexed in C could get value 1.

A cover C can be strengthened by up- and down-lifting [11] as follows:

$$\sum_{t \in C \setminus D} z_t + \sum_{t \notin C} \gamma_t z_t + \sum_{t \in D} \delta_t z_t \leq |C \setminus D| + \sum_{t \in D} \delta_t - 1 \tag{11}$$

where $D \subset C$. In this way a *lifted cover inequality* (LCI) is obtained. A *global LCI* (GLCI) shares with a local one the same form (11). The difference lies in the way up- and down-lifting coefficients γ, δ can be computed: as in local LCIs, one can solve a 0-1 knapsack problem per coefficient, but [9] gives evidence that computing lifting via a multiple 0-1 knapsack is more effective.

3 Computational Experience

We tested formulation (1)–(6) with the min-max objective

$$\min \quad d \qquad (12)$$

The model was solved using IBM Ilog Cplex version 12.6 with standard settings and single processor. We set a cutoff to halt search as soon as

– either the root gap $\frac{d_{UB} - d_{LP}}{d_{LP}}$ goes below 0.2%, where d_{LP} is the value of the LP relaxation at root node and d_{UB} is the value of the best integer solution found;
– or the program has run for 3600 s.

The experiments were run on a 8-core i7 Intel processor 2.597 GHz with 8 GB RAM and Windows 7 Professional 64 bits operating system.

Test bed

The test bed of our experience consists of both artificial instances imitating DNA sequences and biological samples from http://www.embl.org/. Each artificial instance is obtained by a number of random perturbations of a "seed" string of length n with elements chosen in an alphabet A with four symbols. A seed is generated randomly according to a uniform probability of occurrence of each $\alpha \in A$, independently on position.

From a seed, m new strings are then generated to obtain a problem instance: each string is a perturbation of the seed according to a substitution matrix $\mathbf{S}$, whose rows and columns are associated with the symbols of A. The probability of changing a symbol into another is computed after the elements $s_{\alpha\beta}$ of $\mathbf{S}$:

$$p(\alpha, \beta) = \frac{s_{\alpha\beta}}{\sum_{\gamma \in A} s_{\alpha\gamma}}$$

We assumed $s_{\alpha\alpha}$ identical for all $\alpha \in A$ and $s_{\alpha\beta} = 1$ for $\alpha \neq \beta$: therefore the probability that a derived string preserves any given symbol α of the seed is $s_{\alpha\alpha}$ times larger than that of changing it in any other symbol. In our samples (denoted as diag_$s_{\alpha\alpha}$), we adopted $s_{\alpha\alpha} = 1, 3, 5$. Seed positions are identically affected by the changes expressed by $\mathbf{S}$.

Test outcome

We generated a first set of instances (square problems) with $m = n$ and $n = 50, 100, 150, 200, 300, 400$ using a uniform pseudo-random generator. The outcome for such instances is reported in Table 1.

Results in Table 1 highlight the good performance of our model. All instances are solved in extremely manageable times, optimally or with a negligible cutoff gap ($\leq 0.2\%$): on average, we solved problems with 400 strings of length 400 in slightly more than ten minutes. Standard deviation of results testifies a sparse range of difficulty of the problems generated, thus indicating a test with no

Table 1. Computational test (I): simulation of DNA sequences, square problems.

problem		nodes		time (sec)		root gap	
sample	size ($n = m$)	average	standard deviation	average	standard deviation	average	standard deviation
diag_1	50	32071.13	52450.57	59.58	99.15	0.6056	0.1453
	100	18946.63	20771.31	183.88	175.47	0.2603	0.0445
	150	1477.33	1551.04	90.08	62.31	0.1966	0.0246
	200	589.40	737.10	119.87	90.44	0.1877	0.0148
	300	302.33	264.69	342.76	156.35	0.1728	0.0213
	400	152.00	102.57	636.03	230.04	0.1602	0.0402
diag_3	50	19655.97	20251.82	36.10	36.70	0.6720	0.1747
	100	16823.67	19238.87	146.32	140.25	0.2518	0.0568
	150	2154.77	1919.04	105.81	63.87	0.2061	0.0266
	200	737.10	1474.77	123.06	123.57	0.1873	0.0295
	300	192.33	189.39	251.68	114.84	0.1707	0.0297
	400	146.00	115.39	632.05	261.98	0.1661	0.0334
diag_5	50	40820.60	163209.53	67.62	260.50	0.6397	0.1415
	100	13034.30	31779.50	105.99	202.37	0.2507	0.0406
	150	1497.40	1541.88	78.77	50.96	0.1924	0.0257
	200	419.97	533.73	92.10	59.53	0.1874	0.0200
	300	280.03	359.23	280.74	164.38	0.1743	0.0265
	400	174.00	185.45	614.48	363.84	0.1567	0.0387

uncontrolled bias. We note an inverse correlation between branch-and-bound nodes and instance size, to be explained with the different impact of cutoff on the search tree as the solution value increases (rank distance grows in fact quadratically with string length). Such a correlation is however direct when considering CPU time. Finally, the small gaps at root (LP relaxation vs. best integer found, last column of Table 1) indicate that the multi-weighted matching model is a good way to represent the structure of the CSP under rank-distance.

The second set of artificial instances consists of rectangular problems. In a first subset, we keep constant the target set size ($m = 50$) and vary the string lengths from $n = 100$ to $n = 600$, step 100. For each n, we generated and solved 10 random instances. Symmetrically, in another subset we keep constant the string length ($n = 100$) and vary the target set size from $m = 100$ to $m = 1000$, step 100. Also in this case, we solved 10 random repetitions for each m. Results are shown in Tables 2 and 3.

Again, we observe a good model performance: the algorithm was seldom halted for expired time limit (3600 s), and solution time remains manageable also for very large problems (about 45 min in the worst case). As in square problems, we record an inverse correlation of the search tree nodes and m (but not n).

Table 2. Computational test (II): simulation of DNA sequences, rectangular problems with 50 strings and variable length n (average values on 10 repetitions per size).

length (n)	gap closed (%)	average #nodes	average time (sec)	average gap at root (%)
100	100 %	61083.80	166.74	0.7196
200	100 %	82697.70	411.08	0.7784
300	100 %	78889.00	661.49	0.7521
400	80 %	179059.70	1996.86	0.9438
500	50 %	181677.70	2762.45	0.9837
600	70 %	99378.20	1934.09	0.8314

Table 3. Computational test (III): simulation of DNA sequences, rectangular problems with fixed length ($n = 100$) and variable number of strings (average values on 10 repetitions per size). All problems were closed at cutoff value.

#strings (m)	average #nodes	average time (sec)	average gap at root (%)
100	7614	86.74	0.24605
200	850.2	75.96	0.18553
300	151	89.47	0.1675
400	97	161.80	0.17069
500	103	290.41	0.15061
600	123	581.68	0.1733
700	28	408.03	0.13433
800	71	1231.48	0.14502
900	75	1579.70	0.14172
1000	28	1458.41	0.1367

For further validation, we replicated the experiments carried out in [6]. These tests aim at finding rank-central n-nucleotides strings between pairs of sequences. We used the same substrings as [6], extracted from Homo Sapiens V00662, Pan Paniscus D38116, Equus Asinus X97337 (first 200 nucleotides), Rattus Norvegicus X14848, Mus Musculus V00711, Myoxusglis AJ001562, Bos Taurus V00654 (first 150 nucleotides). In all cases we obtained the optimal solution in a fraction of the computational time used by the heuristic [6] to find a non-optimal solution.

4   Conclusions and Future Research

We developed and tested an integer linear programming formulation of the CLOSEST STRING PROBLEM (CSP) under the rank distance defined in [6,7]. Our experiments show that, even for quite large problems, the use of local search heuristics is not justified, since they are outperformed by integer linear programming. In fact, using Cplex 12.6 we could solve problems with up to

400 quaternary strings of 400 elements each in few minutes, and tackle rectangular problems of larger size in very reasonable CPU time. Moreover, replicas of experiments in [6] show that our method finds optimal solutions more quickly than recent heuristics do to find suboptimal solutions.

According to our experiments, room for improvement is to be searched in rectangular problems with either a large target set or very long strings. A possible approach might be that of exploiting valid inequalities. Natural candidates are cover inequalities and their global lifting, that is, as suggested in [9], sequential lifting taking the whole multi-knapsack constraint into account.

References

1. Arbib, C., Labbé, M., Servilio, M.: Scheduling two chains of unit jobs on one machine: a polyhedral study. Networks **58**(2), 103–113 (2011)
2. Arbib, C., Servilio, M., Ventura, P.: Improved integer linear programming formulations for the 0–1 closest string problem (2015, submitted)
3. Balas, E.: Facets of the knapsack polytope. Math. Program. **8**, 146–164 (1975)
4. Chimani, M., Woste, M., Bocker, S.: A closer look at the closest string and closest substring problem. In: Proceedings of the 13th Workshop on Algorithm Engineering and Experiments – ALENE, pp. 13–24 (2011)
5. Della Croce, F., Giarraffa, M.: The selective fixing algorithm for the closest string problem. Comput. Oper. Res. **41**, 24–30 (2014)
6. Dinu, L.P., Ionescu, R.: An efficient rank based approach for closest string and closest substring. PLoS ONE **7**(6), e37576 (2012). doi:10.1371/journal.pone.0037576
7. Dinu, L.P., Popa, A.: On the closest string via rank distance. In: Kärkkäinen, J., Stoye, J. (eds.) CPM 2012. LNCS, vol. 7354, pp. 413–426. Springer, Heidelberg (2012)
8. Hammer, P.L., Johnson, E.L., Peled, U.N.: Facets of regular 0-1 polytopes. Math. Program. **8**, 179–206 (1975)
9. Kaparis, K., Letchford, A.N.: Local and global lifted cover inequalities for the 0-1 multidimensional knapsack problem. Eur. J. Oper. Res. **186**, 91–103 (2008)
10. Levenshtein, V.I.: Binary codes capable of correcting deletions, insertions, and reversals (in Russian). Doklady Akademii Nauk SSSR **163**(4), 845–848 (1965). English translation in Soviet Physics Doklady **10**(8), 707–710 (1966)
11. Roy, T.J., Wolsey, L.A.: Solving mixed integer programming problems using automatic reformulation. Oper. Res. **35**, 45–57 (1987)

Unrelated Parallel Machine Scheduling Problem with Precedence Constraints: Polyhedral Analysis and Branch-and-Cut

Mohammed-Albarra Hassan[1,2]([✉]), Imed Kacem[1], Sébastien Martin[1], and Izzeldin M. Osman[3]

[1] LCOMS EA 7306, Université de Lorraine, 57000 Metz, France
{barra.hassan,imed.kacem,sebastien.martin}@univ-lorraine.fr
[2] University of Gezira, Wadmedani, Sudan
[3] Sudan University of Science and Technology, Khartoum, Sudan
izzeldin@acm.org

Abstract. We consider the problem of unrelated parallel machines with precedence constraints (UPMPC), with the aim of minimizing the makespan. Each task has to be assigned to a unique machine and no preemption is allowed. In this paper, we show the relation between the interval graph and the UPMSPC problem. We propose valid inequalities and study the facial structure of their polytope. Facets are presented to strength the associated integer linear program formulation to help in solving the global problem. We develop a Branch and Cut algorithm for solving the problem and present some experimental results.

Keywords: Polyhedral · Valid inequalities · Unrelated parallel machines · Scheduling · Precedence constraints · Branch-and-Cut

1 Introduction

The problem under consideration is to schedule n jobs on m machines which are arranged in parallel with the aim of minimizing the total completion time. Let J be the set of the jobs and M be the set of the parallel machines. A precedence constraint between two jobs j_1 and j_2 is denoted by $(j_1 \prec j_2)$ and it requires that job j_2 cannot start to be processed until job j_1 will finish its processing. The graph associated with the jobs is denoted by $D = (U, A)$, where U is the set of vertices associated at each job J and A denotes the set of arcs associated with each precedence constraint. We call this graph the precedence graph. We take also the case where $\{u, v, w\} \subseteq U$ such that u before v and v before w then u before w. We consider also the speeds for all machines denoted by s_i, where $i \in M$. Every job $j \in J$ has a processing time p_j and its effective processing time depends on the selected machine i, where $p_{ij}=p_j s_i$. Each machine $i \in M$ cannot process more than one job at a given time. Furthermore, machines have different speeds and preemption of jobs is not allowed. According to the well-known $\alpha|\beta|\gamma$ scheduling problem classification scheme proposed initially by Graham et al. [10], the problem can be denoted as $R|prec|C_{max}$.

© Springer International Publishing Switzerland 2016
R. Cerulli et al. (Eds.): ISCO 2016, LNCS 9849, pp. 308–319, 2016.
DOI: 10.1007/978-3-319-45587-7_27

According to this definition, the problem of unrelated parallel machine with precedence constraints (URPMPC) has many applications in various fields, such as in cloud computing, projects scheduling, textile, semi-conductor manufacturing. The applications occur also in many production and service industries, including telecommunications, health care, and bank service, generally in all situations in which resources have to be allocated to activities over time considering the jobs dependencies. The majority of the literature on parallel machine scheduling considers the case of identical machines. The problem is known to be NP-hard in the strong sense (see Garey and Johnson [10]). Among the numerous papers dealing with such problems, we may cite the following references that concern the minimization of the makespan. Several methods have been proposed to solve this problem. In [11], Martello *et al.* proposed exact and approximation algorithms for optimizing maximum completion time in unrelated parallel machine scheduling without precedence constraints. For the problem of $R||C_{max}$ Ghirardi *et al.* in [8] developed a recovering beam search algorithm. Gacias *et al.* in [3] proposed different methods for solving parallel machine scheduling problem with precedence constraints and setup times between the jobs. They proposed dominance conditions based on the analysis of the problem structure and an extension to setup times of the energetic reasoning constraint propagation algorithm. An exact branch-and-bound procedure and a climbing discrepancy search heuristic based on these components are defined. In [1] Alessandro *et al.* studied the problem of m parallel dedicated machines with a regular criterion. Chain precedence constraints among the tasks, deterministic processing times and processing machine of each task are given which can be viewed as a special case of unrelated machines. They proposed computational complexity results and solution algorithms for some special cases. When the precedence relations among the tasks are given by two chains, they provided efficient solution algorithms for the minimization of the weighted sum of completion times and the number of tardy jobs. In [14] the problem of unrelated parallel machine with setup time was studied. An improved mixed-integer linear formulation was proposed and a Lagrangian heuristic was developed to solve the problem. Kumar [15] proposed approximation algorithms for $R|prec|C_{max}$ and $R|prec|\sum_j W_j C_j$ problems.

However, the literature on parallel machine scheduling with precedence constraints is quite limited. To the best of our knowledge, there are few works on polyhedral study for the problem under consideration. Coll *et al.* [13] proposed an integer model for $R|prec|Cmax$ problem, this model is based on partitioning linear ordering. Some valid inequalities have been derived. Also facet defining inequalities are presented. Mokotoff in [5] dealt with the polyhedral structure of the scheduling problem $R||Cmax$. The authors proposed a mixed integer program and identify some valid inequalities for fixed values of the maximum completion time.

In this paper, we first describe in Sect. 2 an Integer Linear Program (ILP) for the problem. In the following we consider a sub problem, where all valid inequalities remain valid in the general ILP. The facial structure of the polytope associated with the sub problem is investigated in Sect. 3, and facets are presented. The class of these inequalities is based on forbidden subgraphs. A branch and cut algorithm and the associated experimental results are presented in Sect. 4. Concluding remarks and perspectives are made in the last section.

2 Mathematical Formulation

In this section we will recall the Integer Linear Program given in [9] which is compared with two mathematical models found in the literature for solving the same problem and the computational results show that this mathematical model obtained the best results for solving UPMSPC problem. This ILP considers the beginning of the job and the relation between jobs if they processed on the same machine. It also verifies if one job processed before another job or at the same time.

This model is based on interval graph and an m-clique free graph, because the graph induced by each solution must be interval graph, and $m - cliquefree$ graph. A graph $G(V, E)$ is called interval graph if its vertices V can be represented by interval I_V of the real line such that two vertices are adjacent if and only if the corresponding intervals intersect [2]. Let a clique K, in G be a subset of the vertices, $K \subseteq V$, such that every two distinct vertices of K are adjacent. This is equivalent to the condition that the subgraph of G induced by K is complete. Let $I \subseteq E$ be a subset of edges, the graph $G[I]$ is an $m - clique$ free if and only if $G[I]$ does not contain a clique of size strictly greater than m.

In the following we will present the IPL.

For each job we consider a variable defining the beginning of job.

$y_j \in \mathbb{N}^+$ is the beginning of job j for all job j in J.

We consider binary variables for assigning the jobs to machines.

$$x_j^i = \begin{cases} 1 \text{ if job } j \text{ on machine } i, \\ 0 \text{ otherwise,} \end{cases} \quad \forall i \in M, \forall j \in J.$$

For all two jobs sharing a time unit, the associated subgraph must be interval and m-clique free. For this reason we consider the following binary variables to know if two jobs share a time unit.

The variables z_{j_1,j_2}, and $\bar{z}_{j_1,j_2}$ correspond to the edges of the induced interval subgraph.

$$z_{j_1,j_2} = \begin{cases} 1 \text{ if job } j_1 \text{ and job } j_2 \text{ processed at the same time} \\ 0 \text{ otherwise} \end{cases} \quad \forall j_1, j_2 \in J.$$

$$\bar{z}_{j_1,j_2} = \begin{cases} 1 \text{ if job } j_1 \text{ processed before or at the same time with job } j_2 \\ 0 \text{ otherwise} \end{cases} \quad \forall j_1, j_2 \in J.$$

The variable $\bar{z}_{j_1,j_2}$ demonstrate the precedence between jobs. For every $j \in J$:
$C_j \in \mathbb{N}^+$ is the completion time of job j.

$C_{\max} \in \mathbb{N}^+$ is the maximum of C_j.

The URPMPC can be solved by the following ILP, denoted by (P):

$$\min C_{max}$$

$$y_j + \sum_{i \in M} p_{ij} x_j^i \le C_{max}, \qquad\qquad \forall j \in J, \qquad\qquad (1)$$

$$\sum_{i \in M} x_j^i = 1, \qquad\qquad \forall j \in J, \qquad\qquad (2)$$

$$x_{j_1}^i + x_{j_2}^i \le 2 - z_{j_1,j_2} \qquad\qquad \forall j_1, j_2 \in J, \forall i \in M, \qquad\qquad (3)$$

$$y_{j_1} + \sum_{i \in M} p_{ij_1} x_{j_1}^i \le y_{j_2}, \qquad\qquad \forall (j_1, j_2) \in A, \qquad\qquad (4)$$

$$y_{j_1} + \sum_{i \in M} p_{ij_1} x_{j_1}^i \le y_{j_2} + C\bar{z}_{j_2,j_1}, \qquad \forall j_1, j_2 \in J, \qquad\qquad (5)$$

$$\bar{z}_{j_1,j_2} + \bar{z}_{j_2,j_1} \le 1 + z_{j_1,j_2}, \qquad\qquad \forall j_1, j_2 \in J, \qquad\qquad (6)$$

$$\sum_{(j_1,j_2) \in E(\bar{I})} z_{j_1,j_2} - \sum_{(j_1,j_2) \in E \setminus E(\bar{I})} z_{j_1,j_2} \le |E(\bar{I})| - 1, \qquad \forall \bar{I} \subseteq \bar{\mathcal{I}}, \quad (7)$$

$$\sum_{(j_1,j_2) \in E(K)} z_{j_1,j_2} - \sum_{(j_1,j_2) \in E \setminus E(K)} z_{j_1,j_2} \le |E(K)| - 1, \qquad \forall K \subseteq \mathcal{K}, \quad (8)$$

The objective function is to minimize the makespan. Inequalities (1) ensure that the beginning time for each job plus its processing time is less than or equal to the total completion time. Inequalities (2) controls each job to be processed on one machine. Inequalities (3) guarantee that there is no two jobs processed on the same machine at the same time. Inequalities (4) control the precedence constraints between jobs then job j_1 has to be completed before job j_2 can be processed. Inequalities (5) ensure that the starting of any job must be after the finishing of its predecessor. Inequalities (6) ensure that, if the job j_1 processed before or at the same time with job j_2, and job j_2 processed before or at the same time with job j_1 then job j_1 and j_2 will be process at the same time. If we consider a solution given by the vector (z), then the induced subgraph $G = (V, E)$ where for each job $j \in J$ we associate a vertex $v_j \in V$ and for all $z_{j_1 j_2} = 1$ we associate an edge $uv \in E$ must be an interval graph and the clique of maximum size must be less or equal to m. We denote by $\bar{\mathcal{I}}$ the set of all non interval induced subgraph, and by $\mathcal{K}$ the set of all cliques of size greater or equal to $m + 1$. The inequalities (7) ensure that all induced subgraphs are interval graphs. The inequalities (8) ensure that all induced subgraphs have no clique of size greater or equal to $m + 1$. In the following we are interested in the analysis of the polytope associated with the Interval and m-clique free sub problem (IMCFSP). This polytope associated with the ILP obtained by projection to z variables of P.

3   Interval and $m - Cliquefree$ Subgraph Problem

In this section, we present the problem of finding an interval and $m - cliquefree$ subgraph. Let $\mathcal{I} := \{I \subseteq E \mid G[I]$ be the set of edges set inducing interval and *m-clique free subgraph*}. The vector z^I is called the incidence vector associated with I. We define the IMCFSP polytope as follows:

$$P_{\mathcal{I}}(G, m) := conv\{z^I \in \{0, 1\}^{|E|} | I \in \mathcal{I}\},$$

Now, we analyze the dimension of the polytope.

Proposition 1. *Polytope $P_{\mathcal{I}}(G, m)$ is full dimensional.*

In the following we will prove that the trivial inequality is facet.

Proposition 2. *Let $e \in E$. The trivial inequality $z_e \geq 0$ defines a facet of $P_{\mathcal{I}}(G, m)$.*

3.1 Forbidden Subgraphs Inequalities

In this section, we define some families of graphs. In Fig. 1, we can see five forbidden subgraphs [4]. Recall that a hole is a cycle without chord. When the induced subgraph has one of the forbidden graphs, then it is not interval graph. Recall that, clique is new forbidden subgraph. If we have clique of size strictly grater than m, then the subgraph is not m-clique free. We introduced this family of forbidden subgraphs because our sub problem is to find a valid solution, which is interval and m-clique free.

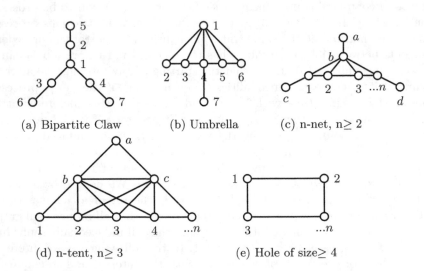

(a) Bipartite Claw   (b) Umbrella   (c) n-net, n$\geq$ 2

(d) n-tent, n$\geq$ 3   (e) Hole of size$\geq$ 4

Fig. 1. Forbidden subgraphs characterization

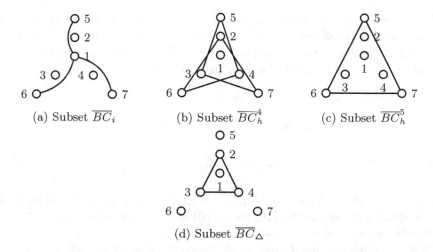

(a) Subset $\overline{BC}_i$　　　　(b) Subset $\overline{BC}_h^4$　　　　(c) Subset $\overline{BC}_h^5$

(d) Subset $\overline{BC}_\triangle$

Fig. 2. Subsets of complementary bipartite claw

Bipartite Claw. In this subsection we give inequalities to avoid the bipartite claw forbidden subgraph. An example is given in Fig. 2d.

We give some notations to help in analyzing the bipartite claw forbidden subgraph.

Consider the complete graph K_7 with seven nodes. We partition this graph to BC and $\overline{BC}$. We denote by BC the set of all edges that construct the bipartite claw as in Fig. 2d. Furthermore, we denote by $\overline{BC}$, the set of edges in the complementary graph. $\overline{BC}$ is partitioned in the following denoted subsets: $\overline{BC}_h^4$ is the set of all edges that formulate a hole of size 4 in bipartite claw. $\overline{BC}_\triangle$ is the set of the three edges, such that when we add one of these edges then we obtain a central triangle. $\overline{BC}_i$ is the set of edges that formulate a triangle with the inner vertex. $\overline{BC}_h^5$ is the set of all edges that formulate a hole of size 5 in bipartite claw. Figure 2 shows these subsets.

- $BC = \{(1,2),(1,3),(1,4),(2,5),(4,7),(3,6)\}$.
- $\overline{BC} = \{(7,1),\ (7,2),\ (7,3),\ (7,5),\ (7,6),\ (6,1),\ (6,2),\ (6,4),\ (6,5),\ (5,1),\ (5,3),\ (5,4),\ (4,2),\ (4,3),\ (3,2)\}$.
- $\overline{BC}_h^4 = \{(3,5),(2,6),(5,4),(2,7),(3,7),(4,6)\}$.
- $\overline{BC}_\triangle = \{(2,3),(2,4),(3,4)\}$.
- $\overline{BC}_i = \{(1,5),(1,6),(1,7)\}$.
- $\overline{BC}_h^5 = \{(5,6),(5,7),(6,7)\}$.

We consider two cases, when $m = 2$, and when $m \geq 3$.

If $m = 2$ then the following inequality is valid.

$$\sum_{e \in BC} z_e \leq 5. \tag{9}$$

Indeed, when we add an edge from $\overline{BC}_\triangle$, by definition, the resulting subgraph will contain a clique of size 3, which is not m-clique free in this case, as well it is $2 - net$. Moreover, when we add an edge $e \in \overline{BC}_h^4$, then we obtain a *hole*. If we add another edge to break this hole then we obtain a clique of size 3.

Proposition 3. *The inequality* (9) *defines a facet, when* $m = 2$.

Now if $m \geq 3$ then the following inequality is valid.

$$\sum_{e \in BC} z_e - \sum_{e \in \overline{BC}} z_e \leq 5. \tag{10}$$

Indeed, if we add one edge may be the resulting graph is interval and m-clique free. Inequality (9) dominates (10) since (10) is a linear combination of (9) and the trivial inequalities $-z_e \leq 0$ where $e \in \overline{BC}$.

Now, we strength this inequality by analyzing when the resulting graph is a valid solution or not. If we add one edge of $\overline{BC}_\triangle$ to the bipartite claw, then the resulting subgraph contains $2 - net$. If we add one edge of $\overline{BC}_h^5$ or $\overline{BC}_h^4$ we obtain a hole of size 5 or 4 respectively. It is clear when we add any one, two or three edges of $\overline{BC}_i$ then the resulting graph become interval and m-clique free.

$$\sum_{e \in BC} 2z_e - \sum_{e \in \overline{BC}_h^4 \cup \overline{BC}_\triangle} z_e - 2 \sum_{e \in \overline{BC}_i} z_e \leq 10 \tag{11}$$

Proposition 4. *Inequality* (11) *defines a facet, when* $m \geq 3$.

Umbrella Inequalities. For the umbrella subgraph as shown in Fig. 3d, let $H = (U, E_u)$ be a graph that formulates the umbrella, and $\overline{E}_u$ be a set of the complementary edges for H. In the following we will give a family of valid inequalities that delete the umbrella subgraphs. To analyze this forbidden subgraph we need the following notations:

Let $E_u^i \subset E_u$ is the set of the inner three edges in umbrella subgraph. $E_u^t \subset \overline{E}_u$ is the set of the edges, such that when we add one of these edges to the umbrella we create new triangle. $E_u^c \subset E_c$ the dashed edges in Fig. 3b. $E_u^a \subset E_u$ be the set of the around edges. $E_u^h \subset \overline{E}_u$ be the set of edges if they connected they will formulate a hole of size 4 or of size 5.

- $E_u^i = \{(1,3),(1,4),(1,5)\}$.
- $E_u^t = \{(1,7),(3,7),(5,7)\}$.
- $E_u^c = \{(2,4),(3,5),(4,6)\}$.
- $E_u^a = \{(1,2),(2,3),(3,4),(4,5),(5,6),(1,6),(4,7)\}$.
- $E_u^h = \{(2,7),(6,7),(2,5),(2,6),(3,6)\}$.

Remark that, the graph induced by $H^u = \{E_u^i \cup E_u^t \cup E_u^c \cup E_u^a \cup E_u^h\}$ is a complete graph.

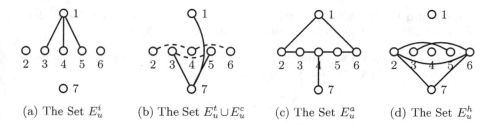

(a) The Set E_u^i (b) The Set $E_u^t \cup E_u^c$ (c) The Set E_u^a (d) The Set E_u^h

Fig. 3. Subsets of umbrella and its complementary

In the umbrella subgraph when $m = 2$, we need the subgraph must be triangle free and hole free. If we remove the edges $(1,5)$, and $(1,3)$, then we have no triangle but we obtain a hole of size 4. Thus in order to be valid, the induced subgraph of umbrella for $m = 2$, at least we need to eliminate 4 edges, because with 7 edges we will have either a hole or a triangle. We can deduce that inequality (12) is valid, for $m = 2$.

$$\sum_{e \in E_u \setminus \{(4,7)\}} z_e + z_{(2,6)} + z_{(2,5)} + z_{(3,6)} + z_{(3,5)} \leq 5. \tag{12}$$

When $m = 3$, to keep all edges of E_u it is necessary to add at least one edge of E_u^t. Moreover, when we add an edge from E_u^c in this case the subgraph contains a clique of size 4. If we add an edge from E_u^h, then the induced subgraph will contain a hole.

Thus, the valid inequalities when $m = 3$ will be:

$$\sum_{e \in E_u^a \setminus \{(4,7)\}} z_e + z_{(2,6)} + z_{(2,5)} + z_{(3,6)} \leq 5. \tag{13}$$

When $m \geq 4$, in order to find a valid solution we can add also the edges from E_u^c. Then, the valid inequalities when $m \geq 4$ will be:

$$\sum_{e \in E_u^a} z_e - \sum_{e \in E_u^t \cup E_u^c} z_e \leq 6. \tag{14}$$

Proposition 5. *Inequality (14) defines a facet if $m \geq 4$.*

Hole Inequalities. It is convenient to define a hole here as an induced subgraph of G isomorphic to C_k for some $k \geq 4$, [2]. The hole C is a forbidden subgraph as in Fig. 1e. Let C denote the set of edges that construct the hole, $C = \{(u_1, u_2), (u_2, u_3), ..., (u_{|C|-1}, u_{|C|}), (u_{|C|}, u_1)\}$. If $(i + k) > |C|$, then $u_{i+k} = u_{i'}$, $i' = (i + k) - |C|$. Let $\overline{C}$ denote the set of all chords of hole C.

Suppose we have a hole of size 4, this graph is non-interval graph. The induced subgraph of hole is valid only if we add at least one chord.

Proposition 6. *For a hole C, the minimum number of necessary chords that will be added to the hole to be an interval graph is $|C| - 3$, when $|C| \geq 4$.*

In the following we will present valid inequalities for the hole forbidden subgraph. If $m = 2$, then inequality (15) is valid.

$$\sum_{e \in C} z_e \leq |C| - 1. \tag{15}$$

Let $e \in C$, if we add one chord to $C \subset \{e\}$, then we will obtain a triangle and it is not valid for $m = 2$. If $m \geq 3$, then Inequality (16) is valid.

$$\sum_{e \in C} (|C| - 3) z_e - \sum_{e \in \overline{C}} z_e \leq (|C| - 1)(|C| - 3). \tag{16}$$

Proposition 7. *Let C be a hole of size greater than 3, then Inequality (16) associated with cycle C defines a facet if $m \geq 3$.*

3.2   Clique Inequalities

In this section, we study the clique subgraph and provide valid inequalities and facets.

Proposition 8. *Let K be a clique of size $m + 1$, such that $K \subseteq V$. Let $e \in E \backslash E(K), e' \in E(K)$, then the graph induced by $E(K) \backslash \{e'\} \cup \{e\}$ is an interval graph and $m - clique$ free.*

Proposition 9. *Let K be a clique of size $m + 1$. The inequality (17) defines a facet.*

$$\sum_{e \in E(K)} z_e \leq |E(K)| - 1. \tag{17}$$

4   Branch and Cut Algorithms

In this section, we present a branch and cut algorithm for URPMPC problem. After illustrating many facets in the previous section, we can apply the cutting plane scheme for these classes of inequalities. Cutting plane algorithms mainly consist in generating constraints by means of a separation procedure. See, for example, (see [12]) for a survey on this domain and (see [7]) for how polyhedral results are used in cutting plane algorithms. The results of the previous sections have allowed us to derive two exact cutting plane algorithms and four heuristic cutting plane algorithms for the sub problem to solve URPMPC. In the following paragraphs we will describe these separation algorithms.

Exact Separation. Let the solution $z^* \in \mathbb{R}$ be a solution of linear relaxation. The separation algorithm consists in finding one bipartite claw violated by z^* then we will add this inequality to the ILP. Remark that, if we select the two first edges $(1,2)$ and $(1,3)$, if $(2z^*_{(1,2)} + 2z^*_{(1,3)} - z^*_{(2,3)} < 2)$ then there does not exist a bipartite claw with in these two edges in this position. With the same argument we test the weight of all partial subgraph to drop non interesting subgraph. This process is used for exact and all heuristic algorithms. The running time of the exact algorithm in the worst case is in $O(n^7)$.

H1-Sep Separation. In this heuristic we start by searching the vertices u_1, u_2, u_3, and u_4 that maximize $(2z_{(1,2)} + 2z_{(1,3)} + 2z_{(1,4)} - z_{(2,4)} - z_{(2,3)} - z_{(3,4)})$. If this value is greater than 4, then we search u_5, u_6, and u_7 such that the BC induced by these vertices is violated by z^*. Using this greedy approach the heuristic running time is in $O(n^4)$.

H2-Sep Separation. This heuristic follows the greedy approach to find a violated BC inequality. Let z^* be the solution of the linear relaxation. We search at each step the best next edge to add in BC. This heuristic has $O(n^2)$ running time. Remark that, we keep the lazy cut for inequalities (7), and (8) presented in [9] to ensure the validity of the optimal solution. We use the same idea for Umbrella separation algorithms.

4.1   Computational Results

To test the efficiency of the inequalities mentioned in Sect. 3.1, we developed the mentioned exact and heuristics separations. All computational results were obtained by using Cplex 12.6 and Java for implementing exact and heuristics algorithms. The ILP with the valid inequalities was tested under the following proposed benchmark of instances.

The processing times are uniformly, distributed between 1 and 100 as it is common in the literature [6]. We generated five different sets of DAG where the graph density is equals 0.15 and calculated as follow $GD = \frac{|E|}{|V|(|V|-1)}$ where E is the set of edges associated with precedence constraints between jobs, and V is the set of vertices associated with jobs, with the following combinations of number of jobs $n = \{10, 12, 14, 16, 18, 20\}$ and the number of machines $m = 3$. The speed of machines generated randomly between 10 and 20. In total 5×6 instances are generated. CPU time required is in seconds. We limit to 3600 s the running time for each instance, and 4.0 GB of RAM.

The result of this test are presented in Table 1. Column "*Cplex*" gives the number of optimal solutions obtained without adding user cuts. Column "*ExactSep*" provides the number of optimal solutions for the exact algorithm. Column "*H1Sep*" and "*H2Sep*" give the numbers of optimal solutions obtained by heuristic *H1Sep* and *H2Sep* respectively. The average of all solved instances is also reported in the same table. We observed that for the instances of $n \geq 16$

Table 1. Number of optimal solutions obtained.

Instance	Cplex	ExactSep	H1Sep	H2Sep
3×10	4/5	4/5	4/5	4/5
3×12	4/5	2/5	3/5	4/5
3×14	1/5	3/5	2/5	2/5
3×16	0/5	2/5	1/5	0/5
3×18	0/5	0/5	1/5	0/5
3×20	0/5	2/5	1/5	0/5
Average	0.16	0.43	0.40	0.26

none of the instances have been solved in the time limit. Whereas, when we use *ExactSep*, and *H1Sep* separation algorithms they are capable to solve more instances. That shows the efficiency of our valid inequalities. We just give a result for small number of machines, but we noticed that Cplex can solve instances with 10 to 12 jobs quickly, because when we add the cuts we increase computational time. However, the exact algorithm and *H1Sep* can solve big number of instances within a reasonable running time.

4.2 Conclusion and Perspectives

In this paper we presented a polyhedral study for this problem. We also proposed families of valid inequalities based on interval and m-clique free subgraph. A polyhedral investigation of the convex hull of these vectors yielded several results on facets for this new polytope. We also designed and implemented a branch-and-cut algorithms based upon families of strong valid inequalities presented in this paper. We separate some forbidden subgraphs. Computational experiments on set of instances have shown that the algorithms are capable to solve many instances to optimality within a reasonable CPU time. Further research in this direction will be helpful to strengthen the integer programming formulations of a large variety of URPMPC problems. In the future work, we will continue on polyhedral study, and will try to find new facets for the polytope associated with this problem. Moreover, we can improve our heuristics. We will work for adding valid inequalities for the other forbidden subgraphs.

References

1. Agnetis, A., Kellerer, H., Nicosia, G., Pacifici, A.: Parallel dedicated machines scheduling with chain precedence constraints. Eur. J. Oper. Res. **221**(2), 296–305 (2012)
2. Schrijver, A.: Combinatorial Optimization: Polyhedra and Efficiency, vol. 24. Springer Science & Business Media, Heidelberg (2003)
3. Gacias, B., Artigues, C., Lopez, P.: Parallel machine scheduling with precedence constraints and setup times. Comput. Oper. Res. **37**(12), 2141–2151 (2010)

4. Lekkeikerker, C., Boland, J.: Representation of a finite graph by a set of intervals on the real line. Fundam. Math. **51**(1), 45–64 (1962)
5. Mokotoff, E., Chrétienne, P.: A cutting plane algorithm for the unrelated parallel machine scheduling problem. Eur. J. Oper. Res. **141**(3), 515–525 (2002)
6. Hall, N.G., Posner, M.E.: Generating experimental data for computational testing with machine scheduling applications. Oper. Res. **49**(7), 854–865 (2011)
7. Aardal, K., Van Hoesel, C.P.M.: Polyhedral techniques in combinatorial optimization II: applications and computations. Stat. Neerl. **53**(2), 131–177 (1999)
8. Ghirardi, M., Potts, C.N.: Makespan minimization for scheduling unrelated parallel machines: a recovering beam search approach. Eur. J. Oper. Res. **165**(2), 457–467 (2005)
9. Hassan, M.-A., Kacem, I., Martin, S., Osman, I.M.: Mathematical formulations for the unrelated parallel machines with precedence constraints. In: Proceedings of the International Conference on Computers & Industrial Engineering, 45th edn. (2015)
10. Graham, R.L., Lawler, E.L., Lenstra, J.K., RinnooyKan, A.H.G.: Optimization and approximation in deterministic sequencing and scheduling: a survey. Ann. Discrete Math. **5**, 287–326 (1979)
11. Martello, S., Soumis, F., Toth, P.: Exact and approximation algorithms for makespan minimization on unrelated parallel machines. Discrete Appl. Math. **75**(2), 169–188 (1997)
12. Jiinger, M., Reinelt, G., Thienel, S.: Practical problem solving with cutting plane algorithms in combinatorial optimization. Comb. Optim. Dimacs **20**, 111–152 (1995)
13. Coll, P.E., Ribeiro, C.C., de Souza, C.C.: Multiprocessor scheduling under precedence constraints: polyhedral results. Discrete Appl. Math. **154**(5), 770–801 (2006)
14. Damodaran, P., Sharma, H.V., Moraga, R.: Scheduling unrelated parallel machines with sequence dependent setup times to minimize makespan. In: Proceedings of the IIE Annual Conference, p. 1. Institute of Industrial Engineers-Publisher (2012)
15. Kumar, V.A., Marathe, M.V., Parthasarathy, S., Srinivasan, A.: Scheduling on unrelated machines under tree-like precedence constraints. Algorithmica **55**(1), 205–226 (2009)

The Multi-terminal Vertex Separator Problem: Polytope Characterization and TDI-ness

Youcef Magnouche[1,2($\boxtimes$)] and Sébastien Martin[1,2]

[1] Lamsade, Université Paris Dauphine, Paris, France
youcef.magnouche@dauphine.fr
[2] LCOMS, Université de Lorraine, Metz, France
sebastien.martin@univ-lorraine.fr
http://www.lamsade.dauphine.fr/
http://lcoms.univ-lorraine.fr/

Abstract. In this paper we discuss a variant of the well-known k-separator problem. Consider the simple graph $G = (V \cup T, E)$ with $V \cup T$ the set of vertices, where T is a set of distinguished vertices called terminals, inducing a stable set and E a set of edges. Given a weight function $w : V \rightarrow \mathbb{N}$, the multi-terminal vertex separator problem consists in finding a subset $S \subseteq V$ of minimum weight intersecting every path between two terminals. We characterize the convex hull of the solutions of this problem in two classes of graph which we call, star trees and clique stars. We also give TDI systems for the problem in these graphs.

Keywords: Vertex separator problem · Total dual integrality · Combinatorial optimization · Polytope characterization

1 Introduction

Let $G = (V \cup T, E)$ be a simple graph with $V \cup T$ the set of vertices, where T is a set of distinguished vertices called terminals, inducing a stable set and E a set of edges. Given a weight function $w : V \rightarrow \mathbb{N}$, the *multi-terminal vertex separator problem* (MTVSP) consists in finding a subset $S \subseteq V$ of minimum weight such that each path between two terminals intersects S. The problem can be solved in polynomial time when $|T| = 2$, [3] but when $|T| \geq 3$, the MTVSP is NP-hard ([6,9]). In this paper we deal with the MTVSP in two specific classes of graph, star trees and clique stars, showing that this problem can be solved in polynomial time for any size of T in these two classes. We show also that the associated polytope is integer and we give a min-max relation for each class. The MTVS problem has applications in different areas like VLSI design, linear algebra, connectivity problems and parallel algorithms. It has also applications in network security, for instance, consider a graph $G = (V \cup T, E)$ representing a telecommunication network, with V the set of routers, T the set of customers and an edge between two vertices represents the possibility of transferring data between each other. We search to set up a monitoring system of minimum cost

© Springer International Publishing Switzerland 2016
R. Cerulli et al. (Eds.): ISCO 2016, LNCS 9849, pp. 320–331, 2016.
DOI: 10.1007/978-3-319-45587-7_28

on some routers, in order to monitor all data exchanged between customers. The set of these routers represents a minimum multi-terminal vertex separator. The MTVSP is a variant of the *k-serparator problem* that consists in partitioning the set of vertices of a graph G, into $k+1$ subsets $\{S, V_1, \ldots, V_k\}$, such that S has a minimum weight and no vertex in V_i is adjacent to a vertex in V_j. Many other variants of the k-serparator problem have been considered in the literature ([2,8]). In [1], the authors discuss the following problem. Given a simple graph $G = (V, E)$ and an integer $\beta(n)$ with $n = |V|$, partition V into three subsets A, B and C such that $|C|$ is minimum, no vertex in A is adjacent to a vertex in B and $\max\{|A|, |B|\} \leq \beta(n)$. In [3] authors consider another variant of the problem. Given a simple graph $G = (V, E)$ with $a, b \in V$ two terminals, the problem here is to partition V into three subsets A, B and C such that $a \in A$, $b \in B$, no edge connecting A and B and the size of the cut induced by C is minimum. They show that this problem can be reduced to a minimum cut problem in an auxiliary graph and then, it can be solved in polynomial time.

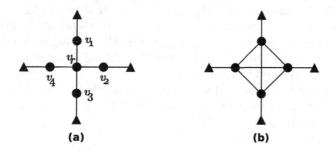

(a)   **(b)**

Fig. 1. Star tree and clique star

The MTVS problem was considered in [6], in which the authors present several valid inequalities and develop a branch-and-cut algorithm to solve the problem. They also present two classes of graph, called star trees, see Fig. 1(a) and clique stars, see Fig. 1(b), on which our work is based. In [10], authors give a linear system for the MTVSP and characterize the class of graph for which it is total dual integral for any size of T, i.e., the dual problem has an integer optimal solution for any integer vertex weight vector. The main motivation of this paper is to derive TDI descriptions for other classes of graph. This is a preliminary work on two specific classes, star trees and clique stars. In this paper we first characterize the polytope of the multi-terminal vertex separators for these two classes of graph and then we give TDI linear systems. The paper is organized as follows, in Sect. 2, we introduce some notations and definitions that we use in the remainder of the paper, in Sect. 3, we characterize the polytope of the multi-terminal vertex separators for two classes of graph, in Sect. 4, we give a total dual integral system for each of these classes of graph.

2 Preliminaries

In this paper, we denote by n the cardinality of V and k the number of terminals in T. A *path* is a set of p distinct vertices $v_1, v_2, \ldots, v_p$ such that for all $i \in \{1, \ldots, p-1\}$, $v_i v_{i+1}$ is an edge. The vertices $v_2, \ldots, v_{p-1}$ are called *internal vertices* of the path. A *terminal path* $P_{tt'}$ is the set of internal vertices of a path P between two terminals $t, t' \in T$, such that $P \cap T = \{t, t'\}$. A terminal path $P_{tt'}$ is *minimal* if there does not exist another terminal path $P_{t_i t_j}$ in the graph, such that $P_{t_i t_j} \subset P_{tt'}$. The *support graph* of an inequality is the graph induced by the vertices of variables having positive coefficient in the inequality. Given a vertex $v \in V \cup T$, we denote by $N(v) \subseteq V \cup T$ the set of vertices adjacent to v. Given a graph H, we denote by $V(H)$ its set of vertices and $E(H)$ its set of edges. Given $x \in \mathbb{R}^V$ and $W \subseteq V \cup T$, we let $x(W) = \sum\limits_{v \in W \cap V} x(v)$. Consider a graph $G = (V \cup T, E)$ and two subgraphs $G_1 = (V_1 \cup T_1, E_1)$, $G_2 = (V_2 \cup T_2, E_2)$ of G. Graph G_1 is said to be *completely included* in G_2, if $V_1 \cup T_1 \subseteq V_2 \cup T_2$.

A star tree $H_k = (V_{H_k} \cup T_{H_k}, E_{H_k})$, Fig. 1(a), where $T_{H_k} = \{t_1, \ldots, t_k\}$, is a tree that is the union of k paths $P_{t_1}, \ldots, P_{t_k}$, such that one end of each P_{t_i} is a common node $v_r \in V_{H_k}$, called the root, and the other end is a terminal t_i.

A clique star $Q_k = (V_{Q_k} \cup T_{Q_k}, E_{Q_k})$, Fig. 1(b), is a graph defined by a clique K_k of k vertices and k disjoint paths $P_{t_1}, \ldots, P_{t_k}$ between all terminals of $T_{Q_k} = \{t_1, t_2, \ldots, t_k\}$ and vertices of K_k.

In the star trees and clique stars the path P_t is refereed as a *branch*. For any star tree H (resp. clique star Q), we denote by $t(H)$ (resp. $t(Q)$) the number of branches of H (resp. Q).

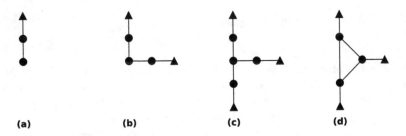

| (a) | (b) | (c) | (d) |

Fig. 2. Star trees and clique stars

Figure 2 gives some star trees and clique stars where the terminals are given by triangles. If $k = 1$, the star tree and the clique star are reduced to a single branch, see Fig. 2(a). If $k = 2$, the star tree and the clique star are reduced to a path between two terminals, see Fig. 2(b).

If $k \geq 3$, the star tree (resp. clique star) with k terminals contains $\binom{k}{k'}$ star trees (resp. clique stars) as subgraphs with $k' \in \{1, \ldots, k\}$ terminals. Let Π (resp. Θ) be the set of all star trees subgraphs of H_k (resp. clique stars subgraphs of Q_k). Note that Π (resp. Θ) contains H_k (resp. Q_k). Let Π^v (resp. Θ^v) be the star trees (resp. clique stars) of Π (resp. Θ) containing vertex $v \in V_{H_k}$ (resp. $v \in V_{Q_k}$).

Let $x \in \{0,1\}^V$ be a vector of variables such that for a vertex $v \in V$, $x(v) = 1$ if v belongs to the separator and $x(v) = 0$ otherwise. The vector x^S is called an incidence vector of the separator S. Consider the vertex weight vector $w \in \mathbb{N}^V$, the MTVSP is equivalent to the following integer linear program P'

$$\min \sum_{v \in V} w(v)x(v) \tag{1}$$

$$x(P_{tt'}) \geq 1 \qquad \forall P_{tt'} \in \Gamma \tag{2}$$
$$x(v) \leq 1 \qquad \forall v \in V \tag{3}$$
$$x(v) \geq 0 \qquad \forall v \in V \tag{4}$$
$$x(v) \ integer \tag{5}$$

where Γ is the set of all terminal paths in G.

Let $P(G,T) = conv(x \in [0,1]^V \,|\, x \ satisfies \ (2))$ be the polytope given by inequalities (2)–(4).

In this paper we consider a star tree H_k and a clique star Q_k satisfying the following hypothesis

1. the number of terminals is at least three, otherwise the linear system (2) and (4) is TDI, [10].
2. each branch of the star tree H_k contains at least one internal vertex, otherwise the linear system (2) and (4) is TDI, [10].

In star trees and clique stars, under the above hypothesis, the polytope $P(G,T)$ is full-dimensional [6].

3   Polytope Characterization

In this section we will characterize the multi-terminal vertex separator's polytope in the star trees and clique stars.

3.1   Star Trees

Proposition 3.11. *In star trees, the polytope $P(G,T)$ is not integral.*

Proof. Consider a star tree $H_k = (V_{H_k} \cup T_{H_k}, E_{H_k})$ with at least one vertex in each branch. Let $\overline{x} \in [0,1]^{V_k}$ be a solution of $P(H_k, T_{H_k})$ defined as follows

- $\overline{x}(v) = 0.5 \qquad \forall v \in N(v_r)$
- $\overline{x}(v) = 0 \qquad \forall v \in V_{H_k} \setminus N(v_r)$

The vector $\overline{x}$ represents a fractional extreme point of $P(H_k, T_{H_k})$, since there is k fractional variables and k terminal path inequalities that are linearly independent and tight by $\overline{x}$. ∎

Consider the following valid inequalities presented in [6]

$$x(V_{H_{k'}} \setminus \{v_r\}) + (k' - 1)x(v_r) \geq k' - 1 \qquad\qquad H_{k'} \in \Pi \qquad (6)$$

We recall that terminal paths are star trees of two terminals and inequalities (2) are included in (6). We notice that all inequalities (6) associated with the star trees of Π with one terminal are dominated by trivial inequalities.

Theorem 3.11. *For any star tree, the polytope given by inequalities (6) and trivial inequalities is integer.*

Proof. Let us assume the contrary and let x^* be a fractional extreme point of the polytope $P(H_k, T_k)$ associated with the star tree H_k, where $|V_{H_k}|$ is minimum (i.e., for all star trees of n' vertices whith $n' < |V_{H_k}|$, the associated polytope is integer). Thus x^* satisfies a unique system of linear independent equalities $\mathcal{A}$

$$x(V_{H_{k'}} \setminus \{v_r\}) + (k' - 1)x(v_r) = k' - 1 \qquad \forall H_{k'} \in \Pi_1 \qquad (7)$$
$$x^*(v) = 1 \qquad \forall v \in V_1 \qquad (8)$$
$$x^*(v) = 0 \qquad \forall v \in V_2 \qquad (9)$$

such that $|\Pi_1| + |V_1| + |V_2| = |V_{H_k}|$, $\Pi_1 \subseteq \Pi$, $V_1 \subseteq V_{H_k}$ and $V_2 \subseteq V_{H_k}$. Moreover we have the following claims

Claim 3.11. *For all $v \in V_{H_k} \setminus \{v_r\}$, $x^*(v) > 0$.*

Proof of claim 3.11. Otherwise, $|V_{H_k}|$ cannot be minimum. ∎

Claim 3.12. *For each branch P_t, $x^*(P_t) \leq 1$.*

Proof of claim 3.12. Otherwise, the variables of internal vertices of P_t must belong to (8) and to no other equality. Thus, $|V_{H_k}|$ cannot be minimum. ∎

Claim 3.13. *For the root vertex v_r, $x^*(v_r) < 1$.*

Proof of claim 3.13. Otherwise, x^* cannot be fractional. ∎

Claim 3.14. *Each branch P_t contains at most one internal vertex.*

Proof of claim 3.14. Otherwise, from Claims 3.11 and 3.12 the variables associated with the internal vertices of P_t are fractional and cannot appear separately in (7) (i.e., if a variable of a vertex $v_i \in P_t$ appears in an equality (7), all the variables associated with the vertices of P_t appear in the same equality). It is easy to construct another feasible solution for the system of equalities $\mathcal{A}$, which contradicts the extremity of x^*. ∎

Claim 3.15. *If there exists a branch P_t such that $x^*(P_t) < 1$, then all support graphs of equalities (7), contain the branch P_t.*

Proof of claim 3.15. Otherwise, let $ax^* = b$ be an equality (7) not containing the variables associated with the vertices of $P_t \setminus \{v_r\}$. Thus, the star tree inequality $x^*(P_t) + ax^* \geq b + 1$ is violated. ∎

Claim 3.16. *For the root vertex v_r, $x^*(v_r) > 0$.*

Proof of claim 3.16. Otherwise, there must exists two branches P_t and $P_{t'}$ such that $x^*(P_t) < 1$ and $x^*(P_{t'}) < 1$. It follows that these two branches must belong to all support graphs of equalities (7). It is easy to construct another solution for the system of equalities $\mathcal{A}$, which contradicts the extremity of x^*. ∎

From the claims 3.11, 3.12, 3.13 and 3.16 we deduce that for all $v \in V_{H_k}$, $0 < x^*(v) < 1$. We distinguish two cases

a. There exists a branch P_t such that $x^*(P_t) < 1$.
   Let $y^* \in \mathbb{R}^V$ defined as follows
   - $y^*(v) = x^*(v)$ for $v \in P_t \setminus \{v_r, t\}$
   - $y^*(v_r) = x^*(v_r) + \epsilon$
   - $y^*(v) = x^*(v) - \epsilon$ for all $v \in V_{H_k} \setminus (P_t \cup \{v_r\})$
   From claim 3.15, the vector y^* satisfies all equalities of $\mathcal{A}$. Contradiction with x^* a fractional extreme point.
b. For each branch P_t, $x^*(P_t) = 1$
   It follows that for each pair of vertices $v_i, v_j \in V_{H_k} \setminus \{v_r\}$, $x^*(v_i) = x^*(v_j) = 1 - x^*(v_r)$. Since all the variables are fractional, the variable associated with each vertex $v \in V_{H_k} \setminus \{v_r\}$ must belong to at least one equality (7)

$$x(V_{H_{k'}} \setminus \{v_r\}) + (k' - 1)x(v_r) = k' - 1 \qquad \text{for } k' \in \{2, \ldots, k\}$$

By the variable changing presented before

$$k' - k'x(v_r) + (k' - 1)x(v_r) = k' - 1$$

Hence, $x(v_r) = 1$ which contradicts the extremity of x^*. ∎

3.2   Clique Stars

Proposition 3.21. *For clique stars, $P(G, T)$ is not integral.* ∎

Consider the following valid inequalities presented in [6]

$$x(Q_{k'}) \geq k' - 1 \qquad\qquad \forall Q_{k'} \in \Theta \qquad (10)$$

We recall that Θ contains all the terminal paths in Q_k since they are clique stars of two terminals and inequalities (2) are included in (10).

Theorem 3.21. *For clique stars, the polytope given by inequalities (10) and trivial inequalities is integer.* ∎

4   TDI-ness

In this section we give a TDI descriptions for the multi-terminal vertex separator problem in star trees and clique stars.

4.1   Star Trees

We first introduce some notations. Consider two star trees H_l^i and H_s^j subgraphs of H_k, such that $s \geq 2$, $l \geq 2$, $i \in \{1, \ldots, \binom{k}{l}\}$ and $j \in \{1, \ldots, \binom{k}{s}\}$. We denote by $H_{l,s}^{i \cap j}$ the star tree subgraph of H_k, whose branches are all those in common with H_l^i and H_s^j. We denote by $H_{l,s}^{i \cup j}$ a star tree subgraph of H_k with $\min\{s + l - t(H_{l,s}^{i \cap j}), s + l - 1\}$ terminals, whose branches belong either to H_l^i or to H_s^j. If $t(H_{l,s}^{i \cap j}) = 0$, $H_{l,s}^{i \cup j}$ is any star tree of $s + l - 1$ branches.

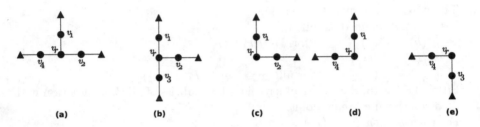

Fig. 3. Star trees, subgraphs of H_4 in Fig. 1(a)

To illustrates these notations, if H_l^i is the graph in Fig. 3(c) and H_s^j the graph in Fig. 3(e), then $H_{l,s}^{i \cup j}$ is the graph in Fig. 3(a) or the graph in Fig. 3(b) and $H_{l,s}^{i \cap j}$ does not exist. If H_l^i is the graph in Fig. 3(a) and H_s^j the graph in Fig. 3(b), then $H_{l,s}^{i \cup j}$ is the graph in Fig. 1(a) and $H_{l,s}^{i \cap j}$ is the graph in Fig. 3(c).

Let P^* be the linear program defined by the variable vector x, the objective function (1), the trivial inequalities (4) and inequalities (6). Let $y \in \mathbb{R}_+^\Pi$ be the dual variable vector associated with inequalities (6). Consider the dual D^* of P^*

$$\max \quad \sum_{H_{k'} \in \Pi} (k' - 1) y_{H_{k'}}$$

$$\sum_{H \in \Pi^v} y_H \leq w(v) \qquad\qquad \forall v \in V \setminus \{v_r\} \qquad (11)$$

$$\sum_{H \in \Pi} (k' - 1) y_H \leq w(v_r) \qquad\qquad (12)$$

$$y_H \geq 0 \qquad\qquad \forall H \in \Pi \qquad (13)$$

We notice that D^* consists in packing star trees of Π in H_k satisfying the capacity w of each vertex. Let $y^* \in \mathbb{R}_+^\Pi$ be an optimal solution of D^*.

The solution y^* is called *maximal optimal* if for each other optimal solution $\overline{y} \in \mathbb{R}_+^\Pi$ there exists $s \in \{1, \ldots, k\}$ satisfying the following conditions

1. $\displaystyle\sum_{i=1}^{f_l} \overline{y}_{H_l^i} = \sum_{i=1}^{f_l} y_{H_l^i}^*, \qquad$ for all $l \in \{s + 1, \ldots, k\}$ and $f_l = \binom{k}{l}$

2. $\displaystyle\sum_{i=1}^{f} \overline{y}_{H_s^i} < \sum_{i=1}^{f} y_{H_s^i}^*, \qquad$ for $f = \binom{k}{s}$

There must exists one maximal optimal solution in D^*, and in the following we suppose that y^* is a maximal optimal.

Lemma 4.11. *For each pair of star trees H_l^i and $H_s^j \in \Pi$, such that $y_{H_l^i}^* > 0$, $y_{H_s^j}^* > 0$ either H_l^i is completely included in H_s^j or H_s^j is completely included in H_l^i.* ∎

Corollary 4.11. *For $s \in \{1, \ldots, k\}$, there exists at most one star tree H_s^j with a value $y_{H_s^j}^* > 0$ over all star trees with s terminals.*

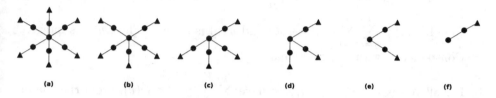

(a)   (b)   (c)   (d)   (e)   (f)

Fig. 4. A maximal optimal solution structure

Figure 4 illustrates the structure of the maximal optimal solution y^* (each star tree is included in another, except H_k, and no more than one star tree with the same number of terminals).

Theorem 4.11. *For star trees, the linear system of P^* is TDI.*

Proof. We should prove that D^* has an integer optimal solution. For this, we need to show the claims below.

Claim 4.11. *If for a star tree $H_s^j \in \Pi$, $\sum_{l=s}^{k} \sum_{i=1}^{\binom{k}{l}} y_{H_l^i}^* < w_v$ for each vertex $v \in V(H_s^j) \setminus \{v_r\}$ then $y_{H_p^i}^* = 0$ for each star tree $H_p^i \in \Pi$ with $p \leq s-1$ terminals.*

Proof of claim 4.11. Let us assume the contrary, then there exists $H_s^j \in \Pi$ such that $\sum_{l=s}^{k} \sum_{i=1}^{\binom{k}{l}} y_{H_l^i}^* < w_v$ for all vertices $v \in V(H_s^j) \setminus \{v_r\}$ and there exists $p \in \{2, \ldots, s-1\}$ and $i \in \{1, \ldots, \binom{k}{p}\}$ such that $y_{H_p^i}^* > 0$ (from Lemma 4.11, H_p^i is a subgraph of H_s^j). We suppose that p is maximum. To prove the claim, we will show that y^* cannot be maximal optimal by constructing another solution $\bar{y} \in \mathbb{R}_+^\Pi$ from y^*. Indeed, proving that $\bar{y}$, obtained by adding $\alpha > 0$ to $y_{H_s^j}^*$ and subtracting $\beta > 0$ from $y_{H_p^i}^*$, is feasible and optimal, will contradicts the maximality of y^*. To guarantee the optimality of $\bar{y}$ we should have $\alpha \times (s-1) = \beta \times (p-1)$. Then if $\alpha = \min\{\frac{y_{H_p^i}^*(p-1)}{(s-1)}, \min_{v \in V_{H_s^j} \setminus \{v_r\}} \{c(v) - \sum_{l=s}^{k} \sum_{i=1}^{\binom{k}{l}} y_{H_l^i}^*\}\}$ then $\alpha > 0$ and $\beta = \frac{\alpha(s-1)}{p-1}$. Since p is maximum, thus $\bar{y}$ must be feasible optimal solution for D^*. Thus our claim holds. ∎

Claim 4.12. *If y^* is fractional then there exists exactly one star tree $H_s^j \in \Pi$ such that $y_{H_s^j}^*$ is fractional.*

Proof of claim 4.12. We suppose that there exists two different star trees H_l^i and H_s^j, such that $y_{H_s^j}^*$ and $y_{H_l^i}^*$ are fractional. We suppose that s is maximum (i.e., for all $p \in \{s+1, \ldots, k\}$, $y_{H_p}^*$ is integer). From Corollary 4.11, $s > l$. We distinguish two cases

a. There exists a vertex $v \in V_{H_s^j} \setminus \{v_r\}$ such that $\sum\limits_{p=s}^{k} \sum\limits_{q=1}^{\binom{k}{p}} y_{H_p^q}^* = w(v)$. Since s is maximum, we know that $y_{H_p}^*$ is integer for any star tree H_p with $p \in \{s+1, \ldots, k\}$, $\sum\limits_{p=s}^{k} \sum\limits_{q=1}^{\binom{k}{p}} y_{H_p^q}^* = w(v)$ and $w(v)$ is integer. Thus, $y_{H_s^j}^*$ is integer. Contradiction with $y_{H_s^j}^*$ fractional.

b. For all vertex $v \in V_{H_s^j} \setminus \{v_r\}$, we have $\sum\limits_{p=s}^{k} \sum\limits_{q=1}^{\binom{k}{p}} y_{H_p^q}^* < w(v)$. From the Claim 4.11, $y_{H_p}^* = 0$ for any star tree $H_p \in \Pi$ with $p \leq s-1$ terminals. Contradiction with $y_{H_l^i}^*$ fractional.

Thus there exists at most one star tree $H_s^j \in \Pi$ such that $y_{H_s^j}^*$ is fractional. ∎

Claim 4.13. *If y^* is fractional then there exists another optimal solution $\bar{y}$ that is integer.*

Proof of claim 4.13. Let $H_s^j \in \Pi$ be the star tree such that $y_{H_s^j}^*$ is fractional. We distinguish three cases

- If $s = 1$ then let $\bar{y} \in \mathbb{R}_+^\Pi$ be the solution obtained from y^* by setting $y_{H_s^j}^* = 0$. The vector $\bar{y}$ represents an integer feasible optimal solution.
- If $s \geq 2$ then it is clear that $(s-1)y_{H_s^j}^*$ is integer. We denote by $\epsilon = y_{H_s^j}^* - \lfloor y_{H_s^j}^* \rfloor$. It follows that $(s-1)\lfloor y_{H_s^j}^* \rfloor + (s-1)\epsilon$ is integer. Thus $(s-1)\epsilon$ is integer. Let $\bar{y} \in \mathbb{R}_+^\Pi$ be another solution obtained from y^* by subtracting ϵ from $y_{H_s^j}^*$ and by adding 1 to $y_{H_{\epsilon \times (s-1)+1}}^*$ for an arbitrary star tree $H_{\epsilon \times (s-1)+1}$.

Thus $\bar{y}$ is an integer optimal solution for D^*. ∎

Then the proof is ended and the linear system of P^* is TDI. ∎

As consequence, we obtain the following *min-max* relation: In star trees, the minimum number of vertices covering all terminal paths is equal to the maximum packing of star trees.

4.2   Clique Stars

For this section we introduce some notations. Consider two clique stars Q_l^i and Q_s^j subgraphs of Q_k such that $s \geq 2$, $l \geq 2$, $i \in \{1, \ldots, \binom{k}{l}\}$ and $j \in \{1, \ldots, \binom{k}{s}\}$.

Let $Q_{l,s}^{i \cap i}$ be the clique star subgraph of Q_k whose branches are all those in common with Q_l^i and Q_s^j. We denote by $Q_{l,s}^{i \cup j}$ the clique star subgraph of Q_k whose branches are all those in common either with Q_l^i or with Q_s^j.

Let P^Q be the linear program defined by the variable vector x, the objective function (1) and inequalities (4) and (10). Let D^Q be the dual of P^Q. We notice that the D^Q consists in packing clique stars of Θ in Q_k satisfying the capacity of each vertex. Let $y \in \mathbb{R}_+^\Theta$ be the dual variables associated with inequalities (10) and y^* the optimal solution of D^Q. The solution y^* is called *maximal optimal* if for each other optimal solution $\bar{y} \in \mathbb{R}_+^\Theta$ there exists $s \in \{1, \ldots, k\}$ satisfying the following conditions

1. $\displaystyle\sum_{i=1}^{f} \bar{y}_{Q_l^i} = \sum_{i=1}^{f} y_{Q_l^i}^*$,   for all $l \in \{s+1, \ldots, k\}$ and $f = \binom{k}{l}$

2. $\displaystyle\sum_{i=1}^{f} \bar{y}_{Q_s^i} < \sum_{i=1}^{f} y_{Q_s^i}^*$,   for $f = \binom{k}{s}$

There must exists one maximal optimal solution in D^Q, and in the following we suppose that y^* is a maximal optimal.

Theorem 4.21. *For clique stars, the linear system of P^Q is TDI.*

Proof. We have the following claims.

Claim 4.21. *For all two different subgraphs Q_l^i and Q_s^j of Q_k, such that $y_{Q_s^j}^* > 0$ and $y_{Q_s^j}^* > 0$, either Q_s^j is completely included in Q_l^i or Q_l^i is completely included in Q_s^j.*

Proof of claim 4.21. We suppose that there exists two subgraphs Q_l^i and Q_s^j of Q_k, such that $y_{Q_l^i}^* > 0$ and $y_{Q_s^j}^* > 0$ and no one is included in the other. There exists $\epsilon > 0$ such that $\bar{y} \in \mathbb{R}_+^\Theta$, obtained from y^* by subtracting ϵ from $y_{Q_l^i}^*$ and from $y_{Q_s^j}^*$ and by adding ϵ to $y_{Q_{l,s}^{i \cup j}}^*$ and to $y_{Q_{l,s}^{i \cap j}}^*$, is feasible and optimal solution for D^*. Thus contradiction with y^* maximal optimal. ∎

Corollary 4.21. *For $s \in \{1, \ldots, k\}$, there exists at most one clique star Q_s^j with a value $y_{Q_s^j}^* > 0$ over all clique stars with s terminals.*

Claim 4.22. *For all Q_s^j subgraph of Q_k, there exists a vertex $v \in V(Q_s^j)$ such that $\displaystyle\sum_{p=s}^{k} \sum_{q=1}^{\binom{k}{p}} y_{Q_p^q}^* = w(v)$.*

Proof of claim 4.22. We suppose there exists Q_s^j subgraph of Q_k, such that for all $v \in V(Q_s^j)$, $\displaystyle\sum_{p=s}^{k} \sum_{q=1}^{\binom{k}{p}} y_{Q_p^q}^* < w(v)$. There must exist $Q_p^q \in \Theta$ subgraph of Q_k such that $2 \leq p < s$ and $y_{Q_p^q}^* > 0$, Otherwise the solution is not optimal. We suppose that p is maximum. There exists $0 < \epsilon \leq y_{Q_p^q}^*$ such that $\bar{y} \in \mathbb{R}_+^\Theta$, obtained from y^* by subtracting ϵ from $y_{Q_p^q}^*$ and by adding ϵ to $y_{Q_s^j}^*$, is feasible and optimal solution for D^*. Thus contradiction with y^* maximal optimal. ∎

Then we deduce an algorithm to solve D^*. We start by packing the clique star of Θ having a maximum number of terminals until the capacity of some vertex is all used. The branches containing a saturated vertex, are we subtract the number of packed clique stars from the capacities of the other vertices. The same operations are repeated until Q_k becomes a branch.

Algorithm 1. An exact algorithm for solving the D^*

Data: The graph $Q_k = (V_{Q_k} \cup T_{Q_k}, E_{Q_k})$, a vector $w \in \mathbb{N}^V$
Result: A maximal optimal solution y^*

1 **begin**
2   Let $Q^{k+1} \leftarrow Q_k$;
3   **for** $(i = k \rightarrow 1)$ **do**
4         $Q^i \leftarrow$ clique star obtained from Q^{i+1} by deleting each branch P_t containing a vertex v with $w(v) = 0$;
5         $y^*_{Q^i} = \min\limits_{\forall v \in V(Q^i)} \{w(v)\}$;
6   **for** $(v \in V(Q^i))$ **do**
7             $w(v) = w(v) - \min\limits_{\forall v \in V(Q^i)} \{w(v)\}$;

Corollary 4.22. *From Claims 4.21 and 4.22, the algorithm* 1 *gives an optimal solution y^* for D^*, and since the capacities are integer, it follows that y^* is integer.*

As consequence, we obtain the following *min-max* relation: In clique stars, the minimum number of vertices covering all terminal paths is equal to the maximum packing of clique stars.

5 Conclusion

In this paper we characterized the polytope of the multi-terminal vertex separators in two classes of graph, the star trees and the clique stars and we showed that the associated linear system is total dual integral. Hence, the multi-terminal vertex separator problem is polynomial in these two classes of graph. It would be interesting to extend the results on other classes of graph, for instance, the terminal cycles [6], the graph composed of a cycle C of k vertices and k disjoint paths between each vertex of C and k terminals, the terminal tree [6], which is a tree with all leaves int T.

References

1. Balas, E., Souza, C.: The vertex separator problem: a polyhedral investigation. Math. Program. **103**(3), 583–608 (2005)
2. Ben-Ameur, W., Mohamed-Sidi, M.-A., Neto, J.: The k-separator problem. In: Du, D.-Z., Zhang, G. (eds.) COCOON 2013. LNCS, vol. 7936, pp. 337–348. Springer, Heidelberg (2013)

3. Ben-Ameur, W., Didi Biha, M.: On the minimum cut separator problem. Networks **59**(1), 30–36 (2012)
4. Chen, J., Liu, Y., Lu, S.: An improved parameterized algorithm for the minimum node multiway cut problem. Algorithmica **55**(1), 1–13 (2009)
5. Cornaz, D., Furini, F., Lacroix, M., Malaguti, E., Mahjoub, A.R., Martin, S.: Mathematical Formulations for the balanced vertex k-separator problem. In: Control, Decision and Information Technologies. IEEE (2014)
6. Cornaz, D., Magnouche, Y., Mahjoub, A.R., Martin, S.: The multi-terminal vertex separator problem: polyhedral analysis and branch-and-cut. In: International Conference on Computers and Industrial Engineering (2015)
7. Cygan, M., Pilipczuk, M., Pilipczuk, M., Wojtaszczyk, J.O.: On multiway cut parameterized above lower bounds. J. ACM Trans. Comput. Theory (TOCT) **5**(1), 290–309 (2013). Article No. 3
8. Biha, M., Meurs, M.J.: An exact algorithm for solving the vertex separator problem. J. Glob. Optim. **49**(3), 425–434 (2010)
9. Garg, N., Vazirani, V.V., Vazirani, M.: Multiway cuts in directed and node weighted graphs. In: Shamir, E., Abiteboul, S. (eds.) ICALP 1994. LNCS, vol. 820, pp. 487–498. Springer, Heidelberg (1994)
10. Jost, V., Naves, G.: The graphs with the max-Mader-flow-min-multiway-cut property. J. Comput. Res. Repository - CORR (2011)
11. Marx, D.: Parameterized graph separation problems. In: Downey, R.G., Fellows, M.R., Dehne, F. (eds.) IWPEC 2004. LNCS, vol. 3162, pp. 71–82. Springer, Heidelberg (2004)
12. Schrijver, A.: Combinatorial Optimization: Polyhedra and Efficiency (2003)

Toward Computer-Assisted Discovery and Automated Proofs of Cutting Plane Theorems

Matthias Köppe[(✉)] and Yuan Zhou

Department of Mathematics, University of California, Davis, CA, USA
{mkoeppe,yzh}@math.ucdavis.edu

Abstract. Using a metaprogramming technique and semialgebraic computations, we provide computer-based proofs for old and new cutting-plane theorems in Gomory–Johnson's model of cut generating functions.

1 Introduction

Inspired by the spectacular breakthroughs of the polyhedral method for combinatorial optimization in the 1980s, generations of researchers have studied the facet structure of convex hulls to develop strong cutting planes. It is a showcase of the power of experimental mathematics: Small examples are generated, their convex hulls are computed (for example, using the popular tool PORTA [9]), conjectures are formed, theorems are proved. Some proofs feature brilliant new ideas; other proofs are routine. Once the theorems have been found and proved, separation algorithms for the cutting planes are implemented. Numerical tests are run, the strength-versus-speed trade-off is investigated, parameters are tuned, papers are written.

In this paper, we ask how much of this process can be automated: In particular, *can we use algorithms to discover and prove theorems about cutting planes?* This paper is part of a larger project in which we aim to automate more stages of this pipeline. We focus on general integer and mixed integer programming, rather than combinatorial optimization, and use the framework of cut-generating functions [10], specifically those of the classic single-row Gomory–Johnson model [14,15]. Cut-generating functions are an attractive framework for our study for several reasons. First, it is essentially dimensionless: Cuts obtained from cut-generating functions can be applied to problems of arbitrary dimension. Second, it may be a way towards effective multi-row cuts, though the computational approaches so far have disappointed. Third, work on new cuts in the single-row Gomory–Johnson model has, with few exceptions, become a routine, but error-prone task that leads to proofs of enormous complexity; see for example [24,25]. Fourth, finding new cuts in the multi-row Gomory–Johnson model has a daunting complexity, and few attempts at a systematic study have been made. Fifth, working on the Gomory–Johnson model is timely because only recently, after decades of theoretical investigations, the first computational tools for cut-generating functions in this model became available in [3] and the software implementation [20].

© Springer International Publishing Switzerland 2016
R. Cerulli et al. (Eds.): ISCO 2016, LNCS 9849, pp. 332–344, 2016.
DOI: 10.1007/978-3-319-45587-7_29

Of course, automated theorem proving is not a new proposition. Probably the best known examples in the optimization community are the proof of the Four Color Theorem, by Appel–Haken [1], and more recently and most spectacularly the proof of the Kepler Conjecture by Hales [18] and again within Hales' Flyspeck project in [19]. In the domains of combinatorics, number theory, and plane geometry, Zeilberger with long-term collaborator Shalosh B. Ekhad have pioneered automated discovery and proof of theorems; see, for example [13]. Many sophisticated automated theorem provers, by names such are HOL light, Coq, Isabelle, Mizar, etc. are available nowadays; see [28] and the references within for an interesting overview.

Our approach is pragmatic. Our theorems and proofs come from a metaprogramming trick, applied to the practical software implementation [20] of computations with the Gomory–Johnson model; followed by computations with semialgebraic cell complexes. As such, all of our techniques are reasonably close to mathematical programming practice. The correctness of all of our proofs depends on the correctness of the underlying implementation. We make no claims that our proofs can be formalized in the sense of the above mentioned formal proof systems that break every theorem down to the axioms of mathematics; we make no attempt to use an automated theorem proving system.

Our software is in an early, proof-of-concept stage of development. In this largely computational and experimental paper we report on the early successes of the software. We computationally verify the results on the `gj_forward_3_slope`[1] (https://github.com/mkoeppe/infinite-group-relaxation-code/search?q=%22def def+gj_forward_3_slope%28%22) and `drlm_backward_3_slope` (https://github. com/mkoeppe/infinite-group-relaxation-code/search?q=%22def+drlm_backward _3_slope%28%22) functions. We find a correction to a theorem by Chen [8] regarding the extremality conditions for his `chen_4_slope` (https://github.com/mk oeppe/infinite-group-relaxation-code/search?q=%22def+chen_4_slope%28%22) family.[2] We find a correction to a result by Miller, Li and Richard [24] on the so-called $CPL_3^=$ functions (`mlr_cpl3_...`). We discover several new parametric families, `kzh_3_slope_param_extreme_1` (https://github.com/mkoeppe/infinite-group-relaxation-code/search?q=%22def+kzh_3_slope_param_extreme_1%28%22) and `kzh_3_slope_param_extreme_2` (https://github.com/mkoeppe/infinite-group-relaxation-code/search?q=%22def+kzh_3_slope_param_extreme_2%28%22), of extreme functions and corresponding theorems regarding their extremality, with automatic proofs.

---

[1] A function name shown in typewriter font is the name of the constructor of this function in the Electronic Compendium, part of the SageMath program [20]. In an online copy of this paper, hyperlinks lead to this function in the GitHub repository.

[2] This is a new result, which should not be confused with our previous result in [22] regarding Chen's family of 3-slope functions (`chen_3_slope_not_extreme`, https://github.com/mkoeppe/infinite-group-relaxation-code/search?q=%22def+ch en_3_slope_not_extreme%28%22).

2 The Gomory–Johnson Model

We restrict ourselves to the single-row (or, "one-dimensional") infinite group problem, which has attracted most of the attention in the past and for which the software [20] is available. It can be written as

$$\sum_{r \in \mathbb{R}} r\, y(r) \equiv f \pmod 1,$$

$$y \colon \mathbb{R} \to \mathbb{Z}_+ \text{ is a function of finite support},$$

(1)

where f is a given element of $\mathbb{R} \setminus \mathbb{Z}$. We study the convex hull $R_f(\mathbb{R}, \mathbb{Z})$ of the set of all functions $y \colon \mathbb{R} \to \mathbb{Z}_+$ satisfying the constraints in (1). The elements of the convex hull are understood as functions $y \colon \mathbb{R} \to \mathbb{R}_+$.

After a normalization, valid inequalities for the convex set $R_f(\mathbb{R}, \mathbb{Z})$ can be described using so-called *valid functions* $\pi \colon \mathbb{R} \to \mathbb{R}$ via $\langle \pi, y \rangle := \sum_{r \in \mathbb{R}} \pi(r) y(r) \geq 1$. Valid functions π are cut-generating functions for pure integer programs. Take a row of the optimal simplex tableau of an integer program, corresponding to a basic variable x_i that currently takes a fractional value:

$$x_i = -f_i + \sum_{j \in N} r_j x_j, \quad x_i \in \mathbb{Z}_+,\, x_N \in \mathbb{Z}_+^N.$$

Then a valid function π for $R_{f_i}(\mathbb{R}, \mathbb{Z})$ gives a valid inequality $\sum_{j \in N} \pi(r_j) x_j \geq 1$ for the integer program. (By a theorem of Johnson [21], this extends easily to the mixed integer case: A function ψ can be associated to π, so that they together form a *cut-generating function pair* (ψ, π), which gives the coefficients of the continuous and of the integer variables.)

In the finite-dimensional case, instead of merely valid inequalities, one is interested in stronger inequalities such as tight valid inequalities and facet-defining inequalities. These rôles are taken in our infinite-dimensional setting by *minimal functions* and *extreme functions*. Minimal functions are those valid functions that are pointwise minimal; extreme functions are those that are not a proper convex combination of other valid functions.

By a theorem of Gomory and Johnson [14], minimal functions for $R_f(\mathbb{R}, \mathbb{Z})$ are classified: They are the subadditive functions $\pi \colon \mathbb{R} \to \mathbb{R}_+$ that are periodic modulo 1 and satisfy the *symmetry condition* $\pi(x) + \pi(f - x) = 1$ for all $x \in \mathbb{R}$.

Obtaining a full classification of the *extreme* functions has proved to be elusive, however various authors have defined parametric families of extreme functions and provided extremality proofs for these families. These parametric families of extreme functions from the literature, as well as "sporadic" extreme functions, have been collected in an electronic compendium as a part of the software [20]; see [22].

We refer the interested reader to the recent surveys [4,5,11] for a more detailed exposition.

3 Examples of Cutting-Plane Theorems in the Gomory–Johnson Model

To illustrate what cutting-plane theorems in the Gomory–Johnson model look like, we give three examples, paraphrased for precision from the literature where they were stated. As we will show later, the last theorem is incorrect.

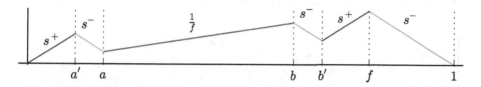

Fig. 1. gj_forward_3_slope (https://github.com/mkoeppe/infinite-group-relaxation-co de/search?q=%22def+gj_forward_3_slope%28 %22)

Theorem 3.1 (reworded from Gomory–Johnson [16, Theorem 8]). *Let $f \in (0,1)$ and $\lambda_1, \lambda_2 \in \mathbb{R}$. Define the periodic, piecewise linear* gj_forward_3_slope *(Fig. 1, https://github.com/mkoeppe/infinite-group-relaxation-code/search?q=% 22def+gj_forward_3_slope(%22) function $\pi \colon \mathbb{R}/\mathbb{Z} \to \mathbb{R}$ as follows. The function π satisfies $\pi(0) = \pi(1) = 0$; it has 6 pieces between 0 and 1 with breakpoints at $0, a', a, b, b', f$ and 1, where $a = \frac{\lambda_1 f}{2}$, $a' = a + \frac{\lambda_2(f-1)}{2}$, $b = f - a$ and $b' = f - a'$. The slope values of π on these pieces are $s^+, s^-, \frac{1}{f}, s^-, s^+$ and s^-, respectively, where $s^+ = \frac{\lambda_1 + \lambda_2}{\lambda_1 f + \lambda_2(f-1)}$ and $s^- = \frac{1}{f-1}$. If λ_1 and λ_2 satisfy that (i) $0 \le \lambda_1 \le \frac{1}{2}$, (ii) $0 \le \lambda_2 \le 1$ and (iii) $0 < \lambda_1 f + \lambda_2(f-1)$, then the function π is an extreme function for $R_f(\mathbb{R}/\mathbb{Z})$.*

Theorem 3.2 (Dey–Richard–Li–Miller [12]; in this form, for the real case, in [22], Theorem 4.1). *Let f and b be real numbers such that $0 < f < b \le \frac{1+f}{4}$. The periodic, piecewise linear* drlm_backward_3_slope *(Fig. 2, https://github.com/mkoeppe/infinite-group-relaxation-code/search?q=%22def+ drlm_backward_3_slope%28%22) function $\pi \colon \mathbb{R}/\mathbb{Z} \to \mathbb{R}$ defined as follows is an extreme function for $R_f(\mathbb{R}/\mathbb{Z})$:*

$$
\pi(x) = \begin{cases}
\frac{x}{f} & \text{if } 0 \le x \le f \\
1 + \frac{(1+f-b)(x-f)}{(1+f)(f-b)} & \text{if } f \le x \le b \\
\frac{x}{1+f} & \text{if } b \le x \le 1+f-b \\
\frac{(1+f-b)(x-1)}{(1+f)(f-b)} & \text{if } 1+f-b \le x \le 1
\end{cases}
$$

Theorem 3.3 (reworded from Chen [8], Theorem 2.2.1). *Let $f \in (0,1)$, $s^+ > 0, s^- < 0$ and $\lambda_1, \lambda_2 \in \mathbb{R}$. Define the periodic, piecewise linear* chen_4_ slope *(Fig. 3, https://github.com/mkoeppe/infinite-group-relaxation-code/ search?q=%22def+chen_4_slope%28%22) function $\pi \colon \mathbb{R}/\mathbb{Z} \to \mathbb{R}$ as follows. The*

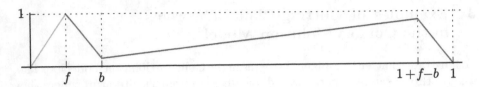

Fig. 2. drlm_backward_3_slope (https://github.com/mkoeppe/infinite-group-relaxati
code/search?q=%22def+drlm_backward_3_slope%28%22)

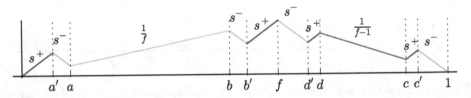

Fig. 3. chen_4_slope (https://github.com/mkoeppe/infinite-group-relaxation-code/sea
rch?q=%22def+chen_4_slope%28%22)

function π *satisfies* $\pi(0) = \pi(1) = 0$; *it has* 10 *pieces between* 0 *and* 1 *with breakpoints at* $0, a', a, b, b', f, d', d, c, c', 1$, *where*

$$a' = \frac{\lambda_1(1-s^-f)}{2(s^+-s^-)}, \quad a = \frac{\lambda_1 f}{2}, \quad c = 1 - \frac{\lambda_2(1-f)}{2}, \quad c' = 1 - \frac{\lambda_2(1-s^+(1-f))}{2(s^+-s^-)}$$

and $b = f - a$, $b' = f - a'$, $d = 1 + f - c$, $d' = 1 + f - c'$. *The slope values of* π
on these pieces are $s^+, s^-, \frac{1}{f}, s^-, s^+, s^-, s^+, \frac{1}{f-1}, s^+$ *and* s^-, *respectively. If the parameters* $f, \lambda_1, \lambda_2, s^+$ *and* s^- *satisfy that*

$$f \geq \tfrac{1}{2}, \quad s^+ \geq \tfrac{1}{f}, \quad s^- \leq \tfrac{1}{f-1}, \quad 0 \leq \lambda_1 < \min\{\tfrac{1}{2}, \tfrac{s^+-s^-}{s^+(1-s^-f)}\}, \text{ and}$$

$$f - \tfrac{1}{s^+} < \lambda_2 < \min\left\{\tfrac{1}{2}, \tfrac{s^+-s^-}{s^-(s^+(f-1)-1)}\right\},$$

then π *is an extreme function for* $R_f(\mathbb{R}, \mathbb{Z})$.

Observation 3.4

(i) *These theorems are about families of periodic, continuous piecewise linear functions* $\pi \colon \mathbb{R} \to \mathbb{R}$ *that depend on a finite number of real parameters in a way that breakpoints and slope values can be written as rational functions of the parameters.*

(ii) *There are natural conditions on the parameters to make the function even constructible; for example, in Theorem 3.2, if* $f < b$ *is violated, then the function is not well-defined. These conditions are inequalities of rational functions of the parameters. Hence the set of parameter tuples such that the construction describes a function is a semialgebraic set.*

(iii) *There are additional conditions on the parameters that ensure that the function is an extreme function. Again, all of these conditions are inequalities of rational functions of the parameters. Hence the set of parameter tuples such that the construction gives an extreme function is a semialgebraic set.*

Remark 3.5. Some families of extreme functions in the literature are defined in more general ways. Some use parameters that are integers (for example, `drlm_2_slope_limit` (https://github.com/mkoeppe/infinite-group-relaxation-code/search?q=%22def+drlm_2_slope_limit%28%22) has integer parameters that control the number of pieces of the function). Others use non-algebraic operations such as the floor/ceiling/fractional part operations to define the breakpoints and slope values of the function (for example, `dg_2_step_mir` (https://github.com/mkoeppe/infinite-group-relaxation-code/search?q=%22def+dg_2_step_mir%28%22)). Another family, `bhk_irrational` (https://github.com/mkoeppe/infinite-group-relaxation-code/search?q=%22def+bhk_irrational%28%22), requires an arithmetic condition, the $\mathbb{Q}$-linear independence of certain parameters, for extremality. These families are beyond the scope of this paper.

4   Semialgebraic Cell Structure of Extremality Proofs

The minimality of a given periodic piecewise linear function can be easily tested algorithmically; see, for example, [4, Theorem 3.11]. Basu, Hildebrand, and Köppe [3] gave the first algorithmic tests for extremality for a given function π whose breakpoints are rational with a common denominator q. The simplest of these tests uses their finite-oversampling theorem (see [5, Theorem 8.6] for its strongest form). Extremality of the function π is equivalent to the extremality of its restriction to the refined grid $\frac{1}{3q}\mathbb{Z}/\mathbb{Z}$ for the finite master group problem. Thus it can be tested by finite-dimensional linear algebra.

The proof of the finite-oversampling theorem in [3] (see also [5, Sect. 7.1] for a more high-level exposition) provides another algorithm, based on the computation of "affine-imposing" ("covered") intervals and the construction of "equivariant" perturbation functions. This algorithm in [3] is also tied to the use of the grid $\frac{1}{q}\mathbb{Z}/\mathbb{Z}$; but it has since been generalized in the practical implementation [20] to give a completely *grid-free algorithm*, which is suitable also for rational breakpoints with huge denominators and for irrational breakpoints.[3]

Observation 4.1. *On inspection of this grid-free algorithm, we see that it only uses algebraic operations, comparisons, and branches* (if-then-else *and* loops), *and then returns either* True *(to indicate extremality) or* False *(non-extremality).*

Enter *parametric analysis* of the algorithm, that is, we wish to run the algorithm for a function from a parametric family and observe how the run of the algorithm and its answer changes, depending on the parameters. It is then a simple observation that for any algorithm of the type described in Observation 4.1,

---

[3] The finiteness proof of the algorithm, however, does depend on the rationality of the data. In this paper we shall ignore the case of functions with non-covered intervals and irrational breakpoints, such as the `bhk_irrational` (https://github.com/mkoeppe/infinite-group-relaxation-code/search?q=%22def+bhk_irrational%28%22) family.

the set of parameters where the algorithm returns *True* must be a union of sets described by equations and inequalities of rational functions in the parameters. If the number of operations (and thus the number of branches) that the algorithm takes is bounded finitely, then it will be a finite union of "cells", each corresponding to a particular outcome of comparisons that led to branches, and each described by finitely many equations and inequalities of rational functions in the parameters. Thus it will be a semialgebraic set.

Within each of the cells, we get the "same" proof of extremality. A complete proof of extremality for a parametric family is merely a collection of cells, with one proof for each of them. This is what we compute as we describe below.

5   Computing One Proof Cell by Metaprogramming

Now we assume that we are given a tuple of *concrete* parameter values; we will compute a semialgebraic description of one cell of the proof, i.e., a cell of parameter tuples for which the algorithm takes the same branches.

It is well known that modern programming languages provide facilities known as "operator overloading" or "virtual methods" that allow us to conveniently write "generic" programs that can be run on objects of various types. For example, the program [20], written in the SageMath system [26], by default works with (arbitrary-precision) rational numbers; but when parameters are irrational algebraic numbers, it makes exact computations in a suitable real number field.

We make use of the same facilities for a metaprogramming technique that transforms the program [20] for testing extremality for a function corresponding to a given parameter tuple into a program that computes a description of the cell that contains the given parameter tuple. No code changes are necessary.

We define a class of elements[4] that support the algebraic operations and comparisons that our algorithm uses, essentially the operations of an ordered field. Each element stores (1) a symbolic expression[5] of the parameters in the problem, for example $x + y$ and (2) a concrete value, which is the evaluation of this expression on the given parameter tuple, for example 13. In the following, we denote elements in the form $x + y\,|_{=13}$. Every algebraic operation ($+$, $-$, $*$, ...) on the elements of the class is performed both on the symbolic expressions and on the concrete values. For example, if one multiplies the element $x\,|_{=7}$ and another element $x + y\,|_{=13}$, one gets the element $x^2 + xy\,|_{=91}$.

When a comparison ($<$, $\leq$, $=$, ...) takes place on elements of the class, their concrete values are compared to compute the Boolean return value of the comparison. For example, the comparison $x^2 + xy\,|_{=91} > 42$ evaluates to *True*. But we now have a constraint on the parameters x and y: The inequality $x^2 + xy > 42$ needs to hold so that our answer *True* is correct. We record this constraint.[6]

---

[4] These elements are instances of the class `ParametricRealFieldElement`. Their **parent**, representing the field, is an instance of the class `ParametricRealField`.

[5] Since all expressions are, in fact, rational functions, we use exact seminumerical computations in the quotient field of a multivariate polynomial ring, instead of the slower and less robust general symbolic computation facility.

[6] This information is recorded in the **parent** of the elements.

After a run of the algorithm, we have a description of the parameter region for which all the comparisons would give the same truth values as they did for the concrete parameter tuple, and hence the algorithm would take the same branches. This description is typically a very long and highly redundant list of inequalities of rational functions in the parameters.

It is crucial to simplify the description. "In theory", manipulation of inequalities describing semialgebraic sets is a solved problem of effective real algebraic geometry. Normal forms such as Cylindrical Algebraic Decomposition (CAD) [6, Chapts. 5 and 11] are available in various implementations, such as in the standalone QEPCAD B [7] or those integrated into CAS such as Maple and Mathematica, underlying these systems' 'solve' and 'assume' facilities. In computational practice, we however observed that these systems are extremely sensitive to the number of inequalities, rendering them unsuitable for our purposes; see [27, Sect. 5] for a study with Maple. We therefore roll our own implementation.

1. Transform inequalities and equations of rational functions into those of polynomials by multiplying by denominators, and bring them in the normal form $p(x) < 0$ or $p(x) = 0$. In the case of inequalities, this creates the extra constraint that the denominator takes the same sign as it does on the test point. So this transformation may break cells into smaller cells.
2. Factor the polynomials $p(x)$ and record the distinct factors as equations and inequalities. In the case of inequalities, this potentially breaks cells into smaller cells. We can ignore the factors with even exponents in inequalities.
3. Reformulation–linearization: Expand the polynomial factors in the standard monomial basis and replace each monomial by a new variable. This gives a linear system of inequalities and equations and thus a not-necessarily-closed polyhedron in an extended variable space. We use this polyhedron to represent our cell. Indeed, its intersection with the algebraic variety of monomial relations is in linear bijection with the semialgebraic cell.
4. All of this is implemented in an incremental way. We use the excellent Parma Polyhedra Library [2] via its SageMath interface written in Cython. PPL is based on the double description method and supports not-necessarily-closed polyhedra. It also efficiently supports adding inequalities dynamically and injecting a polyhedron into a higher space. The latter becomes necessary when a new monomial appears in some constraint. The PPL also has a fast implementation path for discarding redundant inequalities.

In our preliminary implementation, we forgo opportunities for strengthening this extended reformulation by McCormick inequalities, bounds propagation etc., which would allow for further simplification. We remark that all of these polyhedral techniques ultimately should be regarded as a preprocessing of input for proper real-algebraic computation. They are not strong enough on their own to provide "minimal descriptions" for semialgebraic cells. In a future version of our software, we will combine our preprocessing technique with the CAD implementation in Mathematica.

6   Computing the Cell Complex Using Wall-Crossing BFS

Define the graph of the cell complex by introducing a node for each cell and an edge if a cell is obtained from another cell by flipping one inequality. We compute the cell complex by doing a breadth-first search (BFS) in this graph. This is a well-known method for the case of the cells of arrangements of hyperplanes; see [17, Chap. 24] and the references within. The nonlinear case poses challenges due to degeneracy and possible singularities, which we have not completely resolved.

Our preliminary implementation uses a heuristic numerical method to construct a point in the interior of a neighbor cell, which will be used as the next concrete parameter tuple for re-running the algorithm described in Sect. 5. This may fail, and so we have no guarantees that the entire parameter space is covered by cells when the breadth-first search terminates. This is the weakest part of our current implementation.

7   Automated Proofs and Corrections of Old Theorems

Using our implementation, we verified Theorems 3.1 and 3.2, as well as other theorems regarding classical extreme functions from the literature. Figure 4 shows the visualizations of the corresponding cell complexes. Using our implementation we also investigated Theorem 3.3 regarding `chen_4_slope` (https://github.com/mkoeppe/infinite-group-relaxation-code/search?q=%22def+chen_4_slope%28%22) and discovered that it is incorrect. For example, the function with parameters $f = 7/10$, $s^+ = 2$, $s^- = -4$, $\lambda_1 = 1/100$, $\lambda_2 = 49/100$ satisfies the hypotheses of the theorem; however, it is not subadditive and thus not an extreme function. On the other hand, the stated hypotheses are also not necessary for extremality. For example, the function with parameters $f = 7/10$, $s^+ = 2$, $s^- = -4$, $\lambda_1 = 1/10$, $\lambda_2 = 1/10$ does not satisfy the hypotheses, however it is extreme. We omit a statement of corrected hypotheses that we found using our code.

We also investigated another family of functions, the so-called $\mathrm{CPL}_3^=$ functions, introduced by the systematic study by Miller, Li, and Richard [24]. Their method can be regarded as a predecessor of our method, albeit one that led to an error-prone manual case analysis (and human-generated proofs). Though our general method can be applied directly, we developed a specialized version of our code that follows Miller, Li, and Richard's method to allow a direct comparison. This revealed mistakes in [24] (we omit the details in this extended abstract).

8   Computer-Assisted Discovery of New Theorems

In [23], the authors conducted a systematic computer-based search for extreme functions on the grids $\frac{1}{q}\mathbb{Z}$ for values of q up to 30. This resulted in a large catalog of extreme functions that are "sporadic" in the sense that they do not belong to any parametric family described in the literature. Our goal is to automatically embed these functions into parametric families and to automatically prove

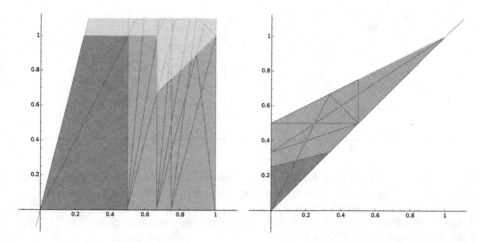

Fig. 4. The cell complexes of two parametric families of functions. *Left*, `gj_forward_3_slope` (https://github.com/mkoeppe/infinite-group-relaxation-code/search?q=%22 def+gj_forward_3_slope%28%22), showing the (λ_1, λ_2)-plane for fixed $f = 4/5$. *Right*, `drlm_backward_3_slope` (https://github.com/mkoeppe/infinite-group-relaxation-co de/search?q=%22def+drlm_backward_3_slope%28%22), showing the parameters $(f, bkpt)$. Cell colors: 'not constructible' (*white*), 'constructible, not minimal' (*yellow*), 'minimal, not extreme' (*green*) or 'extreme' (*blue*). (Color figure online)

theorems about their extremality. In this section, we report on cases that have been done successfully with our preliminary implementation; the process is not completely automatic yet.

We picked an interesting-looking 3-slope extreme function found by our computer-based search on the grid $\frac{1}{q}\mathbb{Z}$. We then introduced parameters f, a, b, v to describe a preliminary parametric family that we denote by `param_3_slope_1`. In the concrete function that we started from, these parameters take the values $\frac{6}{19}, \frac{1}{19}, \frac{5}{19}, \frac{8}{15}$; see Fig. 5 (left). So a denotes the length of the first interval right to f, b denotes the length of interval centered at $(1 + f)/2$ and $v = \pi(f + a)$. By this choice of parameters, the function automatically satisfies the equations corresponding to the symmetry conditions. Next we run the parametric version of the minimality test algorithm. It computes *True*, and as a side-effect computes a description of the cell in which the minimality test is the same.

```
sage: K.<f,a,b,v>=ParametricRealField([6/19,1/19,5/19,8/15])
sage: h = param_3_slope_1(f,a,b,v)
sage: minimality_test(h)
True
sage: K._eq_factor
{-f^2*v + 3*f*b*v + f^2 + f*a - 3*f*b - 3*a*b - f*v + b}
```

In particular, the above line shows that it has discovered one nonlinear equation that holds in the cell corresponding to the minimality proof of the function. We use this equation, quadratic in f and multilinear in the other parameters, to

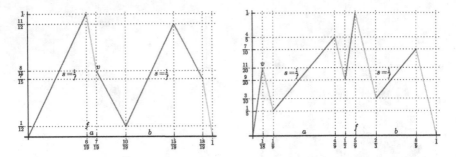

Fig. 5. Two new parametric families of extreme functions. *Left*, `kzh_3_slope_param_extreme_1`   (https://github.com/mkoeppe/infinite-group-relaxation-code/search?q= %22def+kzh_3_slope_param_extreme_1%28%22). *Right*, `kzh_3_slope_param_extreme_2` (https://github.com/mkoeppe/infinite-group-relaxation-code/search?q=%22def+kzh_ 3_slope_param_extreme_2%28%22).

eliminate one parameter.[7] This gives our parametric family `kzh_3_slope_param_extreme_1` (https://github.com/mkoeppe/infinite-group-relaxation-code/search ?q=%22def+kzh_3_slope_param_extreme_1%28%22), which depends only on f, a, and b. The definition of the parametric family is the only input to our algorithm. Cells with respect to this family will be full-dimensional; this helps to satisfy a current implementation restriction of our software. Indeed, re-running the algorithm yields the following simplified description of the cell in which the concrete parameter tuple lies.

```
3*f + 4*a - b - 1 < 0                    -a < 0
-f^2 - f*a + 3*f*b + 3*a*b - b < 0       -f + b < 0
f*a - 3*a*b - f + b < 0                  -f - 3*b + 1 < 0
-f^2*a + 3*f*a*b - 3*a*b - f + b < 0
```

We then compute the cell complex by BFS as described in Sect. 6. By inspection, we observe that the collection of the cells for which the function is extreme happens to be a convex polytope (this is not guaranteed). We discard the inequalities that appear twice and thus describe inner walls of the complex. By inspection, we discard nonlinear inequalities that are redundant. We obtain a description of the union of the cells for which the function is extreme as a convex polytope. We obtain the following:

Theorem 8.1. *Let* $f \in (0,1)$ *and* $a, b \in \mathbb{R}$ *such that*

$$0 \le a, \quad 0 \le b \le f \text{ and } 3f + 4a - b - 1 \le 0.$$

The piecewise linear **kzh_3_slope_param_extreme_1** *(https://github.com/mkoep pe/infinite-group-relaxation-code/search?q=%22def+kzh_3_slope_param_extreme_*

---

[7] We plan to automate this in a future version of our software.

1%28%22) function $\pi \colon \mathbb{R}/\mathbb{Z} \to \mathbb{R}$ *defined as follows is extreme. The function* π *has breakpoints at*

$$0, f, f + a, \frac{1 + f - b}{2}, \frac{1 + f + b}{2}, 1 - a, 1.$$

The values at breakpoints are given by $\pi(0) = \pi(1) = 0$, $\pi(f + a) = 1 - \pi(1 - a) = v$ *and* $\pi(\frac{1+f-b}{2}) = 1 - \pi(\frac{1+f+b}{2}) = \frac{f-b}{2f}$, *where* $v = \frac{f^2 + fa - 3fb - 3ab + b}{f^2 + f - 3bf}$.

A similar process leads to a theorem about the family kzh_3_slope_param_ extreme_2 (https://github.com/mkoeppe/infinite-group-relaxation-code/sear ch?q=%22def+kzh_3_slope_param_extreme_2%28%22) shown in Fig. 5 (right). We omit the statement of the theorem.

References

1. Appel, K., Haken, W.: Every planar map is four colorable. Part I: discharging. Illinois J. Math. **21**(3), 429–490 (1977). http://projecteuclid.org/euclid.ijm/ 1256049011
2. Bagnara, R., Hill, P.M., Zaffanella, E.: The Parma Polyhedra Library: toward a complete set of numerical abstractions for the analysis and verification of hardware and software systems. Sci. Comput. Program. **72**(1–2), 3–21 (2008). http://dx. doi.org/dx.doi.org/10.1016/j.scico.2007.08.001
3. Basu, A., Hildebrand, R., Köppe, M.: Equivariant perturbation in Gomory and Johnson's infinite group problem. I. The one-dimensional case. Math. Oper. Res. **40**(1), 105–129 (2014). doi:10.1287/moor.2014.0660
4. Basu, A., Hildebrand, R., Köppe, M.: Light on the infinite group relaxation I: foundations and taxonomy. 4OR **14**(1), 1–40 (2016). doi:10.1007/s10288-015-0292-9
5. Basu, A., Hildebrand, R., Köppe, M.: Light on the infinite group relaxation II: sufficient conditions for extremality, sequences, and algorithms. 4OR **14**(2), 107–131 (2016). doi:10.1007/s10288-015-0293-8
6. Basu, S., Pollack, R., Roy, M.-F.: Algorithms in Real Algebraic Geometry. Algorithms and Computation in Mathematics, vol. 10. Springer, Berlin (2006). MR2248869 (2007b:14125), ISBN:978-3-540-33098-1; 3-540-33098-4
7. Brown, C.W.: QEPCAD B: a program for computing with semi-algebraic sets using CADs. SIGSAM Bull. **37**(4), 97–108 (2003). doi:10.1145/968708.968710
8. Chen, K.: Topics in group methods for integer programming. Ph.D. thesis, Georgia Institute of Technology, June 2011
9. Christof, T., Löbel, A.: Porta: Polyhedron representation transformation algorithm. http://comopt.ifi.uni-heidelberg.de/software/PORTA/
10. Conforti, M., Cornuéjols, G., Daniilidis, A., Lemaréchal, C., Malick, J.: Cut-generating functions and S-free sets. Math. Oper. Res. **40**(2), 253–275 (2013). http://dx.doi.org/10.1287/moor.2014.0670
11. Conforti, M., Cornuéjols, G., Zambelli, G.: Corner polyhedra and intersection cuts. Surv. Oper. Res. Manag. Sci. **16**, 105–120 (2011)
12. Dey, S.S., Richard, J.-P.P., Li, Y., Miller, L.A.: On the extreme inequalities of infinite group problems. Math. Program. **121**(1), 145–170 (2010). doi:10.1007/ s10107-008-0229-6

13. Ekhad XIV, S.B., Zeilberger, D.: Plane geometry: an elementary textbook, 2050. http://www.math.rutgers.edu/~zeilberg/GT.html

14. Gomory, R.E., Johnson, E.L.: Some continuous functions related to corner polyhedra, I. Math. Program. **3**, 23–85 (1972). doi:10.1007/BF01584976

15. Gomory, R.E., Johnson, E.L.: Some continuous functions related to corner polyhedra, II. Math. Program. **3**, 359–389 (1972). doi:10.1007/BF01585008

16. Gomory, R.E., Johnson, E.L.: T-space and cutting planes. Math. Program. **96**, 341–375 (2003). doi:10.1007/s10107-003-0389-3

17. Goodman, J.E., O'Rourke, J. (eds.): Handbook of Discrete and Computational Geometry. CRC Press Inc., Boca Raton (1997). ISBN:0-8493-8524-5

18. Hales, T.C.: A proof of the Kepler conjecture. Ann. of Math. **162**(3), 1065–1185 (2005). doi:10.4007/annals.2005.162.1065. MR2179728 (2006g:52029)

19. Hales, T.C., Adams, M., Bauer, G., Dang, D.T., Harrison, J., Hoang, T.L., Kaliszyk, C., Magron, V., McLaughlin, S., Nguyen, T.T., et al.: A formal proof of the Kepler conjecture (2015). arXiv preprint arXiv:1501.02155

20. Hong, C.Y., Köppe, M., Zhou, Y.: SageMath program for computation and experimentation with the 1-dimensional Gomory-Johnson infinite group problem (2014). http://github.com/mkoeppe/infinite-group-relaxation-code

21. Johnson, E.L.: On the group problem for mixed integer programming. Math. Program. Study **2**, 137–179 (1974)

22. Köppe, M., Zhou, Y.: An electronic compendium of extreme functions for the Gomory-Johnson infinite group problem. Oper. Res. Lett. **43**(4), 438–444 (2015). doi:10.1016/j.orl.2015.06.004

23. Köppe, M., Zhou, Y.: New computer-based search strategies for extreme functions of the Gomory-Johnson infinite group problem (2015). arXiv:1506.00017 [math.OC]

24. Miller, L.A., Li, Y., Richard, J.-P.P.: New inequalities for finite and infinite group problems from approximate lifting. Naval Res. Logist. (NRL) **55**(2), 172–191 (2008). doi:10.1002/nav.20275

25. Richard, J.-P.P., Li, Y., Miller, L.A.: Valid inequalities for MIPs and group polyhedra from approximate liftings. Math. Program. **118**(2), 253–277 (2009). doi:10.1007/s10107-007-0190-9

26. Stein, W.A., et al.: Sage Mathematics Software (Version 7.1). The Sage Development Team (2016). http://www.sagemath.org

27. Sugiyama, M.: Cut-generating functions for integer linear programming. Bachelor thesis, UC Davis, June 2015. https://www.math.ucdavis.edu/files/1514/4469/2452/Masumi_Sugiyama_ugrad_thesis.pdf

28. Wiedijk, F.: The seventeen provers of the world. http://www.cs.ru.nl/~freek/comparison/comparison.pdf

Approximating Interval Selection on Unrelated Machines with Unit-Length Intervals and Cores

Kateřina Böhmová[1(✉)], Enrico Kravina[1], and Matúš Mihalák[1,2]

[1] Department of Computer Science, ETH Zurich, Zurich, Switzerland
{katerina.boehmova,enrico.kravina}@inf.ethz.ch
[2] Department of Knowledge Engineering,
Maastricht University, Maastricht, Netherlands
matus.mihalak@maastrichtuniversity.nl

Abstract. We consider a scheduling problem with machine dependent intervals, where each job consists of m fixed intervals, one on each of the m machines. To schedule a job, exactly one of the m intervals needs to be selected, making the corresponding machine busy for the time period equal to the selected interval. The objective is to schedule a maximum number of jobs such that no two selected intervals from the same machine overlap. This problem is NP-hard and admits a deterministic 1/2-approximation. The problem remains NP-hard even if all intervals have unit length, and all m intervals of any job have a common intersection. We study this special case and show that it is APX-hard, and design a 501/1000-approximation algorithm.

Keywords: Fixed interval scheduling · Interval selection · Computational complexity · Approximation algorithms

1 Introduction

We study a *fixed-interval* scheduling problem with m machines and n jobs, called INTERVALSELECTION *on unrelated machines*, where each job has on every machine an open interval of the reals (denoting the exact time interval when the job can be processed on the machine). By scheduling a job on a machine, one implicitly *selects* the corresponding interval of the job (and makes the machine unavailable for that time period), a job is scheduled by selecting one of its intervals. The goal is to schedule the maximum number of jobs such that no two selected intervals from the same machine intersect.

If $m = 1$, the problem becomes the classic interval scheduling problem which is solvable in $O(n \log n)$ time by the following greedy algorithm: Scan iteratively the right endpoints of the intervals from left to right, and in each iteration *select* the considered interval, if and only if it does not intersect any of the previously selected intervals. We will refer to this algorithm as the *single-machine greedy*.

Already for $m \geq 2$, the problem is NP-hard [2]. A straightforward generalization of the single-machine greedy has an approximation ratio of 1/2: Consider the machines one by one in an arbitrary order, run the single-machine greedy on

© Springer International Publishing Switzerland 2016
R. Cerulli et al. (Eds.): ISCO 2016, LNCS 9849, pp. 345–356, 2016.
DOI: 10.1007/978-3-319-45587-7_30

the intervals of the currently considered machine, add all the selected intervals to the solution and remove the jobs that correspond to them from all the subsequent machines. We will refer to this algorithms as the *multi-machine greedy*. For $m = 2$, there exists a 2/3-approximation algorithm; this algorithm has recently been generalized for any constant m, achieving approximation ratio $(\frac{1}{2} + \frac{1}{3m(m-1)})$ [3].

There are many variants of the broad class of interval scheduling, see, e.g., a recent surveys by Kolen et al. [7] and by Kovalyov et al. [8]. In a classic variant (with identical machines), each job is identified with exactly one interval, and a job (i.e., the interval) can be scheduled on any of the machines. This problem is a special variant of INTERVALSELECTION where the intervals of each particular job are the very same time interval (on all the machines). This problem is polynomially solvable even in the weighted case [1,4] (and for any m).

On the other hand, INTERVALSELECTION can be seen as a special case of $JISP_k$, the *job interval selection problem*, where each job has exactly k intervals on the real line (and the goal is to schedule a maximum number of jobs). Any instance of $JISP_k$ where the real line can be split into k parts (by $k - 1$ vertical lines) so that every part contains exactly one interval for each job, is also an instance of INTERVALSELECTION, where each part represents one machine. $JISP_k$ for $k \geq 2$ was shown to be NP-hard [6,9] and subsequently even APX-hard [10]. There is a *deterministic* 1/2-approximation algorithm for $JISP_k$ [10] which works similarly as the multi-machine greedy, and a *randomized* $\frac{e-1}{e}$-approximation algorithm [5]. This is the only algorithm that beats the barrier of 1/2 in a general setting. The algorithm is randomized, and there is no standard approach to de-randomize it. Thus, beating the approximation ratio of 1/2 in the *deterministic* case is, in that view, a main open problem.

In this paper, we study a special case of INTERVALSELECTION, where all the m intervals of every job have a point in common. In other words, the intersection of all the m intervals is non-empty. We call such a common point a *core* of the job. We call this special case INTERVALSELECTION *with cores*.

Situations where jobs have cores arise naturally in practice. Our motivation comes from the problem of assigning cars to n users of a car-sharing system with m cars, each at a different location. Assume that every user can reach every car (say, by public transport), and she wants to use it to arrive to a particular place of a fixed-time appointment. Then, depending on the distance of the car to the place of the appointment, she needs to specify different time interval for which she needs each of the cars. Clearly, the time of the appointment naturally induces a time point common to all intervals specified by a particular user, and thus a "core" in the underlying scheduling problem.

INTERVALSELECTION with cores can be solved optimally in a running time exponential in m by a dynamic programming algorithm [11] (and thus in polynomial time, whenever m is a constant). However, for a non-constant m, the problem was shown to be NP-hard even for unit intervals [2] (i.e., when every interval has a length one). One can show that even for the unit intervals case, the multi-machine greedy remains a 1/2-approximation (i.e., there is an instance of the problem on which the algorithm schedules only a half of an optimum number of jobs).

In this paper we show that INTERVALSELECTION with cores and unit intervals can be deterministically approximated strictly better than 1/2, without any restriction on the number of machines m of the given instance. At the same time, we remark that the problem is not an easy one by showing that it is APX-hard. The two main results are stated in the following theorems.

Theorem 1. INTERVALSELECTION *with cores and unit intervals is APX-hard.*

Theorem 2. *There is a deterministic $\frac{501}{1000}$-approximation algorithm for* INTERVALSELECTION *with cores and unit intervals.*

Due to space constraints, the details of the APX-hardness are omitted entirely. We note that the proof uses a new analysis of an existing NP-hardness proof [2]. (We also note that the APX-hardness of the more general problem JISP$_2$ does not carry over to our problem.)

The remaining of the paper describes the algorithm and its analysis. The approximation ratio of our algorithm is at least 501/1000. We believe that this ratio can be further improved by fine-tuning the parameters of the algorithm, and using a more careful analysis. In particular, in most of the subroutines of the algorithm we actually obtain a ratios strictly better than 501/1000. In some cases, we set the constants of the algorithm or in the analysis in such a way that we get exactly 501/1000. Thus, by a slight modification of the parameters, better approximation ratio can be obtained. None the less, we prefer to keep the algorithm and its analysis relatively simple. We also note that our algorithm is, to the best of our knowledge, the first *deterministic* algorithm having an approximation ratio better than 1/2 for any NP-hard variant of INTERVALSELECTION that does not restrict the number of machines m.

1.1 Standard Techniques Fail

Before we describe the algorithm, let us remark that some standard techniques – greedy approach and a shifting technique – fail in the goal of achieving an approximation ratio better than 1/2. Due to space constraints, we omit the details.

2 The Approximation Algorithm

In the following, we use the term *window* to refer to a time interval (independent of machines and jobs). Also, if a job has more cores, we fix any of those (say, the left-most), and refer to it as *the* core of the job.

The approximation algorithm, which we call SPLITANDMERGE, is a recursive divide-and-conquer algorithm. Described on a high level, in every step it either provides a *good enough solution* for the considered (sub)instance (and goes back in the recursion), or identifies a *middle* subinstance – a set of jobs with cores inside a small window W, for which a *good enough solution* can be provided; in this case, the algorithm recursively proceeds with the two subinstances induced by the jobs having cores *left* of W, and *right* of W, respectively, and then merges the three solutions. Here, *good enough solution* needs to schedule close-to-optimum many jobs of the subinstance, and at the same time, it must allow to merge the solutions from the *left*, *middle*, and the *right* subinstances without losing much of the quality.

The algorithm considers two types of a *middle* subinstance. The first type is induced by a window W of size 6 which contains at most m cores. The second type is structurally more involved: the algorithm first runs a greedy algorithm, called *adaptive greedy*, for the considered (sub)instance; then, if there is a window W of size 20 such that the adaptive greedy schedules all jobs with cores in W, then one can identify a sub-window W that induces the *middle* instance.

The algorithm takes special care of middle windows W for which the left or the right subinstance is empty. For this reason, we say that a window W is *on the left border* of an instance, if there is no core strictly to the left of W, and at the same time there is a core at the left endpoint of W. We analogously define W to be *on the right border*. We say that W is *in the interior* of an instance if there is both a core strictly to the left and strictly to the right of W.

The algorithm SPLITANDMERGE works as follows (the auxiliary procedures are subroutines which are described in detail later):

(1) If there are at most m jobs, return a schedule where every job is scheduled (one on each of m machines).

(2a) Else if there is a window of size 6 in the interior of the instance that contains at most m cores, then return the result obtained by the *sparse interior procedure* described in Sect. 2.1.

(2b) Else if one of the two windows of size 3 on the borders of the instance contains at most m cores, then return the result obtained by the *sparse border procedure* described in Sect. 2.2.

(3) Else, run the *adaptive greedy algorithm* on the whole instance to obtain a solution S_{AG}; and mark all the cores of jobs which are not scheduled in S_{AG}.

    (Ia) If there is a window of size 20 in the interior of the instance which does not contain a marked core, then return the result obtained by the *middle splitting procedure* described in Sect. 2.3.

    (Ib) Else if one of the windows of size 9 on the borders of the instance does not contain a marked core, then return the result obtained by the *border splitting procedure* described in Sect. 2.4.

    (II) Else return the solution S_{AG}.

We now describe the *adaptive greedy* algorithm: it consists of m iterations, where m is the number of machines. In each iteration it processes one machine by running the single-machine greedy on it. All jobs scheduled on that machine are made unavailable for subsequent iterations. The order in which the machines are processed is decided in an adaptive (greedy) way. In each iteration, the chosen machine is the one that maximizes the number of selected intervals (among all not yet processed machines). Such a machine is found by running single-machine greedy on each of the remaining unprocessed machines. Obviously, the running time of multi-machine greedy is polynomial in n and m, since it runs the single-machine greedy $(m + (m-1) + \ldots + 2) = O(m^2)$ many times.

The adaptive greedy algorithm has the following guarantee on the number of scheduled jobs:

Lemma 1. *During the first q iterations, the adaptive greedy selects at least $\left(\frac{q}{m} - \frac{q^2}{2m^2}\right) s$ intervals, where s is the optimum of the instance.*

Proof. Let us fix an underlying optimum of size s. Let t_i, $i \in \{1, \ldots, m\}$, denote the number of intervals selected by adaptive greedy during its ith iteration and let s_i, $i \in \{1, \ldots, m\}$, denote the number of intervals of the underlying optimum that are still available in the beginning of the ith iteration.

Observe that an interval I from the optimum can become unavailable for two reasons: either the adaptive greedy selects another interval I' on the same machine that intersects I and that ends earlier than I, or the adaptive greedy scheduled the job of I earlier (by selecting an interval of the job on a different machine). Thus, every chosen interval of the adaptive greedy can make at most two intervals of the optimum unavailable. Therefore, if t_i intervals are selected by the adaptive greedy in iteration i, we have $s_{i+1} \geq s_i - 2t_i$. In the beginning of the ith iteration, there are still $m - i + 1$ machines to be processed, and there are still s_i intervals of the optimum available on those machines. Therefore, there is a machine with at least $\frac{s_i}{m-i+1}$ many intervals of the optimum and from the strategy of adaptive greedy and optimality of single-machine greedy it follows that $t_i \geq \frac{s_i}{m-i+1}$.

Using these recurrence relations we prove the lemma by induction on m and $q \leq m$. We note that for $q = 1$ and arbitrary m, the statement of the lemma holds, since in the first iteration at least $\frac{s}{m}$ intervals are selected. Now let us consider the base case, that is, $m = 1$. Since $q \leq m$, we are in the situation where $q = 1$ and the statement of the lemma follows from the just noted fact.

Next, assuming $m \geq q \geq 2$ and that the statement holds for any instance with $m - 1$ machines, we analyze the number of intervals selected in the first q iterations. After the fist iteration, $t_1 \geq \frac{s_1}{m}$ intervals are selected and the optimum on the remaining machines is of size $s_2 \geq s_1 - 2t_1$. By applying the induction hypothesis we obtain that in the next $q-1$ iterations at least $\left(\frac{q-1}{m-1} - \frac{(q-1)^2}{2(m-1)^2}\right)(s_1 - 2t_1)$ intervals are selected. Therefore, altogether, in the next q iterations, at least $t_1 + \left(\frac{q-1}{m-1} - \frac{(q-1)^2}{2(m-1)^2}\right)(s_1 - 2t_1)$ intervals are selected. For increasing values of $t_1 \geq s_1/m$, this lower bound also increases, and thus the bound is minimized for $t_1 = s_1/m$. Therefore, at least

$$\frac{s_1}{m} + \left(\frac{q-1}{m-1} - \frac{(q-1)^2}{2(m-1)^2}\right)\left(s_1 - 2\frac{s_1}{m}\right) \geq^{(*)} \left(\frac{q}{m} - \frac{q^2}{2m^2}\right)s_1$$

intervals are selected (the inequality $(*)$ can be obtained by a straightforward manipulation of the formula). $\qquad\square$

Observe that for $q = m$, the algorithm schedules at least $\frac{1}{2}s$ many intervals. Thus, this algorithm alone does not achieve a better approximation ratio than $\frac{1}{2}$. However, observe that in the first rounds, the lemma guarantees that the algorithm takes larger fractions of s than in the last rounds. Thus, if we had a good alternative bound on the number of intervals selected by the algorithm in its last rounds, we could obtain a better approximation ratio by simply summing the two different lower bounds on the number of selected intervals, i.e., the lower bound for the first q' rounds of the algorithm, plus the lower bound for the last $m - q'$ rounds of the algorithm. Later on, we will provide exactly such a lower bound on the number of intervals selected in the last rounds of the algorithm.

2.1   The Sparse Interior Procedure

If the *sparse interior procedure* is called, then there must be a window W of size 6 in the interior that contains at most m cores. Let x be the left endpoint of W. The sparse interior procedure creates a *left* subinstance consisting of jobs with cores to the left of $x + 2$, and a *right* subinstance consisting of jobs with cores to the right of $x + 4$. After that, it recursively calls SPLITANDMERGE to obtain solutions to these two subinstances. Finally, the sparse interior procedure merges these two results as follows. We observe that intervals of jobs with cores to the left of $x + 2$ cannot intersect intervals of jobs with cores to the right of $x + 4$ (since all intervals have unit length). This implies that the sparse interior procedure can merge the left and right solutions without any conflicts. It remains to add to this result all the jobs with cores in the window from $x + 2$ to $x + 4$. The sparse interior procedure schedules these jobs on those machines where no job with core in the window from x to $x + 2$ and no job with core in the window from $x + 4$ to $x + 6$ has been scheduled. Since by assumption there are at most m jobs in the window from x to $x + 6$, this can always be done.

2.2   The Sparse Border Procedure

The *sparse border procedure* is similar to the sparse interior procedure. We describe only the sparse border procedure for the right border, since the procedure for the left border is symmetric. Suppose that the window W of size 3 on the right border of the instance contains at most m cores, and let x be the left endpoint of W. The sparse border procedure uses SPLITANDMERGE (recursively) to obtain a solution for the subinstance that consists of the jobs with cores to the left of $x + 2$. We observe that the jobs with cores in the window from $x + 2$ to $x + 3$ cannot conflict with jobs with cores to the left of x. Since there are no more than m jobs with cores in the window from x to $x + 3$, the sparse border procedure can schedule all the jobs with cores in the window from $x + 2$ to $x + 3$ without causing any conflict.

2.3   The Middle Splitting Procedure

The middle splitting procedure is applied if there is a window W of size 20 in the interior of the instance with no marked core inside W. This means that all jobs with core inside W have been scheduled by the adaptive greedy (i.e. they are in the solution S_{AG}).

The procedure identifies a *middle window*, a sub-window of W, to naturally split the instance into a left, middle, and right instance. The procedure recursively calls SPLITANDMERGE to obtain solutions for the left and right instances, and combines them with the "unblocked" intervals of S_{AG} having core in the middle window.

To choose the middle window, we first subdivide the window W in 10 windows of size 2, see Fig. 1. Starting from the left, we call these windows w_5^L, w_4^L, w_3^L, w_2^L, w_1^L, w_1^R, w_2^R, w_3^R, w_4^R, w_5^R. For a window w we denote with $|w|$ the number of cores inside of w. For $i \in \{1,2,3,4,5\}$, we let $r_i = |w_i^L| + |w_i^R|$ and define the windows σ_i as $\sigma_i = \bigcup_{j=1}^{i}(w_j^L \cup w_j^R)$. Finally, we let $\kappa_i := |\sigma_i|$ and observe that $\kappa_i = \sum_{j=1}^{i} r_j$. The middle window is chosen among the four candidates σ_2, σ_3, σ_4, and σ_5. Motivated by the desired approximation guarantee, let $\alpha := \frac{501}{1000}$. For $i \in \{2,3,4\}$ we say that the window σ_i is *valid* if $r_i \le \frac{1-\alpha}{\alpha}\kappa_{i-1}$. If one of σ_2, σ_3 or σ_4 is valid, we choose it as the middle window (if more than one of is valid, any one of them can be chosen). If on the other hand none of these three windows is valid, we choose σ_5 as the middle window.

2.4 The Border Splitting Procedure

The border splitting procedure is applied if there is a window W of size 9 on the left or on the right border of the instance with no marked intervals in it (which means that all jobs whose core is inside W have been scheduled by the adaptive greedy). We only describe the border splitting procedure for W on the right border of the instance, since the case where W lies on the left border of the instance is symmetric.

Let x be the left border of W. We define four windows w_1, w_2, w_3, and w_4 as follows (cf. Fig. 1). Window w_1 has size 3 and ranges from $x + 6$ to $x + 9$. The remaining windows have size 2. The window w_2 ranges from $x + 4$ to $x + 6$, w_3 from $x + 2$ to $x + 4$, and w_4 from x to $x + 2$. For $i \in \{1,2,3,4\}$ we let $r_i = |w_i|$,

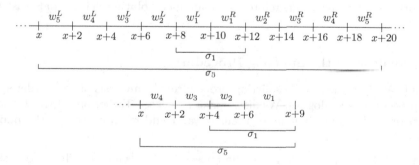

Fig. 1. Finding a middle window for the middle splitting procedure (up) and for the border splitting procedure (down). For brevity, only σ_1 and σ_5 are depicted.

i.e., r_i is the number of cores in the window w_i. For $i \in \{1, 2, 3, 4\}$, we define the windows σ_i as $\sigma_i = \bigcup_{j=1}^{i} w_j$. Finally, let $\kappa_i := |\sigma_i|$. Like in the middle splitting procedure, the border splitting procedure has to choose an appropriate *border window* among the three candidates σ_2, σ_3, and σ_4. Again, we use the constant $\alpha = \frac{501}{1000}$. For $i \in \{2, 3\}$ we say that the window σ_i is *valid* if $r_i \leq \frac{1-\alpha}{\alpha} \kappa_{i-1}$. If one of σ_2 or σ_3 is valid, we choose it as the border window (if more than one of is valid, any one of them can be chosen). Otherwise we choose σ_4 as the border window.

The border window naturally splits the instance into the left subinstance and the middle subinstance. Restricting S_{AG} to jobs with chore inside the border window gives a solution for the middle instance (scheduling all jobs), and we obtain a solution for the left subinstance by recursively calling SPLITAND-MERGE. Afterwards, the algorithm merges the two solutions by discarding intervals of the solution for the middle instance that conflict with the solution for the left subinstance.

3   Analysis

We show that the approximation ratio of SPLITANDMERGE is at least $\frac{501}{1000}$ using an induction on the size of the input instances. The base of the induction consists of instances with at most m intervals which are solved exactly by SPLITANDMERGE. SPLITANDMERGE recursively calls different procedures. In the following subsections we analyze each of the procedures.

3.1   Analysis of the Sparse Interior/Border Procedures

Both the sparse interior procedure and the sparse border procedure divide the instance into a "left", "middle", and "right" subinstances (where the left or the right subinstances may be empty), solve the left and the right subinstances recursively, combine the obtained solutions without any conflicts, and additionally schedules *all* jobs of the middle instance. By induction hypothesis, each of the solutions of the smaller parts achieves a ratio of $\frac{501}{1000}$ over the optimum on the corresponding part. Therefore, the combined solution achieves at least the same ratio.

3.2   Analysis of the Splitting Procedures

We only analyze the middle splitting procedure; the analysis of the border splitting procedure is analogous. We distinguish two cases. Either one of the windows σ_2, σ_3 or σ_4 is valid, or none of them is valid and σ_5 is chosen as the middle window.

Case 1. Let σ_k, $k \in \{2, 3, 4\}$ be the chosen (valid) window. The chosen window induces a left, a middle, and a right subinstance. We show that the algorithm approximates all three subinstances with approximation ratio $\frac{501}{1000}$ and these three solutions can be merged without any conflict, which implies that the merged solution is also a $\frac{501}{1000}$-approximation.

The left and the right subinstances are approximated with ratio $\frac{501}{1000}$ by inductive hypothesis. We now analyze the approximation ratio of the solution for the middle subinstance. Since window σ_k is valid, we know that $r_k \leq \frac{1-\alpha}{\alpha}\kappa_{k-1}$, which we can rewrite as $\alpha \leq \frac{\kappa_{k-1}}{r_k+\kappa_{k-1}}$. Recall that the optimum of the middle subinstance is $r_k + \kappa_{k-1}$. Also, since the intervals inside of σ_{k-1} do not conflict with jobs of the left or right subinstance (because their cores are more than two units away), there are at least κ_{k-1} intervals from the solution of the middle subinstance which are not discarded when merging the subinstances. Therefore, the middle subinstance is approximated with ratio at least $\alpha = \frac{501}{1000}$.

Case 2. Consider the case where none of the three candidate middle windows is valid. First we show that this implies $\kappa_4 > 5m$ as follows. The fact that every window of size 6 contains more than m cores implies that $|w_2^L|+|w_1^L|+|w_1^R| > m$ and that $|w_1^L| + |w_1^R| + |w_2^R| > m$. Adding these two inequalities we conclude that $2r_1+r_2 > 2m$. Furthermore, since the window σ_2 is not valid, we have that $r_2 > \frac{1-\alpha}{\alpha}\kappa_1 = \frac{1-\alpha}{\alpha}r_1$. The last two inequalities can be combined and solved for r_1+r_2 (multiply the second with α and add it to the first) to obtain $\kappa_2 = r_1+r_2 > \frac{2}{1+\alpha}m$. Since σ_3 and σ_4 are also not valid, we obtain $\kappa_4 = (r_4 + \kappa_3) > (\frac{1-\alpha}{\alpha}\kappa_3 + \kappa_3) = (\frac{1-\alpha}{\alpha}+1)(r_3+\kappa_2) > (\frac{1-\alpha}{\alpha}+1)^2\kappa_2 > (\frac{1-\alpha}{\alpha}+1)^2\frac{2}{1+\alpha}m = \frac{2}{(\alpha+1)\alpha^2}m > 5m$.

Now, to bound the number of jobs with cores in σ_5 that "survive" the merging with left and right subinstance solutions, we make two observations. First, notice that on each machine at most four intervals of the middle instance conflict with the left or the right solution (two per merging side). Second, since none of the jobs with cores in σ_4 conflicts with the left or the right solution, all these κ_4 jobs are scheduled in the combined solution. Therefore, the combined solution contains at least $\max(r_5 - 4m, 0) + \kappa_4$ jobs with cores in σ_5

Since the optimum for the middle subinstance is $r_5 + \kappa_4$, we obtain that the approximation ratio for the middle subinstance is at least $\frac{\max(r_5-4m,0)+\kappa_4}{r_5+\kappa_4}$. To show that this ratio is greater than $\frac{501}{1000}$ we distinguish two cases. If $r_5 \leq 4m$ then $r_5 \leq \frac{4}{5}5m \leq \frac{4}{5}\kappa_4$, and we obtain that the approximation ratio of the middle subinstance is at least $\frac{\kappa_4}{r_5+\kappa_4} \geq \frac{\kappa_4}{\frac{4}{5}\kappa_4+\kappa_4} = \frac{5}{9} > \frac{501}{1000}$. If $r_5 \geq 4m$, then the approximation ratio is at least $\frac{r_5-4m+\kappa_4}{r_5+\kappa_4} = 1 - \frac{4m}{r_5+\kappa_4} \geq 1 - \frac{4m}{4m+5m} = \frac{5}{9} > \frac{501}{1000}$.

3.3 Analysis when no Splitting is Necessary

We consider an instance which has been processed by adaptive greedy and which contains at least one marked core (that is, a core of a job that has not been scheduled). We start with the following observations.

Lemma 2. *If there is a marked core at $x \in \mathbb{R}$, then on every machine adaptive greedy has selected an interval completely contained in the window from $x - 2$ to $x + 1$.*

Proof. Consider a fixed machine M. At some point during adaptive greedy, a run of single-machine greedy is performed on M. Furthermore, consider an unscheduled job J with its core at x. On machine M, the interval I of J lies completely between $x-1$ and $x+1$. Since I has not been selected by single-machine greedy, it means that an interval I' has been selected instead. This interval I' has its right endpoint in the interior of I. Hence I' has to lie completely between $x-2$ and $x+1$.   □

Lemma 3. *If there are k marked cores such that the distance between any two of those cores is strictly greater than 2, then on each machine at least k intervals are selected.*

Proof. From Lemma 2 we know that for each unmarked core at x, an interval completely inside the window from $x-2$ to $x+1$ is selected. Now consider two cores, at x_1 and at $x_2 > x_1 + 2$. Since the overlap of the windows $(x_1 - 2, x_1 + 1)$ and $(x_2 - 2, x_2 + 1)$ is less then one unit, the intervals selected for the marked cores x_1 and x_2 must be distinct.   □

Thus, if there are many marked cores far away from each other, then the adaptive greedy scheduled many jobs. We define the *width* of an instance as the distance between the leftmost and the rightmost core.

Lemma 4. *If all windows of size 20 in the interior of the instance and both windows of width 9 at the borders of the instance contain at least one marked core, then adaptive greedy selected on each machine at least $\frac{\omega+2}{22}$ intervals, where ω is the width of the instance.*

Proof. Along with the proof, see the illustration in Fig. 2. Let $x_1 < x_2 < \cdots < x_k$, $k \in \mathbb{N}$ be the positions of the marked cores (there may be multiple cores in each of these positions). For $i \in \{1, 2, \ldots\}$ let w_i be the window from $x_1 + 22(i-1)$ to $x_1 + 22i$. We divide every window w_i into a left part consisting of all the points to the left or on $x_1 + 22(i-1) + 2$, and into a right part consisting of all the points strictly to the right of $x_1 + 22(i-1) + 2$. Let q be the smallest positive integer number such that x_k is in w_q. Since each window of size 20 in the interior contains at least one marked core, there must be a marked core in the right part of every window w_j. We now show that there are at least $\frac{\omega+2}{22}$ marked cores with distance at least 2 from each other, which together with Lemma 3 concludes the proof. We distinguish two cases: either x_k lies in the left part of w_q, or it lies in the right part of w_q.

We first consider the case where x_k lies in the left part of w_q. We know that there is a marked core at x_1. Furthermore we know that there is at least one marked core in the right part of each of the windows $w_1, \ldots, w_{q-1}$. Hence in total there are at least q marked cores with distance at least 2 from each other. By definition of q and since x_k lies in the left part of w_q, we know that the distance from x_1 to x_k is at most $22(q-1) + 2)$. The leftmost core is at most 9 units to the left of x_1, and the rightmost core is at most 9 units to the right of x_k. Therefore the width of the instance is bounded by $\omega \leq 9 + 22(q-1) + 2 + 9$,

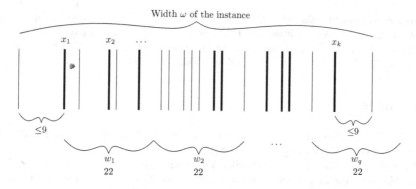

Fig. 2. The vertical lines represent cores of jobs. The bold lines represent marked cores of jobs.

from which it follows that $w + 2 \leq 22q$. Therefore in this case the inequality that we want to show holds.

Next, we consider the case where x_k lies in the right part of w_q. We know that there is a marked core at x_1, and that there is at least one marked core in the right part of each of the windows $w_1, \ldots, w_q$. Hence in total we have at least $q + 1$ marked cores with distance at least 2 from each other. By definition of q we know that the distance from x_1 to x_k is at most $22q$. Accounting for the space of at most 9 on the left and on the right yields a bound on the width of the instance of $w \leq 22q + 18$. It follows that $w + 2 \leq 22(q + 1)$. We conclude that also in this case the inequality that we want to show holds.   □

Let s be the size of an optimum solution. It is easy to see that the width of the instance is at least $\frac{s}{m} - 2$. Therefore we know from Lemma 4 that on each machine at least $\frac{1}{22} \frac{s}{m}$ intervals are selected.

We now apply Lemma 1 for a carefully chosen number of iterations q of the adaptive greedy. For this purpose let μ be the smallest nonnegative number such that $(\frac{21}{22} + \mu)m$ is a natural number. Note that $\mu < \frac{1}{m}$. Now, from Lemma 1 we know that in the first $(\frac{21}{22} + \mu)m$ iterations of adaptive greedy, at least $((\frac{21}{22} + \mu) - \frac{1}{2}(\frac{21}{22} + \mu)^2) s = (\frac{483}{968} + \frac{1}{22}\mu - \frac{1}{2}\mu^2) s$ intervals are selected.

Since on each machine at least $\frac{s}{22m}$ intervals are selected, it follows that during the last $m - (\frac{21}{22} + \mu) m$ iterations of adaptive greedy, at least $\frac{1}{22} \frac{s}{m} (m - (\frac{21}{22} + \mu)m) = s (\frac{2}{968} - \frac{1}{22}\mu)$ intervals are selected.

We sum these two quantities and obtain that in total at least $s (\frac{485}{968} - \frac{1}{2}\mu^2)$ intervals are selected. Since $\mu \leq \frac{1}{m}$, for large m this converges towards $\frac{485}{968} s$. In particular, for $m \geq 150$ we obtain an approximation ratio of at least $\frac{501}{1000}$. It can be shown that for $m < 150$ the algorithm also achieves approximation ratio of $\frac{501}{1000}$. For simplicity, let us omit this here. For the sake of seeing that there is a $\frac{501}{1000}$-approximation algorithm, one can modify the presented algorithm in that it runs the exact (optimum) algorithm from the literature for every $m < 150$, and runs SPLITANDMERGE for $m \geq 150$.

Acknowledgements. Kateřina Böhmová is a recipient of the Google Europe Fellowship in Optimization Algorithms, and this research is supported in part by this Google Fellowship. The project has been partially supported by the Swiss National Science Foundation (SNF) under the grant number 200021‑156620.

References

1. Arkin, E.M., Silverberg, E.B.: Scheduling jobs with fixed start and end times. Discrete Appl. Math. **18**(1), 1–8 (1987)
2. Böhmová, K., Disser, Y., Mihalák, M., Widmayer, P.: Interval selection with machine-dependent intervals. In: Dehne, F., Solis-Oba, R., Sack, J.-R. (eds.) WADS 2013. LNCS, vol. 8037, pp. 170–181. Springer, Heidelberg (2013)
3. Böhmová, K., Kravina, E., Mihalák, M.: Maximization problems competing over a common ground set and their black-box approximation (unpublished manuscript)
4. Bouzina, K.I., Emmons, H.: Interval scheduling on identical machines. J. Glob. Optim. **9**, 379–393 (1996)
5. Chuzhoy, J., Ostrovsky, R., Rabani, Y.: Approximation algorithms for the job interval selection problem and related scheduling problems. In: Proceedings of 42nd IEEE Symposium on Foundations of Computer Science (FOCS), pp. 348–356 (2001)
6. Keil, J.M.: On the complexity of scheduling tasks with discrete starting times. Oper. Res. Lett. **12**(5), 293–295 (1992)
7. Kolen, A.W.J., Lenstra, J.K., Papadimitriou, C.H., Spieksma, F.C.R.: Interval scheduling: a survey. Naval Res. Logistics (NRL) **54**(5), 530–543 (2007)
8. Kovalyov, M.Y., Ng, C., Cheng, T.E.: Fixed interval scheduling: models, applications, computational complexity and algorithms. Eur. J. Oper. Res. **178**(2), 331–342 (2007)
9. Nakajima, K., Hakimi, S.L.: Complexity results for scheduling tasks with discrete starting times. J. Algorithms **3**(4), 344–361 (1982)
10. Spieksma, F.C.R.: On the approximability of an interval scheduling problem. J. Sched. **2**(5), 215–227 (1999)
11. Sung, S.C., Vlach, M.: Maximizing weighted number of just-in-time jobs on unrelated parallel machines. J. Sched. **8**, 453–460 (2005)

Balanced Partition of a Graph for Football Team Realignment in Ecuador

Diego Recalde[1,2(✉)], Daniel Severín[3,4], Ramiro Torres[1], and Polo Vaca[1]

[1] Departamento de Matemática, Escuela Politécnica Nacional, Ladrón de Guevara
E11-253, EC170109 Quito, Ecuador
{diego.recalde,ramiro.torres,polo.vaca}@epn.edu.ec
[2] Research Center on Mathematical Modelling (MODEMAT),
Escuela Politécnica Nacional, Quito, Ecuador
[3] FCEIA, Universidad Nacional de Rosario, Rosario, Argentina
daniel@fceia.unr.edu.ar
[4] CONICET, Rosario, Argentina

Abstract. In the second category of the Ecuadorian football league, a set of football teams must be grouped into k geographical zones according to some regulations, where the total distance of the road trips that all teams must travel to play a Double Round Robin Tournament in each zone is minimized. This problem can be modeled as a k-clique partitioning problem with constraints on the sizes and weights of the cliques. An integer programming formulation and a heuristic approach were developed to provide a solution to the problem which has been implemented in the 2015 edition of the aforementioned football championship.

Keywords: Integer programming models · Graph partitioning · Heuristics · Football

1 Introduction

The regulations of the Ecuadorian football federation (FEF) stipulate that the second category of the Ecuadorian professional football league be conformed by the best two teams from each provincial football association that have their venues in the capital city or in nearby cities in the province. For the design of the first stage of the championship in this category, called the *zonal stage*, the provinces are grouped into 4 geographical zones. For example, during the 2014 edition of this league, 21 provincial associations participated (42 teams total), and there were three zones with 10 teams and one zone with 12. The teams of each zone are divided into two *subgroups* with the same cardinality. The division into subgroups is made at random, satisfying the constraints that two teams of the same provincial association do not belong to the same subgroup, and every subgroup must have the same number of best and second-best teams whenever possible. The teams of every subgroup play a Double Round Robin Tournament, i.e., 8 tournaments of this type. Finally, the best teams of each subgroup and

© Springer International Publishing Switzerland 2016
R. Cerulli et al. (Eds.): ISCO 2016, LNCS 9849, pp. 357–368, 2016.
DOI: 10.1007/978-3-319-45587-7_31

the four second-best teams with the highest scores advance to the next stage of the second category league, called the *national stage*, which is the prelude to the *final stage* of this championship. The FEF managers asked themselves, and the authors of this work, whether the design of the geographical zones for the *zonal stage* of this championship was optimal or not. This question resulted in an interesting mathematical problem, which is addressed in this work.

In the context of this problem and according to FEF regulations, the distance between two provinces is defined as the road trip distance between its capital cities. Thus, the problem proposed by the FEF managers consists in grouping the provinces into k zones such that the number of provinces in each zone differs at most by one, there exists a certain homogeneity of football performance among the teams in each class of the partition, and the total geographical distance between the provinces in each zone is minimized.

From the practical application point of view, and in the most general form, this problem is known in the literature as *sports team realignment*: professional sport teams are grouped in divisions in which a tournament is played to classify one or more teams to another stage in a sports league. Divisions are usually based on geography in order to minimize travel costs. The number of teams in each division needs to be similar to ensure that each team has equal opportunity to become the champion of the division, and therefore, to qualify for another stage in the sport championship. Peculiarly, in the Ecuadorian football realignments, the divisions are composed of provinces instead of teams, which could also be viewed as a territory design problem. This generated a suggestion to FEF of a new form of realignment, according to what is done in other leagues in the region and internationally, which will be explained later.

The sports team realignment problem has been modeled in different ways and for different leagues. A quadratic binary programming model is set up to divide 30 teams, of the National Football League (NFL) in the United States, into 6 compact divisions of 5 teams each [11]. The results, obtained directly from a nonlinear programming solver, are considerably less expensive for the teams in terms of total intradivisional travel, in comparison with the realignment of the 1995 edition of this league. On the other hand, McDonald and Pulleyblank [6] propose a geometric method to construct realignments for several sports leagues in the United States: NHL, MLB, NFL and NBA. The authors claim that with their approach they always find the optimal solution. To prove this, they solve mixed integer programming problems corresponding to practical instances, using CPLEX.

When it is possible to divide the teams into divisions of equal size, the sports team realignment problem can be modeled as a k-way equipartition problem: given an undirected graph with n nodes and edge costs, the problem is to find a k-partition of the set of vertices, each of the same size, such that the total cost of edges which have both endpoints in one of the subsets of the partition is minimized. Mitchell [7] solved the realignment of the NFL optimally for 32 teams and 8 divisions; the problem is modeled as a k-way equipartition problem, and is solved using a branch-and-cut algorithm; the author shows that the 2002 edition of the NFL could have reduced the sum of intradivisional travel distances by 45 %. Later, the same problem was solved using a branch-and-price method [8].

When $n \bmod k \neq 0$, the sports team realignment problem is modeled as a *Clique Partitioning Problem* (CPP) with minimum clique size constraints [9], where the size of the subsets (clusters) satisfies a lower bound of $\lfloor n/k \rfloor$ and the problem is solved with a branch-and-price-and-cut method. Since they consider an undirected complete graph, all the clusters are also cliques and that is why they refer to the problem as a *clique partitioning problem*.

The CPP has been extensively studied in the literature. This graph optimization problem was introduced by Grötschel and Wakabayashi [1] to formulate a clustering problem. They studied this problem from a polyhedral point of view and the theoretical results are used in a cutting plane algorithm that includes heuristic separation routines for some classes of facets. Jaehn and Pesch [2] propose a branch-and-bound algorithm where tighter bounds for each node in the search tree are reported. Ferreira et al. [3] analyze the problem of partitioning a graph satisfying capacity constraints on the sum of the node weights in each subset of the partition. Additionally, constraints on the size of subcliques are introduced for the CPP and the structure of the resulting polytope is studied [5].

This project is part of a cooperation agreement between the Department of Mathematics of Escuela Politécnica Nacional and the football association of Pichincha province (AFNA) in Ecuador. This agreement was based on a previous successful project for scheduling the first division of the professional Ecuadorian football league [10].

In Sect. 2 of this paper, an integer programming formulation for the balanced k-clique partitioning problem is proposed as a base model to solve the practical application in the second category of the Ecuadorian football league. A heuristic algorithm to find feasible solutions for practical instances is shown in Sect. 3. Practical and computational experience based on real-world data is reported in Sect. 4, and the paper ends in Sect. 5 with some concluding remarks.

2   Balanced k-Clique Partitioning Problem with Weight Constraints

By associating provincial capitals with the nodes of a graph, the distance between provinces with costs on the edges, and a measure of football performance of the teams of each province with weights on the nodes, the realignment problem in the second category of the Ecuadorian football league problem can be modeled as a k-Clique Partitioning Problem with constraints on the size (number of nodes in each subset differs at most in one) and weight of the cliques (total sum of node weights in the clique). From now on, we refer to this problem as a *balanced k-clique partitioning problem with weight constraints (BWk-CPP)*.

Let $G = (V, E)$ be an undirected complete graph with node set $V = \{1, \ldots, n\}$, edge set $E = \{\{i, j\} : i, j \in V, i \neq j\}$, cost on the edges $d : E \longrightarrow \mathbb{R}^+$ with $d_{ij} = d_{ji}$, weights on nodes $w : V \longrightarrow \mathbb{R}^+$ and a fixed number k, with $n \geq k \geq 2$. A k-clique partition of G is a collection of k subgraphs

$(V_1, E(V_1)), \ldots, (V_k, E(V_k))$ of G, where $V_i \neq \emptyset$, for all $i = 1, \ldots, k$, $V_i \cap V_j = \emptyset$, for all $i \neq j$, $\cup_{i=1}^{k} V_i = V$, and $E(V_i)$ is the set of edges with end nodes in V_i. Moreover, let $W_L, W_U \in \mathbb{R}^+$, $W_L \leq W_U$, be the lower and upper bounds, respectively, for the weight of each clique. The weight of a clique is the total sum of the node weights in the clique. Then, the balanced k-clique partitioning problem with weight constraints (BWk-CPP) consists of finding a k-clique partition such that

$$W_L \leq \sum_{j \in V_c} w_j \leq W_U, \qquad \forall c = 1, \ldots, k, \qquad (1)$$

$$\|V_i| - |V_j\| \leq 1, \qquad \forall i, j \in \{1, 2, \ldots, k\}, \quad i < j, \qquad (2)$$

and the total edge cost over all cliques is minimized.

Notably, to the extent of our knowledge, balanced k-clique partitioning problems with weight constraints have not been reported in the literature. In fact, some of these problems can be obtained from BWk-CPP by fixing parameters adequately. For instance, if weight constraints (1) are suppressed, the balanced k-clique partitioning problem appears. Similarly, if (1) is removed and $n \bmod k = 0$, i.e., the size of the cliques coincides with n/k, the problem becomes the so-called k-way equipartition problem. Moreover, when $k = 2$ is fixed, n is even and (1) is taken away, the equicut problem arises. It is known that all these problems are $\mathcal{NP}$-hard. Even, if we restrict ourselves to $k = 3$, $W_L = W_U = 0$ and weights $w_i = 0$ for all $i \in V$, the decision problem associated to our problem BWk-CPP is $\mathcal{NP}$-complete.

Proposition 1. *Let $G = (V, E)$ be a complete graph, cost on the edges $d : E \longrightarrow \mathbb{R}^+$ and a value $t \in \mathbb{R}^+$. Deciding if G has a 3-clique partition satisfying (2) with total edge cost at most t is $\mathcal{NP}$-complete.*

Proof. We give a polynomial transformation from the 3-EQUITABLE COLORING PROBLEM which is $\mathcal{NP}$-complete [4], and consists of deciding if an undirected graph $G = (V, E)$ has a partition of V into stable sets V_1, V_2, V_3 of G satisfying (2).

For a given graph $G = (V, E)$, consider the following instance of the decision problem: a complete graph $G' = (V, E')$, costs $d_{ij} = 1$ for all $\{i, j\} \in E$ and $d_{ij} = 0$ for all $\{i, j\} \in E' \backslash E$, and $t = 0$. Clearly, there exists a 3-clique partition of G' with total edge cost zero if and only if V_1, V_2, V_3 are stable sets satisfying (2). □

The last result gives us a remote possibility of finding a polynomial time algorithm to solve the BWk-CPP problem to optimality. It is also known that approaches based on Integer Linear Programming have proven to be one of the best tools to solve these kind of hard problems. An integer programming formulation is provided below.

If $n \bmod k \neq 0$, a set A of zero-weight dummy nodes of cardinality $k - (n \bmod k)$ are included to the set V, i.e., $V := V \cup A$. Consequently, zero-cost dummy edges are included to the set E, i.e., $E := E \cup \{\{i,j\} : i \in A, j \in V, i \neq j\}$. Observe that the inclusion of the set of vertices A to the set V implies that $n \mod k = 0$.

Let x_i^c be the variable that takes the value 1 if the node $i \in V$ belongs to clique c, for all $c = 1, \ldots, k$, and 0 otherwise. Moreover, $x_{ij}^c = 1$ indicates that the edge $\{i,j\} \in E$ is assigned to clique c, for all $c = 1, \ldots, k$, and $x_{ij}^c = 0$ otherwise. Then, the BWk-CPP can be formulated as:

$$\min \sum_{c=1}^{k} \sum_{\{i,j\} \in E} d_{ij} x_{ij}^c \tag{3}$$

$$\sum_{i \in V} x_i^c = \frac{n}{k}, \qquad \forall c = 1, 2, \ldots, k \tag{4}$$

$$\sum_{c=1}^{k} x_i^c = 1, \qquad \forall i \in V \tag{5}$$

$$\sum_{i \in A} x_i^c \leq 1, \qquad \forall c = 1, 2, \ldots, k \tag{6}$$

$$W_L \leq \sum_{i \in V} w_i x_i^c \leq W_U, \qquad \forall\, c = 1, \ldots, k \tag{7}$$

$$\sum_{j \in \delta(i)} x_{ij}^c = \left(\frac{n}{k} - 1\right) x_i^c, \qquad \forall i \in V, c = 1, 2, \ldots, k \tag{8}$$

$$x_i^c, x_{ij}^c \in \{0,1\}, \qquad \forall i \in V, c = 1, 2, \ldots, k, \text{ and } \{i,j\} \in E, \tag{9}$$

where $\delta(i)$ is the set of incident edges to node $i \in V$.

The objective function (3) seeks to minimize the total edge cost of the cliques; constraints (4) build cliques of equal size; constraints (5) ensure that each node belongs to exactly one clique; constraints (6) guarantee that there is at most one dummy node in each clique; constraints (7) impose the weight requirement on each clique; constraints (8) establish that if a node belongs to a subgraph, then it is connected with the other $n/k - 1$ nodes in the subgraph, i.e., it is a clique. Observe that (8) can be obtained as linear combinations of the classical forcing constraints $x_{ij}^c \geq x_i^c + x_j^c - 1$.

The formulation provided above has exactly $k(n + 3) + n$ constraints and $kn(n + 1)/2$ variables. For small size instances ($n \leq 22$; $k = 4$), the last formulation was solved to optimality using the Gurobi solver, in order to provide a solution for the realignment problem in the Ecuadorian Football League. However, a heuristic approach was also explored in order to obtain feasible solutions for practical instances in which the Gurobi solver failed to return an optimal solution quickly.

3   Heuristic Approach

In this section a heuristic approach to solve the problem addressed in this paper is explained. Although it is not intended to compare the performance of this heuristic with state-of-the-art heuristics for Graph Partitioning Problems, it behaves reasonably fast and meets our needs and expectations. It is based on the decomposition of $BWk\text{-}CPP$ in two sub-problems that are much easier to solve. The first sub-problem consists of finding a set of k initial nodes that are the "seeds" of the subsets in the partition. Then, the k subsets are constructed by solving the second sub-problem, where the remaining nodes are assigned to the subsets adequately, so that they become a clique and satisfy the size and weight constraints. Each iteration of the heuristic consists of solving the two sub-problems one after the other, where the first sub-problem provides a different set of seed nodes. The process is repeated a fixed number of iterations and the best solution is stored.

3.1   Seed Nodes Location Phase

For the choice of seed nodes, two variants have been tested:

Variant 1:   Select k nodes randomly, where the probability of choosing a node $i \in V$ is directly proportional to its weight w_i. As in the practical application, the weight of a node corresponds to a measure of football performance of a team, it is more likely that a strong team be chosen as the seed of a group, which is desirable from the practical point of view.

Variant 2:   Perform an iterative location procedure sketched in Algorithm 1, where $\mathcal{S}$ is the set of seed nodes.

Algorithm 1. Seed location, variant 2.

$\mathcal{S} = \emptyset$

$p = 1$

Choose one node $j \in V$ randomly using the variant 1.

Assign node j as the seed of subset V_p and set $\mathcal{S} = \{j\}$.

for $p = 2, \ldots, k$ **do**

   Find the farthest node j to the seed nodes in the subsets $V_1, \ldots, V_{p-1}$ by computing:

$$j = \arg \max_{i \in V \setminus \mathcal{S}} \sum_{s \in \mathcal{S}} d_{si}$$

   and set node j as the seed node of V_p.

   $\mathcal{S} = \mathcal{S} \cup \{j\}$

end for

The second variant chooses, as a seed node, the "farthest node" from the set of seed nodes chosen in a previous iteration. In the context of the application, it makes sense that the seed teams be located in a disperse fashion, ensuring that they will never play together. Recall that a separate tournament is played in each subset of teams (zone).

3.2   Allocation-Completion Phase

After the seed location stage has been performed, and every subset $V_l, l = 1, \ldots, k$ has a seed node, the remaining $n - k$ nodes must be allocated to the sets V_l to conform k cliques that satisfy size and weight requirements and with minimum total cost. In order to find the solution to this problem, the clique condition over the subsets is relaxed and the following model is solved: let $\mathcal{S}$ be the set of seed nodes determined in the last phase; let x_i^c be a variable that is equal to 1 if the remaining node $i \in V \setminus \mathcal{S}$ is assigned to the subset with seed node c, and 0 otherwise. Then, the allocation model can be formulated as:

$$\min \sum_{i \in V \setminus \mathcal{S}} \sum_{c \in \mathcal{S}} d_{ic} x_i^c \tag{10}$$

s.t.

$$\sum_{c \in \mathcal{S}} x_i^c = 1, \qquad\qquad \forall\, i \in V \setminus \mathcal{S}, \tag{11}$$

$$\sum_{i \in A} x_i^c \leq 1, \qquad\qquad \forall c \in \mathcal{S} \tag{12}$$

$$\sum_{i \in V \setminus \mathcal{S}} x_i^c = \frac{n}{k} - 1, \qquad\qquad \forall\, c \in \mathcal{S}, \tag{13}$$

$$W_L \leq w_c + \sum_{i \in V \setminus \mathcal{S}} w_i x_i^c \leq W_U, \qquad\qquad \forall c \in \mathcal{S}, \tag{14}$$

$$x_i^c \in \{0, 1\}, \qquad\qquad \forall\, i \in V \setminus \mathcal{S},\, c \in \mathcal{S} \tag{15}$$

The expression (10) aims to minimize the total edge cost between nodes and seed nodes. Constraints (11) ensure that all the remaining nodes in $V \setminus \mathcal{C}$ are assigned to exactly one subset; constraints (12) guarantee again that there is at most one dummy node in each clique; constraints (13) impose the number of nodes on each subset (every subset already has the seed node). Finally, constraints (14) guarantee weight requirements. This model has $n + 2k$ constraints and $(n - k)k$ variables, which are far fewer constraints and variables than the original model.

After obtaining a node partition $\mathcal{P} = \{V_1, \ldots, V_k\}$ of V, by solving the seed location and allocation phases sequentially, we can easily construct a feasible clique partition for the $BWk\text{-}CPP$ by completing the remaining edges $\{i, j\}$ (completion step), where $i, j \in V_l \setminus \{s\}$, for all $l = 1, \ldots, k$ and i, j are not seed nodes. This phase is completed by computing the balanced k-clique partitioning cost $z(\mathcal{P})$ induced by $\mathcal{P}$ as $z(\mathcal{P}) = \sum_{l=1}^{k} \sum_{\substack{i,j \in V_l \\ i<j}} d_{ij}$.

Finally, the location and allocation-completion phases are integrated in the following routine: a complete graph with weights on the nodes and costs on the edges, a fixed number of iterations $N > 0$, and the number of cliques $k \geq 2$ in which the graph must be partitioned are given as inputs. In the main loop, seed location and allocation-completion phases are performed in cascade a fixed number of iterations. At the end, the feasible balanced k-clique partition induced by a partition $\mathcal{P}^*$ with minimum value $z(\mathcal{P}^*)$ is returned. Figure 1 depicts how the allocation-completion phase works.

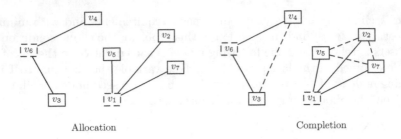

Allocation   Completion

Fig. 1. Given a complete graph with seven nodes and $k = 2$, the allocation-completion phase is performed where $\mathcal{S} = \{v_1, v_6\}$ is the set of seed nodes.

3.3   Local Search

After a feasible solution of the *BWk-CPP* is found by using the heuristic method explained before, a local search algorithm is performed with the hope to find a partition of less total cost. Given a balanced k-clique partition induced by $\mathcal{P} = \{V_1, \ldots, V_k\}$, a feasible solution reachable from $\mathcal{P}$ is made up by exchanging the seed node $s \in V_l$ with a neighbor node $j \in V_l$, for every $l = 1, \ldots, k$, one exchange on each iteration. Note that this procedure returns a different set of seed nodes $\mathcal{S}'$. Using this new set of seed nodes, the allocation-completion phase is performed to obtain a new partition $\mathcal{P}' = \{V_1', \ldots, V_k'\}$. If the total cost $z(\mathcal{P}')$ is smaller than $z(\mathcal{P})$, then the current best solution is updated to the clique partition induced by $\mathcal{P}'$. The procedure stops when all possible changes have been done in every subset V_l, $l = 1, \ldots, k$.

4   Practical Experience and Computational Results

For the 2014 edition of the second category of the Ecuadorian professional football league, the realignment of provinces in the first stage (*zonal stage*) was made using an empirical method. This empirical solution is presented in Fig. 2, where the zones, the provinces in each zone, and the total distance of road trips are depicted. The value shown in the figure is the distance traveled by all the teams to play a Double Round Robin Tournament in each zone. This value is computed by multiplying the total cost of the clique induced by every subgroup by four, and adding these quantities for all the cliques. The latter is due to the fact that in a Round Robin Tournament, every pair of teams play once at a home venue (home game) and once visiting the other team (away game). Thus, for every edge $\{i, j\}$ associated with teams with venues in the cities (nodes) i and j, respectively, in an away game every team travels twice the distance d_{ij} between the capital cities: once to go the game, and once to get back home.

On the other hand, the optimal solution for the 2014 edition of the championship, which was provided by the mathematical programming approach, is depicted in Fig. 3. The node weights were computed considering historical performance of the teams, in a very similar way to the approach of the South American

Fig. 2. Empirical solution, 2014 edition. Total road trip distance: 39830.4 km.

Fig. 3. Optimal solution, 2014 edition. Total road trip distance: 38298 km.

Football Confederation for ranking teams. The bounds on the node weights were fixed to $W_L = \mu(n/k) - \sigma$ and $W_U = \mu(n/k) + \sigma$, where μ and σ are the mean and standard deviation of the node weights, respectively.

The optimal solution showed that the first stage of the second category football championship could have reduced the total road trip distance by 1532.4 km, which represents a difference of 4 % with respect to the empirical solution. It is important to remark that the evaluation of the empirical versus the mathematical programming method is made assuming that only one double Round Robin Tournament is played in each zone. The latter is because the realignment in the Ecuadorian football league considers provinces instead of teams, as it was stated in the introduction of this article. Even though the reduction in travel distance was small, FEF managers were pleased to know that it was possible to improve the realignment. This allowed this mathematical method to be implemented in practice for the 2015 edition of the championship.

Once the problem was mathematically modeled and solved, AFNA managers suggested the inclusion of additional constraints for the realignment of the 2015 edition of the championship, which are detailed as follows.

4.1 Historically Strong and Weak Teams

The teams representing the provinces of Guayas, Manabí and Pichincha are considered to be the strongest teams because of the good results obtained during the last editions of the championship. Thus, for the 2015 edition it was required that these provinces belong to different zones in the realignment. This task was done by fixing adequately the variables corresponding to these teams.

On the other hand, the five teams representing the provinces of the Amazon region of Ecuador have obtained poor results in previous editions of the tournament. Therefore, in this case the requirement was that at most three Amazon provinces belong to the same zone: $\sum_{i \in Amazon} x_i^c \leq 3$, for all $c = 1, \ldots, k$.

Taking into account these restrictions, which were added to the model (3)–(9) and to the heuristic algorithm, the optimal realignment for the 2015 edition of

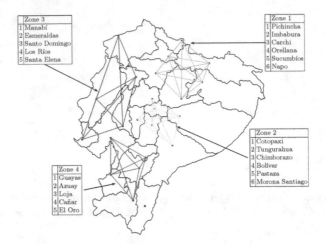

Fig. 4. 2015 edition: optimal realignment. Total road trip distance: 45354 km.

the championship was computed. The proposal of optimal realignment, including this solution, was presented under the auspices of AFNA to the FEF; this solution, which was used in practice, appears in Fig. 4.

4.2   A Proposal to Change the Design of the Zonal Stage

As it was stated in the introduction of this work, in the zonal stage, the teams in each zone are divided into two subgroups, where a Double Round Robin Tournament is played in each one of them. A question immediately arises: why not make the realignment in order to obtain the subgroups directly? Unfortunately, this change in the realignment method requires a change in the regulations of the FEF that has not been implemented yet. Nevertheless, we wanted to show the realignment of teams for the 2015 edition of the championship and to compare its benefits against the realignment of provinces. In this case, the nodes in the graph correspond to the exact position of the venues of the teams, instead of assuming that the venue is the capital of the province. Notice that now we have the usual realignment problem reported in the literature.

For the 2015 edition of the football league, the total road trip distance of all the teams during the zonal stage, considering every subgroup in each zone (8 subgroups, 44 teams in total), was 94736 km. The optimal solution, under this proposal, reported a total road trip of 86192 km. In conclusion, if this new proposal had been implemented, the second category football championship could have reduced the total road trip distance by 8544 km, which represents a difference of approximately 9 %. This reduction is significant if we consider that Ecuador has only a total area of 283 561 km^2.

A proposal including the team realignment explained in this subsection was presented to FEF managers in mid-2015. As of the submission of this paper, the FEF regulations have not yet been reformed to put this new realignment methodology into practice.

4.3   Computational Results

The performance of the Gurobi solver and the heuristics are shown in Table 1. Three instances are presented: the 2014 edition, the 2015-proposal related to the new design of the zonal stage explained in Sect. 4.2, and the 2015 edition which was put into practice. The data corresponding to these instances is accessible at [12].

The integer programming formulation, and the heuristic method were implemented in the C++ programming language on a Core i7 PC with 8 GB RAM running Ubuntu 12.04 LTS and using GUROBI v6.5.0 as an IP solver. The running time in any case was limited to 14400 s and the number of iterations of the heuristic algorithm was set to at most 5000 iterations because beyond this number of iterations, no improvement was observed.

Table 1. Computational results

Instance	n	k	Gurobi solver			Heuristic approach				
							Variant 1		Variant 2	
			cost	t (s)	Gap (%)	# iterations	cost	t (s)	cost	t (sec)
2014	21	4	38298	11.3	0	500	38298	3.9	38298	2.1
2015-proposal	44	8	**86192**	**14400**	12.6	5000	**87296**	**177.4**	87903.0	54.5
2015	22	4	45354	13.2	0	1000	45354	7.8	45354	1.5

5   Concluding Remarks

The realignments obtained by the optimization method were presented to the AFNA managers, who were pleased by the reduction of the total distance travelled by the teams. They emphasized the potential of mathematical tools, as opposed to the empirical process, in which political and subjective issues may affect the ultimate decision. Under the auspices of AFNA, the solution was presented to the managers of FEF to be considered for the design of leagues in the future. In fact, this methodology was used for designing the 2015 edition of the zonal league.

Acknowledgments. We thank Patricio Torres, authority of the team Liga Deportiva Universitaria, and Galo Barreto, former manager of AFNA, for their support to this project. This research was partially supported by the PACK-COVER MATH-AmSud cooperation project.

References

1. Grötschel, M., Wakabayashi, Y.: A cutting plane algorithm for a clustering problem. Math. Program. **45**, 59–96 (1989)
2. Jaehn, F., Pesch, E.: New bounds and constraint propagation techniques for the clique partitioning problem. Discrete Appl. Math. **161**, 2025–2037 (2013)

3. Ferreira, C., Martin, A., de Souza, C., Weismantel, R., Wolsey, L.: The node capacitated graph partitioning problem: a computational study. Math. Program. **81**,
   229–256 (1998)
4. Furmanczyk, H., Kubale, M.: Equitable coloring of graphs. In: Graph Colorings,
   Providence, Rhode Island, pp. 35–53. American Mathematical Society (2004)
5. Labbé, M., Özsoy, F.A.: Size-constrained graph partitioning polytopes. Discrete
   Math. **310**, 3473–3493 (2010)
6. McDonald, B., Pulleyblank, W.: Realignment in the NHL, MLB, NFL, and NBA.
   J. Quant. Anal. Sports **10**, 225–240 (2014)
7. Mitchell, J.: Branch-and-cut for the k-way equipartition problem. Technical report,
   Department of Mathematical Sciences, Rensselaer Polytechnic Institute (2001)
8. Ji, X., Mitchell, J.E.: Finding optimal realignments in sports leagues using a
   branch-and-cut-and-price approach. Int. J. Oper. Res. **1**, 101–122 (2005)
9. Ji, X., Mitchell, J.E.: Branch-and-price-and-cut on the clique partitioning problem
   with minimum clique size requirement. Discrete Optim. **4**, 87–102 (2007)
10. Recalde, D., Torres, R., Vaca, P.: Scheduling the professional ecuadorian football
    league by integer programming. Comput. Oper. Res. **40**, 2478–2484 (2013)
11. Saltzman, R., Bradford, R.M.: Optimal realignments of the teams in the National
    Football League. Eur. J. Oper. Res. **93**, 469–475 (1996)
12. Test instances for the football team realignment in Ecuador. http://www.math.
    epn.edu.ec/~recalde/#TestInstances/

On a General Framework for Network Representability in Discrete Optimization

(Extended Abstract)

Yuni Iwamasa[✉]

Department of Mathematical Informatics,
Graduate School of Information Science and Technology,
University of Tokyo, Tokyo 113-8656, Japan
yuni_iwamasa@mist.i.u-tokyo.ac.jp

Abstract. In discrete optimization, representing an objective function as an s-t cut function of a network is a basic technique to design an efficient minimization algorithm. A network representable function can be minimized by computing a minimum s-t cut of a directed network, which is a very easy and fastly solved problem. Hence it is natural to ask what functions are network representable. In the case of pseudo Boolean functions (functions on $\{0,1\}^n$), it is known that any submodular function on $\{0,1\}^3$ is network representable. Živný-Cohen-Jeavons showed by using the theory of expressive power that a certain submodular function on $\{0,1\}^4$ is not network representable.

In this paper, we introduce a general framework for the network representability of functions on D^n, where D is an arbitrary finite set. We completely characterize network representable functions on $\{0,1\}^n$ in our new definition. We can apply the expressive power theory to the network representability in the proposed definition. We prove that some ternary bisubmodular function and some binary k-submodular function are not network representable.

Keywords: Network representability · Valued constraint satisfaction problem · Expressive power · k-submodular function

1 Introduction

The minimum s-t cut problem is one of the most fundamental and efficiently solved problems in discrete optimization. Thus, representing a given objective function by the s-t cut function of some network leads to an efficient minimization algorithm. This idea goes back to a classical paper by Ivănescu [10] in 60's, and revived in the context of computer vision in the late 80's. Efficient image denoising and other segmentation algorithms are designed by representing the energy functions as s-t cut functions. Such a technique (*Graph Cut*) is now popular in computer vision; see [5,14] and references therein. An s-t cut function is a representative example of *submodular functions*. Mathematical modeling

© Springer International Publishing Switzerland 2016
R. Cerulli et al. (Eds.): ISCO 2016, LNCS 9849, pp. 369–380, 2016.
DOI: 10.1007/978-3-319-45587-7_32

and learning algorithms utilizing submodularity are now intensively studied in machine learning community; see e.g. [1]. Hence efficient minimization algorithms of submodular functions are of great importance, but it is practically impossible to minimize very large submodular functions arising from machine learning by using generic polynomial time submodular minimization algorithms such as [7, 11,15,17]. Thus, understanding efficiently minimizable subclasses of submodular functions and developing effective uses of these subclasses for practical problems have been being important issues.

What (submodular) functions are efficiently minimizable via a network representation and minimum cut computation? Ivǎnescu [10] showed that all submodular functions on $\{0,1\}^2$ are network representable, and Billionet-Minoux [2] showed that the same holds for all submodular functions on $\{0,1\}^3$. Note that it is meaningful to investigate network representability of functions having a few variables, since they can be used as building blocks for large network representations. Kolmogorov-Zabih [14] introduced a formal definition of the network representability, and showed that network representable functions are necessarily submodular. Are all submodular functions network representable? This question was negatively solved by Živný-Cohen-Jeavons [19]. They showed that a certain submodular function on $\{0,1\}^4$ is *not* network representable. In proving the non-existence of a network representation, they utilized the theory of expressive power developed in the context of valued constraint satisfaction problems.

In this paper, we initiate a network representation theory for functions on D^n, where D is a general finite set beyond $\{0,1\}$. Our primary motivation is to give a theoretical basis for applying network flow methods to multilabel assignments (e.g. Potts model) in practical area. Our main target as well as our starting point is network representations of *k-submodular functions* [8] which have recently been gained attention as a promising generalization of submodular functions on $\{0,1,2,\ldots,k\}^n$ [6,12]. Iwata-Wahlström-Yoshida [12] considered a network representation of k-submodular functions for design of FPT algorithms. Independently, Ishii [9] considered another representation, and showed that all 2-submodular (bisubmodular) functions on $\{0,-1,1\}^2$ are network representable. In this paper, by generalizing and abstracting their approaches, we present a unified framework for network representations of functions on D^n. Features of the proposed framework as well as results of this paper are summarized as follows:

- In our network representation, to represent a function on D^n, each variable in D is associated with several nodes. More specifically, three parameters (k,ρ,σ) define one network representation. The previous network representations (by Kolmogorov-Zabih, Ishii, and Iwata-Wahlström-Yoshida) can be viewed as our representations for special parameters.
- We completely characterize network representable functions on $\{0,1\}^n$ under our new definition; they are network representable in the previous sense or they are monotone (Theorems 7 and 8). The minimization problem of monotone functions is trivial. This means that it is sufficient only to consider the original network representability for functions on $\{0,1\}^n$.

- An important feature of our framework is its compatibility with the expressive power theory, which allows us to prove that a function cannot admit any network representation.
- As application of above, we prove that some bisubmodular function on $\{0, -1, 1\}^3$ and some k-submodular function on $\{0, 1, 2, \ldots, k\}^2$ are *not* network representable for a certain parameter (Theorems 9 and 10). This answers negatively an open problem raised by [12].

The proofs of theorems had to be omitted due to space constraints. They will be included in the full version of this paper.

Organization. In Sect. 2, we introduce a submodular function, an s-t cut function, and a k-submodular function. We also introduce the network representation of submodular functions by Kolmogorov-Zabih [14]. Furthermore we explain concepts of expressive power and weighted polymorphism, which play key roles in proving the non-existence of a network representation. In Sect. 3, we explain the previous network representations of k-submodular functions. Then we introduce a framework for the network representability of functions on D^n, and discuss its compatibility with the expressive power theory. We also present our results on the network representability in our framework.

Notation. Let $\mathbf{Q}$ and $\mathbf{Q}_+$ denote the sets of rationals and nonnegative rationals, respectively. In this paper, functions can take the infinite value $+\infty$, where $a < +\infty$ and $a + \infty = +\infty$ for $a \in \mathbf{Q}$. Let $\overline{\mathbf{Q}} := \mathbf{Q} \cup \{+\infty\}$. For a function $f : D^n \to \overline{\mathbf{Q}}$, let $\mathrm{dom}\, f := \{x \in D^n \mid f(x) < +\infty\}$. For a positive integer k, let $[k] := \{1, 2, \ldots, k\}$, and $[0, k] := [k] \cup \{0\}$. By a (directed) network $(V, A; c)$, we mean a directed graph (V, A) endowed with rational nonnegative edge capacity $c : A \to \mathbf{Q}_+ \cup \{+\infty\}$. A subset $X \subseteq V$ is also regarded as a characteristic function $X : V \to \{0, 1\}$ defined by $X(i) = 1$ for $i \in X$ and $X(i) = 0$ for $i \notin X$. A function $\rho : F \to E$ with $F \supseteq E$ is called a *retraction* if it satisfies $\rho(a) = a$ for $a \in E$. $\rho : F \to E$ is extended to $\rho : F^n \to E^n$ by defining $(\rho(x))_i := \rho(x_i)$ for $x \in F^n$ and $i \in [n]$.

2 Preliminaries

2.1 Submodularity

A *submodular function* is a function f on $\{0, 1\}^n$ satisfying the following inequalities

$$f(x) + f(y) \geq f(x \wedge y) + f(x \vee y) \qquad (x, y \in \{0, 1\}^n),$$

where binary operations $\wedge, \vee$ are defined by

$$(x \wedge y)_i := \begin{cases} 1 & \text{if } x_i = y_i = 1, \\ 0 & \text{if } x_i = 0 \text{ or } y_i = 0, \end{cases} \qquad (x \vee y)_i := \begin{cases} 1 & \text{if } x_i = 1 \text{ or } y_i = 1, \\ 0 & \text{if } x_i = y_i = 0, \end{cases}$$

for $x = (x_1, x_2, \ldots, x_n)$ and $y = (y_1, y_2, \ldots, y_n)$.

The *s-t cut function* of a network $G = (V \cup \{s, t\}, A; c)$ is a function C on 2^V defined by

$$C(X) := \sum_{(u,v) \in A,\ u \in X \cup \{s\},\ v \notin X \cup \{s\}} c(u, v) \qquad (X \subseteq V).$$

For $X \subseteq V$, we call $X \cup \{s\}$ an *s-t cut*. An *s-t* cut function is submodular. In particular, an *s-t* cut function can be efficiently minimized by max-flow min-cut algorithm. The current fastest one is $O(|V||A|)$ time algorithm by Orlin [16].

Let us introduce a class of functions on $[0, k]^n$, which also plays key roles in discrete optimization. A *k-submodular function* is a function f on $[0, k]^n$ satisfying the following inequalities

$$f(x) + f(y) \geq f(x \sqcap y) + f(x \sqcup y) \qquad (x, y \in [0, k]^n),$$

where binary operations $\sqcap, \sqcup$ are defined by

$$(x \sqcap y)_i := \begin{cases} x_i & \text{if } x_i = y_i, \\ 0 & \text{if } x_i \neq y_i, \end{cases} \quad (x \sqcup y)_i := \begin{cases} y_i \ (\text{resp. } x_i) & \text{if } x_i = 0 \ (\text{resp. } y_i = 0), \\ x_i & \text{if } x_i = y_i, \\ 0 & \text{if } 0 \neq x_i \neq y_i \neq 0, \end{cases}$$

for $x = (x_1, x_2, \ldots, x_n)$ and $y = (y_1, y_2, \ldots, y_n)$. A *k*-submodular function was introduced by Huber-Kolmogorov [8] as an extension of submodular functions. In $k = 1$, a *k*-submodular function is submodular, and in $k = 2$, a *k*-submodular function is called *bisubmodular*, which domain is typically written as $\{0, -1, 1\}^n$ (see [4]). It is not known whether a *k*-submodular function can be minimized in polynomial time under oracle model for $k \geq 3$. By contrast, it is known that a *k*-submodular function can be minimized in polynomial time in valued constraint satisfaction problem model for all k [13].

In the following, we denote the set of all submodular functions having at most n variables as $\Gamma_{\text{sub},n}$, and let $\Gamma_{\text{sub}} := \bigcup_n \Gamma_{\text{sub},n}$. We also denote the set of all bisubmodular functions (resp. *k*-submodular functions) having at most n variables as $\Gamma_{\text{bisub},n}$ (resp. $\Gamma_{k\text{sub},n}$).

2.2   Network Representation over $\{0, 1\}$

A function $f : \{0, 1\}^n \to \overline{\mathbf{Q}}$ is said to be *network representable* if there exist a network $G = (V, A; c)$ and a constant $\kappa \in \mathbf{Q}$ satisfying the following:

- $V \supseteq \{s, t, 1, 2, \ldots, n\}$.
- For all $x = (x_1, x_2, \ldots, x_n) \in \{0, 1\}^n$, it holds that

$$f(x) = \min\{C(X) \mid X : s\text{-}t \text{ cut}, X(i) = x_i \text{ for } i \in [n]\} + \kappa.$$

This definition of the network representability was introduced by Kolmogorov-Zabih [14]. A network representable function has the following useful properties:

Property 1: A network representable function f can be minimized via computing a minimum s-t cut of a network representing f.

Property 2: The sum of network representable functions f_1, f_2 is also network representable, and a network representation of $f_1 + f_2$ can easily be constructed by combining networks representing f_1, f_2.

By the Property 1, a network representable function can be minimized efficiently, provided a network representation is given. By the Property 2, it is easy to construct a network representation of a function f if f is the sum of "smaller" network representable functions. Hence it is meaningful to investigate network representability of functions having a few variables. For example, by the fact that all submodular functions on $\{0, 1\}^2$ are network representable, we know soon that the sum of submodular functions on $\{0, 1\}^2$ is also network representable. This fact is particularly useful in computer vision application. Moreover, thanks to extra nodes, a function obtained by a *partial minimization* (defined in Sect. 2.3) of a network representable function is also network representable.

2.3 Expressive Power

It turned out that the above definition of network representability is suitably dealt with in the theory of *expressive power*, which has been developed in the literature of valued constraint satisfaction problems [18]. In this subsection, we introduce the concepts concerning expressive power.

Let D be a finite set, called a *domain*. A *cost function* on D is a function $f : D^r \to \overline{\mathbf{Q}}$ for some positive integer $r = r_f$, called the *arity* of f. A set of cost functions on D is called a *language* on D. A cost function $f_= : D^2 \to \overline{\mathbf{Q}}$ defined by $f_=(x, y) = 0$ if $x = y$ and $f_=(x, y) = +\infty$ if $x \neq y$, is called the *weighted equality relation* on D. A *weighted relational clone* [3] on D is a language Γ on D such that

- $f_= \in \Gamma$,
- for $\alpha \in \mathbf{Q}_+$, $\beta \in \mathbf{Q}$, and $f \in \Gamma$, it holds that $\alpha f + \beta \in \Gamma$,
- an addition of $f, g \in \Gamma$ belongs to Γ, and
- for $f \in \Gamma$, a partial minimization of f belongs to Γ.

Here an *addition* of two cost functions f, g is a cost function h obtained by

$$h(x_1, \ldots, x_n) = f(x_{\sigma_1(1)}, \ldots, x_{\sigma_1(r_f)}) + g(x_{\sigma_2(1)}, \ldots, x_{\sigma_2(r_g)}) \quad (x_1, \ldots, x_n \in D)$$

for some $\sigma_1 : [r_f] \to [n]$ and $\sigma_2 : [r_g] \to [n]$. A *partial minimization* of f of arity $n + m$ is a cost function h of arity n obtained by

$$h(x_1, \ldots, x_n) = \min_{x_{n+1}, \ldots, x_{n+m} \in D} f(x_1, \ldots, x_n, x_{n+1}, \ldots, x_{n+m}) \quad (x_1, \ldots, x_n \in D).$$

For a language Γ, the *expressive power* $\langle \Gamma \rangle$ of Γ is the smallest weighted relational clone (as a set) containing Γ [18]. A cost function f is said to be *representable by a language* Γ if $f \in \langle \Gamma \rangle$.

By using these notions, Živný-Cohen-Jeavons [19] noted that the set of network representable functions are equal to the expressive power of $\Gamma_{\text{sub},2}$.

Lemma 1 ([19]). *The set of network representable functions coincides with* $\langle \Gamma_{\mathrm{sub},2} \rangle$.

The previous results for network representability and nonrepresentability are summarized as follows.

Theorem 2. *The following hold:*

[14] $\langle \Gamma_{\mathrm{sub},2} \rangle \subseteq \Gamma_{\mathrm{sub}}$.
[2] $\langle \Gamma_{\mathrm{sub},2} \rangle = \langle \Gamma_{\mathrm{sub},3} \rangle$.
[19] $\langle \Gamma_{\mathrm{sub},2} \rangle \not\supseteq \Gamma_{\mathrm{sub},4}$.

When proving $\langle \Gamma_{\mathrm{sub},2} \rangle \not\supseteq \Gamma_{\mathrm{sub},4}$, Živný-Cohen-Jeavons [19] actually found a 4-ary submodular function f such that $f \in \langle \Gamma_{\mathrm{sub},2} \rangle$.

2.4   Weighted Polymorphism

How can we prove $f \notin \langle \Gamma \rangle$? We here introduce an algebraic machinery, called a *weighted polymorphism*, proving this. A function $\varphi : D^k \to D$ is called a k-ary *operation* on D. For $x^1 = (x_1^1, \ldots, x_n^1), \ldots, x^k = (x_1^k, \ldots, x_n^k) \in D^n$, we define $\varphi(x^1, \ldots, x^k)$ by $\varphi(x^1, \ldots, x^k) := \left(\varphi(x_1^1, \ldots, x_1^k), \ldots, \varphi(x_n^1, \ldots, x_n^k) \right) \in D^n$. A k-ary *projection* $e_i^{(k)}$ for $i \in [k]$ on D is defined by $e_i^{(k)}(x) = x_i$ for $x = (x_1, x_2, \ldots, x_k) \in D^k$. A k-ary operation φ is called a *polymorphism* of Γ if for all $f \in \Gamma$ and for all $x^1, x^2, \ldots, x^k \in \mathrm{dom}\, f$, it satisfies $\varphi(x^1, x^2, \ldots, x^k) \in \mathrm{dom}\, f$. Let $\mathrm{Pol}^{(k)}(\Gamma)$ be the set of k-ary polymorphisms of Γ, and let $\mathrm{Pol}(\Gamma) := \bigcup_k \mathrm{Pol}^{(k)}(\Gamma)$. Note that for any Γ, all projections are in $\mathrm{Pol}(\Gamma)$. Let us define a weighted polymorphism. A function $\omega : \mathrm{Pol}^{(k)}(\Gamma) \to \mathbf{Q}$ is called a k-ary *weighted polymorphism* of Γ [3] if it satisfies the following:

- $\sum_{\varphi \in \mathrm{Pol}^{(k)}(\Gamma)} \omega(\varphi) = 0$.
- If $\omega(\varphi) < 0$, then φ is a projection.
- For all $f \in \Gamma$ and for all $x^1, x^2, \ldots, x^k \in \mathrm{dom}\, f$,

$$\sum_{\varphi \in \mathrm{Pol}^{(k)}(\Gamma)} \omega(\varphi) f(\varphi(x^1, x^2, \ldots, x^k)) \leq 0.$$

Let $\mathrm{wPol}^{(k)}(\Gamma)$ be the set of k-ary weighted polymorphisms of Γ, and let $\mathrm{wPol}(\Gamma) := \bigcup_k \mathrm{wPol}^{(k)}(\Gamma)$. Here the following lemma holds:

Lemma 3 ([3]). *Suppose that Γ is a language on D and f is a cost function on D. If there exist some $\omega \in \mathrm{wPol}^{(k)}(\Gamma)$ and $x^1, x^2, \ldots, x^k \in \mathrm{dom}\, f$ satisfying*

$$\sum_{\varphi \in \mathrm{Pol}^{(k)}(\Gamma)} \omega(\varphi) f(\varphi(x^1, x^2, \ldots, x^k)) > 0,$$

then it holds that $f \notin \langle \Gamma \rangle$.

Thus we can prove nonrepresentability by using Lemma 3.

3   General Framework for Network Representability

3.1   Previous Approaches of Network Representation over D

Here we explain previous approaches of network representation for functions for a general finite set D. Ishii [9] considered a method of representing a bisubmodular function, which is a function on $\{0, -1, 1\}^n$, by a skew-symmetric network. A network $G = (\{s^+, s^-, 1^+, 1^-, \ldots, N^+, N^-\}, A; c)$ is said to be *skew-symmetric* if it satisfies that if $(u, v) \in A$, then $(\overline{v}, \overline{u}) \in A$ and $c(u, v) = c(\overline{v}, \overline{u})$. Here define $\overline{u}$ by $\overline{u} := i^+$ if $u = i^-$ and $\overline{u} := i^-$ if $u = i^+$. An s^+-s^- cut X is said to be *transversal* if $X \not\supseteq \{i, \overline{i}\}$ for every $i \in [n]$. The set of transversal s-t cuts is identified with $\{0, -1, 1\}^N$ by $X \mapsto x_i := X(i^+) - X(i^-)$ for $i \in [N]$. Ishii gave a definition of the network representability for a function on $\{0, -1, 1\}^n$ as follows:

> A function $f : \{0, -1, 1\}^n \to \overline{\mathbf{Q}}$ is said to be *skew-symmetric network representable* if there exist a skew-symmetric network $G = (V, A; c)$ and a constant $\kappa \in \mathbf{Q}$ satisfying the following:
> - $V \supseteq \{s^+, s^-, 1^+, 1^-, 2^+, 2^-, \ldots, n^+, n^-\}$.
> - For all $x = (x_1, x_2, \ldots, x_n) \in \{0, -1, 1\}^n$,
>
> $$f(x) = \min\{C(X) \mid X : \text{transversal } s\text{-}t \text{ cut},$$
> $$X(i^+) - X(i^-) = x_i \text{ for } i \in [n]\} + \kappa.$$

In a skew-symmetric network, the minimal minimum s^+-s^- cut is transversal [9]. Hence a skew-symmetric network representable function can be minimized efficiently via computing a minimum s^+-s^- cut. Here the following holds:

Lemma 4 ([9]). *Skew-symmetric network representable functions are bisubmodular.*

Moreover Ishii proved the following theorem:

Theorem 5 ([9]). *All binary bisubmodular functions are skew-symmetric network representable.*

This representation has both Properties 1 and 2. Therefore a bisubmodular function given as the sum of binary bisubmodular functions is skew-symmetric network representable. Thanks to extra nodes, a bisubmodular function given as partial minimization of a skew-symmetric network representable function is also skew-symmetric network representable.

Iwata-Wahlström-Yoshida [12] considered another method of representing a k-submodular function by a network. For an s-t cut X, let $\underline{X} := \{s\} \cup \bigcup_{i \in [n]} \{i^l \mid X \cap \{i^1, i^2, \ldots, i^k\} = i^l$ for some $l \in [k]\}$. For $x = (x_1, x_2, \ldots, x_n) \in [0, k]^n$, let $X_x := \{s\} \cup \bigcup_{x_i \neq 0} \{i^l \mid x_i = l\}$.

A function $f : [0, k]^n \to \overline{\mathbf{Q}}$ is said to be k-*network representable* if there exist a network $G = (V, A; c)$ and a constant $\kappa \in \mathbf{Q}$ satisfying the following:
- $V = \{s, t\} \cup \{i^l \mid (i, l) \in [n] \times [k]\}$.
- The s-t cut function C of G satisfies

$$C(X) \geq C(\underline{X}) \qquad (X : s\text{-}t \text{ cut}).$$

- For all $x = (x_1, x_2, \ldots, x_n) \in [0, k]^n$, it holds that

$$f(x) = C(X_x) + \kappa.$$

k-network representable functions can be minimized via computing a minimum s-t cut by definition, and constitute an efficiently minimizable subclass of k-submodular functions, as follows.

Lemma 6 ([12]). *k-network representable functions are k-submodular.*

Iwata-Wahlström-Yoshida constructed networks representing *basic k-submodular functions*, which are special k-submodular functions. This method also has both Properties 1 and 2. Therefore a k-submodular function given as the sum of basic k-submodular functions is k-network representable.

As was seen in Sect. 2.3, network representable functions on $\{0, 1\}^n$ are considered as the expressive power of $\Gamma_{\text{sub},2}$, and hence we can apply the expressive power theory to network representability. However Ishii and Iwata-Wahlström-Yoshida network representation methods cannot enjoy the expressive power theory, since (i) the set of network representable functions under Iwata-Wahlström-Yoshida method is not a weighted relational clone, and (ii) the concept of expressive power only focuses on the representability of functions on the same domain. We introduce, in the next subsection, a new network represetability definition for resolving (i), and in Sect. 3.4, we also introduce an extension of expressive power for resolving (ii).

3.2   Definition

By abstracting the previous approaches, we here develop a unified framework for network representability over D. The basic idea is the following: Consider networks having nodes $i^1, i^2, \ldots, i^k$ for each $i \in [n]$, where $|D| \leq 2^k$. We associate one variable x_i in D with k nodes $i^1, i^2, \ldots, i^k$. The k nodes have 2^k intersection patterns with s-t cuts. We specify a set of $|D|$ patterns, which represents D, for each i. The cut function restricted to cuts with specified patterns gives a function on D^n. To remove effect of irrelevant patterns in minimization, we fix a retraction from all patterns to specified patterns, and consider networks with the property that the retraction does not increase cut capacity. Now functions represented by such networks are minimizable via minimum s-t cut with retraction.

A formal definition is given as follows. Let k be a positive integer, and E a subset of $\{0, 1\}^k$. We consider a node i^l for each $(i, l) \in [n] \times [k]$. For a retraction $\rho : \{0, 1\}^k \to E$, a network $G = (V, A; c)$ is said to be (n, ρ)-*retractable* if G satisfies the following:

- $V \supseteq \{s, t\} \cup \{i^l \mid (i, l) \in [n] \times [k]\}$.
- For all $x = (x_1^1, \ldots, x_1^k, x_2^1, \ldots, x_2^k, \ldots, x_n^1, \ldots, x_n^k) \in \{0, 1\}^{kn}$,

$$C_{\min}(x) \geq C_{\min}(\rho(x_1^1, \ldots, x_1^k), \ldots, \rho(x_n^1, \ldots, x_n^k)),$$

where

$$C_{\min}(x) := \min\{C(X) \mid X : s\text{-}t \text{ cut}, X(i^l) = x_i^l \text{ for } (i, l) \in [n] \times [k]\}.$$

Let σ be a bijection from D to E. A function $f : D^n \to \overline{\mathbf{Q}}$ is said to be (k, ρ, σ)-*network representable* if there exist an (n, ρ)-retractable network $G = (V, A; c)$ and a constant $\kappa \in \mathbf{Q}$ satisfying that

$$f(x) = C_{\min}(\sigma(x_1), \sigma(x_2), \ldots, \sigma(x_n)) + \kappa$$

for all $x = (x_1, x_2, \ldots, x_n) \in D^n$. A (k, ρ, σ)-network representable function can be minimized efficiently via computing a minimum s-t cut.

The network representability in the sense of Kolmogorov-Zabih is the same as the $(1, \mathrm{id}, \mathrm{id})$-network representability, where $\mathrm{id} : \{0, 1\} \to \{0, 1\}$ is the identity map. Let $\rho_k : \{0, 1\}^k \to \{0, 1\}^k$ and $\sigma_k : [0, k] \to \{0, 1\}^k$ be maps defined by

$$\rho_k(x) := \begin{cases} x & \text{if } x = (0, \ldots, 0, \overset{i}{1}, 0, \ldots, 0) \text{ for some } i \in [k], \\ (0, \ldots, 0) & \text{otherwise,} \end{cases}$$

$$\sigma_k(x) := \begin{cases} (0, \ldots, 0, \overset{i}{1}, 0, \ldots, 0) & \text{if } x = i \in [k], \\ (0, \ldots, 0) & \text{if } x = 0. \end{cases}$$

Then skew-symmetric network representability is a special class of the $(2, \rho_2, \sigma_2)$-network representability, and k-network representability is a special class of the (k, ρ_k, σ_k)-network representability.

The (k, ρ, σ)-network representability possesses both Properties 1 and 2. Furthermore a function given as a partial minimization of a (k, ρ, σ)-network representable function is also (k, ρ, σ)-network representable.

3.3   Results on Network Representability

In our network representation, one variable is associated with "several" nodes even if $D = \{0, 1\}$. Hence the set of network representable functions on $\{0, 1\}^n$ in our sense may be strictly larger than that in the original. The following theorem says that additional network representable functions are only monotone.

Theorem 7. *If a function f on $\{0, 1\}^n$ is (k, ρ, σ)-network representable for some k, ρ, σ, then f is $(1, \mathrm{id}, \mathrm{id})$-network representable, or monotone.*

The minimization of a monotone function is trivial. Therefore it is sufficient only to consider $(1, \mathrm{id}, \mathrm{id})$-network representability (original network representability) for functions on $\{0, 1\}^n$.

We give a more precise structure of network representable functions on $\{0,1\}^n$. Let $\rho_1 : \{0,1\}^2 \to \{0,1\}^2$, $\rho_2 : \{0,1\}^2 \to \{0,1\}^2$, and $\sigma_0 : \{0,1\} \to \{0,1\}^2$ be functions defined by

$$\rho_1(x) = \begin{cases} (1,0) & \text{if } x = (1,0), \\ (0,1) & \text{otherwise,} \end{cases} \qquad \rho_2(x) = \begin{cases} (0,1) & \text{if } x = (0,1), \\ (1,0) & \text{otherwise,} \end{cases}$$

$$\sigma_0(x) = \begin{cases} (1,0) & \text{if } x = 1, \\ (0,1) & \text{if } x = 0. \end{cases}$$

Then the following holds:

Theorem 8. *A function f on $\{0,1\}^n$ is (k, ρ, σ)-network representable for some k, ρ, σ if and only if f is $(1, \mathrm{id}, \mathrm{id})$-network representable, $(2, \rho_1, \sigma_0)$-network representable, or $(2, \rho_2, \sigma_0)$-network representable.*

We next present network nonrepresentability results for functions on D^n, especially, k-submodular functions. These results will be proved via the theory of expressive power. We have seen in Theorem 5 that all binary bisubmodular functions are $(2, \rho_2, \sigma_2)$-network representable. We show that the same property does not hold for ternary bisubmodular functions.

Theorem 9. *Some ternary bisubmodular function is not $(2, \rho_2, \sigma_2)$-network representable.*

We also know that all binary basic k-submodular functions are (k, ρ_k, σ_k)-network representable [12], and their sum is efficiently minimizable. A natural question raised by [12] is whether all binary k-submodular functions are k-network representable or not. We answer this question negatively.

Theorem 10. *Some binary k-submodular function is not (k, ρ_k, σ_k)-network representable for all $k \geq 3$.*

Theorems 7, 8, 9 and 10 are consequences of Theorems 13, 14, 15 and 16 in the next subsection.

3.4   Extended Expressive Power

To incorporate the theory of expressive power into our framework, we introduce a way of handling languages on D from a language Γ on another domain F, which generalizes previous arguments. Let k be a positive integer with $|D| \leq |F|^k$. Let E be a subset of F^k with $|E| = |D|$, $\rho : F^k \to E$ a retraction, and $\sigma : D \to E$ a bijection. We define $\langle \Gamma \rangle^k$ by

$$\langle \Gamma \rangle^k := \{f \in \langle \Gamma \rangle \mid \text{The arity } r_f \text{ of } f \text{ is a multiple of } k\}.$$

Regard $\langle \Gamma \rangle^k$ as a language on F^k. A language $\langle \Gamma \rangle_\rho^k$ on E is defined by

$$\langle \Gamma \rangle_\rho^k := \{f|_{E^{r_f}} \mid f \in \langle \Gamma \rangle^k, \ f(x) \geq f(\rho(x)) \text{ for } x \in \mathrm{dom}\, f\}.$$

A function f is *representable by* (Γ, ρ, σ) if there exists $g \in \langle \Gamma \rangle_\rho^k$ such that $f(x) = g(\sigma(x))$ for $x \in D^n$. We define a language $\langle \Gamma \rangle_{(\rho,\sigma)}^k$ on D as the set of functions representable by (Γ, ρ, σ). By comparing these notions to our network representations, we obtain an generalization of Lemma 1.

Lemma 11. *The set of (k, ρ, σ)-network representable functions coincides with $\langle \Gamma_{\mathrm{sub},2} \rangle_{(\rho,\sigma)}^k$.*

The following theorem enables us to deal with our network representability on D^n from the theory of expressive power.

Theorem 12. *For a language Γ on F, $\langle \Gamma \rangle_{(\rho,\sigma)}^k$ is a weighted relational clone on D.*

Let Γ be a language on F. A function f on D is called *representable by Γ* if $f \in \langle \Gamma \rangle_{(\rho,\sigma)}^k$ for some positive integer k, $\rho : F^k \to E$, and $\sigma : D \to E$. The set of cost functions on D representable by a language Γ is denoted by $\overline{\langle \Gamma \rangle}_D$. Notice that $\overline{\langle \Gamma \rangle}_D$ is not a weighted rational clone in general. By using these notations, Theorems 7 and 8 are reformulated as follows. Here let Γ_{mono} be the set of monotone functions.

Theorem 13. $\langle \Gamma_{\mathrm{sub},2} \rangle \subsetneq \overline{\langle \Gamma_{\mathrm{sub},2} \rangle}_{\{0,1\}} \subsetneq \langle \Gamma_{\mathrm{sub},2} \rangle \cup \Gamma_{\mathrm{mono}}$.

Theorem 14. $\overline{\langle \Gamma_{\mathrm{sub},2} \rangle}_{\{0,1\}} = \langle \Gamma_{\mathrm{sub},2} \rangle \cup \langle \Gamma_{\mathrm{sub},2} \rangle_{(\rho_1,\sigma_0)}^2 \cup \langle \Gamma_{\mathrm{sub},2} \rangle_{(\rho_2,\sigma_0)}^2$.

Theorem 9 is rephrased as $\Gamma_{\mathrm{bisub},3} \not\subseteq \langle \Gamma_{\mathrm{sub},2} \rangle_{(\rho_2,\sigma_2)}^2$. We prove a stronger statement such that $\Gamma_{\mathrm{bisub},3}$ is not included even in the set of $(\Gamma_{\mathrm{sub}}, \rho_2, \sigma_2)$-representable functions.

Theorem 15. $\Gamma_{\mathrm{bisub},3} \not\subseteq \langle \Gamma_{\mathrm{sub}} \rangle_{(\rho_2,\sigma_2)}^2$.

Theorem 10 is rephrased as $\Gamma_{k\mathrm{sub},2} \not\subseteq \langle \Gamma_{\mathrm{sub},2} \rangle_{(\rho_k,\sigma_k)}^k$. Again we prove a stronger statement such that $\Gamma_{k\mathrm{sub},2}$ is not included even in the set of $(\Gamma_{\mathrm{sub}}, \rho_k, \sigma_k)$-representable functions.

Theorem 16. $\Gamma_{k\mathrm{sub},2} \not\subseteq \langle \Gamma_{\mathrm{sub}} \rangle_{(\rho_k,\sigma_k)}^k$ *for all $k \geq 3$.*

We prove Theorem 15 (resp. Theorem 16) by finding some ternary bisubmodular function (resp. binary k-submodular function) and some weighted polymorphism of $\langle \Gamma_{\mathrm{sub}} \rangle_{(\rho_2,\sigma_2)}^2$ (resp. $\langle \Gamma_{\mathrm{sub}} \rangle_{(\rho_k,\sigma_k)}^k$) satisfying the condition in Lemma 3.

Acknowledgments. We thank Hiroshi Hirai and Magnus Wahlström for careful reading and numerous helpful comments. This research is supported by JSPS Research Fellowship for Young Scientists and by the JST, ERATO, Kawarabayashi Large Graph Project.

References

1. Bach, F.: Learning with submodular functions: A convex optimization perspective. Found. Trends Mach. Learn. **6**(2–3), 145–373 (2013)
2. Billionnet, A., Minoux, M.: Maximizing a supermodular pseudoboolean function: a polynomial algorithm for supermodular cubic functions. Discrete Appl. Math. **12**, 1–11 (1985)
3. Cohen, D.A., Cooper, M.C., Creed, P., Jeavons, P.G., Živný, S.: An algebraic theory of complexity for discrete optimization. SIAM J. Comput. **42**(5), 1915–1939 (2013)
4. Fujishige, S.: Submodular Functions and Optimization, 2nd edn. Elsevier, Amsterdam (2005)
5. Greig, G., Porteous, B., Seheult, A.: Exact minimum a posteriori estimation for binary images. J. R. Stat. Soc. Ser. B **51**, 271–279 (1989)
6. Gridchyn, I., Kolmogorov, V.: Potts model, parametric maxflow and k-submodular functions. In: Proceedings of International Conference on Computer Vision (ICCV 2013), pp. 2320–2327 (2013)
7. Grötschel, M., Lovász, L., Schrijver, A.: Geometric Algorithms and Combinatorial Optimization. Springer, Berlin (1988)
8. Huber, A., Kolmogorov, V.: Towards minimizing k-submodular functions. In: Mahjoub, A.R., Markakis, V., Milis, I., Paschos, V.T. (eds.) ISCO 2012. LNCS, vol. 7422, pp. 451–462. Springer, Heidelberg (2012)
9. Ishii, Y.: On a representation of k-submodular functions by a skew-symmetric network. Master thesis, The University of Tokyo (2014). (in Japanese)
10. Ivănescu, P.L.: Some network flow problems solved with pseudo-boolean programming. Oper. Res. **13**(3), 388–399 (1965)
11. Iwata, S., Fleischer, L., Fujishige, S.: A combinatorial strongly polynomial algorithm for minimizing submodular functions. J. ACM **48**(4), 761–777 (2001)
12. Iwata, Y., Wahlström, M., Yoshida, Y.: Half-integrality, LP-branching and FPT algorithms. SIAM J. Comput. (to appear)
13. Kolmogorov, V., Thapper, J., Živný, S.: The power of linear programming for general-valued CSPs. SIAM J. Comput. **44**(1), 1–36 (2015)
14. Kolmogorov, V., Zabih, R.: What energy functions can be minimized via graph cuts? IEEE Trans. Pattern Anal. Mach. Intell. **26**(2), 147–159 (2004)
15. Lee, Y.T., Sidford, A., Wong, S.C.: A faster cutting plane method and its implications for combinatorial and convex optimization. In: Proceedings of the 56th Annual IEEE Symposium on Foundations of Computer Science (FOCS 2015) (2015). arXiv:1508.04874v2
16. Orlin, J.B.: Max flows in $O(nm)$ time, or better. In: Proceedings of the 45th Annual ACM Symposium on the Theory of Computing (STOC 2013), pp. 765–774 (2013)
17. Schrijver, A.: A combinatorial algorithm minimizing submodular functions in strongly polynomial time. J. Comb. Theory Ser. B **80**, 346–355 (2000)
18. Živný, S.: The Complexity of Valued Constraint Satisfaction Problems. Cognitive Technologies. Springer, Heidelberg (2012)
19. Živný, S., Cohen, D.A., Jeavons, P.G.: The expressive power of binary submodular functions. Discrete Appl. Math. **157**(15), 3347–3358 (2009)

A Compact Representation for Minimizers of k-Submodular Functions (Extended Abstract)

Hiroshi Hirai and Taihei Oki[⊠]

Department of Mathematical Informatics,
Graduate School of Information Science and Technology,
The University of Tokyo, Tokyo 113-8656, Japan
{hirai,taihei_oki}@mist.i.u-tokyo.ac.jp

Abstract. k-submodular functions were introduced by Huber and Kolmogorov as a generalization of bisubmodular functions. This paper establishes a compact representation of minimizers of k-submodular functions by posets with inconsistent pairs (PIPs), and completely characterizes the class of PIPs (elementary PIPs) corresponding to minimizers of k-submodular functions. Our representation coincides with Birkhoff's representation theorem if $k = 1$ and with signed-poset representation by Ando and Fujishige if $k = 2$. We also give algorithms to construct the elementary PIP representing the minimizers of a k-submodular function f for three cases: (i) a minimizing oracle of f is available, (ii) f is network-representable, and (iii) f is the objective function of the relaxation of multiway cut problem. Furthermore, we provide an efficient algorithm to enumerate all maximal minimizers from the PIP representation. Our results are applied to obtain all maximal persistent assignments in labeling problems arising from computer vision.

1 Introduction

The minimizer set of a submodular function forms a distributive lattice. By celebrated Birkhoff's representation theorem, the minimizer set is compactly written by a poset (partially ordered set). Applications of this fact include DM-decomposition of matrices and further generalizations [12]. In this paper we shall consider the minimizer sets of k-submodular functions. A *k-submodular function* was introduced by Huber and Kolmogorov [8] as a generalization of submodular and bisubmodular functions. 1-submodular functions and 2-submodular functions are identical to submodular functions and bisubmodular functions, respectively.

The main result of this paper is to establish a compact representation for minimizers of k-submodular functions by *posets with inconsistent pairs* (PIPs) [2,3,13]. This is a generalization of the compact representation of minimizer sets of submodular functions and Ando–Fujishige's signed-poset representation for bisubmodular functions [1]. In our representation, each minimizer corresponds to some special ideal of the PIP, called a *consistent ideal*. We also characterize PIPs arising from k-submodular functions. Such PIPs are called *elementary*.

© Springer International Publishing Switzerland 2016
R. Cerulli et al. (Eds.): ISCO 2016, LNCS 9849, pp. 381–392, 2016.
DOI: 10.1007/978-3-319-45587-7_33

Our representation is compact since the number of elements in an elementary PIP is at most kn, where n is the number of variables of the corresponding k-submodular function. Moreover, we present three algorithms to obtain the elementary PIP representing the minimizer set of a k-submodular function f for three cases: (1) a minimizing oracle of f is given, (2) f is network-representable, and (3) f is the relaxed multiway cut function. We also give a fast algorithm to enumerate all maximal consistent ideals of an elementary PIP.

The rest of this paper is organized as follows. Section 2 is preliminaries. In Sect. 3, we introduce our compact representation for minimizers of k-submodular functions. In Sect. 4, we present our algorithms. Finally in Sect. 5, we describe applications to *k-submodular relaxation* [6,9].

Omitted proofs and algorithms are given in the full version.

2   Preliminaries

For a nonnegative integer n, we denote $\{1, 2, \ldots, n\}$ by $[n]$ (with $[0] := \varnothing$).

2.1 k-Submodular Function

Let S_k be a finite set of $k + 1$ elements $\{0, 1, \ldots, k\}$ and $\preceq$ a partial order on S_k defined by $a \preceq b \overset{\text{def}}{\Longleftrightarrow} a \in \{0, b\}$ for each $a, b \in S_k$. For a subset $X \subseteq S_k{}^n$, the subposet $(X, \preceq)$ of $(S_k{}^n, \preceq)$ is simply denoted by X. For every $x = (x_1, x_2, \ldots, x_n) \in S_k{}^n$, the subset $\{i \in [n] \mid x_i \neq 0\}$ of $[n]$ is called *support* of x and denoted by $\operatorname{supp} x$.

A *k-submodular function* is a function $f \colon S_k{}^n \to \bar{\mathbb{R}} := \mathbb{R} \cup \{+\infty\}$ satisfying the following inequality

$$f(x) + f(y) \geq f(x \sqcap y) + f(x \sqcup y) \tag{1}$$

for all $x, y \in S_k{}^n$. Here the binary operation $\sqcap$ on $S_k{}^n$ is given by

$$(x \sqcap y)_i := \begin{cases} \min\{x_i, y_i\} & (x_i \text{ and } y_i \text{ are comparable on } \preceq), \\ 0 & (x_i \text{ and } y_i \text{ are incomparable on } \preceq), \end{cases} \tag{2}$$

for every $x, y \in S_k{}^n$ and $i \in [n]$. The other operation $\sqcup$ in (1) is dually defined by changing min to max in (2).

It is not known whether k-submodular functions for $k \geq 3$ can be minimized in polynomial time on the standard oracle model. However, some special classes of k-submodular functions are efficiently minimizable. For example, Kolmogorov, Thapper and Živný [11] showed that a sum of low-arity k-submodular functions can be minimized in polynomial time through linear programming. Note that it is assumed that each low-arity function is given as the table storing all function values; hence the total input size is $\mathrm{O}(mk^r)$, where m is the number of low-arity k-submodular functions and r is the maximum arity. We can also minimize a nonnegative combination of (binary) *basic k-submodular functions*, introduced by Iwata, Wahlström and Yoshida [9], by computing a minimum (s,t)-cut on a directed network; see Sect. 4.2.

A nonempty subset of $S_k{}^n$ is said to be $(\sqcap, \sqcup)$-*closed* if it is closed under the operations $\sqcap$ and $\sqcup$. From (1), the following obviously holds.

Lemma 1. *The minimizer set of a k-submodular function is $(\sqcap, \sqcup)$-closed.*

2.2 Median Semilattice and PIP

A key tool for providing an efficient representation of $(\sqcap, \sqcup)$-closed sets is a correspondence between *median semilattices* and *PIPs*, which was established by Barthélemy and Constantin [3].

A *median semilattice* [16] is a meet-semilattice $\mathcal{L} = (L, \leq)$ satisfying following conditions:

(MS1) every principal ideal is a distributive lattice.
(MS2) for all $x, y, z \in L$, if $x \vee y$, $y \vee z$ and $z \vee x$ exist, then $x \vee y \vee z$ exists in L.

Note that every distributive lattice is a median semilattice. We denote the set of join-irreducible elements of $\mathcal{L}$ by $\mathcal{J}(\mathcal{L})$. The minimum element of $\mathcal{L}$ is not join-irreducible.

Next we introduce a *poset with inconsistent pairs* (PIP). This notion was independently introduced in different contexts with different names: "event structure" in [13], "site" in [3], and "PIP" in [2] which we use. Let $(P, \leq)$ be a finite poset and $\#$ a symmetric binary relation on P. A PIP is a triplet $(P, \leq, \#)$ satisfying the following:

(IC1) for all $p, q \in P$ with $p \# q$, there is no $r \in P$ with $p \leq r$ and $q \leq r$.
(IC2) for all $p, q, p', q' \in P$, if $p' \leq p, q' \leq q$ and $p' \# q'$, then $p \# q$.

The relation $\#$ is called an *inconsistent relation*. Each unordered pair $\{p, q\}$ of P is called *inconsistent* if $p \# q$. Note that any inconsistent pair of P is incomparable and has no common upper bound in P. An inconsistent pair $\{p, q\}$ of P is said to be *minimally inconsistent* if $p' \leq p, q' \leq q$ and $p' \# q'$ imply $p = p'$ and $q = q'$ for all $p', q' \in P$. If $\{p, q\}$ is minimally inconsistent, the $p \# q$ is particularly denoted by $p \,\#\!\!\#\, q$. We can check the following properties about the minimally inconsistent relation:

(MIC1) for all $p, q \in P$ with $p \,\#\!\!\#\, q$, there is no $r \in P$ with $p \leq r$ and $q \leq r$.
(MIC2) for all $p, q, p', q' \in P$ with $p' \leq p$ and $q' \leq q$, if $p' \,\#\!\!\#\, q'$ and $p \,\#\!\!\#\, q$, then $p' = p$ and $q' = q$.

Indeed, PIPs can also be defined as a triplet $(P, \leq, \#\!\!\#)$, where $(P, \leq)$ is a poset and $\#\!\!\#$ is a binary symmetric relation on P satisfying the conditions (MIC1) and (MIC2). In this definition, the inconsistent relation $\#$ on P is obtained by

$$p \# q \;\overset{\text{def}}{\Longleftrightarrow}\; \text{there exist } p', q' \in P \text{ with } p' \leq p, q' \leq q \text{ and } p' \,\#\!\!\#\, q' \tag{3}$$

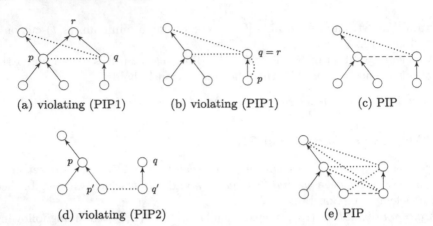

(a) violating (PIP1)   (b) violating (PIP1)   (c) PIP

(d) violating (PIP2)   (e) PIP

Fig. 1. Examples of PIPs and non-PIP structures. Solid arrows indicate the orders between elements (drawn from the smaller elements to the larger ones). Dotted lines and dashed lines represent the inconsistent relations. In (a), (b) and (d), labels indicate where the violations of (PIP1) and (PIP2) are. In (c) and (e), the minimally inconsistent relations are drawn by dashed lines.

for every $p, q \in P$. Since both definitions of PIPs are fundamentally equivalent, we make use of a convenient one. For a PIP $\mathcal{P} = (P, \leq, \#)$, every ideal of $(P, \leq)$ is called a *consistent ideal* of $\mathcal{P}$ if it contains no (minimally) inconsistent pair. We denote the family of consistent ideals of $\mathcal{P}$ by $\mathcal{C}(\mathcal{P})$. Figure 1 shows examples of PIPs and non-PIP structures.

The following theorem associates median semilattices with PIPs. Namely, there is a one-to-one correspondence between median semilattices and PIPs.

Theorem 2 ([3]). *The following hold.*

(1) *Let* $\mathcal{L} = (L, \leq)$ *be a median semilattice and* $\#$ *a symmetric binary relation on* L *defined by*

$$x \mathrel{\#} y \overset{\text{def}}{\iff} x \vee y \text{ does not exist on } L$$

for every $x, y \in L$. *Then* $\mathcal{P} := (\mathcal{J}(\mathcal{L}), \leq, \#)$ *forms a PIP with inconsistent relation* $\#$. *The poset* $(\mathcal{C}(\mathcal{P}), \subseteq)$ *is isomorphic to* $\mathcal{L}$, *where the isomorphism is given by* $I \mapsto \bigvee_{x \in I} x$ *(with* $\varnothing \mapsto \min \mathcal{L}$*).*

(2) *Let* $\mathcal{P}$ *be a PIP. The poset* $\mathcal{L} := (\mathcal{C}(\mathcal{P}), \subseteq)$ *forms a median semilattice. Let* $\mathcal{P}'$ *be the PIP obtained from* $\mathcal{L}$ *in the same way as defined in (1). Then* $\mathcal{P}'$ *is isomorphic to* $\mathcal{P}$, *where the isomorphism is given by* $I \mapsto \max I$.

3 $(\sqcap, \sqcup)$-Closed Set and Elementary PIP

Our starting point for a compact representation of $(\sqcap, \sqcup)$-closed sets is the following.

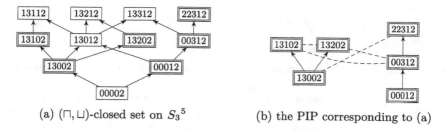

(a) $(\sqcap, \sqcup)$-closed set on S_3^5 (b) the PIP corresponding to (a)

Fig. 2. Example of a $(\sqcap, \sqcup)$-closed set and the corresponding PIP. Join-irreducible elements are surrounded by double-lined frames. In (b), only minimally inconsistent relations are drawn.

Lemma 3. *Any $(\sqcap, \sqcup)$-closed set on S_k^n is a median semilattice.*

Let $\smile$ be a symmetric binary relation on S_k^n defined by

$$x \smile y \overset{\text{def}}{\iff} x \vee y \text{ does not exist on } S_k^n$$

for every $x, y \in S_k^n$. Note that if $x \not\smile y$ then it holds $x \vee y = x \sqcup y$ for every $x, y \in M$. From Theorem 2 (1) and Lemma 3, we obtain the following.

Theorem 4. *Let $M \subseteq S_k^n$ be a $(\sqcap, \sqcup)$-closed set. Then $\mathcal{P} := (\mathcal{J}(M), \preceq, \smile)$ forms a PIP with inconsistent relation $\smile$. The poset $(\mathcal{C}(\mathcal{P}), \subseteq)$ is isomorphic to M, where the isomorphism is given by $I \mapsto \bigvee_{x \in I} x$ (with $\varnothing \mapsto \min M$).*

Figure 2 shows an example of a $(\sqcap, \sqcup)$-closed set and the corresponding PIP. From Theorem 4, it will be turned out that the set $\mathcal{J}(M)$ of join-irreducible elements of every $(\sqcap, \sqcup)$-closed set $M \subseteq S_k^n$ does not lose any information about the structure of M. To reconstruct each element in M, the minimum element of M is needed besides from $\mathcal{J}(M)$. The other elements in M can be obtained as the join of one or more join-irreducible elements of M. Therefore M can be completely reconstructed from the pair $(\mathcal{J}(M), \min M)$. Furthermore, the following proposition guarantees the compactness of this representation.

Proposition 5. *Let M be a $(\sqcap, \sqcup)$-closed set on S_k^n. The number of join-irreducible elements of M is at most kn.*

From Theorem 4, any $(\sqcap, \sqcup)$-closed set can be represented by a PIP. However, not all PIPs correspond to some $(\sqcap, \sqcup)$-closed sets. A natural question then arises: *What class of PIPs represents $(\sqcap, \sqcup)$-closed sets?* The main result (Theorem 7) of this section answers this question. In what follows, we frequently denote the PIP $(\mathcal{J}(M), \preceq, \smile)$ by $\mathcal{J}(M)$ for brevity.

Definition 6. *Let $\mathcal{P} = (P, \leq, \#)$ be a PIP with poset $(P, \leq)$ and minimally inconsistent relation $\#$ on P. We call $\mathcal{P}$ elementary if it satisfies the following conditions:*

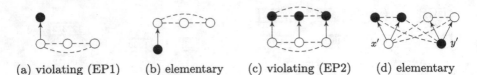

(a) violating (EP1)   (b) elementary   (c) violating (EP2)   (d) elementary

Fig. 3. Examples of elementary PIPs and non-elementary PIPs. In all diagrams, the drawn PIPs satisfy the condition (EP0) with $n = 2$. The partition of elements are illustrated by colors (white and black). Non-minimally inconsistent relations are not drawn in each diagram.

(EP0) P is the disjoint union of $P_1, P_2, \ldots, P_n$ such that for all distinct $x, y \in P$, it holds $x \mathbin{\#} y$ if and only if $\{x, y\} \subseteq P_i$ for some $i \in [n]$.

(EP1) for any distinct $i, j \in [n]$, if $|P_i| \geq 2$ and $|P_j| = 1$, there are no elements $x \in P_i$ and $y \in P_j$ with $x < y$.

(EP2) for any distinct $i, j \in [n]$, if $|P_i| \geq 2$ and $|P_j| \geq 2$, either of the following holds:

(EP2-1) every $x \in P_i$ and $y \in P_j$ are not comparable.

(EP2-2) there exist $x' \in P_i$ and $y' \in P_j$ such that $x' < y$ and $y' < x$ for all $x \in P_i \setminus \{x'\}$ and $y \in P_j \setminus \{y'\}$.

Examples of elementary PIP and non-elementary PIPs are illustrated in Fig. 3.

Theorem 7. *The following hold:*

(1) For every $(\sqcap, \sqcup)$-closed set M, the set $\mathcal{J}(M)$ of join-irreducible elements of M forms an elementary PIP, and $(\mathcal{C}(\mathcal{J}(M)), \subseteq)$ is isomorphic to $(M, \preceq)$.

(2) For every elementary PIP $\mathcal{P}$, there is a $(\sqcap, \sqcup)$-closed set M such that $(M, \preceq)$ is isomorphic to $(\mathcal{C}(\mathcal{P}), \subseteq)$.

In case of $k = 2$, the condition (EP2) in Definition 6 is equivalent to the skew-symmetricity of the corresponding Hasse diagram, hence our representation for $(\sqcap, \sqcup)$-closed sets coincides with one by Ando–Fujishige [1].

4 Algorithms

4.1 By a Minimizing Oracle

We can obtain the elementary PIP for the minimizer set of a k-submodular function $f \colon S_k{}^n \to \bar{\mathbb{R}}$ by a minimizing oracle, which returns a minimizer of f and its restrictions. Before describing the algorithm, we introduce some additional notations. Let $\mathcal{D}(f)$ be the minimizer set of f. For $i \in [n]$ and $l \in S_k$, we define a new k-submodular function $f|_{(i,l)} \colon S_k{}^n \to \bar{\mathbb{R}}$ from f by

$$f|_{(i,l)}(x_1, \ldots, x_i, \ldots, x_n) := f(x_1, \ldots, \overset{i}{\smile}{l}, \ldots, x_n)$$

for every $x \in S_k{}^n$. Namely, $f|_{(i,l)}$ is a function obtained by fixing the i-th variable of f to l.

Algorithm 1. Obtain the minimum minimizer of a k-submodular function

---

Input : A k-submodular function $f\colon S_k{}^n \to \bar{\mathbb{R}}$
Output: The minimum minimizer $\min \mathcal{D}(f)$ of f
1: **function** GETMINIMUMMINIMIZER(f)
2:     $x \leftarrow$ GETMINIMIZER(f)
3:     **for** $i \in \operatorname{supp} x$ **do**
4:         **if** $\min f|_{(i,0)} = \min f$ **then**
5:             $x_i \leftarrow 0$
6:     **return** x

The minimum minimizer. We first present an algorithm to obtain the minimum minimizer of a k-submodular function in Algorithm 1, where GETMINIMIZER is the minimizing oracle. This algorithm correctly returns the minimum element of $\mathcal{D}(f)$ since $\min f|_{(i,0)}$ is equal to $\min f$ if $(\min \mathcal{D}(f))_i = 0$ and otherwise it holds $\min f|_{(i,0)} > \min f$. The algorithm calls the oracle at most $n + 1$ times and the time complexity is $\mathrm{O}(n\gamma)$, where γ is the time required by a single oracle call.

Join-irreducible minimizers. Algorithm 2 shows a procedure to collect all join-irreducible minimizers of a k-submodular function. Let x be the minimum minimizer of f. The function $f_x\colon S_k{}^n \to \bar{\mathbb{R}}$ used in the algorithm is defined by $f_x(y) := f((y \sqcup x) \sqcup x)$ for every $y \in S_k{}^n$. Since $((y \sqcup x) \sqcup x)_i$ is equal to y_i if $x_i = 0$ and to x_i if $x_i \neq 0$, we can regard f_x as a function obtained by fixing each i-th variable of f to x_i if $x_i \neq 0$. Note that the minimum values of f and f_x are same. The correctness of this algorithm is based on the following proposition: the set of join-irreducible minimizers of f coincides with the set

$$\left\{ \min \mathcal{D}\left(f_x|_{(i,l)}\right) \middle| i \in [n] \setminus \operatorname{supp} x,\ l \in [k],\ \min f_x|_{(i,l)} = \min f \right\}. \tag{4}$$

Each join-irreducible minimizer is collected according to (4) one-by-one. The time complexity is $\mathrm{O}\left(kn^2\gamma\right)$ since the algorithm calls Algorithm 1 at most $nk + 1$ times. Consequently, if a minimizing oracle is available, the entire minimizer set can also be obtained in polynomial time.

Theorem 8. *On the minimizing oracle model, we can obtain the elementary PIP for the minimizer set of a k-submodular function $f\colon S_k{}^n \to \bar{\mathbb{R}}$ in $\mathrm{O}\left(kn^2\gamma\right)$ time, where γ is the time required by a single oracle call.*

4.2   On Network-Representable k-Submodular Functions

Iwata, Wahlström and Yoshida [9] introduced *basic k-submodular functions*, a special class of k-submodular functions. They showed a reduction of the minimization problem of a nonnegative combination of (binary) basic k-submodular functions to the minimum cut problem on a directed network. We describe their method and present an algorithm to obtain the elementary PIP that represents the minimizer set.

Algorithm 2. Collect all join-irreducible minimizers of a k-submodular function

Input : A k-submodular function $f\colon S_k{}^n \to \bar{\mathbb{R}}$
Output: The set $\mathcal{J}(\mathcal{D}(f))$ of all join-irreducible minimizers of f
1: **function** GetJoinIrreducibleMinimizers(f)
2:     $x := $ GetMinimumMinimizer(f)
3:     $f_x := $ the function obtained by fixing the i-th variable of f to x_i for all $i \in \operatorname{supp} x$
4:     $J \leftarrow \varnothing$
5:     **for** $i \in [n] \setminus \operatorname{supp} x$ **do**
6:         **for** $l \leftarrow 1$ **to** k **do**
7:             **if** $\min f_x|_{(i,l)} = \min f$ **then**
8:                 $J \leftarrow J \cup \left\{ \text{GetMinimumMinimizer}(f_x|_{(i,l)}) \right\}$
9:     **return** J

Let n and k be nonnegative integers. We consider a directed network $N = (V, A, c)$ with vertex set V, edge set A and nonnegative edge capacity c. Suppose that V consists of source s, sink t and other vertices v_i^l, where $i \in [n]$ and $l \in [k]$. Let $U_i := \{v_i^1, v_i^2, \ldots, v_i^k\}$ for $i \in [n]$. An (s,t)-cut of N is a subset X of V such that $s \in X$ and $t \notin X$. We call an (s,t)-cut X *legal* if $|X \cap U_i| \leq 1$ holds for every $i \in [n]$. There is a natural bijection ψ from $S_k{}^n$ to the set of legal (s,t)-cuts of N defined by

$$\psi(x) := \{s\} \cup \{v_i^{x_i} \mid i \in \operatorname{supp} x\}$$

for every $x \in S_k{}^n$. For an (s,t)-cut X of N, let $\check{X}$ denote a legal (s,t)-cut obtained by removing vertices in $X \cap U_i$ from X for every $i \in [n]$. The *capacity* of X is defined as $d(X) := \sum_{e \in \delta X} c(e)$, where δX is the set of edges from X to $V \setminus X$. We say that a network N *represents* a function $f\colon S_k{}^n \to \bar{\mathbb{R}}$ if it satisfies the following:

(NR1) there exists a constant $K \in \mathbb{R}$ such that $f(x) = d(\psi(x)) + K$ for all $x \in S_k{}^n$.
(NR2) it holds $d(\check{X}) \leq d(X)$ for all (s,t)-cuts X of N.

From (NR1), the minimum value of $f + K$ is equal to the capacity of a minimum (s,t)-cut of N. For every minimum (s,t)-cut X of N, $\check{X}$ is also a minimum (s,t)-cut since N satisfies the condition (NR2). Therefore $\psi^{-1}(\check{X})$ is a minimizer of f, and a minimum (s,t)-cut can be computed by maximum flow algorithms. Indeed, Iwata, Wahlström and Yoshida [9] showed that a nonnegative combination of basic k-submodular functions are representable by such networks.

Now we shall consider obtaining all minimizers of a k-submodular function $f\colon S_k{}^n \to \bar{\mathbb{R}}$ represented by a network N. The minimizer set of f is isomorphic to the family of legal minimum (s,t)-cuts of N ordered by inclusion, where the isomorphism is ψ. As for the family of (not necessarily legal) minimum (s,t)-cuts, it is well-known that the family forms a distributive lattice. Thus, by Birkhoff's representation theorem, the family is efficiently representable by a poset. Picard and Queyranne [15] showed how to obtain the poset from the residual graph corresponding to a maximum (s, t)-flow of N. Our result (Theorem 9) is based on their algorithm. For an (s, t)-flow φ of N, the *residual graph* corresponding to φ is a directed graph $G_\varphi = (V, A_\varphi)$, where

$$A_\varphi := \{a \in A \mid \varphi(a) < c(a)\} \cup \{(u, v) \in V \times V \mid (v, u) \in A \text{ and } 0 < \varphi(v, u)\}.$$

Theorem 9. *Let $N = (V, A, c)$ be a network representing a k-submodular function $f \colon S_k^{\,n} \to \bar{\mathbb{R}}$ and G_φ denote the residual graph corresponding to a maximum (s, t)-flow φ of N. Let $\mathcal{V}$ be the set consisting of sccs of G_φ other than the following:*

(1) sccs reachable from s.
(2) sccs reachable to t.
(3) sccs reachable to an scc containing two or more elements in U_i for some $i \in [n]$.
(4) sccs reachable to distinct sccs Y and Z such that $|Y \cap U_i| = |Z \cap U_i| = 1$ for some $i \in [n]$.

Let $\le$ be a partial order on $\mathcal{V}$ defined by

$$X \le Y \overset{\text{def}}{\Longleftrightarrow} X \text{ is reachable from } Y \text{ on } G_\varphi$$

for every $X, Y \in \mathcal{V}$. In addition, let $\#$ be a symmetric binary relation on $\mathcal{V}$ defined as follows:

$$X \mathbin\# Y \overset{\text{def}}{\Longleftrightarrow} X \ne Y \text{ and there exists } i \in [n] \text{ such that } |X \cap U_i| = |Y \cap U_i| = 1$$

for every $X, Y \in \mathcal{V}$. Then $\mathcal{P} = (\mathcal{V}, \le, \#)$ forms an elementary PIP, where $\#$ is the minimally inconsistent relation of $\mathcal{P}$. The consistent ideal family of $\mathcal{P}$ is isomorphic to the set of minimizers of f, where the isomorphism is $\psi^{-1} \circ \tau_N$.

By using an efficient technique (like dynamic programming) for obtaining $\mathcal{V}$, the following theorem holds.

Theorem 10. *Let $f \colon S_k^{\,n} \to \bar{\mathbb{R}}$ be a k-submodular function represented by a network N with m edges. We can obtain the elementary PIP for the minimizer set of f in $O(\mathrm{MF}(kn, m) + n\,(k + m))$ time.*

Here $\mathrm{MF}(n', m')$ denotes the time complexity to find a maximum (s, t)-flow of a directed network with n' nodes and m' edges. The state-of-the art algorithm for the maximum (s, t)-flow problem on a directed network is [14] by Orlin, where $\mathrm{MF}(n', m') = O(n'm')$. Therefore the time complexity for obtaining $\mathcal{V}$ from G_φ is much less than one for computing G_φ from the network N.

4.3   Multiway Cut Problem

Let $N = (V, E, c, S)$ be an undirected network with vertex set V, edge set E, nonnegative edge capacity c and nonempty subset $S = \{s_1, s_2, \ldots, s_k\}$ of V, called *terminals*. A *multiway cut* $\mathcal{X}$ is a partition of V such that each part contains exactly one of the terminals. The *capacity* of $\mathcal{X}$ is defined as the sum of capacities of all edges whose ends belong to different parts in $\mathcal{X}$. The *multiway cut problem* (MCP) in N is the problem of finding a multiway cut with minimum capacity. This problem is known to be NP-hard if $k \geq 3$.

There is a natural relaxation, implicitly in [4] of MCP, which turned out to be a k-submodular relaxation [6,9]. For every $X, Y \subseteq V$, let $d(X, Y)$ denote the sum of capacities of edges between X and Y. $d(X, V \setminus X)$ is denoted by $d(X)$. For each $s \in S$, an *s-isolating cut* is a cut X of N such that $X \cap S = \{s\}$. A *semi-multiway cut* $\mathcal{Y}$ of N is a subpartition $\{X_1, X_2, \ldots, X_k\}$ of V such that X_l is an s_l-isolating cut for each $l \in [k]$. The *capacity* of $\mathcal{Y}$ is defined as $\frac{1}{2} \sum_{l=1}^{k} d(X_l)$. If $\mathcal{Y}$ forms a multiway cut, the capacities of $\mathcal{Y}$ as a semi-multiway cut and as a multiway cut are equal. A semi-multiway cut is said to be *minimum* if its capacity is the minimum in all semi-multiway cuts of N. The relaxation problem (RMCP) of MCP is the problem of finding a minimum semi-multiway cut of N. A standard uncrossing argument gives a simple solution of this problem as follows.

Lemma 11. *For $l \in [k]$, let Y_l be the minimal minimum s_l-isolating cut. Then $\{Y_1, Y_2, \ldots, Y_k\}$ is a minimum semi-multiway cut. In particular, a minimum semi-multiway cut is precisely a subpartition consisting of minimum s-isolating cuts over $s \in S$.*

Here an *s-isolating cut* is said to be *minimum* if its capacity is minimum in all *s*-isolating cuts of N. In particular, RMCP can be solved by k minimum-cut computations.

The objective function of RMCP is a representative example of k-submodular functions. Let $n := |V|$ and $m := |E|$. Observe that S_k^{n-k} and the set of all semi-multiway cuts are in one-to-one correspondence by $S_k^{n-k} \ni x \mapsto \{\{s_l\} \cup \{u \mid x_u = l\}\}_{l \in [k]}$. Therefore, finding a minimum semi-multiway cut is also viewed as the minimization of function $g_N : S_k^{n-k} \to \mathbb{R}$ defined by $x \mapsto \frac{1}{2} \sum_{l=1}^{k} d(\{s_l\} \cup \{u \mid x_u = l\})$.

Lemma 12 ([6,9]). *g_N is k-submodular.*

From Lemma 11 we have a minimizing oracle of g_N, and obtain the PIP representation of the minimizer set of g_N (= the set of minimum semi-multiway sets) by the algorithm in Sect. 4.1. Furthermore, g_N is network-representable [9] in the sense of the previous subsection, though the representing network is different from N. Therefore we also obtain the PIP representation for g_N in $O(\mathrm{MF}(kn, km) + n(k + m))$ time if we use the network construction of [9].

We proved a more efficient algorithm is possible.

Theorem 13. *Let $N = (V, E, c, S)$ be an undirected network with vertex set V, edge set E, nonnegative edge capacity c and terminal set S. We can obtain the elementary PIP representing the minimum semi-multiway cuts of N in $O(\log k \cdot \mathrm{MF}(n, m))$ time, where $n := |V|, m := |E|$ and $k := |S|$.*

4.4   Enumeration of All Maximal Consistent Ideals

The compact representation for $(\sqcap, \sqcup)$-closed sets by elementary PIPs is kind of a data compression. Hence it is natural to consider an efficient way to extract elements of the original $(\sqcap, \sqcup)$-closed sets. This corresponds to an enumeration of consistent ideals of elementary PIPs. Indeed, PIPs and their consistent ideal families can be regarded as special cases of Boolean 2-CNFs and their satisfiability instances, respectively. Thus we can enumerate all consistent ideals in output-polynomial time [5] (i.e. the algorithm stops in time polynomial in the length of the input and output).

As described in Sect. 5, maximal consistent ideals are of special interest. Now we consider enumerating them. This can also be done in output-polynomial time by using the algorithm of [10] in $O(k^3 n^3)$ time per one output. We developed a faster algorithm. Due to the space limit, we state only the following theorem; see the full version for detail.

Theorem 14. *Let* $\mathcal{P} = (P, \leq, \#)$ *be an elementary PIP with n parts and* $G = (P, A)$ *a directed graph representing the poset* $(P, \leq)$. *We can enumerate all maximal consistent ideals of* $\mathcal{P}$ *in* $O(|P| + |A| + nN)$ *time, where N is the number of maximal consistent ideals of* $\mathcal{P}$.

Our algorithm is efficient since the running time is proportional to the length of the input and output.

5   Application

For a function $g \colon [k]^n \to \bar{\mathbb{R}}$, a k-submodular function $f \colon S_k{}^n \to \bar{\mathbb{R}}$ is called a *k-submodular relaxation* [6,9] of g if it satisfies $f(x) = g(x)$ for all $x \in [k]^n$. Iwata, Wahlström and Yoshida [9] investigated k-submodular relaxations as a key tool for designing efficient FPT algorithms. Gridchyn and Kolmogorov [6] applied k-submodular relaxations to labeling problems in computer vision. Hirai and Iwamasa [7] characterized the class of functions which admit k-submodular relaxations.

The most important property of k-submodular relaxations is the following, called *persistency* [6,9].

Lemma 15 ([6,9]). *For every minimizer $x \in S_k{}^n$ of f, there exists a minimizer $y \in [k]^n$ of g such that $x_i \neq 0$ implies $x_i = y_i$ for each $i \in [n]$.*

Namely, each minimizer of f gives us partial information about a minimizer of g. In particular, minimizers that contain more nonzero elements have more information. Indeed, the following lemma holds.

Lemma 16. *Let M be a $(\sqcap, \sqcup)$-closed set. The supports of all maximal (with respect to $\preceq$) elements in M are all the same.*

From this lemma, it will be turned out that all maximal minimizers of f have the same amount of information about minimizers of g. As described in Sect. 4.4, we can enumerate all maximal minimizers efficiently. We hope that this enumeration algorithm will be applied to FPT algorithms and labeling problems arising from computer vision.

Acknowledgement. This work was partially supported by JSPS KAKENHI Grant Numbers 25280004, 26330023, 26280004, and by the JST, ERATO, Kawarabayashi Large Graph Project.

References

1. Ando, K., Fujishige, S.: ⊔, ⊓-closed families and signed posets. Technical report, Forschungsinstitut für Diskrete Mathematik, Universität Bonn (1994)
2. Ardila, F., Owen, M., Sullivant, S.: Geodesics in CAT(0) cubical complexes. Adv. Appl. Math. **48**(1), 142–163 (2012)
3. Barthélemy, J.P., Constantin, J.: Median graphs, parallelism and posets. Discrete Math. **111**(1–3), 49–63 (1993)
4. Dahlhaus, E., Johnson, D.S., Papadimitriou, C.H., Seymour, P.D., Yannakakis, M.: The complexity of multiterminal cuts. SIAM J. Comput. **23**(4), 864–894 (1994)
5. Feder, T.: Network flow and 2-satisfiability. Algorithmica **11**, 291–319 (1994)
6. Gridchyn, I., Kolmogorov, V.: Potts model, parametric maxflow and k-submodular functions. In: IEEE International Conference on Computer Vision (ICCV 2013), pp. 2320–2327 (2013)
7. Hirai, H., Iwamasa, Y.: On k-submodular relaxation. SIAM J. Discrete Math. (2016, to appear)
8. Huber, A., Kolmogorov, V.: Towards minimizing k-submodular functions. In: Mahjoub, A.R., Markakis, V., Milis, I., Paschos, V.T. (eds.) ISCO 2012. LNCS, vol. 7422, pp. 451–462. Springer, Heidelberg (2012)
9. Iwata, Y., Wahlström, M., Yoshida, Y.: Half-integrality, LP-branching and FPT algorithms. SIAM J. Comput. (2016, to appear)
10. Kavvadias, D.J., Sideri, M., Stavropoulos, E.C.: Generating all maximal models of a Boolean expression. Inf. Process. Lett. **74**, 157–162 (2000)
11. Kolmogorov, V., Thapper, J., Živný, S.: The power of linear programming for general-valued CSPs. SIAM J. Comput. **44**, 1–36 (2015)
12. Murota, K.: Analysis, Matrices and Matroids for Systems. Springer, Berlin (2010)
13. Nielsen, M., Plotkin, G., Winskel, G.: Petri nets, event structures and domains, part I. Theoret. Comput. Sci. **13**, 85–108 (1981)
14. Orlin, J.B.: Max flows in O(nm) time, or better. In: Proceedings of the 45th Annual ACM Symposium on Theory of Computing (STOC 2013), pp. 765–774 (2013)
15. Picard, J.C., Queyranne, M.: On the structure of all minimum cuts in a network and applications. In: Rayward-Smith, V.J. (ed.) Combinatorial Optimization II. Mathematical Programming Studies, vol. 13, pp. 8–16. Springer, Berlin (1980)
16. Sholander, M.: Medians and betweenness. Proc. Am. Math. Soc. **5**(5), 801–807 (1954)

Optimization Models for Multi-period Railway Rolling Stock Assignment

Susumu Morito[✉], Yuho Takehi, Jun Imaizumi, and Takayuki Shiina

Department of Industrial Management and Systems Engineering, Waseda University,
3-4-1 Ohkubo, Shinjuku, Tokyo 169-8555, Japan
{morito,tshiina}@waseda.jp, yuhotakehi@akane.waseda.jp, jun@toyo.jp

Abstract. It is necessary for railway companies to construct daily schedules of assigning rolling stocks to utilization paths. A utilization path consists of a series of trains that a particular rolling stock performs in a day. A mixed integer programming model based on Lai et al. [1] is presented and is shown that straightforward applications of the model result in too much computational time and also inappropriate assignment schedules due to *end effects*. We show that the model can be modified to alleviate these difficulties, and also show that the repeated applications of the optimization model in the rolling horizon allow to generate a feasible assignment schedule for a longer period of time thus indicating the feasibility of the optimization approach.

Keywords: Railway rolling stock · Inspection requirements · Utilization path · Mixed integer programming

1 Introduction

In order to operate trains, it is necessary for a railway company to assign rolling stocks to trains. A series of trains that a particular rolling stock executes during a day is called a **utilization path**. A rolling stock assignment problem, then, is a generalized multi-period assignment problem of making an assignment schedule for rolling stocks to a given set of utilization paths for a specified time horizon. Important considerations of the rolling stock assignment problem are inspection requirements. This paper considers two types of inspections, namely, **daily inspections** and **monthly inspections**. Normally, these inspections must be performed at a specified set of stations or train bases.

2 A Rolling Stock Assignment Problem

2.1 Problem Statement

We list assumptions of the rolling stock assignment problem.

© Springer International Publishing Switzerland 2016
R. Cerulli et al. (Eds.): ISCO 2016, LNCS 9849, pp. 393–402, 2016.
DOI: 10.1007/978-3-319-45587-7_34

1. A finite horizon model is considered in which a period corresponds to a day.
2. The starting station and the ending station of each utilization path are given.
3. Each utilization path is assigned to a rolling stock.
4. No deadhead is allowed, and thus the ending station of the utilization path in day k must coincide with the starting station of the utilization path in day $k + 1$ for each rolling stock.
5. Initial conditions with regard to the utilization path assigned to each rolling stock in day 0 are given, together with the accumulative operating times for daily and monthly inspections, and the accumulative operating mileage for monthly inspection at the end of day 0 for each rolling stock.
6. For each rolling stock, a daily inspection must be performed at least every other day. On the other hand, a monthly inspection should be performed at least once in 30 days, or within 30,000 km, whichever comes first.
7. Monthly inspections are performed at specific stations (including nearby train bases), and utilization paths which perform monthly inspections are known in advance together with the locations where inspections are performed.
8. Daily inspections can be performed at one of several stations that can perform daily inspections.
9. There exists an upper limit for the number of inspections that can be performed in a day at a particular station.
10. Costs of a daily and a monthly inspections are known, and the total inspection cost is minimized.

2.2   Related Studies

Lai et al. [1] studied rolling stock assignment at Taiwan High Speed Rail, which forms the basis of this study. Their problem is basically the same as ours except that deadheads are allowed possibly at very high cost. They presented a mixed integer programming formulation, which again forms the basis of this study, and detailed information concerning its application to their problems.

Maroti and Kroon [2,3] consider adjustments of a given roling stock schedule so that train units that require maintenance in the forthcoming one to three days can reach the maintenance facility in time. They considered a multicommodity type model for the maintenance routing problem.

2.3   Typical Instances

Typical instances considered in this paper are (1) a Taiwan High Speed Rail instance of Lai et al. consisting of 30 utilization paths and 30 sets of rolling stocks, and (2) an instance of a bullet-train line (Shinkansen) of Japan consisting of 26 utilization paths and 26 sets of rolling stocks.

Out of 30 utilization paths of the Taiwan instance, 24 paths are operation utilization paths which connect all operational trains in the timetable, 2 paths are maintenance spare utilization paths for major maintenance performed, say, once in 3 years, 2 paths are operation spare utilization paths for unexpected disruptions from defective trains, and finally 2 monthly inspection utilization paths in which only monthly inspections are performed. On the other hand, all of 26 utilization paths of the Japan instance are all operation utilization paths, among which 4 operation utilization paths include monthly inspections to be performed at specific stations. No maintenance utilization paths and operational spare utilization paths are considered in the Japan instance.

3   Formulation

We first define notations, and then present the mixed integer programming formulation.

Sets

E @Set of stations (including nearby train bases) where daily inspections can be performed
K @Set of days in the planning horizon
N @Set of utilization paths
N_e @Set of utilization paths which end at depot e
V @Set of rolling stocks

Constants

$G_i(A_i)$ @Starting (Ending) station of utilization path i
L_i @Operating mileage of utilization path i in kilometer
T_i @Operating time of utilization path i in day
Q @Cost of a monthly inspection
F @Cost of a daily inspection
I_{1D} @Upper bound of accumulative operating time for daily inspection in day
I_{2D} @Upper bound of accumulative operating time for monthly inspection in day
I_{2L} @Upper bound of accumulative operating mileage for monthly inspection in km
M @Arbitrarily large numbers
U_e @Capacity of monthly inspection at station e
W @Weight used in the objective function

Variables

x_i^{kv} @Variable that takes value 1 when rolling stock v is assigned to utilization path i in day k, and 0 otherwise
y_e^{kv} @Variable that takes value 1 when rolling stock v performs monthly inspection at station e in day k, and 0 otherwise
z^{kv} @Variable that takes value 1 when rolling stock v performs daily inspection in day k, and 0 otherwise

$D_{1v}^k(D_{2v}^k)$ @Accumulative operating time of rolling stock v for daily (monthly) inspection in day

L_{2v}^k @Accumulative operating mileage of rolling stock v for monthly inspection in kilometer

$$\text{minimize} \quad Q\sum_{e\in E}\sum_{k\in K}\sum_{v\in V} y_e^{kv}$$

$$+ F\sum_{k\in K}\sum_{v\in V} z^{kv}$$

$$+ W\sum_{k\in K}\sum_{v\in V}(D_{1v}^k + D_{2v}^k + L_{2v}^k) \tag{1}$$

$$\text{s.t.} \quad \sum_{i\in N} x_i^{kv} = 1 \qquad \forall k\in K, \forall v\in V \tag{2}$$

$$\sum_{v\in V} x_i^{kv} = 1 \qquad \forall k\in K, \forall i\in I \tag{3}$$

$$D_{1v}^k \le I_{1D} \qquad \forall k\in K, \forall v\in V \tag{4}$$

$$D_{1v}^k \le D_{1v}^{k-1} + \sum_{i\in N} T_i x_i^{kv}$$

$$- M\sum_{e\in E} y_e^{kv} - M z^{kv}$$

$$\forall k\in K, \forall v\in V \tag{5}$$

$$D_{2v}^k \le I_{2D} \qquad \forall k\in K, \forall v\in V \tag{6}$$

$$D_{2v}^k \le D_{2v}^{k-1} + \sum_{i\in N} T_i x_i^{kv} - M z^{kv}$$

$$\forall k\in K, \forall v\in V \tag{7}$$

$$L_{2v}^k \le I_{2L} \qquad \forall k\in K, \forall v\in V \tag{8}$$

$$L_{2v}^k \ge L_{2v}^{k-1} + \sum_{i\in N} L_i x_i^{kv} - M z^{kv} \qquad \forall k\in K, \forall v\in V \tag{9}$$

$$\sum_{i\in N_e} x_i^{kv} \ge y_e^{kv}$$

$$\forall k\in K, \forall v\in V, \forall e\in E \tag{10}$$

$$\sum_{i\in N} x_i^{kv} \ge z^{kv} \qquad \forall k\in K, \forall v\in V \tag{11}$$

$$\sum_{v\in V} y_e^{kv} \le U_e \qquad \forall k\in K, \forall v\in E \tag{12}$$

$$\sum_{i\in N} A_i x_i^{(k-1)v} = \sum_{i\in N} G_i x_i^{kv} \qquad \forall k\in K, \forall v\in V \tag{13}$$

$$x_i^{kv} \in \{0,1\} \;\; \forall i\in N, \forall k\in K, \forall v\in V \tag{14}$$

$$y_e^{kv} \in \{0,1\} \;\; \forall k\in N, \forall v\in V, \forall e\in E \tag{15}$$

$$z^{kv} \in \{0,1\} \qquad \forall k \in K, \forall v \in V \tag{16}$$

$$D_{1v}^k \geq 0 \qquad \forall k \in K, \forall v \in V \tag{17}$$

$$D_{2v}^k \geq 0 \qquad \forall k \in K, \forall v \in V \tag{18}$$

$$L_{2v}^k \geq 0 \qquad \forall k \in K, \forall v \in V \tag{19}$$

The major difference between the above model and the original model of Lai et al. [1] is the treatment of deadheads. The above model prohibits deadheads with Constraint (13), but other than that, the above model is basically identical to their model.

4   Evaluation of the Mixed Integer Programming Model

Detailed information of input data for Lai's experiments is given in Lai et al. [1]. They indicate that the problem with the planning horizon of 28 days (4 weeks) was solved in CPU 6 h, whereas the manual assignment took 8 h. They showed the results of 4-week assignments obtained manually and by the optimization model.

We evaluated the mixed integer programming model on their instance (some undisclosed information is estimated) using a standard commercial solver. Experiments are performed on a PC with Intel(R) Core i7-4770 CPU 3.40 GHz and 8 GB memory using AMPL-Gurobi version 6.0.2 on Windows 7 Professional.

Table 1 shows, for two instances of the planning horizon of 7 days and 14 days, upper and lower bounds, duality gap (i.e., (upper bound − lower bound)/lower bound) * 100), the numbers of monthly and daily inspections, CPU times in seconds to solve the linear programming relaxation optimally and also when the best integer solution is obtained, together with the cutoff time when the computations are aborted because of too much CPU time. We managed to obtain a feasible integer solution when the planning horizon was 7 days, but failed to get even a feasible integer solution when the planning horizon was extended to 14 days. Obviously, no feasible integer solutions could be obtained when the planning horizon is more than 14 days.

The main reason for difficulties in solving longer horizon problems is estimated to be the limited number of utilization paths in which monthly inspections can be performed. This could be verified by solving problems under the

Table 1. Evaluation of model/instance of Lai et al. [1] (CPU time in seconds)

Planning horizon	Upper bound	Lower bound	GAP (%)	# of Monthly inspections	# of Daily inspections	LP time	IP time	Cutoff time
7	12423	6864	44.8	2	83	55	3585	7749
14	–	31469	–	–	–	765	–	8682

artificial condition that monthly inspections could be performed in all utilization paths. Under this fictitious assumption, we managed to obtain feasible solutions for the problem with the planning horizon of 28 days, even though the optimality could not be reached at all. The duality gaps and CPU time are also very large. Another difficulties were found from the (feasible) solution obtained under the artificial condition that monthly inspections could be performed in any of the utilization paths.

Figure 1 is an assignment schedule for a 21-day problem, where monthly inspections are performed in colored (yellow) cells. Note that all monthly inspections are performed during day 1 through day 13, and no monthly inspections are performed after day 13. Also, the right-most column of the figure shows the cumulative operating mileage of each rolling stock at the end of the planning horizon of 21 days. Note that the cumulative operating mileages at the end of the horizon are very close to the upper limit of 30,000 km. This is because the problem is formulated as a finite horizon problem, and also because the objective function leads to less number of inspections during the planning horizon. This is sometimes called the **End Effects**.

It is obvious that the model cannot obtain a feasible assignment schedule for the "next" planning horizon starting from day 22, as ending operating mileages of most rolling stocks are too close to the upper limit of 30,000 km.

Figure 1 is a result for the Taiwan instance, but similar results are obtained for the Japan instance also. In the next section, we try to modify the model to reduce end effects and also to speed up the computations.

	1	2	3	4	5	6	7	8	9	10	11	12	13	14	15	16	17	18	19	20	21	走行距離
1	23	24	17	17	17	18	4	5	9	16	18	19	6	3	4	5	9	16	24	25	3	24450
2	6	2	26	23	18	19	2	16	22	23	22	23	22	23	22	23	22	23	18	4	5	25532
3	19	3	4	5	20	21	17	17	17	17	25	3	4	5	16	17	17	18	4	5	9	25989
4	13	14	3	19	10	23	22	23	18	4	5	15	2	6	10	1	2	15	2	20	21	27650
5	9	6	10	1	2	9	3	4	3	7	1	10	1	2	3	4	3	19	16	25		25406
6	12	1	2	9	3	4	3	7	13	14	3	4	5	15	9	9	16	17	17	17	17	26090
7	17	25	13	26	23	17	18	19	7	1	7	8	10	1	2	6	6	26	13	14	10	29785
8	16	22	13	14	7	11	12	8	26	8	2	20	21	18	19	10	11	12	11	12	11	23550
9	18	19	9	7	1	2	9	3	4	5	9	6	3	4	5	3	4	3	6	9	15	25780
10	26	8	7	8	6	13	26	11	12	11	12	1	7	8	6	20	21	25	9	26	1	25050
11	25	26	11	12	8	26	11	12	1	3	4	5	26	11	12	8	7	1	26	13	14	26000
12	4	5	6	2	9	3	19	2	3	19	10	11	12	13	14	16	18	19	10	23	22	20769
13	22	23	22	11	12	13	14	6	2	26	8	16	18	19	7	11	12	8	20	21	24	28985
14	2	9	16	18	19	7	1	9	15	15	6	9	15	7	13	14	10	11	12	11	12	29979
15	21	17	25	3	4	5	15	10	8	6	16	17	24	17	17	25	3	4	5	6	20	27853
16	13	15	20	21	22	8	6	13	10	13	14	2	9	26	1	15	13	10	8	10	8	26000
17	3	4	5	6	26	1	10	1	6	9	20	21	17	25	13	2	20	21	22	1	16	29757
18	5	16	24	22	11	12	8	20	21	18	19	26	11	12	8	26	8	7	1	2	6	27850
19	14	20	21	24	24	22	13	14	16	22	23	18	19	9	26	13	14	2	15	13	2	26750
20	1	7	1	20	21	24	24	18	19	2	13	7	8	10	11	12	13	14	7	8	7	28855
21	24	18	19	10	13	14	7	13	14	10	11	12	13	14	20	21	24	22	23	24	18	29000
22	10	11	12	13	14	6	16	22	11	12	13	14	16	22	23	18	19	6	16	18	19	28647
23	11	12	8	15	15	10	23	24	24	24	17	22	23	24	18	19	26	13	14	7	13	28740
24	20	21	18	4	5	16	25	26	23	25	26	13	14	16	24	22	23	24	25	3	4	28711
25	8	10	23	25	16	25	20	21	25	20	21	25	20	21	25	7	1	9	3	19	26	29591
26	7	13	14	16	25	20	21	25	20	21	24	24	25	20	21	24	25	20	21	22	23	27938

Fig. 1. End effects (Color figure online)

5 Modifications to the Mixed Integer Programming Model

It appears necessary to remove or at least reduce end effects. One way to reduce end effects is forcing to perform monthly inspections more uniformly throughout the planning horizon. This can be achieved by setting some limits on the number of monthly inspections.

From a series of experiments in which utilization paths that can perform monthly inspections are artificially adjusted, we found that "restrictive" instances in which the number of utilization paths that can perform a monthly inspection is very small is difficult to solve, and tends to take longer CPU time. At the same time, we recognized a rather puzzling phenomenon that a feasible assignment schedule that was found by solving the problem in which a monthly inspection can be performed in a moderate number of utilization paths, could not be found when a more relaxed problem in which a monthly inspection can be performed in all utilization paths. That is, a feasible solution that is found in a more restricted problem could not be found when the more relaxed problem is solved. The optimization algorithm and the software seem to have difficulties to find feasible (not necessarily, optimal) solutions to problems with too much freedom, just like they have difficulties to solve very restrictive problems. We experienced similar phenomena several times for very restrictive problems to which feasible solutions are difficult to obtain and also for problems with too much freedom.

Upon various experiments, we come to think that setting some forms of **lower and upper limits for the number of monthly inspections** in the assignment schedule helps reduce not only end effects but also CPU time. Many alternative forms of setting limits can be considered, and after a series of experiments, we found the following ways of setting limits are effective.

(1) **Lower Limit**: For each subperiod of 7 days, the lower limit on the number of monthly inspections is set to 6. Subperiods are considered in an overlapping fashion such as "from day 1 to day 7", "from day 2 to day 8", "from day 3 to day 9", etc. Some informal explanation for the magic number of 6 is the fact that the number of monthly inspections per rolling stock per 7-day subperiod is $7/30$, recalling that a monthly inspection should be performed within 30 days from the previous inspection. Since the number of rolling stocks is 26 in the Japanese instance (the number happens to be the same in the Taiwan instance, if we disregard 2 major maintenance utilization paths and 2 operation spare utilization paths), the lower bound becomes $26 * 7/30 - 6.06 > 6$.

(2) **Upper Limit**: For each day in the planning horizon, the maximum number of monthly inspections is 2.

Adding these constraints concerning lower and upper limits on the number of monthly inspections gives equalizational effects and thus alleviates end effects. Interestingly, the equalization with the added constraints also helps reduce CPU time, and for the Japan instance, the problems for which feasible solutions could be obtained have expanded from the planning horizon of 7 days to 14 days.

6   Applications of the Mixed Integer Programming Model: A Rolling Horizon Approach

The modified model now allows us to generate more "**reasonable**" assignment schedules at least for the planning horizon of 7 days, and possibly more. This, however, does not mean that a 7-day period optimization model can be used in reality, as the process may not be repeated due to infeasibility. It is then necessary to observe that generation of the optimization-based assignment schedule can be repeated for some reasonable period of time.

In this section, we report results of simulation in a rolling horizon fashion to see how long the optimization can be repeated to come up with an assignment schedule for an extended period of time. Two alternative methods are considered to repeat the optimization process, depending on whether planning horizons of two "adjacent" optimization models time-wise overlap or not.

(1) **Non-overlapping Horizon**: Solve first the optimization model for the specified planning horizon. Using the ending conditions of the optimization as initial conditions, solve the optimization model for the next planning horizon. If the length of the planning horizon is one week, solve the optimization model for week 1. Using the final conditions of week 1 as initial conditions, solve the optimization model for week 2, etc.

(2) **Overlapping Horizon**: Solve first the optimization model for the specified planning horizon as in the non-overlapping case. However, the obtained assignment schedule is not used in its entirety, and only the earlier portion of the schedule is adopted. Using the final conditions of the adopted portion as the initial conditions, resolve the optimization model, and so force. For example, if the length of the planning horizon is one week, and only the first 3 days of the resultant schedule are adopted, check the ending conditions of day 3, and using these conditions as initial conditions, the optimization problem is solved again for the next one week, i.e., from day 4 to day 10. The non-overlapping horizon can be viewed as a special case of the overlapping horizon in which the length of the optimization planning horizon coincides with the length of the adopted part of the assignment schedule.

Table 2 summarizes, for the Japan instance, the maximum period of time for which a feasible assignment schedule can be generated, under the non-overlapping horizon (Case No.1–No.3) and also for the overlapping horizon (Case No.4–No.6). The table should be read as follows: For the planning horizon of 7 days under the non-overlapping horizon, the optimization yielded a feasible assignment schedule only for the first 7 days only, and the optimization model became infeasible for the next 7 days due to the ending conditions of consecutive mileage at the end of day 7. Similarly, when the planning horizon is 5 days, the optimization could not be repeated at all. However, when the planning horizon is reduced to 3 days, the optimization model can be run twice to generate a feasible schedule for the total of 6 days, but no further.

The results for the overlapping horizon cases can be read in a similar fashion where the length of overlapping horizons is shown. For example, when the planning horizon is 7 days and the length of overlapping horizons is 3 days, the first

Table 2. Simulation results for non-overlapping and overlapping horizons

Case no	Planning horizon	Overlapping period	# of Times optimization model can be solved	Maximum period feasible schedule is obtained
1	7 days	none	1	7 days
2	5 days	none	1	5 days
3	3 days	none	2	6 days
4	7 days	2 days	2	12 days
5	7 days	3 days	3	15 days
6	7 days	4 days	≥10	≥30 days

	1	2	3	4	5	6	7	8	9	10	11	12	13	14	15	16	17	18	19	20	21	22	23	24	25	26	27	28	29	30	走行距離
1	2	6	2	10	13	14	22	23	18	16	22	8	15	6	7	15	20	21	25	15	6	13	14	2	22	8	9	23	24	25	16050
2	5	3	4	4	5	26	21	22	23	26	8	10	8	15	26	1	6	7	20	21	22	19	3	19	19	16	18	24	22	8	23762
3	16	17	22	19	7	11	26	11	12	25	26	23	21	17	17	8	15	6	13	14	2	12	23	18	14	9	2	8	16	22	24367
4	17	22	23	2	16	25	3	19	26	21	24	17	13	14	15	12	1	15	26	11	12	18	4	5	8	2	3	17	25	20	24103
5	14	9	20	14	6	10	23	18	4	17	17	24	6	7	1	6	7	1	22	13	14	24	22	11	2	10	23	3	4	5	21153
6	22	23	24	26	11	12	14	3	19	7	1	20	2	20	21	3	19	2	11	12	23	2	9	10	26	23	17	9	15	15	29609
7	25	20	21	18	4	5	9	10	8	6	6	15	17	22	8	7	23	22	12	23	18	3	19	26	11	12	13	6	6	7	26953
8	3	19	3	12	1	2	4	5	2	4	5	9	3	19	6	17	25	26	15	6	15	1	26	13	9	3	19	18	19	9	14710
9	4	5	9	1	3	4	8	16	22	24	25	6	1	16	22	11	12	23	5	9	16	14	16	25	6	26	8	14	20	21	1566
10	21	25	16	16	22	8	17	17	17	8	20	21	7	13	14	22	13	14	14	26	11	23	18	4	21	24	22	15	26	13	25017
11	8	16	18	8	2	9	20	21	24	15	15	7	26	1	9	21	24	18	23	18	4	4	5	20	7	1	15	16	17	17	27792
12	26	11	12	5	9	7	15	15	20	20	21	25	9	10	13	19	10	11	18	4	5	9	2	3	16	17	25	1	7	1	26653
13	23	18	19	23	18	19	7	1	7	13	14	2	11	12	23	5	2	9	2	2	9	6	15	15	3	19	7	11	12	11	7042
14	11	12	11	3	19	16	1	7	1	10	13	14	24	18	19	18	4	5	9	16	24	26	1	16	1	15	6	26	13	14	2296
15	13	14	10	24	17	17	16	25	6	19	7	1	19	26	11	4	5	3	4	5	3	16	24	24	10	13	14	19	3	19	9322
16	9	2	15	13	14	6	24	24	25	1	16	22	22	11	12	9	9	16	3	19	26	21	25	9	17	25	26	13	14	16	4234
17	10	8	26	17	24	24	12	13	14	22	23	18	5	3	4	24	18	4	16	17	17	17	17	17	25	20	21	10	11	12	8019
18	7	1	6	6	15	3	6	6	15	5	10	13	20	21	24	10	11	12	19	3	19	25	10	8	5	6	20	7	1	6	26975
19	18	4	5	25	10	1	18	4	5	23	18	19	23	24	25	26	8	10	17	25	10	7	13	14	12	11	12	25	2	10	15329
20	20	21	17	9	20	21	11	12	11	18	19	3	16	25	2	16	17	17	6	7	1	5	20	21	23	18	4	21	18	4	17475
21	6	15	7	15	26	13	19	20	21	11	12	11	12	23	18	2	16	24	24	24	25	15	6	7	13	14	10	4	5	3	10677
22	1	26	8	7	23	18	25	26	13	12	11	12	25	9	16	14	26	8	10	8	7	20	21	22	4	5	16	5	9	2	17045
23	15	10	1	22	8	20	13	14	3	3	4	5	4	5	3	23	22	13	8	10	8	10	8	6	20	21	24	12	23	24	9935
24	19	7	13	21	25	15	2	2	9	9	3	4	18	4	5	20	21	25	1	20	21	11	12	23	24	22	11	2	10	23	19963
25	12	13	14	11	12	23	10	8	10	2	2	26	10	8	20	13	14	20	7	1	20	8	7	1	15	7	1	20	21	18	4807
26	24	24	25	20	21	22	5	9	16	14	9	16	14	2	10	25	3	19	21	22	13	22	11	12	18	4	5	22	8	26	21154

Fig. 2. Feasible assignment obtained by rolling schedule with overlapping horizon (Color figure online)

4 days of the assignment schedule obtained by the first optimization is adopted, and the remaining schedule of 3 days are discarded. Based on the ending conditions of day 4, the optimization model is run again for 7 days, to come up with the total of a feasible assignment schedule of 11 days. This way, we could repeat solving optimization model three times when the length of the overlapping period is 3 days. As we increased the length of the overlapping period to 4 days (i.e., discard the last 4 days of the assignment schedule of 7 days), the optimization could be repeated at least 10 times to come up with the assignment schedule of at least 30 days. Generally, we can observe that the shorter the planning horizon, the longer the period for which a feasible schedule is obtained.

Finally, Fig. 2 shows a feasible assignment schedule obtained in Case No.6 of Table 2, in which we can observe monthly inspections as indicated in color (yellow) moving from the lower left corner toward the upper right corner, reflecting the fact that the initial cumulative mileages of rolling stocks are higher as we go down the figure.

7   Conclusions

The rolling stock assignment problem is considered, and it is shown that the mixed integer programming model of Lai et al. [1] have difficulties in generating reasonable assignment schedules when the planning horizon becomes longer. In particular, we pointed out that end effects of the finite horizon model gave difficulties of generating unreasonable assignment schedules. Rather simple modifications of the model by adding constraints limiting the number of monthly inspections are shown to at least reduce the difficulties due to end effects and also help reduce CPU time of the optimization.

Simulation experiments were performed to verify that the optimization model can be run several times in a rolling horizon fashion with the non-overlapping as well as the overlapping horizons. This study showed the feasibility of using mixed integer programming optimization for the rolling stock assignment problems.

Acknowledgments. This work was supported by JSPS Kakenhi Grant No. 26350437. The authors would like to express sincere appreciations to Professor Y.C. Lai of National Taiwan University and Mr. Satoshi Kato of Railway Technical Research Institute.

References

1. Lai, Y.C., Zeng, W.W., Liu, K.C., Wang, S.W.: Development of rolling stock assignment system for Taiwan high speed rail. In: Proceedings of 5th International Seminar on Railway Operations Modelling and Analysis, Copenhagen (2013)
2. Maroti, G., Kroon, L.F.: Maintenance routing for train units: the transition model. Transp. Sci. **39**, 518–525 (2005)
3. Maroti, G., Kroon, L.: Maintenance routing for train units: the interchange model. Comput. Oper. Res. **34**, 1121–1140 (2007)

Sum-of-Squares Rank Upper Bounds
for Matching Problems

Adam Kurpisz, Samuli Leppänen$^{(\boxtimes)}$, and Monaldo Mastrolilli

IDSIA, 6928 Manno, Switzerland
{adam,samuli,monaldo}@idsia.ch

Abstract. The matching problem is one of the most studied combinatorial optimization problems in the context of extended formulations and convex relaxations. In this paper we provide upper bounds for the rank of the Sum-of-Squares (SoS)/Lasserre hierarchy for a family of matching problems. In particular, we show that when the problem formulation is strengthened by incorporating the objective function in the constraints, the hierarchy requires at most $\lceil \frac{k}{2} \rceil$ rounds to refute the existence of a perfect matching in an odd clique of size $2k + 1$.

1   Introduction

A matching in a simple graph $G = (V, E)$ is a subset of edges $M \subseteq E$ such that every vertex of V is incident to at most one edge in M. The problem of finding a matching of maximum possible cardinality is known to admit a polynomial time algorithm first given by Edmonds [7], and it is has been extensively studied in combinatorial optimization and mathematics. The MAXIMUM MATCHING problem can be formulated as an integer linear program (ILP) in the form $\max_{x \in \{0,1\}^E} \{\sum_{e \in E} x_e \mid x(\delta(v)) \leq 1, \forall v \in V\}$, where $\delta(v)$ denotes the set of edges incident to a vertex v in G and $x(\delta(v)) = \sum_{i \in \delta(v)} x_i$. Of particular interest is the *matching polytope*, which is the convex hull of the feasible points of the ILP when the graph G is complete with n vertices.

Interestingly, despite the fact that the matching problem can be solved in polynomial time, the matching polytope cannot be described using a polynomial number of linear inequalities and hence the problem cannot be solved using a single linear program (LP) of polynomial size (however, the decision version of the matching problem can be solved using a polynomial-sized LP and thus the optimization version is solvable to arbitrary precision using a sequence of LPs [3]). In his seminal work, Yannakakis [26] showed that no *symmetric* LP of polynomial size can describe the matching polytope, and later Rothvoß [21] showed this for any LP of polynomial size. In light of these negative results for LPs, it is natural to ask whether the matching problem can be expressed compactly in a framework such as semidefinite programming (SDP) that is more

Supported by the Swiss National Science Foundation project 200020-144491/1 "Approximation Algorithms for Machine Scheduling Through Theory and Experiments".

© Springer International Publishing Switzerland 2016
R. Cerulli et al. (Eds.): ISCO 2016, LNCS 9849, pp. 403–413, 2016.
DOI: 10.1007/978-3-319-45587-7_35

powerful than linear programming but still allows efficient optimization. Recently Braun et al. [5] proved that no symmetric polynomial size SDP captures the matching polytope. This result was proved by showing that among all symmetric SDP relaxations for the matching problem, the Sum-of-Squares (SoS)/Lasserre SDP hierarchy is *optimal*. More precisely, for every constant t, the t-round SoS SDP relaxation yields at least as good an approximation as any SDP relaxation of size $n^{O(t)}$. The result in [5] follows by appealing to a result by Grigoriev [10] that shows that $\Omega(n)$-rounds of the SoS SDP hierarchy cannot refute the existence of a perfect matching in an odd clique of size $n = 2k + 1$.

A particular systematic way of studying LP and SDP formulations of combinatorial problems are the so-called *lift-and-project* methods, which produce a sequence of convex relaxations converging to the integral polytope of the problem. Typically in such methods a starting formulation of a problem is lifted by adding new variables and constraints, and after optimizing over the relaxation, the solution is projected back to the space of the integral polytope. Some of the most studied lift-and-project methods include the Lovász-Schrijver (LS) [15], Sherali-Adams (SA) [22] and the SoS [12,17,19,23] hierarchies. It is known that the SoS hierarchy produces stronger relaxations than the LS and SA hierarchies (see for example [13]). Common to these hierarchies is that they are parameterized by their *level* t, which is a positive integer less than or equal to n, the number of decision variables. Using for example the ellipsoid method, solving the relaxations is possible in polynomial time if $t = O(1)$ and hence their study is also of interest in the context of approximation algorithms. Recent results show that lift-and-project methods might produce the best possible mathematical models of a given size for certain class of problems. For some constraint satisfaction problems, like MAX CUT, Chan et al. [6] proved that the SA hierarchy is at least as good as any LP of the same size. On the other hand, Lee et al. [14] proved that for approximating maximum constraint satisfaction problems, SDPs of polynomial-size are equivalent in power to those arising from constant level SoS relaxations.

Since the matching problem admits a polynomial time algorithm, but cannot be solved using a single small LP or symmetric SDP, it is of interest to study how lift and project methods perform when applied to the problem. We call the *rank* of a lift-and-project method the smallest level t such that the method at that level exactly captures the integral polytope of the underlying problem. In [9] Goemans and Tunçel proved that the rank of the LS procedure for the matching polytope of K_{2k+1} is at least $2k - 1$ and at most $2k^2 - 1$, and in [24] Stephen and Tunçel proved that the rank is exactly k for a stronger semidefinite variant of the LS procedure. For the SA hierarchy, Mathieu and Sinclair [16] showed that the rank is exactly $2k - 1$. Recently Worah [25] studied the performance of the SA_* hierarchy (a variant of the SA hierarchy) when applied to the matching polytope. He showed that the SA_* rank of the matching polytope is at most k. The first result for the SoS hierarchy was given by Grigoriev [10], who showed that the rank is $\Omega(k)$. In [1] Au and Tunçel sketch a proof that the rank of the matching polytope of K_{2k+1} for the SoS hierarchy is at least $\lfloor \frac{k}{2} \rfloor$ and at most k (the full proof is postponed for the future in a subsequent paper [2]).

In this paper we provide upper bounds for the SoS rank of the polytopes of a general family of matching problems. More precisely, we consider the r-UNIFORM HYPERGRAPH MAXIMUM b-MATCHING, where the problem is to find a maximal cardinality subset of hyperedges in an r-uniform hypergraph[1] under the constraint that no vertex can be incident to more than b edges. We show that when the natural polytope of the problem is strengthened by requiring that the objective function attains a specific value (i.e., by adding to the initial formulation the constraint $\sum_{e \in E} x_e = c$ for appropriate constant c), the SoS rank has an upper bound of $\max\left\{b, \frac{1}{2}\lfloor\frac{b|V|}{r}\rfloor\right\}$. In the case $r = 2, b = 1$ corresponding to the MAXIMUM MATCHING problem, the problem formulation we use is the same used by Grigoriev [10] when the graph is K_{2k+1}. When combining our result with the solution of Au and Tunçel [1], we show that the SoS rank of the considered problem is $\lceil\frac{k}{2}\rceil$ or $\lfloor\frac{k}{2}\rfloor$ (when k is even the characterization is tight).

We obtain our result by showing that at the level given above the SoS hierarchy is able to detect a contradiction between the matching constraints and the constraint that the objective function attains a specific superoptimal value. Alternative ways to obtain upper bounds for the hierarchy include the *Decomposition Theorem* [11] (see also [20]) and a recent result [8] for unconstrained quadratic optimization problems. We remark that our technique is more specialized than the Decomposition Theorem and yields an upper bound for the matching problems that is tighter by a factor of 2 than what one obtains by applying the Decomposition Theorem.

2   The SoS Hierarchy

In this section we provide a definition of the SoS hierarchy [12,19] when applied to feasibility problems with 0/1-variables. Although feasibility testing and optimization are equivalent in their complexity up to logarithmic factors, the feasibility testing formulation usually produces tighter SoS relaxations since the objective function is incorporated in the constraints. For convenience, our presentation follows the "pseudoexpectation" notation (see for example [4,18]).

Consider the following feasibility problem

$$p(x) = c,$$
$$q_i(x) \geq 0, \ \forall i = 1, ..., m,$$
$$x_i^2 = x_i, \ \forall i = 1, ..., n, \tag{1}$$

where $p(x)$ is a linear function, $c \in \mathbb{R}$ and $q_i(x)$ are polynomials of degree at most 2.

Let $\mathbb{R}[x]$ denote the ring of polynomials with real coefficients in variables $x = (x_1, ..., x_n) \in \mathbb{R}^n$, and $\mathbb{R}[x]_d$ the set of polynomials of $\mathbb{R}[x]$ of degree at most d.

---

[1] An r-uniform hypergraph is given by a set of vertices V and set of hyperedges E where each hyperedge $e \in E$ is incident to exactly r vertices.

Lasserre [12] proposed the following hierarchy of relaxations of (1) parameterized by an integer t: find a linear map $\tilde{\mathbb{E}} : \mathbb{R}[x]_{2t+2} \to \mathbb{R}$ that satisfies

$$\tilde{\mathbb{E}}(1) = 1,$$
$$\tilde{\mathbb{E}}(u^2(x)) \geq 0, \quad \forall u \in \mathbb{R}[x]_{t+1},$$
$$\tilde{\mathbb{E}}(u^2(x)(p(x) - c)) = 0, \quad \forall u \in \mathbb{R}[x]_t,$$
$$\tilde{\mathbb{E}}(u^2(x)q_i(x)) \geq 0, \quad \forall u \in \mathbb{R}[x]_t, i = 1, ..., m,$$
$$\tilde{\mathbb{E}}(u^2(x) \cdot (x_i^2 - x_i)) = 0, \quad \forall u \in \mathbb{R}[x]_t, i = 1, ..., n. \quad (2)$$

The linear map $\tilde{\mathbb{E}}(\cdot)$ is usually called *degree-$2t+2$ pseudoexpectation operator*. For consistency, we refer to it as the level-t pseudoexpectation operator. In (2) one looks for a linear operator over a finite dimensional space whose dimension depends on the fixed constant t. It can be shown that (2) corresponds to solving a semidefinite programming problem that is solvable in time $(m + 1)n^{O(t)}$ to arbitrary precision using for example the ellipsoid method.

To see that (2) is a relaxation of (1), consider a feasible point x^* to (1) and the pseudoexpectation defined by $\tilde{\mathbb{E}}(f(x)) = f(x^*)$ which merely evaluates any given polynomial f at the point x^*. Then $\tilde{\mathbb{E}}(\cdot)$ is linear and satisfies the conditions of (2) since x^* is a feasible solution. Furthermore, it can be shown (see for example [13]) that any feasible solution to the relaxation (2) gives a feasible solution to (1) when the parameter $t = n$.

Next, we discuss some properties of the pseudoexpectation operator and present the essential ingredients for our main result.

Lemma 1. *Assume $\tilde{\mathbb{E}}(\cdot)$ is a solution to (2), and the problem has the equality constraint $q(x) = 0$ (i.e., two constraints of the form $q(x) \geq 0$ and $-q(x) \geq 0$) of degree at most 2. Then $\tilde{\mathbb{E}}(uq) = 0$ for every $u \in \mathbb{R}[x]_{2t}$.*

Proof. It is sufficient to prove that the claim is true for monomials. Let $r(x) = x_1^{i_1} \cdots x_n^{i_n}$ be any monomial such that $r \in \mathbb{R}[x]_{2t}$. We partition the product r into two parts, v and w, such that $v = x_1^{k_1} \cdots x_n^{k_n}$ and $w = x_1^{l_1} \cdots x_n^{l_n}$, $r = vw$ and $v^2, w^2 \in \mathbb{R}[x]_{2t}$. Then, by definition since $\tilde{\mathbb{E}}(\cdot)$ is feasible, we have

$$0 = \tilde{\mathbb{E}}((v + w)^2 q) = \tilde{\mathbb{E}}(v^2 q) + 2\tilde{\mathbb{E}}(vwq) + \tilde{\mathbb{E}}(w^2 q) = 2\tilde{\mathbb{E}}(rq)$$

which proves the claim.   □

The above lemma together with the requirement $\tilde{\mathbb{E}}(x_i^2) = \tilde{\mathbb{E}}(x_i)$ (originating from the constraints of the relaxation (2)) allows us to *linearize* monomials, in other words, we can write for any polynomial $u \in \mathbb{R}[x]_{2t}$ and index $i = 1, ..., n$ that $\tilde{\mathbb{E}}(x_i^2 u) = \tilde{\mathbb{E}}(x_i u)$. Hence, in what follows we only consider multilinear polynomials (i.e., polynomials consisting of monomials of the form $\prod_{i \in I} x_i$ for some $I \subseteq \{1, ..., n\}$).

In the following lemma we show that under certain conditions any feasible solution to the relaxation (2) gives a feasible solution to (1) for parameter t potentially much smaller than n. We apply the lemma to a family of matching problems, but provide a more general statement in order to highlight the underlying assumptions.

Lemma 2. *Let $\mathcal{P}$ be a feasibility problem of the form (1) with the function $p(x) = \sum_{i=1}^{n} x_i$. If the constraints of $\mathcal{P}$ imply that for the level-t pseudoexpectation operator the equation $\tilde{\mathbb{E}}(\prod_{i \in I} x_i) = 0$ holds for every $I \subseteq [n]$ of size $|I| \geq 2t + 1$, and $c > 2t$, then there is no feasible solution to the SoS relaxation at level t.*

Proof. Assume that there exists a level-t pseudoexpectation operator $\tilde{\mathbb{E}}(\cdot)$ for $\mathcal{P}$. We prove by induction that $\tilde{\mathbb{E}}(\prod_{i \in S} x_i) = 0$ for any $S \subseteq [n]$, $|S| \leq 2t$. Indeed, assume $\tilde{\mathbb{E}}(\prod_{i \in S'} x_i) = 0$ for $|S'| = r$ and consider some $S \subseteq E$ such that $|S| = r - 1$. Consider the relaxation of the equation $\sum_{i=1}^{n} x_i = c$. We have by Lemma 1

$$0 = \tilde{\mathbb{E}}\left(\prod_{i \in S} x_i \left(\sum_{j \in [n]} x_j - c \right) \right) = \tilde{\mathbb{E}}\left(\prod_{i \in S} x_i \sum_{j \in S} x_j + \prod_{i \in S} x_i \sum_{j \in [n] \setminus S} x_j - c \prod_{i \in S} x_i \right)$$

$$= \tilde{\mathbb{E}}\left((|S| - c) \prod_{i \in S} x_i \right) + \tilde{\mathbb{E}}\left(\prod_{i \in S} x_i \sum_{j \in [n] \setminus S} x_j \right) = (|S| - c)\tilde{\mathbb{E}}\left(\prod_{i \in S} x_i \right)$$

where the last equality follows from the induction hypothesis. Since $|S| - c \neq 0$, we get that $\tilde{\mathbb{E}}\left(\prod_{i \in S} x_i \right) = 0$. In particular this implies that $\tilde{\mathbb{E}}(x_i) = 0$ which contradicts the assumption that $\tilde{\mathbb{E}}\left(\sum_{i \in [n]} x_i - c \right) = 0$. $\qquad \square$

Lemma 3. *If $u \in \mathbb{R}[x]_{t+1}$ and $\tilde{\mathbb{E}}(\cdot)$ is a level-t pseudoexpectation operator such that $\tilde{\mathbb{E}}(u^2) = 0$, then $\tilde{\mathbb{E}}(uw) = 0$ for every $w \in \mathbb{R}[x]_{t+1}$.*

The proof of Lemma 3 can be found for example in the appendix of [4].

Lemma 4. *Let $u \in \mathbb{R}[x]_{t+1}$ be any monomial and $\tilde{\mathbb{E}}(\cdot)$ a level-t pseudoexpectation operator such that $\tilde{\mathbb{E}}(u) = 0$, then $\tilde{\mathbb{E}}(uw) = 0$ for every w such that $uw \in \mathbb{R}[x]_{2t+2}$.*

Proof. It is sufficient to prove the claim for monomials $w = \prod_{i \in I} x_i$. Write $w = \prod_{k \in K} x_k \prod_{j \in J} x_j$ for some disjoint K, J such that $K \cup J = I$, $u \prod_{k \in K} x_k \in \mathbb{R}[x]_{t+1}$ and $\prod_{j \in J} x_j \in \mathbb{R}[x]_{t+1}$. Then, since by the linearization property it holds $\tilde{\mathbb{E}}(u) = \tilde{\mathbb{E}}(u^2) = 0$, we have by Lemma 3 that $\tilde{\mathbb{E}}(u \prod_{k \in K} x_k) = 0$. Using the linearization again, we have $\tilde{\mathbb{E}}(u \prod_{k \in K} x_k) = \tilde{\mathbb{E}}(u^2 \prod_{k \in K} x_k^2) = 0$ and thus again by Lemma 3, $\tilde{\mathbb{E}}\left(u \prod_{k \in K} x_k \prod_{j \in J} x_j \right) = 0$. $\qquad \square$

3   SoS Rank Upper Bound for the r-Uniform Hypergraph b-Matching Problem

Given an r-uniform hypergraph $G = (V, E)$ with n vertices and m hyperedges, each incident to exactly r vertices, a b-*matching* M in G is a set of edges such that at most b edges share a common vertex. The r-UNIFORM HYPERGRAPH

MAXIMUM b-MATCHING problem in G consists of finding a b-matching of maximum cardinality. Below is a natural integer linear programing formulation of the problem:

$$\max \sum_{e \in E} x_e$$

$$\text{s.t.} \sum_{e \in \delta(v)} x_e \leq b, \qquad \text{for each } v \in V,$$

$$x_e \in \{0, 1\}, \qquad \text{for each } e \in E. \tag{3}$$

We consider a slightly modified formulation, where the objective function is incorporated in the constraints. For this formulation, which is the same as used by Grigoriev [10] in the case $r = 2, b = 1$ and for the complete graph, we obtain an upper bound for the SoS rank. This modification replaces the optimization problem with the problem of finding a feasible solution. Then, the question of whether or not the SoS hierarchy captures exactly the integral polytope is replaced by the question of whether or not the hierarchy is able to detect that there is no solution with a superoptimal value. Hence, we formulate the problem (3) as

$$\sum_{e \in E} x_e = c,$$

$$\sum_{e \in \delta(v)} x_e \leq b, \qquad \text{for each } v \in V,$$

$$x_e \in \{0, 1\}, \qquad \text{for each } e \in E, \tag{4}$$

where $c \in \mathbb{R}$.

In what follows we show an upper bound for the SoS rank of (4). Informally, we show that when the constant c is set such that (4) does not have a feasible solution, after certain level the SoS hierarchy detects that the constraint $\sum_{e \in E} x_e - c = 0$ is inconsistent with the other constraints.

Then, we discuss the special case when $b = 1$ and $r = 2$, which is the usual MAXIMUM MATCHING problem. More precisely, we show that for the complete graph K_{4k+1}, the SoS relaxation of (4) has rank exactly k. We do this by appealing to the result of Au and Tunçel [1] to argue that the rank of the SoS relaxation of (4) is at least k and by our main result at most k.

3.1   Upper Bound for the SoS Rank for the Matching Problem (4)

In this section we prove the upper bound of $\max\left\{b, \frac{1}{2}\lfloor \frac{b|V|}{r} \rfloor\right\}$ for the rank of SoS relaxation of (4).

Lemma 5. *For every SoS relaxation of (4) of level t at least b, the following holds:*

$$\tilde{\mathbb{E}}\left(\prod_{e \in I} x_e\right) = 0$$

for every $v \in V$ and every $I \subseteq \delta(v)$ such that $b < |I| \leq 2t + 2$.

Proof. First consider any vertex v and a set J of b edges incident to v. Then

$$0 \geq \tilde{\mathbb{E}}\left(\left(\prod_{e \in J} x_e\right)^2 \left(\sum_{f \in \delta(v)} x_f - b\right)\right) = \tilde{\mathbb{E}}\left(\sum_{f \in \delta(v) \setminus J} x_f \prod_{e \in J} x_e + \sum_{f \in J} x_f \prod_{e \in J} x_e - b \prod_{e \in J} x_e\right)$$

$$= \tilde{\mathbb{E}}\left(\sum_{f \in \delta(v) \setminus J} \prod_{e \in J \cup \{f\}} x_e + \sum_{f \in J} \prod_{e \in J} x_e x_f - b \prod_{e \in J} x_e\right) = \sum_{f \in \delta(v) \setminus J} \tilde{\mathbb{E}}\left(\prod_{e \in J \cup \{f\}} x_e\right).$$

Here we used $\tilde{\mathbb{E}}\left(\sum_{f \in J} \prod_{e \in J} x_e x_f\right) = \tilde{\mathbb{E}}\left(\sum_{f \in J} \prod_{e \in J} x_e\right) = \tilde{\mathbb{E}}\left(b \prod_{e \in J} x_e\right)$, and since $\tilde{\mathbb{E}}\left(\prod_{e \in J \cup \{f\}} x_e\right) = \tilde{\mathbb{E}}\left(\prod_{e \in J \cup \{f\}} x_e^2\right) \geq 0$, by the linearization property of the pseudoexpectation operator, it follows that $\tilde{\mathbb{E}}\left(\prod_{e \in J \cup \{f\}} x_e\right) = 0$.

The claim follows then by Lemma 4 and by noting that any set $I \subseteq \delta(v)$ of size $|I| > b$ contains a subset H of size $b + 1$ for which $\tilde{\mathbb{E}}\left(\prod_{e \in H} x_e\right) = 0$ by the above reasoning. □

Lemma 6. *For an r-uniform hypergraph $G = (V, E)$ the maximum b-matching is at most of size $\lfloor \frac{b|V|}{r} \rfloor$.*

Proof. Let V_M denote the vertices incident to some edge in a matching M. Let $\delta_M(v)$ denote the set of edges in the matching incident to a vertex $v \in V_M$. Each vertex $v \in V_M$ can be incident to at most b edges, so the number of edges can be counted as

$$|M| = \frac{1}{r} \sum_{v \in V_M} |\delta_M(v)| \leq \frac{|V_M| b}{r} \leq \frac{|V| b}{r}.$$

Since $|M|$ is an integer, we get that $|M| \leq \lfloor \frac{|V| b}{r} \rfloor$. It is easy to check that the bound is tight at least when the graph is complete and $b = r$. □

Using the above lemma we can show that for the matching problems considered in this paper the pseudoexpectation has the property required by Lemma 2. The following is the main result of the paper.

Theorem 1. *There is no feasible pseudoexpectation operator for the SoS relaxation of level greater than or equal to $t = \max\left\{b, \frac{1}{2}\lfloor \frac{b|V|}{r} \rfloor\right\}$ for (4) when c is not integral or greater than $2t + 1$.*

Proof. By Lemma 6, the size of a maximum b-matching in any r-uniform hypergraph $G = (V, E)$ is at most $\lfloor \frac{b|V|}{r} \rfloor$, so in any set of edges $S \subseteq E$ of size $\lfloor \frac{b|V|}{r} \rfloor + 1$ there must be a subset I such that $I \subseteq \delta(v)$ for some v and $|I| = b + 1$.

Then, since $t \geq b$ by Lemma 5, $\tilde{\mathbb{E}}\left(\prod_{e \in I} x_e\right) = 0$. Furthermore, $|S| \leq 2t + 1$ and so Lemma 4 implies that $\tilde{\mathbb{E}}\left(\prod_{e \in S} x_e\right) = 0$. Applying Lemma 2 proves the claim. □

3.2   Exact SoS Rank for the Matching Polytope of K_{4k+1}.

In [1] Au and Tunçel show that the SoS rank of (3) for K_{4k+1} is at least k. More precisely, they show that the following operator defines a SoS relaxation of (3) of level $k-1$

$$\tilde{\mathbb{E}}\left(\prod_{e \in I} x_e\right) = \begin{cases} \prod_{i=0}^{|I|-1} \frac{1}{4k-2i}, \text{whenever the } I \subseteq E \text{ is a matching} \\ 0, \quad \text{otherwise.} \end{cases} \tag{5}$$

In the following we show that the pseudoexpectation operator (5) is also a feasible solution to the SoS relaxation of the feasibility matching polytope (4) for a constant $c = 2k + \frac{1}{2}$, when $b = 1$ and $r = 2$. Note that this choice of c implies that every constraint of the form $\sum_{e \in \delta(v)} x_e \leq 1$ has to be satisfied with equality. Conversely, if we replace the inequalities with equalities, the constraint $\sum_{e \in E} x_e = 2k + \frac{1}{2}$ is implied.

Thus, if we are able to show that the pseudoexpectation (5) is feasible when the inequalities in (3) are replaced by equalities, then the pseudoexpectation is also feasible for (4) when $c = 2k + \frac{1}{2}$. Therefore, we need to show that $\tilde{\mathbb{E}}\left(u^2(x)(\sum_{e \in \delta(v)} x_e - 1)\right) = 0$ for every $v \in V$ and for every polynomial $u \in \mathbb{R}[x]_{k-1}$ for the pseudoexpectation defined in (5).

Lemma 7. *For the pseudoexpectation operator* (5)

$$\tilde{\mathbb{E}}\left(u^2(\sum_{e \in \delta(v)} x_e - 1)\right) = 0, \text{for every} v \in V \text{ for every} u \in \mathbb{R}[x]_{k-1}.$$

Proof. We prove that $\tilde{\mathbb{E}}\left(\prod_{i \in S} x_i(\sum_{e \in \delta(v)} x_e - 1)\right) = 0$ for any monomial $\prod_{i \in S} x_i \in \mathbb{R}[x]_{2k-2}$, which implies the claim. Consider any set $S \subseteq E, |S| \leq 2k-2$. Suppose S does not form a matching in K_{4k+1}. Then for every $v \in V$

$$\tilde{\mathbb{E}}\left(\prod_{i \in S} x_i(\sum_{e \in \delta(v)} x_e - 1)\right) = \tilde{\mathbb{E}}\left(\sum_{e \in \delta(v)} \prod_{i \in S \cup \{e\}} x_i\right) - \tilde{\mathbb{E}}\left(\prod_{i \in S} x_i\right) = 0$$

since for every $e \in \delta(v)$, $S \cup \{e\}$ does not form a matching.

Next suppose that S forms a matching in K_{4k+1}. This implies that for every $v \in V$, $|S \cap \delta(v)| \leq 1$. Assume first that the intersection is nonempty and let $S \cap \delta(v) = \{e'\}$. Then

$$\tilde{\mathbb{E}}\left(\prod_{i \in S} x_i(\sum_{e \in \delta(v)} x_e - 1)\right) = \tilde{\mathbb{E}}\left(\sum_{e \in \delta(v) \setminus \{e'\}} \prod_{i \in S \cup \{e\}} x_i\right) + \tilde{\mathbb{E}}\left(\prod_{i \in S \cup \{e'\}} x_i\right) - \tilde{\mathbb{E}}\left(\prod_{i \in S} x_i\right)$$

which is 0 since for every $e \in \delta(v) \setminus \{e'\}$, $S \cup \{e\}$ does not form a matching and thus $\tilde{\mathbb{E}}\left(\prod_{i \in S \cup \{e\}} x_i\right) = 0$. Furthermore, $S \cup \{e'\} = S$, so $\tilde{\mathbb{E}}\left(\prod_{i \in S \cup \{e'\}} x_i\right) = \tilde{\mathbb{E}}(\prod_{i \in S} x_i)$.

Finally, assume $S \cap \delta(v) = \emptyset$, and let $M \subseteq \delta(v)$ such that $e \in M$ if and only if $S \cup \{e\}$ forms a matching in K_{4k+1}. We get

$$\tilde{\mathbb{E}}\left(\prod_{i \in S} x_i \left(\sum_{e \in \delta(v)} x_e - 1 \right) \right) = \tilde{\mathbb{E}}\left(\sum_{e \in M} \prod_{i \in S \cup \{e\}} x_i \right) + \tilde{\mathbb{E}}\left(\sum_{e \in \delta(v) \setminus M} \prod_{i \in S \cup \{e\}} x_i \right) - \tilde{\mathbb{E}}\left(\prod_{i \in S} x_i \right)$$

which is 0 since for every $e \in \delta(v) \setminus M$, $\tilde{\mathbb{E}}\left(\prod_{i \in S \cup \{e\}} x_i \right) = 0$, and

$$\tilde{\mathbb{E}}\left(\sum_{e \in M} \prod_{i \in S \cup \{e\}} x_i \right) = (4k - 2|S|) \prod_{i=0}^{|S|} \frac{1}{4k - 2i} = \prod_{i=0}^{|S|-1} \frac{1}{4k - 2i} = \tilde{\mathbb{E}}\left(\prod_{i \in S} x_i \right).$$

Here we used the fact that $|M| + 2|S| = |\delta(v)| = 4k$. Note that the level bound for the relaxation is needed here, since it allows us to assume that $|S| \leq 2k - 2$. A larger level would permit the case $|S| = 2k$, and the last step of the proof would fail. □

The above lemma shows that (5) is a feasible level $k - 1$ pseudoexpectation operator for the SoS relaxation of (4) when the graph is K_{4k+1}. On the other hand, from Theorem 1 we get that when $b = 1$, $r = 2$, the rank is at most $\frac{1}{2} \lfloor \frac{|V|}{2} \rfloor = k$. We remark that for the graph K_{4k+3}, Lemma 7 is still true, but the upper bound for the rank from Theorem 1 is $k + 1$, implying that the rank is either k or $k + 1$.

4   An Open Question

It is known that the convergence of the SoS hierarchy to the convex hull of integral solutions for a given problem is sensitive to the problem formulation. In this paper we proved that for the complete graph K_{4k+1} the rank of the SoS relaxation of the formulation (4) for $b = 1, r = 2$ is exactly k. On the other hand, the following theorem is shown in [1].

Theorem 2 ([1]). *For the complete graph K_{2k+1} the rank of the SoS relaxation of the formulation (3) for $b = 1, r = 2$ is at least $\lfloor \frac{k}{2} \rfloor$ and at most k.*

It is natural that the stronger formulation (4) produces a potentially tighter relaxation than the relaxation of (3). However, we are able to show the following, which suggests that the SoS ranks of (4) and (3) might not be very different.

Lemma 8. *There is no feasible symmetric pseudoexpectation operator for the SoS relaxation of level $t = 1$ for (3) with $b - 1$ and $r = 2$ for the complete graph K_5 with the objective value greater than 2.*

In the above lemma the symmetry of the solution refers to the situation where the image of each monomial $\tilde{\mathbb{E}}(\prod_{i \in I} x_i)$ is either only dependent on the size of the set I or zero. In this sense, the solution (5) is symmetric.

Open question: Is the SoS rank of (3) different from the SoS rank of (4) with the parameters $b = 1, r = 2$ in the case of the complete graph?

References

1. Au, Y.H., Tunçel, L.: Complexity analyses of Bienstock–Zuckerberg and Lasserre relaxations on the matching and stable set polytopes. In: Günlük, O., Woeginger, G.J. (eds.) IPCO 2011. LNCS, vol. 6655, pp. 14–26. Springer, Heidelberg (2011)
2. Au, Y., Tunçel, L.: A comprehensive analysis of polyhedral lift-and-project methods (2013). CoRR, abs/1312.5972
3. Avis, D., Bremner, D., Tiwary, H.R., Watanabe, O.: Polynomial size linear programs for non-bipartite matching problems and other problems in P (2014). CoRR, abs/1408.0807
4. Barak, B., Kelner, J.A., Steurer, D.: Rounding sum-of-squares relaxations. In: STOC, pp. 31–40 (2014)
5. Braun, G., Brown-Cohen, J., Huq, A., Pokutta, S., Raghavendra, P., Roy, A., Weitz, B., Zink, D.: The matching problem has no small symmetric SDP. In: SODA, pp. 1067–1078 (2016)
6. Chan, S.O., Lee, J.R., Raghavendra, P., Steurer, D.: Approximate constraint satisfaction requires large LP relaxations. In: FOCS, pp. 350–359. IEEE Computer Society (2013)
7. Edmonds, J.: Paths, trees and flowers. Can. J. Math. **17**, 449–467 (1965)
8. Fawzi, H., Saunderson, J., Parrilo, P.A.: Sparse sums of squares on finite abelian groups and improved semidefinite lifts. Math. Program., 1–43 (2016)
9. Goemans, M.X., Tunçel, L.: When does the positive semidefiniteness constraint help in lifting procedures? Math. Oper. Res. **26**(4), 796–815 (2001)
10. Grigoriev, D.: Linear lower bound on degrees of Positivstellensatz calculus proofs for the parity. Theoret. Comput. Sci. **259**(1–2), 613–622 (2001)
11. Karlin, A.R., Mathieu, C., Nguyen, C.T.: Integrality gaps of linear and semidefinite programming relaxations for Knapsack. In: Günlük, O., Woeginger, G.J. (eds.) IPCO 2011. LNCS, vol. 6655, pp. 301–314. Springer, Heidelberg (2011)
12. Lasserre, J.B.: Global optimization with polynomials and the problem of moments. SIAM J. Optim. **11**(3), 796–817 (2001)
13. Laurent, M.: A comparison of the Sherali-Adams, Lovász-Schrijver, and Lasserre relaxations for 0–1 programming. Math. Oper. Res. **28**(3), 470–496 (2003)
14. Lee, J.R., Raghavendra, P., Steurer, D.: Lower bounds on the size of semidefinite programming relaxations. In: Servedio, R.A., Rubinfeld, R. (eds.) STOC, pp. 567–576. ACM (2015)
15. Lovász, L., Schrijver, A.: Cones of matrices and set-functions and 0–1 optimization. SIAM J. Optim. **1**(12), 166–190 (1991)
16. Mathieu, C., Sinclair, A.: Sherali-Adams relaxations of the matching polytope. In: STOC, pp. 293–302 (2009)
17. Nesterov, Y.: Global Quadratic Optimization via Conic Relaxation, pp. 363–384. Kluwer Academic Publishers, Dordrecht (2000)
18. O'Donnell, R., Zhou, Y.: Approximability and proof complexity. In: SODA, pp. 1537–1556 (2013)
19. Parrilo, P.: Structured semidefinite programs and semialgebraic geometry methods in robustness and optimization. Ph.D. thesis, California Institute of Technology (2000)
20. Rothvoß, T.: The Lasserre Hierarchy in Approximation Algorithms – Lecture Notes for the MAPSP 2013 Tutorial, June 2013
21. Rothvoß, T.: The matching polytope has exponential extension complexity. In: STOC, pp. 263–272 (2014)

22. Sherali, H.D., Adams, W.P.: A hierarchy of relaxations between the continuous and convex hull representations for zero-one programming problems. SIAM J. Discrete Math. **3**(3), 411–430 (1990)
23. Shor, N.: Class of global minimum bounds of polynomial functions. Cybernetics **23**(6), 731–734 (1987)
24. Stephen, T., Tunçel, L.: On a representation of the matching polytope via semidefinite liftings. Math. Oper. Res. **24**(1), 1–7 (1999)
25. Worah, P.: Rank bounds for a hierarchy of Lovász and Schrijver. J. Comb. Optim. **30**(3), 689–709 (2015)
26. Yannakakis, M.: Expressing combinatorial optimization problems by linear programs. J. Comput. Syst. Sci. **43**(3), 441–466 (1991)

A Novel SDP Relaxation for the Quadratic Assignment Problem Using Cut Pseudo Bases

Maximilian John$^{(\boxtimes)}$ and Andreas Karrenbauer

Max Planck Institute for Informatics, Saarbrücken, Germany
{maximilian.john,andreas.karrenbauer}@mpi-inf.mpg.de

Abstract. The quadratic assignment problem (QAP) is one of the hardest combinatorial optimization problems. Its range of applications is wide, including facility location, keyboard layout, and various other domains. The key success factor of specialized branch-and-bound frameworks for minimizing QAPs is an efficient implementation of a strong lower bound. In this paper, we propose a lower-bound-preserving transformation of a QAP to a different quadratic problem that allows for small and efficiently solvable SDP relaxations. This transformation is self-tightening in a branch-and-bound process.

Keywords: Quadratic assignment · Semidefinite program · Lower bound · Branch and bound

1 Introduction

Assignment problems are some of the best-studied problems in combinatorial optimization, the task being to find a one-to-one correspondence of n items and n locations, i.e., an

$$x \in \Pi_n := \left\{ X \in \mathbb{Z}^{n \times n} : \sum_{i=1}^{n} x_{ik} = 1 \, \forall k \in [n] \text{ and } \sum_{k=1}^{n} x_{ik} = 1 \, \forall i \in [n] \right\},$$

such that the total cost of the assignment is minimized. If the objective function is linear, i.e., of the form $\sum_{i,k} c_{ik} x_{ik}$, the optimum can be computed efficiently due to Birkhoff's theorem [1], e.g., with the Hungarian method [2] in $\mathcal{O}(n^3)$, or with the bipartite matching algorithm of Duan and Su [3] for integer costs of at most C in $O(n^{5/2} \log C)$. However, the linear objective function restricts the modeling power because it does not account for the interaction between the items nor for the interactions between the locations. Therefore, Koopmans and Beckmann investigated a variant of a quadratic objective function of the form $\sum_{i,j,k,\ell} c_{ijk\ell} x_{ik} x_{j\ell}$ that also considers pair-wise dependencies of the input objects [4]. In their variant, the cost factors into dependencies between items and dependencies between locations, respectively. That is, $c_{ijk\ell} = f_{ij} \cdot d_{k\ell}$, or $C = F \otimes D$ in matrix notation using the Kronecker product. This variant of the QAP offers

© Springer International Publishing Switzerland 2016
R. Cerulli et al. (Eds.): ISCO 2016, LNCS 9849, pp. 414–425, 2016.
DOI: 10.1007/978-3-319-45587-7_36

various practical applications such as the facility location problem [5] or the keyboard layout problem [6]. Moreover, it generalizes the traveling salesman problem [7] and several further real-life combinatorial problems such as the wiring problem [8] or hospital layout [9,10]. However, the QAP is very hard to solve even for small instances ($n \geq 30$). For example, the problem library QAPLIB [11] still contains decades-old unsolved instances and ones that were solved only recently by newly proposed techniques and/or the usage of massive computational power [12]. In the time of writing, its smallest unsolved instance consists of just 30 items and locations. On the theoretical side, Queyranne showed that the QAP is NP-hard to approximate within any constant factor, even if the cost can be factorized to a symmetric block diagonal flow matrix and a distance matrix describing the distances of a set of points on a line [13]. Detailed surveys on the Quadratic Assignment Problem can be found in [14–16].

A systematic approach for solving a QAP is to compute relaxations in a branch-and-bound framework. One of the earliest published lower bounds, the Gilmore-Lawler bound [17,18] for the Koopmans-Beckmann variant, reduces the problem to a linear assignment problem. However, it deteriorates quickly with increasing instance sizes [19]. On the other hand, already the first level of the reformulation linearization technique (RLT) by Frieze and Yadegar [20] produces strong lower bounds. But this comes at the expense of introducing n^4 many binary variables $y_{ijk\ell}$, i.e., one for each quadratic term occurring in the objective function. Thus, it takes a lot of resources (both in terms of CPU and RAM) to solve LP-relaxations for instances of practical input size. In contrast, the formulation of Kaufman and Broeckx [21] only contains $\mathcal{O}(n^2)$ variables, and thus, its LP-relaxation can be solved efficiently. Moreover, the primal heuristics of state-of-the-art MIP-solvers are able to quickly produce strong incumbents with this formulation. However, the lower bounds obtained by relaxing the integrality constraints of this formulation are very weak such that they often do not even surpass the trivial lower bound of $\sum_{i,j} \min\{c_{ijk\ell} : k, \ell \in [n]\}$ in reasonable time, which makes it impractical to use this formulation alone to close the gap between upper and lower bounds in a branch-and-bound process.

Furthermore, there are various relaxations of the QAP as a semidefinite program (SDP). For example, SDP-relaxations for the non-convex constraint $Y = X \otimes X$ were introduced in [22,23]. Recent approaches (e.g., [24]) have shown that these approaches can often efficiently produce good lower bounds for the QAP and beat common linear relaxations. We follow a different approach since we do not derive our SDP from this formulation, but transform the QAP to a different quadratic problem beforehand. In that sense, our approach is somewhat orthogonal to recent other SDP relaxations.

1.1 Our Contribution

In this paper, we propose a novel SDP derived from a lower-bound-preserving transformation of a QAP instance to an auxiliary quadratic minimization problem with only $\mathcal{O}(n \log n)$ variables. SDP-relaxations with that few variables can be solved efficiently with modern interior point methods for conic optimization problems. Moreover, it is straight forward to integrate our relaxation in a branch-and-bound framework. While branching on single assignment variables typically results in very unbalanced branch-and-bound trees, our approach avoids this by design. To this end, we introduce the concept of cut pseudo bases, which has not been used — to the best of our knowledge — in this context before. Our goal was to develop an approach that still works with limited computational resources, e.g., on a laptop, for the cases when the lower bounds provided by Kaufman-Broeckx are too weak and when it is already infeasible to solve the LP-relaxation of RLT1. Furthermore, we present experimental results for instances with $n \geq 25$ in which we outperform both lower bounds mentioned above in terms of efficiency and effectiveness. The bounds produced by our SDP always exceed — just by construction — the trivial lower bound mentioned above.

2   A Novel Lower Bound Using SDP

Let n denote, throughout this paper, the respective number of items and locations. We assume for the sake of presentation that n is a power of 2. This is not a restriction because we can pad n with dummy items and locations. Moreover, the dummy items can be projected out easily in an implementation so that this also does not harm its performance.

The derivation is done in two steps. First, we design a new quadratic program that lower bounds the QAP and allows for a balanced branching tree. In the second step, we relax the new problem to an SDP.

Concerning the goal of achieving balanced branching trees, we revisit the well-known problem of branching on single assignment variables. Setting x_{ik} to 1 means fixing item i to location j, which is a very strong decision that affects all other variables in the i-th row or k-th column, forcing them to 0. On the other hand, if we set x_{ik} to 0, we just decide not to fix i to j. However, there are still $n - 1$ other possible locations for i, so we basically did not decide much. This yields highly imbalanced branching trees as it is much more likely to prune in the 1-branches of a branch-and-bound process. This undesirable effect can be avoided by the well-known idea of generalized upper-bound branching (see Sect. 7 of [25]). Inspired by this, we consider a similar approach illustrated in the following IP formulation with n auxiliary z-variables:

$$\text{minimize} \quad \sum_{i,j,k,\ell=1}^{n} c_{ijk\ell} x_{ik} x_{jl}$$

$$\text{s.t.} \quad \sum_{i=1}^{n} x_{ik} = 1 \qquad \qquad \forall k \in [n]$$

$$\sum_{k=1}^{n/2} x_{ik} = z_i \qquad \sum_{k=n/2+1}^{n} x_{ik} = 1 - z_i \quad \forall i \in [n]$$

$$x_{ik} \in \{0,1\} \qquad \qquad \forall i,k \in [n]$$

$$z_i \in \{0,1\} \qquad \qquad \forall i \in [n].$$

If we branch on the z-variables instead of the assignment variables, our branching tree is much more likely to be balanced because either choice is equally strong. However, it is not sufficient to branch only on these z-variables because too many degrees of freedom still remain open even after all z-variables are set. If we want to completely determine the x-variables and thus be able to project them out, we should introduce further binary z-variables. To this end, we introduce *cut pseudo bases*.

2.1 Introduction of Cut Pseudo Bases

The key idea of cut pseudo bases is the usage of cuts in the complete graph with the locations as nodes. Consider a balanced subset of the nodes, i.e., one of size $n/2$. Instead of assigning an item to a certain location, we now assign it to one of the "halves" of the location space. We repeat this cutting of the location space until we reach a state where — after a finite number of assignments — every item can be uniquely mapped to a single location. Moreover, we cut the space in a balanced way, i.e., we require that each side of the cut is equally large. Let us formalize these requirements.

Definition 1. *A set of cuts over the location space such that*

- *all cuts are balanced,*
- *all singleton locations can be expressed by a linear combination of cuts, and*
- *it is inclusion-wise minimal*

is called a cut pseudo base.

Clearly, the size of a cut pseudo base is $\log_2 n$ when n is a power of 2 and thus $\lceil \log_2 n \rceil$ in general by the padding argument. To illustrate the concept of a pseudo base, consider the following example.

Example 1. Enumerate the $n = 2^k$ locations by $0, \ldots, n-1$, and consider the binary decomposition of these numbers. For every bit $b = 0, \ldots, k-1$, we define a cut that separates all locations with numbers differing in the b-th bit. Then, this collection of cuts forms a cut pseudo base.

Note that any arbitrary cut pseudo base can be transformed to the binary decomposition pseudo base by a permutation of the locations. Hence, we will employ this cut pseudo base as a reference throughout this paper for the sake of presentation and simplicity.

2.2 Exchanging Assignment Variables by Cut Variables

The cut pseudo bases introduced in the previous subsection are balanced by definition, meaning that assigning an item to one side of a cut in the pseudo base is just as effective as assigning it to the other side. However, the decision of whether a particular item should be assigned to a fixed location is highly unbalanced, as we have already discussed. Hence, our goal is to get rid of the assignment variables and introduce the cut variables instead. This will also benefit the number of binary variables which decreases from n^2 assignment variables to $n \log_2 n$ cut variables. Let $(S_b)_{b \in [B]}$ be an arbitrary but fixed cut pseudo base. Note that $B = \lceil \log_2 n \rceil$, otherwise, (S_b) cannot be a cut pseudo base. We introduce the variables $z_i^b \in \{0, 1\}$ for every cut S_b, indicating whether i is assigned to the 0- or 1-side of the cut S_b, i.e., to the outside or the inside, respectively. We relate them to the assignment variables in the following manner.

$$x_{ij} = 1 \Leftrightarrow \forall b \in [B] \ j \in z_i^b\text{-side of cut } S_b$$

We consider an arbitrary cut b in the following and omit the superscript b to simplify the notation and thereby improve readability. Observe that

$$z_i z_j + z_i(1 - z_j) + (1 - z_i)z_j + (1 - z_i)(1 - z_j) = 1$$

holds for any $z_i, z_j \in \mathbb{R}$ and that for a binary solution exactly one of the four terms is 1, and the others vanish.

Thus, we obtain for any assignment x and the corresponding binary z-variables that

$\sum_{\substack{i,j \\ k,\ell}} c_{ijk\ell} x_{ik} x_{j\ell}$

$= \sum_{i,j} [z_i z_j + z_i(1 - z_j) + (1 - z_i)z_j + (1 - z_i)(1 - z_j)] \sum_{k,\ell} c_{ijk\ell} x_{ik} x_{j\ell}$

$\geq \sum_{i,j} z_i z_j \qquad\qquad \cdot \min\{\sum_{k,\ell} c_{ijk\ell} x_{ik} x_{j\ell} : x \in \Pi_{ij}^{(11)}\}$

$+ \sum_{i,j} z_i(1 - z_j) \qquad \cdot \min\{\sum_{k,\ell} c_{ijk\ell} x_{ik} x_{j\ell} : x \in \Pi_{ij}^{(10)}\}$

$+ \sum_{i,j} (1 - z_i)z_j \qquad \cdot \min\{\sum_{k,\ell} c_{ijk\ell} x_{ik} x_{j\ell} : x \in \Pi_{ij}^{(01)}\}$

$+ \sum_{i,j} (1 - z_i)(1 - z_j) \cdot \min\{\sum_{k,\ell} c_{ijk\ell} x_{ik} x_{j\ell} : x \in \Pi_{ij}^{(00)}\}$

where $\Pi_{ij}^{(11)}$ denotes the set of all assignments in which i and j are both assigned inside the cut, where $\Pi_{ij}^{(10)}$ denotes the set of all assignments in which i is assigned inside the cut and j is assigned to the outside, and so on. In the following, we argue that this is indeed a valid lower bound. To this end, let $c_{ij}^{(\alpha\beta)} := \min\{\sum_{k,\ell} c_{ijk\ell} x_{ik} x_{j\ell} : x \in \Pi_{ij}^{(\alpha\beta)}\}$ denote the optimum objective values of the corresponding optimization problems for $\alpha, \beta \in \{0, 1\}$, and observe that $c_{ij}^{(\alpha\beta)}$ only contributes to the right-hand side if $z_i = \alpha$ and $z_j = \beta$.

This yields an objective function that is free of x-variables. Furthermore, the minimum of the original objective taken over all $x \in \Pi$ is bounded from below by the minimum over all z that determine an assignment.

At first glance, it seems that we have to solve $4n^2$ QAPs to compute the coefficients for the new objective function. However, a close inspection of the subproblems reveals that $c_{ij}^{(\alpha\beta)}$ is determined by the minimum $c_{ijk\ell}$ over all k, ℓ such that the b-th bits of k and ℓ are α and β, respectively. This can be computed efficiently for each pair ij by a single scan over all $c_{ijk\ell}$. Note that in the Koopmans-Beckmann variant of a QAP, we have $c_{ijk\ell} = f_{ij} \cdot d_{k\ell}$, and thus, it suffices to scan over the distance pairs $d_{k\ell}$ of the locations k and ℓ. Furthermore, such a single scan can also take additional constraints into account, e.g., excluded pairs due to a branching process. Hence, the lower bound of our approach is self-tightening in a branch-and-bound process. In every branching step, we can update our cost estimation for this particular setting of excluded pairs.

2.3 Towards an SDP

In order to obtain a reasonable SDP relaxation, we apply the typical transformation to map $\{0, 1\}$-variables to $\{-1, 1\}$-variables. That is, we use the linear transformation $z_i = \frac{1+y_i}{2}$. This implies that $1 - z_i = \frac{1-y_i}{2}$. Plugging this into

$$c_{ij}^{(11)} z_i z_j + c_{ij}^{(10)} z_i (1 - z_j) + c_{ij}^{(01)} (1 - z_i) z_j + c_{ij}^{(00)} (1 - z_i)(1 - z_j)$$

yields

$$\sum_{\alpha,\beta=0}^{1} c_{ij}^{(\alpha\beta)} \cdot \frac{1-(-1)^\alpha y_i}{2} \cdot \frac{1-(-1)^\beta y_j}{2} = \sum_{\alpha,\beta=0}^{1} c_{ij}^{(\alpha\beta)} \cdot \frac{1-(-1)^\alpha y_i - (-1)^\beta y_j + (-1)^{\alpha+\beta} y_i y_j}{4}$$

$$= \frac{c_{ij}^{(11)} + c_{ij}^{(10)} + c_{ij}^{(01)} + c_{ij}^{(00)}}{4} + \frac{c_{ij}^{(11)} + c_{ij}^{(10)} - c_{ij}^{(01)} - c_{ij}^{(00)}}{4} \cdot y_i$$

$$+ \frac{c_{ij}^{(11)} - c_{ij}^{(10)} + c_{ij}^{(01)} - c_{ij}^{(00)}}{4} \cdot y_j + \frac{c_{ij}^{(11)} - c_{ij}^{(10)} - c_{ij}^{(01)} + c_{ij}^{(00)}}{4} \cdot y_i y_j.$$

We separate and symmetrize the constant, linear, and quadratic terms such that we can write the total sum over all i, j in matrix-vector notation as

$$y^T C y + c^T y + \gamma$$

with

$$C_{ij} := \frac{c_{ij}^{(11)} + c_{ji}^{(11)} - c_{ij}^{(10)} - c_{ji}^{(10)} - c_{ij}^{(01)} - c_{ji}^{(01)} + c_{ij}^{(00)} + c_{ji}^{(00)}}{8}$$

$$c_i := \frac{1}{4} \sum_{j=1}^{n} c_{ij}^{(11)} + c_{ji}^{(11)} + c_{ij}^{(10)} - c_{ji}^{(10)} - c_{ij}^{(01)} + c_{ji}^{(01)} - c_{ij}^{(00)} - c_{ji}^{(00)}.$$

$$\gamma := \frac{1}{4} \sum_{i=1}^{n} \sum_{j=1}^{n} c_{ij}^{(11)} + c_{ij}^{(10)} + c_{ij}^{(01)} + c_{ij}^{(00)}.$$

To relax the quadratic part in the objective using a semidefinite matrix, we use a standard fact about the trace, i.e., $y^T C y = tr(y^T C y) = tr(C y y^T)$. Thus, we replace the quadratic term $y^T C y$ in the objective function by the Frobenius product[1] $C \bullet Y$ and hope that $Y = y y^T$. However, such a rank-1-constraint is not convex, and we relax it to $Y \succcurlyeq y y^T$, which means that $Y - y y^T$ is positive semidefinite. Since the latter is a Schur complement, this condition is equivalent to

$$\begin{pmatrix} 1 & y^T \\ y & Y \end{pmatrix} \succcurlyeq 0.$$

---

[1] The Frobenius product $A \bullet B := tr(A^T B) = \sum_{i,j} a_{ij} b_{ij}$ is the standard inner product on the space of $n \times n$ matrices used in semi-definite programming.

To accomplish this, we could augment the matrix Y by a 0-th row and column, or we could also use an item that has already been fixed w.r.t. the side of the cut b under consideration, e.g., use one of the dummy items introduced to fill up the number of items to a power of 2. That is, if $y_i = 1$ is already fix for some item i, we may require $Y_{ij} = y_j$ for all j and $Y \succeq 0$. The former constraint can be written as $e_i e_j^T \bullet Y - e_j^T y = 0$, which modern SDP solvers such as Mosek [26] directly allow without a transformation to an equivalent block-diagonal pure SDP formulation. If $y_i = -1$, we obtain the constraints $e_i e_j^T \bullet Y + e_j^T y = 0$ for all items j instead.

In the following, we list further constraints that we may add to the SDP to improve the strength of the lower bound on the QAP. Recall that we omitted any superscripts to identify the cut under consideration. However, we will argue with the complete cut pseudo base in the following, so we use Y^b for the matrix corresponding to cut b and y^b to identify the linear terms corresponding to this cut. Similarly, we shall use C^b, c^b, and γ^b to denote the corresponding parts in the objective function. We emphasize again that the cut pseudo base in use is fixed and contains $B = \lceil \log 2(n) \rceil$ many cuts.

Domain of Y. We make sure that every $y_i^b \in \{-1, 1\}$. For the linear variables, we relax this constraint to $y_i^b \in [-1, 1]$, but in the SDP, we can require something stronger. By using the fact that $\left(y_i^b\right)^2 = 1$, we can add the constraint $Y_{ii}^b = 1$ for all $b \in [B], i \in [n]$. Formally, we do this by the SDP constraint $E_i \bullet Y^b = 1$ where E_i has a 1 on index (i, i) and 0s everywhere else.

Injectivity of the Assignment. We ensure that the assignment is injective, i.e., that no two different keys are assigned to the same spot. In terms of y variables, we require for all distinct i and j that y_i^b be different from y_j^b for at least one b. We have $y_i^b = y_j^b$ if and only if the corresponding entry in Y, namely Y_{ij}^b, is 1. Hence, we add the constraint

$$\sum_{b=1}^{B} Y_{ij}^b \leq B - 1 \Leftrightarrow \left(\sum_{b=1}^{B} \frac{1}{2} Y_{ij}^b + \frac{1}{2} Y_{ji}^b\right) \leq B - 1.$$

Note that in an integer optimal solution, the constraints above already ensure all the properties, we want to have. However, we have found that it is beneficial for the relaxed SDP to add the following constraint.

Zero Row Sums. In the original formulation, injectivity implies that the number of keys assigned to one side of a cut is as large as the number of keys assigned to the opposite side. Recall that we are assuming $n = 2^k$, and we have a cut pseudo base. Hence, the implication above indeed holds. In terms of y variables, this can be modeled as the constraint

$$\sum_{j=1}^{n} y_j^b = 0$$

or as

$$\sum_{j=1}^{n} Y_{ij}^{b} = \sum_{j=1}^{n} y_i^b y_j^b = y_b^i \cdot \sum_{j=1}^{n} y_j^b \overset{!}{=} 0$$

in the SDP for an arbitrary fixed $i \in [n]$. Hence, taking the row sum of Y^b in this case yields the term we are looking for.

Total Entry Sum. We have observed that we can condense the zero-sum-constraints to a single one by exploiting the positive semidefiniteness of Y.

Lemma 1. *Let $Y \in \mathbb{R}^{n \times n}$ be positive semidefinite. If $\mathbb{1}\mathbb{1}^T \bullet Y = 0$, then for any $i \in [n]$, it holds that $\sum_{j=1}^{n} Y_{ij}^b = 0$.*

Proof. Observe that
$$0 = \mathbb{1}\mathbb{1}^T \bullet Y = \mathbb{1}^T Y \mathbb{1}.$$

Since Y is positive semidefinite, $\mathbb{1}$ is an eigenvector of Y with eigenvalue 0, which implies that $Y\mathbb{1} = 0\mathbb{1} = \mathbf{0}$ and proves the claim.

Hence, instead of imposing n constraints for every single row of Y, we have shown that one constraint is enough to fix all row sums to 0.

2.4   Alternative Objective Functions for the SDP

In the previous subsection, we first fixed some cut b and then derived a lower bound on the minimum QAP objective value by minimizing an SDP relaxation. That is, we obtain a valid lower bound by solving an SDP with the objective function $C^b \bullet Y^b + (c^b)^T y^b + \gamma^b$, subject to the constraints mentioned above. However, considering only one cut of the pseudo base in the objective could be weak because costs could be evaded by charging them to the other cuts of the pseudo base that are not accounted for in the objective.

Averaging over the Cut Pseudo Base. Since the lower bound holds for arbitrary cuts b, it also holds for the average over all cuts in the cut pseudo base, i.e., the objective becomes

$$\frac{1}{B} \sum_{b=1}^{B} C^b \bullet Y^b + (c^b)^T y^b + \gamma^b.$$

There is no need to add further auxiliary variables or constraints that may harm the numeric stability of an SDP-solver.

Taking the Maximum. An even stronger lower bound is obtained by taking the maximum over the cuts of the pseudo base because the arithmetic mean never exceeds the maximum. The standard way to model the maximum over the cut pseudo base is to introduce a new linear variable - say z - and add $\log_2 n$ many constraints, ensuring that z is at least the cost of each cut in the pseudo base. However, the Mosek solver (v7.1.0.53) often stalled with this objective function, in contrast to the averaging objective.

We are now ready to plug this SDP into a branch-and-bound framework of the QAP. Recall that we branch on the cut variables. However, as soon as we encounter an integral solution, there is a one-to-one-correspondence between the cut variables and the original assignment variables. This means, the cost of every incumbent is calculated with the original formulation. Hence, the framework will produce an optimal solution for any general QAP.

3   Evaluation

We compare our approach to two classical linearizations, the Kaufman-Broeckx linearization [21] and the first level of the Reformulation Linearization Technique (RLT1) [20]. We use the commercial state-of-the-art solver Gurobi (v6.5.1) [27] to solve the linearizations, and we use Mosek (v7.1.0.53) [26] as the SDP solver. All three approaches are embedded in a branch-and-bound framework. We will report the best known lower bound produced by running the branch-and-bound process for one hour.

We ran experiments on a compute server restricted to one Intel (R) Xeon (R) E5-2680 2.50 GHz core and a limited amount of 8 GB RAM running Debian GNU/Linux 7 with kernel 3.18.27.1. The code was compiled with gcc version 4.7.2 using the -O3 flag. The instances are taken from the QAPLIB homepage [11]. The names of the instances are formed by the name of the author (first three letters), the number of items, followed by a single letter identifier. The test instances cover a wide range of QAP applications including keyboard assignment, hospital layout and several further graph problems.

Figure 1 shows the average lower bound of the different approaches after one hour of computation time. We decided to average the lower bounds of a certain instance set because the single test cases within that set were similar and all approaches behaved consistently there. One can see that RLT1 performs quite well if we have enough computation power to compute bounds there (see tai or had, for example). However, many test instances are too large for our computing resources to compute even the RLT1 root relaxation. In these *hard* cases, our approach outperforms both linear relaxations by several orders of magnitude.

The nug instances are a special instance set because this set contains test cases of increasing size. Therefore, the behavior of the three approaches varies throughout the different test cases and the average reported in Fig. 1 cannot reflect the overall behavior for nug. To this end, we report a detailed description of the whole nug test set in Fig. 2. This also shows how our framework scales with increasing n. At the beginning, for small n, RLT1 can solve the instances

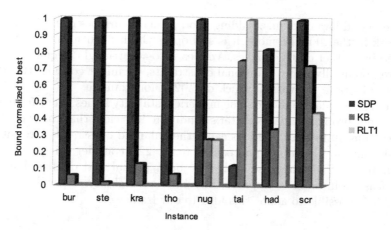

Fig. 1. Averaged QAPLIB instances after one hour of computation (Color figure online)

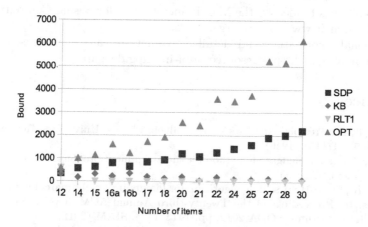

Fig. 2. The Nugent instances with 12 to 30 items after one hour of computation, RLT1 fails to solve $n \geq 15$ within this time and resource limit.

even to optimality, which meets our expectations. For instances of small size, the advantages of our approach in efficiency are just too small to make up for the loss of precision caused by the distance approximations. The trend changes as soon as n grows above 16. The RLT1 formulations are already too large to solve the root relaxation after one hour of computation with a single thread and the bounds of the Kaufman-Broeckx relaxation are lower than ours. This confirms the use-case that we proposed in the introduction of this paper.

4   Future Work

We plan to work on several heuristics to produce better incumbents during the branching, e.g., using randomized rounding. The current version fully focuses on producing good lower bounds. Adding good primal heuristics could improve the framework further. Furthermore, it is still open whether our formulation can be improved by further cutting planes. An idea is to add triangle inequalities. Since these are potentially many additional constraints, we might consider the idea of dynamic constraint activation, which does [28], for example.

Moreover, we plan to tackle some numerical stability issues in our framework. Although our problem satisfies Slater's condition, it is clear that the primal SDP never contains an interior point (because the balanced constraints force at least one eigenvalue to 0). Recent approaches (e.g., [22]) consider this problem and reformulate the SDP they use such that both primal and dual problems are strictly feasible in order to improve the numerical stability. We will investigate whether it is possible to adapt this idea to our framework.

5   Acknowledgments and Supplementary Material

This research was funded by the Max Planck Center for Visual Computing and Communication (www.mpc-vcc.org).

Additional information and detailed evaluation data are released on the project homepage at http://resources.mpi-inf.mpg.de/qap/.

References

1. Birkhoff, D.: Tres observaciones sobre el algebra lineal. Univ. Nac. Tucuman Rev. Ser. A **5**, 147–151 (1946)
2. Kuhn, H.W.: The hungarian method for the assignment problem. Naval Res. Logistics Q. **2**, 83–97 (1955)
3. Duan, R., Su, H.H.: A scaling algorithm for maximum weight matching in bipartite graphs. In: Proceedings of the Twenty-third Annual ACM-SIAM Symposium on Discrete Algorithms, SODA 2012, pp. 1413–1424. SIAM (2012)
4. Koopmans, T., Beckmann, M.J.: Assignment problems and the location of economic activities. Cowles Foundation Discussion Papers 4, Cowles Foundation for Research in Economics, Yale University (1955)
5. Nugent, C., Vollman, T., Ruml, J.: An experimental comparison of techniques for the assignment of facilities to locations. Oper. Res. **16**, 150–173 (1968)
6. Burkard, R., Offermann, J.: Entwurf von Schreibmaschinentastaturen mittels quadratischer Zuordnungsprobleme. Z. Oper. Res. **21**, 121–132 (1977)
7. Burkard, R.E., Çela, E., Pardalos, P.M., Pitsoulis, L.S.: The Quadratic Assignment Problem. Springer, Heidelberg (1998)
8. Steinberg, L.: The backboard wiring problem: a placement algorithm. SIAM Rev. **3**, 37–50 (1961)
9. Krarup, J., Pruzan, P.M.: Computer-aided layout design. In: Balinski, M.L., Lemarechal, C. (eds.) Mathematical Programming in Use. Mathematical Programming Studies, vol. 9, pp. 75–94. Springer, Heidelberg (1978)
10. Elshafei, A.N.: Hospital layout as a quadratic assignment problem. Oper. Res. Q. (1970–1977) **28**, 167–179 (1977)
11. Burkard, R.E., Karisch, S.E., Rendl, F.: Qaplib - a quadratic assignment problem-library. J. Glob. Optim. **10**, 391–403 (1997)

12. Anstreicher, K., Brixius, N., Goux, J.P., Linderoth, J.: Solving large quadratic assignment problems on computational grids. Math. Program. **91**, 563–588 (2014)
13. Queyranne, M.: Performance ratio of polynomial heuristics for triangle inequality quadratic assignment problems. Oper. Res. Lett. **4**, 231–234 (1986)
14. Pardalos, P.M., Rendl, F., Wolkowicz, H.: The quadratic assignment problem: a survey and recent developments. In: Proceedings of the DIMACS Workshop on Quadratic Assignment Problems. DIMACS Series in Discrete Mathematics and Theoretical Computer Science, vol. 16, pp. 1–42. American Mathematical Society (1994)
15. Commander, C.W.: A survey of the quadratic assignment problem, with applications. Morehead Electron. J. Appl. Math. **4**, 1–15 (2005). MATH-2005-01
16. Loiola, E.M., de Abreu, N.M.M., Boaventura-Netto, P.O., Hahn, P., Querido, T.: A survey for the quadratic assignment problem. Eur. J. Oper. Res. **176**, 657–690 (2007)
17. Gilmore, P.C.: Optimal and suboptimal algorithms for the quadratic assignment problem. SIAM J. Appl. Math. **10**, 305–313 (1962)
18. Lawler, E.L.: The quadratic assignment problem. Manage. Sci. **9**, 586–599 (1963)
19. Li, Y., Pardalos, P.M., Ramakrishnan, K.G., Resende, M.G.C.: Lower bounds for the quadratic assignment problem. Ann. Oper. Res. **50**, 387–410 (1994)
20. Frieze, A., Yadegar, J.: On the quadratic assignment problem. Discrete Appl. Math. **5**, 89–98 (1983)
21. Kaufman, L., Broeckx, F.: An algorithm for the quadratic assignment problem using Benders' decomposition. Eur. J. Oper. Res. **2**, 204–211 (1978)
22. Zhao, Q., Karisch, S.E., Rendl, F., Wolkowicz, H.: Semidefinite programming relaxations for the quadratic assignment problem. J. Comb. Optim. **2**, 71–109 (1998)
23. Povh, J., Rendl, F.: Copositive and semidefinite relaxations of the quadratic assignment problem. Discret. Optim. **6**, 231–241 (2009)
24. Peng, J., Mittelmann, H., Li, X.: A new relaxation framework for quadratic assignment problems based on matrix splitting. Math. Program. Comput. **2**, 59–77 (2010)
25. Wolsey, L.A.: Integer Programming. Wiley-Interscience Series in Discrete Mathematics and Optimization. Wiley, New York (1998). A Wiley-Interscience Publication
26. ApS, M.: The MOSEK C optimizer API manual Version 7.1 (Revision 52) (2016)
27. Gurobi Optimization, I.: Gurobi optimizer reference manual (2016)
28. Rendl, F., Rinaldi, G., Wiegele, A.: Solving max-cut to optimality by intersecting semidefinite and polyhedral relaxations. Math. Program. **121**, 307–335 (2008)

The Maximum Matrix Contraction Problem

Dimitri Watel[1,2(✉)] and Pierre-Louis Poirion[1,3]

[1] CEDRIC-CNAM, 292 Rue du Faubourg Saint Martin, 75003 Paris, France
dimitri.watel@ensiie.fr, pierre-louis.poirion@ensta-paristech.fr
[2] ENSIIE, 1 Square de la Résistance, Evry, France
[3] ENSTA Paristech, Evry, France

Abstract. In this paper, we introduce the *Maximum Matrix Contraction problem*, where we aim to contract as much as possible a binary matrix in order to maximize its density. We study the complexity and the polynomial approximability of the problem. Especially, we prove this problem to be NP-Complete and that every algorithm solving this problem is at most a $2\sqrt{n}$-approximation algorithm where n is the number of ones in the matrix. We then focus on efficient algorithms to solve the problem: an integer linear program and three heuristics.

Keywords: Complexity · Approximation algorithm · Linear programming

1 Introduction

In this paper, we are given a two dimensional array in which some entries contain a dot and others are empty. Two lines i and $i+1$ of the grid can be contracted by shifting up every dot of line $i+1$ and of every line after. Two columns j and $j+1$ of the grid can be contracted by shifting left the corresponding dots. However, such a contraction is not allowed if two dots are brought into the same entry. The purpose is maximize the number of neighbor pairs of dots (including the diagonal ones). An illustration is given in Fig. 1.

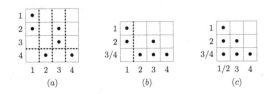

Fig. 1. In Fig. 1.a, we give a 4×4 grid containing 6 dots. Valid contractions are represented by dotted lines and columns. It is not allowed to contract lines 1 and 2 because the two dots (1;1) and (2;1) would be brought into the same entry. Figure 1.b is the result of the contraction of lines 3 and 4 and Fig. 1.c is the contraction of columns 1 and 2. The number of neighbor pairs in each grid is respectively 4, 7 and 10.

© Springer International Publishing Switzerland 2016
R. Cerulli et al. (Eds.): ISCO 2016, LNCS 9849, pp. 426–438, 2016.
DOI: 10.1007/978-3-319-45587-7_37

Motivation. This problem has an application in optimal sizing of wind-farms [1] where we must first define, from a given set of wind-farms location, the neighborhood graph of this set, i.e. the graph such that two wind farms are connected if and only if their corresponding entries in the grid are neighbors. More precisely, given a set of points in the plane, we consider a first grid-embedding such that any two points are at least separated by one vertical line and one horizontal line. Then, we consider the problem of deciding which lines and columns to contract such that the derived embedding maximize the density of the grid, i.e., the number of edges in the corresponding neighbor graph.

Contributions. In this paper, we formally define the grid contraction problem as a binary matrix contraction problem in which every dot is a 1 and every other entry is 0. We study the complexity and the polynomial approximability of the problem. Especially, we prove this problem to be NP-Complete. Nonetheless, every algorithm solving this problem is at most a $2\sqrt{n}$-approximation algorithm where n is the number of 1 in the matrix. We then focus on efficient algorithms to solve the problem. We first investigate the mathematical programming formulation of MMC. We give two formulations: a straightforward non-linear program and a linear program. Secondly, we describe three polynomial heuristics for the problem. We finally give numerical tests to compare the performances of the linear program and each algorithm.

In Sect. 2, we give a formal definition of the problem. In Sect. 3, we prove that the corresponding decision problem is NP-complete, then we give, in Sect. 4 some results about approximability of the problem. In Sect. 5 we derive a linear integer program for the model and run some experiments, then in the next section, we present and compare the three different heuristics.

2 Problem Definition

The following definitions formalize the problem we want to solve with binary matrices. A binary matrix is a matrix with entries from $\{0, 1\}$. Such a matrix modelizes a grid in which each dot is a 1 in the matrix.

Definition 1. *Let M be a binary matrix with p lines and q columns. For each $i \in [\![1; p-1]\!]^1$ and each $j \in [\![1; q-1]\!]$, we define the line contraction matrix L_i and the column contraction matrix C_j by*

$$
L_i = \begin{pmatrix}
1 & 0 & \cdots & 0 & 0 & 0 & 0 & 0 & \cdots & 0 \\
0 & 1 & \cdots & 0 & 0 & 0 & 0 & 0 & \cdots & 0 \\
\vdots & \vdots & \ddots & \vdots & \vdots & \vdots & \vdots & \vdots & \ddots & \vdots \\
0 & 0 & \cdots & 1 & 0 & 0 & 0 & 0 & \cdots & 0 \\
0 & 0 & \cdots & 0 & 1 & 1 & 0 & 0 & \cdots & 0 \\
0 & 0 & \cdots & 0 & 0 & 0 & 1 & 0 & \cdots & 0 \\
0 & 0 & \cdots & 0 & 0 & 0 & 0 & 1 & \cdots & 0 \\
\vdots & \vdots & \ddots & \vdots & \vdots & \vdots & \vdots & \vdots & \ddots & \vdots \\
0 & 0 & \cdots & 0 & 0 & 0 & 0 & 0 & \cdots & 1 \\
0 & 0 & \cdots & 0 & 0 & 0 & 0 & 0 & \cdots & 0
\end{pmatrix}
\qquad
C_j = \begin{pmatrix}
1 & 0 & \cdots & 0 & 0 & 0 & 0 & \cdots & 0 & 0 \\
0 & 1 & \cdots & 0 & 0 & 0 & 0 & \cdots & 0 & 0 \\
\vdots & \vdots & \ddots & \vdots & \vdots & \vdots & \vdots & & \vdots & \vdots \\
0 & 0 & \cdots & 1 & 0 & 0 & 0 & \cdots & 0 & 0 \\
0 & 0 & \cdots & 0 & 1 & 0 & 0 & \cdots & 0 & 0 \\
0 & 0 & \cdots & 0 & 1 & 0 & 0 & \cdots & 0 & 0 \\
0 & 0 & \cdots & 0 & 0 & 1 & 0 & \cdots & 0 & 0 \\
0 & 0 & \cdots & 0 & 0 & 0 & 1 & \cdots & 0 & 0 \\
\vdots & \vdots & \ddots & \vdots & \vdots & \vdots & \vdots & & \vdots & \vdots \\
0 & 0 & \cdots & 0 & 0 & 0 & 0 & \cdots & 1 & 0
\end{pmatrix}.
$$

---

[1] The meaning of $[\![p; q]\!]$ is the list $[p, p+1, \ldots, q]$.

The size of L_i is $p \times p$ and the size of C_j is $q \times q$.

Definition 2. *Let M be a binary matrix of size $p \times q$, $I = [i_1, i_2, \ldots, i_{|I|}]$ a sublist of $[\![1; p-1]\!]$ and $J = [j_1, j_2, \ldots, j_{|I|}]$ a sublist of $[\![1; q-1]\!]$. We assume I and J are sorted. We define the* contraction $C(M, I, J)$ *of the lines I and the columns J of M by the following matrix*

$$C(M, I, J) = \left(\prod_{k=1}^{|I|} L_{i_k} \right) \cdot M \cdot \left(\prod_{k=|J|}^{1} C_{j_k} \right).$$

Example 1. Let M be the matrix corresponding to the grid of Fig. 1.a. The following contraction gives the grid 1.c:

$$C(M, [3], [1]) = L_3 \cdot M \cdot C_1 = \begin{pmatrix} 1&0&0&0 \\ 0&1&0&0 \\ 0&0&1&1 \\ 0&0&0&0 \end{pmatrix} \cdot \begin{pmatrix} 1&0&0&0 \\ 1&0&1&0 \\ 0&0&1&0 \\ 0&1&0&1 \end{pmatrix} \cdot \begin{pmatrix} 1&0&0&0 \\ 1&0&0&0 \\ 0&1&0&0 \\ 0&0&1&0 \end{pmatrix} = \begin{pmatrix} 1&0&0&0 \\ 1&1&0&0 \\ 1&1&1&0 \\ 0&0&0&0 \end{pmatrix}$$

Definition 3. *A contraction $C(M, I, J)$ is said* valid *if and only if $C(M, I, J)$ is a binary matrix.*

Example 2. The following contraction is not valid:

$$C(M, [], [1, 2]) = M \cdot C_2 \cdot C_1 = \begin{pmatrix} 1&0&0&0 \\ 1&0&1&0 \\ 0&0&1&0 \\ 0&1&0&1 \end{pmatrix} \cdot \begin{pmatrix} 1&0&0&0 \\ 0&1&0&0 \\ 0&1&0&0 \\ 0&0&1&0 \end{pmatrix} \cdot \begin{pmatrix} 1&0&0&0 \\ 1&0&0&0 \\ 0&1&0&0 \\ 0&0&1&0 \end{pmatrix} = \begin{pmatrix} 1&0&0&0 \\ 2&0&0&0 \\ 1&0&0&0 \\ 1&1&0&0 \end{pmatrix}$$

Definition 4. *Let M be a binary matrix of size $p \times q$. The* density *is the number of neighbor pairs of 1 in the matrix (including the diagonal pairs). This value may be computed with the following formula:*

$$d(M) = \frac{1}{2} \cdot \sum_{i,j} \left(M_{i,j} \cdot \left(\sum_{\delta=-1}^{1} \sum_{\gamma=-1}^{1} M_{i+\delta, j+\gamma} \right) - 1 \right)$$

where we define that $M_{i,j} = 0$ if $(i, j) \notin [\![1; p-1]\!] \times [\![1; q-1]\!]$

Problem 1. Maximum Matrix Contraction problem (MMC). *Given a binary matrix M of size $p \times q$ such that n entries equal 1 and $p \cdot q - n$ entries equal 0, the Maximum Matrix Contraction problem consists in the search for two sublists I of $[\![1; p-1]\!]$ and J of $[\![1; q-1]\!]$ such that the contraction $C(M, I, J)$ is valid and maximizes $d(C(M, I, J))$.*

We study in the next two sections the complexity and the approximability of this problem.

3   Complexity

This section is dedicated to proving the NP-Completeness of the problem.

Theorem 1. *The decision version of (MMC) is NP-Complete.*

Proof. Let M be an instance of MMC. Given an integer K, a sublist I of $[\![1; p-1]\!]$ and a sublist J of $[\![1; q - 1]\!]$, we can compute in polynomial time the matrix $C(M, I, J)$ and check if the contraction is valid and if $d(C(M, I, J)) \geq K$. This proves the problem belongs to NP.

In order to prove the NP-Hardness, we describe a polynomial reduction from the NP-Complete Maximum Clique problem [2]. Lets $G(V, E)$ be an instance of the Maximum Clique problem, we build an instance M of MMC with $p = q = (4|V| + 6)$. We arbitrarily number the nodes of $G : V = \{v_1, v_2, \ldots v_{|V|}\}$.

Let l_i and c_i be respectively the $6+4(i-1)+1$-th line and the $6+4(i-1)+1$-th column. We associate the four lines $l_i, l_i + 1, l_i + 2, l_i + 3$ and the four columns $c_i, c_i + 1, c_i + 2, c_i + 3$ to v_i. The key idea of the reduction is that each node v is associated with two 1 of the matrix. If we choose the node v to be in the clique, then, firstly, the two 1 associated with v are moved next to each other and this increases the density by one; and secondly, for every node w such that $(v, w) \notin E$, the two 1 associated to cannot be moved anymore.

A complete example is given in Fig. 2. For each node v_i, we set $M_{l_i, c_i} = M_{l_i+2, c_i+2} = 1$. If the nodes v_i and v_j are not linked with an edge, we set $M_{l_i, c_j} = M_{l_i+1, c_j+1} = 1$. If, on the contrary, there is an edge (v_i, v_j), then the intersections of the lines of v_i and the column of v_j is filled with 0. Finally, we add some 1 in the six first columns and the six first lines of the matrix such that only the contractions of the line l_i and the column c_i for $i \in [\![1; n]\!]$ are valid.

The initial density in this matrix is $d_0 = 11 + 6|V| + (|V|(|V| - 1) - 2|E|)$. Note that, in order to add one to the density of the matrix, the only way is to choose a node v_i and contract the column c_i and the line l_i. If the column c_i is contracted and if $(v_i, v_j) \notin E$, the two entries M_{l_j, c_i} and M_{l_j+1, c_i+1} are moved on the same column. Similarly, if the line l_i is contracted, the two entries M_{l_i, c_j} and M_{l_i+1, c_j+1} are moved on the same line. This prohibits the contraction of the line l_j and the column c_j. Consequently, in order to add C to the density, we must find a clique of size C in the graph and contract every line and column associated with the nodes of that clique.

Thus, there is a clique of size K if and only if there is a feasible solution for M of density $d_0 + K$. This concludes the proof of NP-Completeness.   ⊓

The Maximum Clique problem cannot be approximated to within $|V|^{\frac{1}{2}-\varepsilon}$ in polynomial time unless P = NP [3]. Unfortunately, the previous reduction cannot be used to prove a negative approximability result occurs for MMC. Indeed, the density of any feasible solution of the MMC instance we produce is between $d_0 + 1$ and $d_0 + |V|$, with $d_0 = O(|V|^2 - |E|)$. Consequently, the optimal density is at most $(1 + 1/|V|)$ times the worst density. A way to prove a higher inapproximability ratio for MMC would be to modify the reduction such that the gap between the optimal solution and another feasible solution increases.

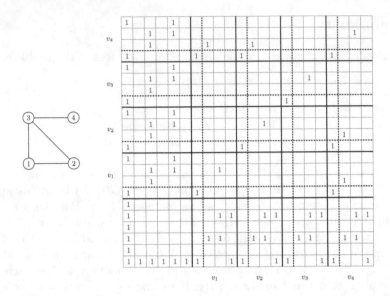

Fig. 2. This figure illustrates, on the left, a graph in which we search for a maximum clique and, on the right, the matrix obtained built with the reduction. We do not show the 0 entries of the matrix for readability. The dotted lines and columns represent the valid contractions.

In the next section, we prove that a $n^{\frac{1}{2}-\varepsilon}$ harness ratio would almost tight the approximability of MMC as there exists a $2\sqrt{n}$-approximation algorithm.

4   Approximability

In this section we define the notion of maximal feasible solution and prove that every algorithm returning a maximal feasible solution is a $2\sqrt{n}$-approximation where n is the number of 1 in the matrix.

Definition 5. *We say a feasible solution is* maximal *if it is not strictly included in another feasible solution. In other words, when all the lines and columns of that solution are contracted, no contraction of any other line or column is valid.*

Lemma 1. *Let M be an instance of MMC, (I, J) be a maximal feasible solution and $M' = C(M, I, J)$ then $2\sqrt{n} \leq d(M') \leq 4n$.*

Proof. A 1 in a matrix cannot have more than 8 neighbors, thus the density of M' is no more than $4n$.

Let p' and q' be respectively $p - |I|$ and $q - |J|$. For each line $i \in [\![1, p'-1]\!]$ of M', there is a column j such that $M'_{i,j} = M'_{i+1,j} = 1$, otherwise we could contract line i and (I, J) would not be maximal. Similarly for each column $j \in [\![1, q'-1]\!]$. Thus $d(M') \geq p' + q'$ where $p' \times q'$ is the size of M'. From the inequality of arithmetic and geometric means, we have $p' + q' \geq 2\sqrt{p' \cdot q'}$ and, as M' contains n entries such that $M'_{i,j} = 1$, $p' \cdot q' \geq n$. Thus $p' + q' \geq 2\sqrt{n}$.   □

From the upper bound and the lower bound given in the previous lemma, we can immediately prove the following theorem.

Theorem 2. *An algorithm returning any maximal solution of an instance of MMC is a $2\sqrt{n}$-approximation.*

Theorem 2 proves a default ratio for every algorithm trying to solve the problem. Note that there are instances in which the ratio between an optimal density and the lowest density of a maximal solution is $O(\sqrt{n})$. An example is given in the external report in [4]. In Sect. 6, we describe three natural heuristics to solve the problem. We show in [4] that their approximability ratio is $O(\sqrt{n})$ by exhibiting a worst case instance.

Determining if MMC can be approximated to within a constant factor is an open question. As it was already pointed at the end of Sect. 3, the problem may possibly be not approximable to within $n^{\frac{1}{2}-\varepsilon}$ and this would almost tight the approximability of MMC.

The next two sections focus on efficient algorithms to solve the problem. The next section is dedicated to the mathematical programming methods.

5   Linear Integer Programming

For $i \in [\![1; p-1]\!]$ (resp. $j \in [\![1; q-1]\!]$), let x_i (resp. y_j) be the binary variable such that $x_i = 1$ (resp. $y_j = 1$) if and only if line i is contracted, i.e. $i \in I$ (resp. column j is contracted, i.e. $j \in J$). From the definitions of Sect. 2, we can model the MMC problem by the following non-linear binary program:

$$(*)\begin{cases} \max_{x,y} \quad d(A) \\ A = \prod_{i=1}^{p-1}((L_i - I_p)x_i + I_p)M \prod_{j=q-1}^{1}((C_j - I_q)y_j + I_q) \\ A_{i,j} \leq 1, \quad \forall(i,j) \in [\![1; p-1]\!] \times [\![1; q-1]\!] \\ x_i, y_j \in \{0,1\} \end{cases}$$

where I_p denotes the identity matrix of size p and where the formula of $d(A)$ is the one given in Definition 4.

Although this formulation is very convenient to write the mathematical model, it is intractable as we would need to add an exponential number of linearizations: for all subset $I, J \subseteq [\![1; p-1]\!] \times [\![1; q-1]\!]$ we would need a variable $x_I = \prod_{i \in I} x_i$ and $y_I = \prod_{j \in J} y_j$.

We now present a linear integer programming model for the MMC problem: instead of linearizing the products $\prod_{i \in I} x_i$ and $\prod_{j \in J} y_j$, we cut the product

$$A = \prod_{i=1}^{p-1}((L_i - I_p)x_i + I_p)M \prod_{j=q-1}^{1}((C_j - I_q)y_j + I_q) \text{ in } T = p+q-1 \text{ time-steps.}$$

More precisely, define $A^{(1)} = M$; for all $t = 2, ..., p$, we define by $A^{(t)}$ the matrix which is computed after deciding the value of y_j for $j \geq p - t + 1$; similarly, for all $p + 1 \leq t \leq T$, $A^{(t)}$ is determined by the value of y_j for all j and by the value of x_i for $i \geq q - t + p$. We obtain the following program:

$$(P) \begin{cases} \max_{x,y} & d(A^{(T)}) \\ & A^{(t+1)} = ((L_{p-t} - I_p)x_{p-t} + I_p)A^{(t)} & \forall 1 \leq t \leq p - 1 \\ & A^{(t+1)} = A^{(t)}((C_{q-t+p} - I_q)y_{q-t+p} + I_q) & \forall p \leq t \leq T \\ & A_{i,j}^{(t)} \leq 1, \quad \forall (i,j,t) \in [1; p-1] \times [1; q-1] \times [2; T] \\ & x_i, y_j \in \{0, 1\} \end{cases}$$

We can easily linearize the model above by introducing, for all $(i, j, t) \in [1; p - 1] \times [1; q-1] \times [2; T]$ $r_{i,j,t} = A_{i,j}^{(t)} * x_{p-t}$ if $1 \leq t \leq p-1$ and $r_{i,j,t} = A_{i,j}^{(t)} * y_{q-t+p}$ if $p + 1 \leq t \leq T$, noticing that the variables $A_{i,j}^{(t)}, x_t, y_t$ are all binary. Finally, after linearizing the product $A_{i,j}^{(T)} A_{k,l}^{(T)}$ in the objective function, $d(A^{(T)})$, we obtain a polynomial size integer programming formulation of the MMC problem.

5.1   Numerical Results

We test the proposed model using IBM ILOG CPLEX 12.6. The experiments are performed on an Intel i7 CPU at 2.70 GHz with 16.0 GB RAM. The models are implemented in Julia using JuMP [5]. The algorithm is run on random squared matrices. Given a size p and a probability r, we produce a random binary matrix M of size $p \times p$ such that $Pr(M_{i,j} = 1) = r$. The expected value of n is then $r \cdot p^2$. We test the model for $n \in \{6, 9, 12\}$ for a probability $r \in \{0.1, 0.15, 0.2, 0.25, 0.3\}$ and we report the optimal value d^* and the running time. For each value of p and r, 10 random instances are created, whose averages are reported on Table 1.

We notice that the integer programming model is not very efficient to solve the problem. For $p = 15$, in most of the cases, CPLEX needs to run more than 2 hours to solve the model.

Table 1. Test of random instances for the integer linear program model.

	r									
	0.1		0.15		0.20		0.25		0.3	
	d^*	time (s)	d^*	time (s)	d^*	time (s)	d^*	time (s)	d^*	time (s)
(p,q)=(6,6)	6.0	0.3	4.1	0.26	12.1	0.2	15.3	0.28	22.0	0.15
(p,q)=(9,9)	15.1	5.3	22.1	5.1	32.3	7.8	36.5	7.0	44.5	3.4
(p,q)=(12,12)	30.8	171.6	48.0	281.2	55.0	183.0	64.4	101.0	71.0	95.1

6   Polynomial Heuristics

In this section, we describe three heuristics for MMC: a first-come-first-served algorithm and two greedy algorithms.

6.1   The LCL Heuristic

This algorithm is a first-come-first-served algorithm. It is divided into two parts: the Line-Column (LC) part and the Column-Line (CL) part.

The LC part computes and returns a maximal feasible solution M^{LC} by, firstly, contracting a maximal set of lines I^{LC} and, then, by contracting a maximal set of columns J^{LC}. The algorithm builds I^{LC} as follows: it checks for each line from $p - 1$ down to 1 if the contraction of that line is valid. In that case, the contraction is done and the algorithm goes on. J^{LC} is built the same way.

The CL part computes and returns a maximal feasible solution M^{CL} by starting with the columns and ending with the lines. The LCL algorithm then returns the solution with the maximum density.

The advantage of such an algorithm is its small time complexity.

Theorem 3. *The time complexity of the LCL algorithm is* $O(p \cdot q)$.

Proof. The four sets I^{LC}, J^{LC}, I^{CL} and J^{CL} can be implemented in time $O(p \cdot q)$ using an auxiliary matrix M'. The proof is given for the first one, the implementation of the three other ones is similar. At first, we copy M into M'. For each line i from $p - 1$ to 1 of M', we check with $2q$ comparisons if there is a column j such that $M'_{i,j} = M'_{i+1,j} = 1$. In that case, we do nothing. Otherwise, we add i to I^{LC} and we replace line i with the sum of the i-th and the $i + 1$-th lines.

Finally, given a matrix M and a set of lines I, one can compute $C(M, I, \emptyset)$ in time $O(p \cdot q)$ by, firstly, computing in time $O(p)$ an array A of size p such that A_i is the number of lines in I strictly lower than i and, secondly, returning a matrix C of size $p - |I| \times q$ such that $C_{i-A_i,j} = M_{i,j}$.   □

Remark 1. Note that, if there is at most one 1 per line of the matrix of the matrix, the LCL algorithm is asymptotically a 4-approximation when n approaches infinity. Indeed, the LC part returns a line matrix in which each entry is a 1. The density of this solution is $n - 1$. As the maximum density is $4n$ by Lemma 1, the ratio is $4\frac{n}{n-1}$. On the contrary, an example given in the external report [4] proves that this algorithm is, in the worst case, at least a $O(\sqrt{n})$-approximation.

6.2   The Greedy Algorithm

The greedy algorithm tries to maximize the density at each iteration. The algorithm computes $d(C(M, \{i\}, \emptyset))$ and $d(C(M, \emptyset, \{j\}))$ for each line i and each column j if the contraction is valid. It then chooses the line or the column maximizing the density. It starts again until the solution is maximal.

Theorem 4. *The time complexity of the Greedy algorithm is $O(p^2 \cdot q^2)$.*

Proof. There are at most $p \cdot q$ iterations. At each iteration, we compute one density per line i and one density per column j. The density of $C(M, \{i\}, \emptyset)$ is the density of M plus the number of new neighbor pairs of 1 due to the contraction of lines i and $i + 1$. The increment can be computed in time $O(q)$ as there are at most three new neighbors for each of the q entries of the four lines $i - 1$ to $i + 2$. Similarly, the density of $C(M, \emptyset, \{j\})$ can be computed in time $O(p)$. Thus one iteration takes $O(p \cdot q)$ iterations. □

Remark 2. We prove in [4] that the greedy algorithm is at least a $O(\sqrt{n})$-approximation algorithm.

6.3   The Neighborization Algorithm

The neighborization algorithm is a greedy algorithm trying to maximize, at each iteration, the number of couple of entries that can be moved next to each other with a contraction. This algorithm is designed to avoid the traps in which the LCL algorithm and the Greedy algorithm fall in by never contracting lines and columns that could prevent some 1 entries to gain a neighbor.

We define a function N from $([\![0; p - 1]\!] \times [\![0; q - 1]\!])^2$ to $\{0, 1\}$. For each couple $c = ((i, j), (i', j'))$ such that $M_{i,j} = 0$ or $M_{i',j'} = 0$, $N(c) = 0$. Otherwise, $N(c) = 1$ if and only if there is a sublist of lines I and a sublist of columns J such that $C(M, I, J)$ is valid and such that the two entries are moved next to each other with this contraction. Finally, we define $N(M)$ as the sum of all the values $N((i, j), (i', j'))$. The algorithm computes $N(C(M, \{i\}, \emptyset))$ and $N(C(M, \emptyset, \{j\}))$ for each line i and each column j if the contraction is valid. It chooses the line or the column maximizing the result and starts again until the solution is maximal.

Theorem 5. *The time complexity of the Greedy algorithm is $O(n^2 \cdot p^3 \cdot q^3 \cdot (p+q))$.*

Proof. Let M be a binary matrix, we first determine the time complexity we need to compute $N(M)$. Let $((i, j), (i', j'))$ be two coordinates such that $M_{i,j} = M_{i',j'} = 1$. We assume $i < i'$ and $j < j'$. The two entries may be moved next to each other if $i' - i - 1$ of the $i' - i$ lines and $j' - j - 1$ of the $j' - j$ columns between the two entries may be contracted and this can be done in time $O(p \cdot q \cdot (j' - j) \cdot (i' - i)) = O(p^2 \cdot q^2)$. As there are at most n^2 entries satisfying $M_{i,j} = M_{i',j'} = 1$, we need $O(n^2 \cdot p^2 \cdot q^2)$ operations to compute $N(M)$.

As there are at most $p \cdot q$ iterations. At each iteration, we computes one value per line i and one value per column j in time $O(n^2 \cdot p^2 \cdot q^2)$. The time complexity is then $O(n^2 \cdot p^3 \cdot q^3 \cdot (p + q))$. □

Remark 3. We prove in [4] that the neighborization algorithm is at least a $O(\sqrt{n})$-approximation algorithm.

6.4   Numerical Results

In this last subsection, we give numerical results of the three algorithms in order to evaluate their performances.

The experiments are performed on an Intel(R) Core(TM) i7-4810MQ CPU @ 2.80GHz processor with 8Go of RAM. The algorithms are implemented with Java 8[2]. The algorithms are run on random squared matrices. Given a size p and a probability r, we produce a random binary matrix M of size $p \times p$ such that $Pr(M_{i,j} = 1) = r$. The expected value of n is then $r \cdot p^2$. Before executing each algorithm, we first reduce the size of each instance by removing every column and line with no 1.

Small instances. We first test the three algorithms on small instances on which we can compute an exact brute-force algorithm. This algorithm exhaustively enumerates every subset of lines and columns for which the contraction is valid and returns the solution with maximum density. The results are summarized on Tables 2 and 3.

We can observe from Table 2 that the running time first increases when r grows and then decreases. Similarly, the number of times the heuristics return an optimal solution first decreases and then increases. The first behavior is explained by the fact that the size of instances with small values of n can be reduced. On the other hand, if r is high, the number of lines and columns of which the contraction is not valid increases and, then, the search space of the algorithms is shortened. Considering the running times, as it was predicted by the time complexities, the LCL and the greedy heuristics are the fastest algorithms. We can observe that the neighborization algorithm can be slower than the exact algorithm on small instances because the running time of the former is more influenced by n than the latter. However, we do not exclude the fact the implementation of the neighborization algorithm may be improved. Considering the quality of the solutions returned by the algorithms, according to Tables 2 and 3, the neighborization heuristic shows better performances than the greedy and the LCL algorithms.

Big instances. We then test the two fastest algorithms LCL and Greedy on bigger instances. The results are given on Table 4. Four interesting differences with Table 2 emerges from Table 4. Firstly, the LCL algorithm is faster than the greedy algorithm. This is coherent with the time complexities. Secondly, the LCL algorithm does not follow the same behavior as the exact algorithm and the neighborization heuristic for small instances: the running time increases with r even if the search space is shortened. Indeed, contrary to the three other algorithms, the implementation does not depend on this search space. Thirdly, the running time of the greedy algorithm first increases with r, then decreases and and finally slowly increases again. This last increase is due to the computation time of the density and the line and columns that can be contracted. Finally, the solution returned by the LCL algorithm seems to be better for small values

---

[2] The implementations can be found at https://github.com/mouton5000/MMCCode.

Table 2. This table details the results for each algorithm. For each values of p and r, the algorithms are executed on 50 instances. We give for each heuristic the mean running time in milliseconds, the mean ratio between the optimal density d^* and returned density d and the number of instances for which the ratio is 1.

p	r	Exact time (ms)	LCL time (ms)	$\frac{d^*}{d}$	$d = d^*$	Greedy time (ms)	$\frac{d^*}{d}$	$d = d^*$	Neigh. time (ms)	$\frac{d^*}{d}$	$d = d^*$
5	0.01	< 1 ms	< 1 ms	1	50	< 1 ms	1	50	< 1 ms	1	50
	0.02	< 1 ms	< 1 ms	1	50	< 1 ms	1	50	< 1 ms	1	50
	0.03	< 1 ms	< 1 ms	1	50	< 1 ms	1	50	< 1 ms	1	50
	0.04	< 1 ms	< 1 ms	1.00	49	< 1 ms	1	50	< 1 ms	1	50
	0.05	< 1 ms	< 1 ms	1.00	48	< 1 ms	1	50	< 1 ms	1	50
	0.1	< 1 ms	< 1 ms	1.00	46	< 1 ms	1.00	49	< 1 ms	1	50
	0.2	< 1 ms	< 1 ms	1.00	45	< 1 ms	1.00	46	< 1 ms	1	50
	0.3	< 1 ms	< 1 ms	1.00	43	< 1 ms	1.00	45	2.52	1.00	49
10	0.01	< 1 ms	< 1 ms	1.00	48	< 1 ms	1	50	< 1 ms	1	50
	0.02	< 1 ms	< 1 ms	1.02	46	< 1 ms	1.00	46	< 1 ms	1	50
	0.03	< 1 ms	< 1 ms	1.04	37	< 1 ms	1.00	41	1.22	1.00	49
	0.04	< 1 ms	< 1 ms	1.02	35	< 1 ms	1.00	39	1.92	1.00	49
	0.05	< 1 ms	< 1 ms	1.10	28	< 1 ms	1.00	26	1.98	1.00	46
	0.1	2.60	< 1 ms	1.00	19	< 1 ms	1.00	21	15.50	1.00	34
	0.2	< 1 ms	< 1 ms	1.00	23	< 1 ms	1.00	23	66.42	1.00	40
	0.3	< 1 ms	< 1 ms	1.00	31	< 1 ms	1.00	34	66.64	1.00	42
15	0.01	< 1 ms	< 1 ms	1.16	33	< 1 ms	1.00	43	< 1 ms	1	50
	0.02	< 1 ms	< 1 ms	1.06	21	< 1 ms	1.00	25	1.64	1.00	40
	0.03	< 1 ms	< 1 ms	1.08	17	< 1 ms	1.00	17	4.36	1.00	40
	0.04	3.76	< 1 ms	1.02	11	< 1 ms	1.00	15	14.84	1.00	34
	0.05	9.40	< 1 ms	1.02	18	< 1 ms	1.00	14	38.96	1.00	33
	0.1	295.74	< 1 ms	1.00	6	< 1 ms	1.00	8	355.54	1.00	19
	0.2	28.24	< 1 ms	1.00	14	< 1 ms	1.00	18	892.10	1.00	33
	0.3	< 1 ms	< 1 ms	1.00	30	< 1 ms	1.00	37	541.58	1.00	45
20	0.01	< 1 ms	< 1 ms	1.18	23	< 1 ms	1.00	31	1.04	1.00	45
	0.02	59.06	< 1 ms	1.14	10	< 1 ms	1.00	15	21.24	1.00	29
	0.03	431.60	< 1 ms	1.04	9	< 1 ms	1.00	6	119.82	1.00	20
	0.04	2275.64	< 1 ms	1.00	2	< 1 ms	1.00	5	273.82	1.00	19
	0.05	10223.92	< 1 ms	1.00	3	< 1 ms	1.00	4	622.92	1.00	8
	0.1	44268.36	< 1 ms	1.00	7	< 1 ms	1.00	2	3809.98	1.00	17
	0.2	424.84	< 1 ms	1.00	15	< 1 ms	1.00	11	5302.22	1.00	33
	0.3	< 1 ms	< 1 ms	1.00	34	< 1 ms	1.00	46	1553.86	1.00	49

Table 3. Each entry of this table details, for each couple of heuristics, the number of instances of Table 2 (there are 1600 instances) for which the line heuristic gives a strictly better results than the column heuristic.

	LCL	Greedy	Neigh
LCL	–	366	70
Greedy	426	–	86
Neigh	629	587	–

Table 4. This table details the results for the LCL algorithm and the greedy algorithm. For each values of p and r, the algorithms are executed on 50 instances. We give for each heuristic the mean running time in milliseconds and how many times the returned density is strictly better than the density returned by the other algorithm.

		LCL		Greedy				LCL		Greedy	
p	r	time (ms)	$d_L > d_G$	time (ms)	$d_L < d_G$	p	r	time (ms)	$d_L > d_G$	time (ms)	$d_L < d_G$
	0.01	< 1 ms	49	17.78	1		0.01	12.00	50	2832.52	0
	0.02	< 1 ms	48	22.58	2		0.02	14.04	21	1890.40	29
	0.03	< 1 ms	43	21.82	5		0.03	16.34	1	1099.38	49
200	0.04	< 1 ms	31	19.26	18	1000	0.04	17.72	1	553.90	49
	0.05	< 1 ms	21	16.76	29		0.05	18.74	5	233.70	45
	0.1	1.28	10	5.18	40		0.1	24.72	0	7.82	0
	0.2	1.92	0	< 1 ms	0		0.2	41.50	0	12.62	0
	0.3	2.58	0	< 1 ms	0		0.3	59.18	0	18.36	0
	0.01	3.28	50	382.06	0		0.01	53.54	49	22068.00	1
	0.02	3.58	44	321.30	6		0.02	59.96	0	10664.44	50
	0.03	3.92	17	237.06	33		0.03	65.66	0	3914.08	50
500	0.04	4.56	10	164.82	40	2000	0.04	71.68	6	1049.00	44
	0.05	4.88	4	104.48	46		0.05	76.36	0	186.04	10
	0.1	6.80	0	4.70	2		0.1	100.16	0	28.88	0
	0.2	10.66	0	3.34	0		0.2	167.42	0	50.46	0
	0.3	15.06	0	4.58	0		0.3	237.54	0	72.88	0

of r and, on the other hand, the greedy algorithm returns better densities for middle values. The two algorithms are equivalent for high values of r because those instances can probably not be contracted.

7   Conclusion

In this paper, we introduced the Maximum Matrix Contraction problem (MMC). We proved this problem is NP-Complete. However, we also proved that every algorithm which solves this problem is an $O(\sqrt{n})$-approximation algorithm. Considering that the NP-Completeness was derived from the Maximum Clique problem, and that this problem cannot be polynomially approximated to within $n^{\frac{1}{2}-\varepsilon}$, MMC is very likely to not being approximable to within the same ratio. Such a result would almost tight the approximability of MMC.

Moreover, we studied four algorithms to solve the problem, an integer linear program, a first-come-first-served algorithm and two greedy algorithms, and gave numerical results. It appears firstly that integer linear programming is not adapted to MMC while the three other heuristics returns really good quality solutions in short amount of time even for large instances. Those results seems to disconfirm the $n^{\frac{1}{2}-\varepsilon}$ inapproximability ratio. It would be interesting to deepen the study in order to produce a constant-factor polynomial approximation algorithm or a polynomial-time approximation scheme if such an algorithm exists.

References

1. Pillai, A., Chick, J., Johanning, L., Khorasanchi, M., de Laleu, V.: Offshore wind farm electrical cable layout optimization. Eng. Optim. **47**(12), 1689–1708 (2015)
2. Karp, R.M.: Reducibility among combinatorial problems. In: Miller, R.E., Thatcher, J.W., Bohlinger, J.D. (eds.) Complexity of Computer Computations, pp. 85–103. Springer, Heidelberg (1972)
3. Håstad, J.: Clique is hard to approximate within $n(1 - \epsilon)$. Acta Math. **182**(1), 105–142 (1999)
4. Watel, D., Poirion, P.: The Maximum Matrix Contraction problem: Appendix. Technical report CEDRIC-16-3645, CEDRIC laboratory, CNAM, France (2016)
5. Lubin, M., Dunning, I.: Computing in operations research using Julia. INFORMS J. Comput. **27**(2), 238–248 (2015)

Integrated Production Scheduling and Delivery Routing: Complexity Results and Column Generation

Azeddine Cheref[1,2]($\boxtimes$), Christian Artigues[1], Jean-Charles Billaut[2],
and Sandra Ulrich Ngueveu[1]

[1] LAAS-CNRS, Université de Toulouse, INP, CNRS, Toulouse, France
{cheref,artigues,ngueveu}@laas.fr
[2] Laboratoire d'Informatique de l'Université de Tours,
64 avenue Jean Portalis, 37200 Tours, France
jean-charles.billaut@univ-tours.fr

Abstract. In this paper, we study an integrated production scheduling and delivery routing problem. The manufacturer has to schedule a set of jobs on a single machine without preemption and to deliver them to multiple customers. A single vehicle with limited capacity is used for the delivery. For each job are associated: a processing time, a size and a specific customer location. The problem consists then to determine the production sequence, to constitute batches and to find the best delivery sequence for each batch. The objectives of the proposed problems are to find a coordinated production and a delivery schedule that minimizes the total completion time (makespan) or the sum of the delivery times of the products. We present complexity results for particular cases and a column generation scheme to solve a relaxed version of the problem, leading to a lower bound of high quality. Some computational results show the good performances of the method.

Keywords: Integrated production and distribution · Complexity · Column generation

1 Introduction

This paper considers an integrated model of scheduling and delivery, where jobs are scheduled on a single machine and finished products are delivered from the manufacturer to multiple customer locations. The relationship between production and distribution being strong, an increasing amount of research has been devoted to this field during the last years. The problem has been largely analysed and reviewed in [4], where the author proposes a classification scheme for a variety of issues reflected by these models.

In this paper, jobs are scheduled on a single machine and preemption is not allowed. Different processing times, sizes and delivery destinations are associated to the jobs. Distribution is performed by a single vehicle with a limited capacity

© Springer International Publishing Switzerland 2016
R. Cerulli et al. (Eds.): ISCO 2016, LNCS 9849, pp. 439–450, 2016.
DOI: 10.1007/978-3-319-45587-7_38

and can be seen as a variant of the multitrip vehicle routing problem [15], in which deliveries are ensured by a single vehicle and with a constrained batching phase. Delivery costs are not taken into account in this paper but the total time required to complete the production and the delivery of the products and the total delivery time are both meaningful indicators of the overall efficiency of the delivery process. Therefore, the objectives of the proposed problems are to find a coordinated production and delivery schedule that minimize the total completion time (makespan) or the sum of the delivery times of the products.

We review below a few relevant papers.

In [13], a similar model is considered but the size of jobs is not included and the authors propose a polynomial time algorithm in the case of a fixed number of distinct destinations. In [6], the authors consider the problem of minimizing the makespan on a single machine scheduling problem with a unique capacitated vehicle and a no wait constraint. No wait constraints implies that the batch must be delivered at its completion time. In [8], an heuristic method is proposed for minimizing the makespan when lifespan constraints are introduced for the products. Most of the models presented in the literature explicitly take into account transportation times to reach the customer's location, but there are no proper routing decisions, since the number of distinct customers is typically very small. Hence, the focus of the analysis is often on scheduling and batching. In [10], the authors consider the problem in which the delivery dates are fixed in advance and in [5], there are various destinations but a batch must contain jobs of the same destination. Complexity results are given by [3] for the problem with a single vehicle, a storage area and one or two customers. In [12], the authors minimize the makespan for the one machine scheduling problem with pickup and delivery in which a single vehicle travels between the machines and the warehouse, whereas in [17], the authors study a similar problem in which three different locations and two vehicles are considered. The first vehicle transports unprocessed jobs between the warehouse and the factory and the second one transports finished jobs between the factory and the customer. Some models in the literature treat a coordinating problem in which the customer sequence is fixed. For example, in [1,16], the authors minimize the total satisfied demand in a single round trip, the authors consider that the products expire in a constant time after their completion time and a time window delivery for each product.

The problem is formally defined in Sect. 2. We present in Sect. 3 some complexity results for particular cases and in Sect. 4 a column generation scheme to solve a relaxed version of the problem, leading to a lower bound of high quality. Computational results are given in Sect. 5.

2   Problem Definition and Notations

We consider a set of n jobs $J = \{J_1, J_2, \ldots, J_n\}$ to be processed on a single machine and delivered to a set of n corresponding customers. Each job J_j, $j = 1, \ldots, n$, requires a certain processing time p_j. Delivery is performed by a single vehicle with capacity c. As mentioned before, there is a set of n customer locations and each job J_j, $j = 1, \ldots, n$, is additionally characterized by its location j and its size s_j where $0 \leq s_j \leq c$. We denote by t_{ij} the transportation time from location i to location j and D_j the arrival time (decision variable) to the location j, i.e. the delivery time of J_j. We use M to denote the machine and, by analogy with vehicle routing problems, we refer to the machine location as the depot.

The vehicle loads a certain number of jobs which have been processed and starts the round trip to deliver them at their respective locations. The set of jobs delivered during a single round trip is called a *batch*. The problem is then to determine the scheduling sequence, cluster the jobs into batches and determine the best route for each batch. Using the notation introduced by [11], the general problem considered here with one machine, several customers, one vehicle and a limited capacity is denoted by $1 \rightarrow D, k \geq 1 | v = 1, c | D_{max}$ for the makespan objective and $1 \rightarrow D, k \geq 1 | v = 1, c | \sum D_j$ for the total completion time ($1 \rightarrow D$ means "one machine to delivery", k is the number of customers, v is the number of vehicles, c indicates that a capacity is considered). An illustration for the problem with $n = 7$ jobs and n customers is given in Fig. 1.

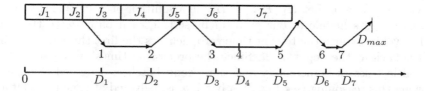

Fig. 1. An illustration for problems $1 \rightarrow D, k \geq 1 | v = 1, c | D_{max}$ or $\sum D_j$

We first discuss the complexity of special cases of the problem. Then we propose extended formulations and a column generation framework for the general case.

3   Particular Cases

In this section, we consider two special cases of the problem for both objectives functions. We give complexity results for the single customer case and some remarks for the fixed-batches case.

3.1   One Customer Case

In [3], the authors prove that the problem $1 \rightarrow D, k = 1 | v = 1, c | D_{max}$ is equivalent to the NP-hard Bin Packing problem when the processing times $p_j = 0$ for all j. Note that this reasoning becomes invalid for the total delivery time objective. However, we prove that problem $1 \rightarrow D, k = 1 | v = 1, c | \sum_j D_j$ is strongly NP-hard by reduction from 3-PARTITION problem. For our purpose, we introduce the 3-PARTITION problem.

3-PARTITION. Given $3h$ integers $a_1, \ldots, a_{3h}$, so that $\sum a_i = hb$, and such that $b/4 < a_i < b/2$ for all i, is it possible to partition them into h disjoint sets each summing up to b?

Theorem 1. *Problem* $1 \rightarrow D, k = 1|v = 1, c = z| \sum_j D_j$ *is NP-hard in the strong sense.*

Proof. Given a 3-PARTITION instance, we construct an instance for our problem as follows:

$$n = 3h \ jobs, c = b, t_{M1} = t_{1M} = t \ \text{and} \ t > 0$$
$$\text{For each job } J_j : p_j = 0, s_j = a_j$$
$$\text{Sum of the delivery times } y = 3th^2$$

From there, the problem consists in determining whether a solution exists such that $\sum_j D_j \le y$.

$\rightarrow$ If there is 3-PARTITION, then there exists a feasible schedule to our problem with $\sum_j D_j \le y$. Let $H_1, H_2, \ldots, H_h$ be a solution of 3-PARTITION. Then, we construct a schedule to our problem by setting each batch b_i to the triple H_i. The vehicle starts the tour at time zero, delivers the first three jobs at time t and is back at the depot at $2t$. Since the processing times are equal to zero, the vehicle restarts immediately and the second batch is delivered at time $3t$. Following this reasoning (see Fig. 2), a batch b_i is delivered at time $(2i-1)t$ and the total delivery time $\sum_j D_j = 3 \sum_{i=1}^{h} (2i-1)t = y$.

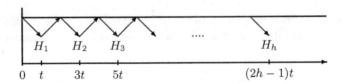

Fig. 2. 3-PARTITION solution

$\leftarrow$ Suppose that a schedule S for our problem exists in which $\sum_j D_j \le y$. According to the generated instance, the number of batches in S cannot be smaller than h, the number of jobs in each batch cannot exceed 3 and the vehicle is never idle at the depot in an optimal solution. We denote by h' the number of batches in S. Firstly, we suppose that $h' = h$ and denote by S_1 the corresponding schedule. This implies that each batch $b_1, b_2, \ldots, b_h$ of S_1 contains exactly three jobs and for each batch $\sum_{j \in b_i} s_j = b$. Thus, $b_1, \ldots, b_h$ define a solution of the 3-PARTITION problem. Suppose now that there exists a schedule S_2 for which $\sum_j D_j \le y$ and $h' > h$. Let n_1 the number of jobs in $b_1, \ldots, b_h$ and n_2 the number of jobs in $b_{h+1}, \ldots, b_{h'}$. We denote by $\sigma_j^{S_1}$ and $\sigma_j^{S_2}$ the jobs scheduled at position j in the schedule S_1 and S_2 respectively. Due to the fact that each batch in S_1 contains three jobs, one can see that $\sum_{j=1}^{n_1} D_{\sigma_j^{S_2}} \geq \sum_{j=1}^{n_1} D_{\sigma_j^{S_1}}$. In the schedule S_2, the remaining n_2 jobs are delivered after the time $(2h - 1)t$ which represents the delivery time of $J_{\sigma_{n_1}^{S_2}}$ and the last job $J_{\sigma_n^{S_1}}$ in S_1. So, $\sum_{j=n_1+1}^{n} D_{\sigma_j^{S_2}} > \sum_{j=n_1+1}^{n} D_{\sigma_j^{S_1}}$ which implies $\sum_j D_j > y$ on the solution S_2.

Remark 1. Problem $1 \to D, k = 1 | v = 1, c | \sum_j D_j$ is polynomially solvable when all jobs have the same size [13]. The authors propose a polynomial time algorithm with a complexity in $O(n^2)$ to solve the problem.

3.2   Fixed-Batch Case

In this case, we consider that the jobs are already clustered into batches and that, for each one, the delivery route is known. We consider below the makespan criterion and the sum of delivery times criterion.

$$1 \to D, k = 1 | v = 1, c, fixed - batches | D_{max} \text{ problem}$$

Proposition 1. *Problem* $1 \to D, k \geq 1 | v = 1, c = z, fixed - batches | D_{max}$ *is polynomially solvable.*

If we consider a batch as a job, this problem becomes equivalent to the well-known polynomial two-machine flow shop problem with makespan criterion [9]. In the resulting problem, we consider the duration of the batch on the machine as the processing time of the corresponding job on the first machine and the duration of the route of the batch as the processing time on the second machine.

$$1 \to D, k = 1 | v = 1, c, fixed - batches | \sum_j D_j \text{ problem}$$

Proposition 2. *Problem* $1 \to D, k \geq 1 | v = 1, c = z, fixed - batches | \sum_j D_j$ *is NP-hard.*

We consider the case in which each batch contains exactly one job and we denote by C'_j the time at which the vehicle is back at the depot after the delivery of job J_j. The delivery time D_j of a job J_j is then $D_j = C'_j - t_{jM}$ and $\sum_{j=1}^{n} D_j = \sum_{j=1}^{n} C'_j - \sum_{j=1}^{n} t_{jM}$, with $\sum_j^n t_{jM}$ a constant. This problem is equivalent to solving the NP-hard two-machine flow shop problem with the sum of completion times criterion [7], in which the processing times of a job J_j on the first machine is equal to p_j, equal to $t_{Mj} + t_{jM}$ on the second machine and a completion time on the second machine equal to C'_j.

4   General Case

For the general case, we first establish the following fundamental dominance property. As there are no release dates for the jobs, from any solution, the machine sequence can obviously be reordered according to the routing sequence and the jobs can be scheduled at the earliest without increasing the objective function for both the D_{max} and $\sum_j D_j$ criteria. The property then follows:

Property 1. There exists an optimal solution satisfying the following conditions:

- Jobs are processed on the machine without idle time,
- Production sequence and routing sequence are the same.

This implies that when the round trip of a batch is given, the sequence on the machine can be deduced. For both objective functions, one can see that the optimal solution of the problem can be obtained by combining batches and the problems are able to be modeled as a set covering problem. Hence, extended formulations and column generation approaches can be considered for both problems. Note that if there was no machine scheduling phase, the problem would resort to the multi-trip travelling salesman problem. Below, we detail these approaches for each criterion. Note that column generation and branch-and-price are techniques of choice for the related multi-trip vehicle routing problem [2,15]. However, we have in our case a single vehicle and a preliminary constrained batching phase due to the machine sequencing sub-problem. It is thus relevant to wonder whether a column generation approach can still be successfully applied or not.

4.1 $1 \rightarrow D, K \geq 1 | V = 1, C | D_{max}$ Problem

We introduce in this section a set covering formulation for the master problem. A column represents a batch and its position on the delivery sequence. The set of feasible batches is denoted as β. For each batch $b \in \beta$, its duration on the machine and its round trip duration are known. We denote by $P_{b,1} = \sum_{j \in b} p_j$ the duration of the batch b on the machine and, $P_{b,2}$ the duration of the round trip that delivers the batch b. Since the jobs contained in a batch are known, $a_{i,b}$ takes the value 1 if the job J_i is in the batch b and 0 otherwise. A unique set of variables $x_{b,k}$ is used to minimize the D_{max} objective. Variables $x_{b,k} \in \{0,1\}$ indicates if the batch b at the position k is selected.

$$\min D_{max} \tag{1}$$

$$D_{max} \geq \sum_{k=1}^{l} \sum_{b \in \beta} P_{b,1} x_{b,k} + \sum_{k=l}^{n} \sum_{b \in \beta} P_{b,2} x_{b,k}, \qquad \forall l \in \{1, ..., n\} \tag{2}$$

$$\sum_{b \in \beta} x_{b,k} \leq 1, \qquad \forall k \in \{1, ..., n\} \tag{3}$$

$$\sum_{b \in \beta} (a_{i,b} \sum_{k=1}^{n} x_{b,k}) = 1, \qquad \forall i \in \{1, ..., n\} \tag{4}$$

$$\sum_{b \in \beta} x_{b,k} \geq \sum_{b \in \beta} x_{b,k+1}, \qquad \forall k \in \{1, ..., n-1\} \tag{5}$$

$$\sum_{b \in \beta} x_{b,k} = 1, \qquad\qquad \forall k \in \{1, ..., \delta\} \qquad (6)$$

$$x_{b,k} \in \{0,1\} \qquad \forall k \in \{1, ..., n\}, \forall b \in \beta \qquad (7)$$

The first set of constraints (2) is equivalent to the *fixed-batches* case presented above. It ensures that the processing of a batch on the machine starts after the completion of the previous one and that the vehicle starts the delivery of a batch after its completion on the machine, and the end of the previous tour. Constraints (3) state that a position can contain at most one batch and constraints (4) ensure that each job is contained in exactly one selected batch. Constraints (5) are symmetry breaking constraints that enforce that all selected columns (batches) appear consecutively at the first positions. Constraints (6) sets a minimum number of batches using a lower bound δ of the number of round trips to deliver all the jobs. To obtain δ, we use the `First Fit Decreasing` rule which is a $3/2$ approximation for the Bin Packing problem. Let τ the number of bins obtained by the `FFD` algorithm. A lower bound of the minimum number of batches δ is then equal to $2/3\tau$.

Following a standard column generation scheme, the master problem is restricted to a subset of variables (columns) $\tilde{\beta} \subseteq \beta$ and a pricing problem is needed to find new non basic variables that can improve the solution for the LP relaxation of model (1–7).

The Pricing Problem. This sub-problem searches for an element of $\beta \setminus \tilde{\beta}$ such that the reduced cost of the new column is negative. We denote by $\bar{c}_{b,k}$ the reduced cost of $x_{b,k}$ and, one has:

$$\bar{c}_{b,k} \le 0 \Leftrightarrow \sum_{i=1}^{n} \underbrace{\frac{p_i \sum_{l=k}^{n} \alpha_l - \gamma_i}{\sum_{l=1}^{k} \alpha_l}}_{l_i} a_{i,b} + P_{b,2} < \underbrace{\frac{\beta_k - \sigma_{k-1} + \sigma_k + \xi_k}{\sum_{l=1}^{k} \alpha_l}}_{r_k}$$

where $\alpha_l, \beta_k, \gamma_i, \sigma_k$ and ξ_k denote the dual values associated to the constraints (2), (3), (4), (5), (6) respectively. Note that the dual values ξ_k exist only for $k \le \delta$, σ_{k-1} for $k \ge 2$ and σ_k for $k \le n - 1$.

We define an auxiliary directed graph $G = (V, A)$ in which $V = N \cup \{v_s, v_d\}$ where nodes $N = \{v_1, \ldots, v_n\}$ represent the locations $1, \ldots, n$ and nodes $\{v_s, v_d\}$ the duplicated depot. The set of arcs is $A = \{(v_i, v_j)|v_i \in N \cup \{v_s\}, v_j \in N \cup \{v_d\}, v_i \ne v_j\}$. Finally, the sub-problem consists in solving an elementary shortest path problem with resource constraints (ESPPRC) on graph G, in which the distance value d_{ij} of an arc (v_i, v_j) is equal to $l_i + t_{ij}$ for $v_i \in V - \{v_s\}$, $d_{sj} = t_{Mj}$ for each arc (v_s, v_j), and $d_{sd} = r_k$ for arc (v_s, v_d). The constraints concern the capacity of the vehicle. A column x_{bk} corresponding to an elementary shortest path in G is introduced in the restricted master problem only if the length of such path is smaller than r_k. From the resulting path and given that the jobs sequence on the machine and the delivery sequence are the same, the round trip length and the batch processing on the machine can be obtained.

An exact method is used to solve the elementary shortest path problem with resource constraints. The interested reader will find more details about the used algorithm in [14].

Starting from the first position, the sub-problem searches for a new column to add by scanning all positions and stops when a column with negative reduced cost is found. The new column is then added. We propose below a heuristic to initiate the column generation process, taking account of this particularity.

Initial Solution Heuristic. In order to accelerate the sub-problem solution phase by using a minimal number of positions, we propose the following initial solution heuristic:

(1) The first step is to assign jobs into batches according to their sizes. The `First Fit Decreasing` rule is used for this purpose.
(2) Given a constitution of the batches, the route is determined using the `nearest neighbor search` rule.
(3) As soon as the duration of the batch on the machine and the duration of the routes are known, the `Johnson's` rule (known to solve the two-machine flow-shop problem to optimality) is used to optimally order the batches.

4.2 $1 \to D, K \geq 1 | V = 1, C | \sum_j D_j$ Problem

In this part, a formulation for the general problem with a cumulative objective function $\sum_j D_j$ is suggested. In order to obtain the delivery time of each job, the new columns of the master problem must take the departure time of a batch into account. Hence, we define binary variables y_{bkt}, which take the value 1 if the batch b is delivered at position k and starts to deliver it at time t if selected, 0 otherwise. For each position k, the departure time is given by variable $S_k \geq 0$. As the considered objective needs the exact delivery time of each job, we denote by R_{ib} the time between the departure time of the vehicle and the arrival time to location i. This value is known once a batch b is given. Let T be an upper bound on the latest possible departure time of the vehicle for the last batch. As long as the triangle inequality holds, this is given by:

$$T = \sum_{i=1}^{n}(p_i + 2t_{0,i})$$

The other notations remain the same as those used for the formulation with the D_{max} objective function. Let β' denote the set of feasible batches.

$$\text{Minimize} \quad \sum_{b \in \beta'} \sum_{k=1}^{n} \sum_{t=1}^{T} \sum_{i \in b} a_{i,b}(t + R_{i,b}) y_{b,k,t} \tag{8}$$

$$S_{k'} \geq \sum_{k=1}^{l} \sum_{b \in \beta'} \sum_{t=1}^{T} P_{b,1} y_{b,k,t} + \sum_{k=l}^{k'-1} \sum_{b \in \beta'} \sum_{t=1}^{T} P_{b,2} y_{b,k,t},$$

$$\forall k' \in \{1, ..., n\}, \qquad \forall l \in \{1, ..., k'\} \quad (9)$$

$$\sum_{b \in \beta'} \sum_{t=1}^{T} y_{b,k,t} \leq 1, \qquad \forall k \in \{1, ..., n\} \quad (10)$$

$$\sum_{b \in \beta'} \sum_{t=1}^{T} (a_{i,b} \sum_{k=1}^{n} y_{b,k,t}) \geq 1, \qquad \forall i \in \{1, ..., n\} \quad (11)$$

$$\sum_{b \in \beta'} \sum_{t=1}^{T} t y_{b,k,t} \geq S_k, \qquad \forall k \in \{1, ..., n\} \quad (12)$$

$$\sum_{b \in \beta'} \sum_{t=1}^{T} y_{b,k,t} \geq \sum_{b \in \beta'} \sum_{t=1}^{T} y_{b,k+1,t}, \quad \forall k \in \{1, ..., n-1\} \quad (13)$$

$$\sum_{b \in \beta'} \sum_{t=1}^{T} y_{b,k,t} = 1, \qquad \forall k \in \{1, ..., \delta\} \quad (14)$$

$$y_{b,k,t} \in \{0,1\} \quad \forall b \in \beta', \forall k \in \{1, ..., n\}, \forall t \in \{1, ..., T\} \quad (15)$$

The delivery time of a job J_i is given by the addition of the departure time of the batch which contains it and the transportation time between the depot and the location i (9). The other constraints are similar to those used for the previous formulation.

We denote by $\bar{c}'_{bkt}$ the reduced cost of variable y_{bkt}. Let $\tilde{\beta}' \subseteq \beta'$. The sub-problem searches for an element of $\beta' \setminus \tilde{\beta}'$ such that the reduced cost of a new column is negative.

$$\bar{c}'_{bkt} \leq 0 \Leftrightarrow \underbrace{\sum_{i=1}^{n} (t + R_{i,b})}_{d_i} a_{i,b} + \underbrace{\sum_{i=1}^{n} (p_{i,1} (\sum_{k'=1}^{n} \sum_{l=k}^{k'} \alpha_{l,k'}) - \gamma_i)}_{l'_i} a_{i,b} +$$

$$+ P_{b,2} \underbrace{(\sum_{k'=k+1}^{n} \sum_{l=1}^{k} \alpha_{l,k'})}_{q} \leq \underbrace{\beta_k + t\delta_k - \sigma_{k-1} + \sigma_k + \xi_k}_{r'_{kt}}$$

where $\alpha_{lk'}, \beta_k, \gamma_i, \delta_k, \sigma_k$ and ξ_k the dual values associated to the constraints (9), (10), (11), (12), (13) and (14) respectively. The dual values ξ_k exists only for $k \leq \delta$, σ_{k-1} for $k \geq 2$ and σ_k for $k \leq n-1$.

An auxiliary graph G' is defined in the same way as the one defined above so that the sub-problem consists in finding an elementary shortest path with resource constraints (ESPPRC) on graph G' in which the distance d_{ij} of an arc (v_i, v_j) is equal to $l'_i + q \cdot t_{ij}$. However, a new label is introduced in order to store the elapsed time R_{ib} between the departure time of the vehicle and the arrival time to location i. To optimally solve the (ESPPRC), Lozano et al. algorithm is also used.

5   Computational Results

The results are performed on a set of data generated as follows. For each job, processing times and the size follow a sets of discrete uniform distribution $\mathcal{U}(1, 100)$ and $\mathcal{U}(1, 10)$ respectively. For a location j, integer coordinates (X_j, Y_j) are randomly generated in the interval $[1, 40]$ and the distances between the locations are obtained by computing the classical euclidean distance.

$$t_{i,j} = t_{j,i} = E\left(\sqrt{(X_i - X_j)^2 + (Y_i - Y_j)^2}\right)$$

Note that the processing times, locations and sizes of the jobs are generated independently of each other. Capacity c of the vehicle is fixed to 20 and 5 instances are generated for each number of jobs $n \in \{20, 30, 40, 50, 60, 70, 80, 90, 100\}$. Hence, the experiments are performed on 45 instances.

The column generation process could be carried out efficiently only for the $D_{\max}$ criterion. Indeed, for the $\sum D_j$ model, the need to explicitly represent time makes the convergence much slower as the number of variables becomes huge in the master problem.

Table 1. Computational results for $1 \rightarrow D, k \geq 1|v = 1, c|D_{max}$ problem

n	$LB(sec)$	#col	%GAP_{init}	%GAP_{UB}
20	1.1	203	21.14	4.62
30	4.6	363	13.67	1.83
40	12.1	595	9.94	0.99
50	46.4	954	10.34	0.88
60	86.4	1136	10.21	0.69
70	215.1	1670	7.30	0.57
80	383.7	1963	7.67	0.42
90	804.8	2637	5.44	0.39
100	1181.9	3041	5.06	–

The experiments have been implemented for the 45 instances on a Xeon 3.20 GHz computer with 8 GB using ILOG CPLEX 12.6 to solve the LPs. We evaluate and compare the solutions obtained by the column generation which represents a lower bound (LB) for the problem with the integer solution obtained by branch and bound on the generated columns, which represents an upper bound (UB). In order to obtain an upper bound in a reasonable time, the subproblem add a single column at each time. Therefore, to obtain a good upper bound, all columns with negative reduced costs are integrated to the master problem which has the advantage of providing good upper bounds and the disadvantage of a larger execution times.

In Table 1, we were interested on the aggregate results for each value of n. The statistics take into account the average CPU times for the column generation (column $LB(sec)$). The number of columns generated during the process is given in column $\#col$. The gap between the initial solution and the relaxed solution is given in column ($\%GAP_{init}$). Finally, the gap between UB solution and the relaxed solution is given by column ($\%GAP_{UB}$). The results show a gap lower than 1% for instances with $n \geq 40$, which proves the very good quality of the bounds. The computational times remain lower than 1200 s for instances with up to 100 jobs.

6   Conclusions

In this paper we presented an integrated production scheduling and delivery routing problem that can be seen as a variant of the multi-trip traveling salesman problem with a constrained batching phase due to machine sequencing constraints. We presented complexity results for particular cases and an efficient column generation scheme for the makespan criterion. In the near future, we shall focus on the implementation of a branch-and-price algorithm to close the remaining gap. We will also focus on finding a better decomposition scheme for the sum of deliveries criterion.

Acknowledgement. This work was supported by the financial support of the ANR ATHENA project, grant ANR-13-BS02-0006 of the French Agence Nationale de la Recherche.

References

1. Armstrong, R., Gao, S., Lei, L.: A zero-inventory production and distribution problem with a fixed customer sequence. Ann. Oper. Res. **159**(1), 395–414 (2008)
2. Azi, N., Gendreau, M., Potvin, J.Y.: An exact algorithm for a vehicle routing problem with time windows and multiple use of vehicles. Eur. J. Oper. Res. **202**(3), 756–763 (2010)
3. Chang, Y.C., Lee, C.Y.: Machine scheduling with job delivery coordination. Eur. J. Oper. Res. **158**(2), 470–487 (2004)
4. Chen, Z.L.: Integrated production and outbound distribution scheduling: review and extensions. Oper. Res. **58**(1), 130–148 (2010)
5. Chen, B., Lee, C.Y.: Logistics scheduling with batching and transportation. Eur. J. Oper. Res. **189**(3), 871–876 (2008)
6. Gao, S., Qi, L., Lei, L.: Integrated batch production and distribution scheduling with limited vehicle capacity. Int. J. Prod. Econ. **160**, 13–25 (2015)
7. Garey, M.R., Johnson, D.S., Sethi, R.: The complexity of flowshop and jobshop scheduling. Math. Oper. Res. **2**(1), 117–129 (1976)

8. Geismar, H.N., Laporte, G., Lei, L., Sriskandarajah, C.: The integrated production and transportation scheduling problem for a product with a short lifespan. J. Comput. **20**(1), 21–33 (2008)
9. Johnson, S.M.: Optimal two-and-three-stage production schedules with set-up times included. Nav. Res. Logistics **1**(1), 61–68 (1954)
10. Hall, N.G., Lesaoana, M., Potts, M.C.: Scheduling with fixed delivery dates. Oper. Res. **49**(1), 134–144 (2004)
11. Lee, C.Y., Chen, Z.L.: Machine scheduling with transportation considerations. J. Sched. **4**(1), 3–24 (2001)
12. Li, C.L., Ou, J.: Machine scheduling with pickup and delivery. Nav. Res. Logistics **52**(7), 617–630 (2005)
13. Li, C.L., Vairaktarakis, G., Lee, C.L.: Machine scheduling with deliveries to multiple customer locations. Eur. J. Oper. Res. **164**(1), 39–51 (2005)
14. Lozano, L., Duque, D., Medaglia, A.L.: An exact algorithm for the elementary shortest path problem with resource constraints. Transp. Sci. **50**(1), 348–357 (2015)
15. Lysgaard, J., Sanne, W.: A branch-and-cut-and-price algorithm for the cumulative capacitated vehicle routing problem. Eur. J. Oper. Res. **236**(3), 800–810 (2014)
16. Viergutz, C., Knust, S.: Integrated production and distribution scheduling with lifespan constraints. Ann. Oper. Res. **213**(1), 293–318 (2014)
17. Wang, X., Cheng, T.C.E.: Production scheduling with supply and delivery considerations to minimize the makespan. Eur. J. Oper. Res. **194**(3), 743–752 (2009)

Author Index

Printed in the United States
By Bookmasters

Printed in the United States
By Bookmasters